PLASTIC BLOW MOLDING HANDBOOK

PLASTIC BLOW MOLDING HANDBOOK

Edited by

Norman C. Lee, P.E.

Sponsored by the Society of Plastic Engineers
and its Blow Molding Division

VAN NOSTRAND REINHOLD
New York

Library of Congress Catalog Card Number 89-22674
ISBN 0-442-20752-2

Printed in the United States of America

Van Nostrand Reinhold
115 Fifth Avenue
New York, New York 10003

Van Nostrand Reinhold International Company Limited
11 New Fetter Lane
London EC4P 4EE, England

Van Nostrand Reinhold
480 La Trobe Street
Melbourne, Victoria 3000, Australia

Nelson Canada
1120 Birchmount Road
Scarborough, Ontario
Canada M1K 5G4

16 15 14 13 12 11 10 9 8 7 6 5 4 3 2 1

Library of Congress Cataloging-in-Publication Data

Plastic blow molding handbook / edited by Norman C. Lee.
p. cm.
ISBN 0-442-20752-2
1. Plastics—Molding—Handbooks, manuals, etc. I. Lee, Norman C., 1934-
TP1150.P54 1990
668.4'12—dc20 89-22674
CIP

Contents

Foreword

Over the years, numerous handbooks and design guides on the subject of plastics have been published. None of these dealt in any depth with the subject of this handbook—blow molding. The recent growth of blow molding as an economically feasible process has been rapid in many areas. This growth, coupled with the lack of technical publications relating to blow molding, prompted the Board of Directors of the Blow Molding Division of the Society of Plastic Engineers to undertake the assimilation of available information and the editing of this milestone publication.

We believe that this *Plastic Blow Molding Handbook* will provide the reader with a greater understanding of the unique process characteristics of blow molding, enable the reader to apply proven techniques in developing new products and applications for blow molding, and will serve as a valuable reference for all who are interested in the plastics industry.

Our thanks are heartily extended to the various authors for their contributions to this pioneering effort in blow molding.

J. H. Moran
Chairman
Blow Molding Division
Society of Plastic Engineers

Preface

The blow molding of plastic articles has in the past had an aura of the mystic around it. As a result, little comprehensive work on the subject has been published. Advances in the technology of polymeric materials, machine controls, computer science, and management techniques have made it necessary to correct the myths and magic.

The purpose of this handbook is to provide background, insight, and reference for the practice of blow molding plastic products.

This book's scope includes a history of blow molding from its earliest known use to the present time, and lists in chronological order significant events that have affected the industry. The book describes in detail the various processes that compose the basic blow molding field. It also gives an overview of the auxiliary machines, processes, and equipment to provide the reader with a general understanding of the support equipment that is required as well as of what can be achieved after molding. Information is provided on plant operation, process control, and quality control, highlighting those aspects that apply to the scope of this book. Comprehensive coverage of the applications of polymeric materials is also included. No attempt has been made to cover financial considerations, however, since this work would not be considered a reference source for that aspect of plant operation. The reader who is interested in more detail than is provided here should consult other, more in-depth works.

The achievement of such an extensive work has taken the efforts of many contributors, some twenty two in all, from a cross section of the plastic industry, encompassing many specialties and specialized knowledge, know-how, and skill. To all of these people, on behalf of myself and the S.P.E. Blow Molding Board of Directors, I wish to express many thanks. I would also like to acknowledge the companies that employ these individuals, since this work would not have been possible without their support. Individuals and companies are acknowledged for their specific contributions at the end of chapters.

A special thanks to Van Nostrand Reinhold's staff, especially Gail Nalven and Rima Weinberg for their untiring assistance and patience in bringing this work to a reality. I also wish to acknowledge the professionalism of Stefania Taflinska, editorial supervisor, Trumbull Rogers, editorial consultant, and William L. Broecker, copy editor, for their work.

Contributors

Samuel L. Belcher, P.E., President, Sabel Plastechs, Inc., Cincinnati, Ohio.

Ross H. Dean, President, Fibrcon, Inc., Whitefish Bay, Wisconsin.

Thomas E. Douglas, Senior Staff Engineer, Zarn, Inc., Reidsville, North Carolina.

Tod F. Eberle, P.E., Market Development Engineering Manager, Johnson Controls, Inc., Plastic Container Division, Manchester, Michigan.

*Stanley E. Eppert, Jr.**, Product Development Engineer, Borg Warner Chemicals, Inc., Washington, West Virginia.

Paul E. Geddes, President, B & G Machinery Co., Grand Rapids, Michigan.

Denes Hunkar, President, Hunkar Laboratories, Inc., Cincinnati, Ohio.

Christopher Irwin, Development Manager, Johnson Controls, Inc., Manchester, Michigan.

Robert R. Jackson, President, Jackson Machinery Company, Cedarburg, Wisconsin.

W. Michael Jaycox, Quality Control Manager, The Dial Corporation, Phoenix, Arizona.

Michael L. Kern, Engineer, Borg Warner Chemicals, Inc.**, Washington, West Virginia.

Norman C. Lee, P.E., Vice President, Research and Development, Zarn, Inc., Reidsville, North Carolina.

---

* Currently Engineer, Himont USA, Wilmington, Delaware

** Now Product Optimization Engineer, G.E. Plastics

James P. Parr[+], Technical Service Engineer, Blow Molding, Hoechst Celanese Corp., Houston, Texas.

Don Peters, Principal Blow Molding Engineer, Phillips 66 Company, Bartlesville, Oklahoma.

Ciro L. Petrucelli, President, Springfield Mold Works, Inc., Westfield, Massachusetts.

Eckard F.H. Raddatz, Manager of Blow Molding Technical Services, Hoechst A.G., Frankfurt, West Germany.

Donatas Satas, Consultant, Satas and Associates, Warwick, Rhode Island.

Virginia A. Sharkey, Advertising and Public Relations Manager, Cumberland Engineering, South Attleboro, Massachusetts.

Max Suit, Technical Consultant, Wentworth Mould and Die Co. Ltd., Hamilton, Ontario, Canada.

John A. Szajna[++], Manager of Molds and Sampling, Continental Plastics Containers, Oak Grove, Illinois.

Timothy Wagner, President, Wagner CAD/CAM Software Engineering, Glastonbury, Connecticut.

Dan Weissmann, Ph.D., Group Leader, Monsanto Chemical Company, Springfield, Massachusetts.

E. Dieter Wunderlich, Vice President, FGH Systems, Denville, New Jersey.

[+] Currently Staff Specialist, Application Technology, Exxon Chemical Company, Houston, Texas

[++] Retired

CHAPTER

1

What Is Blow Molding?

CHRISTOPHER IRWIN

Blow molding is a process used to produce hollow objects from thermoplastic. Blow molded articles have a "reentrant curve" feature. The most common example is a bottle with the top opening or finish much smaller in size than the body.

The basic process has two fundamental phases. First, a preform (or parison) of hot plastic resin in a somewhat tubular shape is created. Second, a pressurized gas, usually air, is used to expand the hot preform and press it against a female mold cavity. The pressure is held until the plastic cools (see Fig. 1-1). This action identifies another common feature of blow molded articles. Part dimensional detail is better controlled on the outside than on the inside, where material wall thickness can alter the internal shape.

In the process approaches used today the method of creating the preform defines the basis of the process. Most machines are either extrusion-based or injection-based; that is, the preform is either an extruded tube or an injection-molded tube. Several special processes, however, have elements of both. For example, most multilayer blow molded articles are produced on extrusion-based machinery, but injection-based machinery is also used. Likewise, most biaxial oriented articles (e.g., PET [polyethylene terephthalate] beverage bottles) are produced on injection-based machinery, but extrusion-based machinery is also used. The neck ring process is both injection- and extrusion-based at the same time. The dip and displacement processes

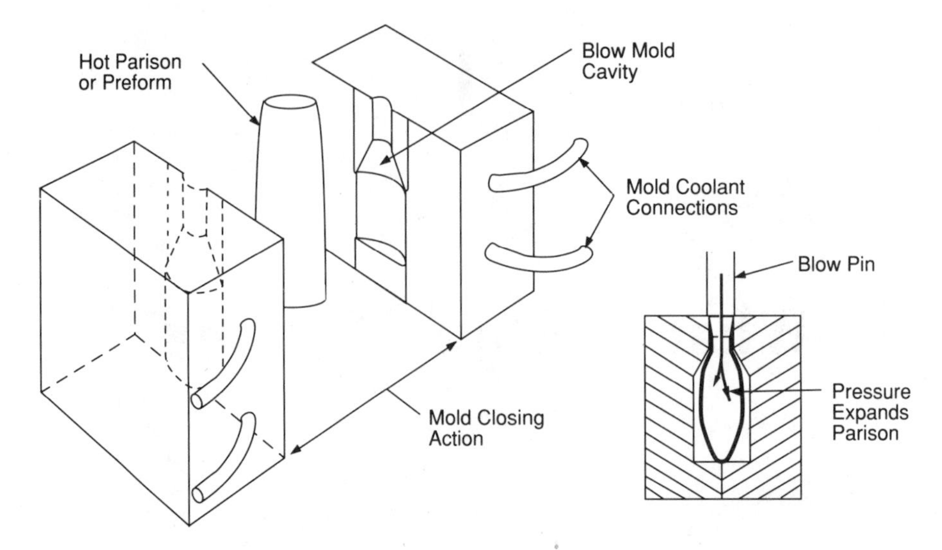

Figure 1-1 Basic blow molding process.

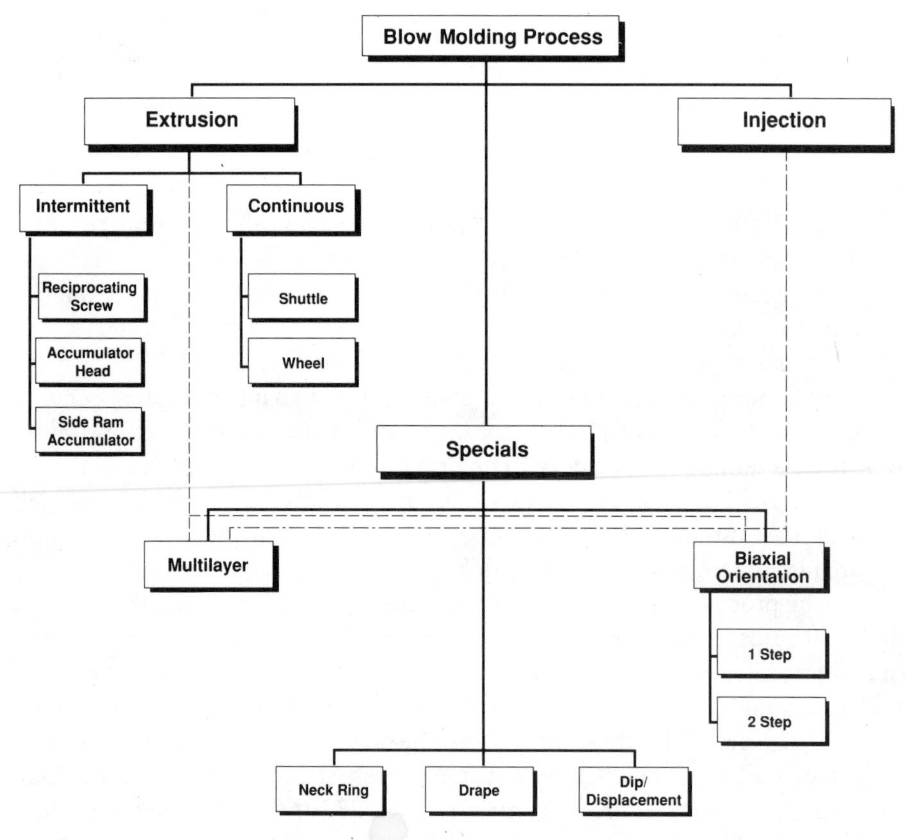

Figure 1-2 Interrelationship of current blow molding processes. Multilayer and biaxial orientation methods have roots in both extrusion and injection approaches.

Figure 1-3 Baby rattles molded from cellulose nitrate about 1890. [16]

mimic injection blow molding, and the drape process has elements of the extrusion approach. Figure 1-2 defines the interrelationship of the processes used today. More detail is given on each approach in later sections.

Blow molding is a multibillion dollar business. In the late 1980s annual worldwide consumption of plastic resin by blow molding processes was about ten billion pounds. Packaging is the primary application for this resin and was used to produce approximately seventy billion bottles, drums, tanks, and other containers.

However, the bottle is not the only shape. A blow molding renaissance is occurring in which engineers and designers are discovering and promoting blow molding for a wide variety of industrial or technical applications. Toy wheels, automobile seat backs, ductwork, surf boards, bellows, fuel tanks, flower pots, automobile bumpers, double-walled tool cases, and cabinet panels are just a few examples of the many creative designs being developed.

Interestingly, the first attempts to blow mold a true thermoplastic, cellulose nitrate, over a century ago were also nonpackaging applications—

mostly novelties and toys. An early celluloid baby rattle (circa 1890) is shown in Figure 1-3 [16]. Packaging applications did not seriously evolve until the late 1950s, when high-density polyethylene became available.

HISTORY OF THE EARLY BLOW MOLDING INDUSTRY

In the United States, John H. Breck, Inc. introduced Banish shampoo in a high-density polyethylene (HDPE) bottle. It had no other purposes but to hold and convey the product to the consumer. In Europe, a supplier of fine foods introduced a rigid polyvinyl chloride (PVC) bottle to hold a liquid aromatic. It had the look and feel of fine glass [1,2].

These two events took place in 1958. Both bottles were more expensive than their glass counterparts, nevertheless they took the plastic bottle from its "cute" squeeze bottle status (a prior niche) to a position as a forthright conveyor of product, and that began the serious challenge to metal and glass packaging. This change marked the beginning of a period of unprecedented growth.

In that same year Kautex (Germany) and Blow-O-Matic (Denmark) displayed the first commercially available blow molding machines, at the National Plastics Exposition in Chicago [3]. Both of these machines were the continuous extrusion type for polyethylene. The Blow-O-Matic machine, however, was forced from the market three years later because of infringement of the Kautex design.

In 1956 Phillips Petroleum Co. and Hoechst began commercial production of HDPE [4,5]. In the United States the price was $0.47 a pound [6]. By 1958 several other resin suppliers (Allied, Celanese, W. R. Grace, Hercules, and others) had entered the U.S. market and had driven the price down to $0.43 a pound. By 1959 the price had dropped again, to $0.38 a pound [7].

The explosive growth of the blow molding industry was caused by several factors, but the most significant was the development of high-density polyethylene. The process and the resin go hand in hand, and a vast majority of all blow molding still is based on this material.

What were some of the early events in machinery, processing, and resin development that led to the Breck Banish shampoo bottle of 1958?

It is obvious that blow molding of plastic resins traces its roots to the blow molding of glass. Reliefs on the walls of Egyptian royal tombs of 1800 B.C. record the art of blow molding and illustrate glass blowers in action. A glass urn has been preserved that shows where the parison, the partially shaped mass of molten glass, was flashed—blown to final shape and trimmed—outside the mold and later ground smooth [8]. The art of glass blowing reached Central Europe at the beginning of the Christian era and spread throughout the Roman Empire.

The first reference to a thermoplastic material (other than glass) is in U.S. Patent 8180 by S. T. Armstrong dated June 24, 1851 (see Fig. 1-4; for

UNITED STATES PATENT OFFICE.

SAML. T. ARMSTRONG, OF NEW YORK, N.Y.

IMPROVEMENT IN MAKING GUTTA-PERCHA HOLLOW WARE.

Specification forming part of Letters Patent No. **8,180**, dated June 24, 1851.

To all whom it may concern:

Be it known that I, S.T. ARMSTRONG, of the city, county, and State of New York, have invented a new and useful Improvement in the Process of Manufacturing Hollow Ware or Articles of Gutta-Percha; and I do hereby declare that the following is a full, clear, and exact description of the principle or character which distinguishes it from all other things before known, and of the method of making, constructing, and using the same.

My improved process is applicable to the making of all kinds of hollow articles which can be formed in molds—such as bottles or articles which may be made hollow—of gutta-percha or gutta-percha compounded with other substances.

After the gutta-percha alone or compounded with other substances has been properly cleansed and prepared in any known or appropriate manner, it is to be formed into a pipe or tube, in the manner of making lead pipe, and for this purpose I use any of the known machines for making lead pipe. The gutta-percha during this part of the process should be kept at a temperature of about 150° of Fahrenheit's scale, which degree of heat is best preserved by applying heat to the exterior of the cylinder of the machine, as is sometimes practiced in the manufacture of lead pipes. As the gutta-percha pipe issues from the die of the machine in a heated state it is plastic and adhesive, so that the end can be closed by pressing it together. I then cut off a piece of the length required and insert it in a mold such as is used for molding glass, with the closed end downward, and after opening the upper end I insert the end of a metal pipe connected with a hydraulic apparatus and force in water under sufficient pressure to expand the gutta-percha until the external surface is brought in contact with the entire surface of the mold. I continue the pressure of water until the gutta-percha is cooled and set, and then I remove the pressure and take the article out of the mold, when it will be found to have taken the exact form of the mold, of whatever figure it may be.

If the form of the article admit of it, the mold may be made in a single piece, or it may be made in two or more parts, depending on the form of the article to be molded; and if the article to be produced is to be made with a neck, when the piece of gutta-percha pipe is put into it in the upper or open end is to be prepared for the reception of the water-pipe by first inserting a conical plug into it until the external surface is forced out against the mold and the inside is made sufficiently large to receive the water-pipe; but if the article to be produced be without a neck, then the gutta-percha pipe is to be cut of greater length, and the open end is bound around the nozzle of the water-pipe to prevent the escape of water when pressure is applied. When the article has been formed the surplus is then cut off and the edge properly trimmed.

Bottles, vases, tumblers, powder-flasks, and such articles can be made in this way to great advantage and of great beauty, as the mold can be ornamented and chased in any way to suit fancy, and, however figured, if sufficient pressure be applied the gutta-percha will receive the impression of the entire figure. As the gutta-percha is cooled by forcing in the water to expand it, it will be set in the mold and retain the form thereof.

Many articles which are not required to be, but which admit of being, made hollow—such as ornamental figures—may be advantageously molded by my improved process, and, if required, after such articles have been molded the nozzle through which the water was introduced can be properly trimmed and closed up.

What I claim as my invention, and desire to secure by Letters Patent in the process above described, is—

The method, substantially as described, of molding articles of gutta-percha or the compounds of gutta-percha with other substances by first making the same in the form of a pipe, and while in a partially heated and plastic state giving to it the form required in a mold by forcing a liquid inside to expand the gutta-percha, as described.

S. T. ARMSTRONG.

Witnesses:
CHAS. J. GILBERT,
CALLS. BROWN.

Figure 1-4

early patent information, see the chapter appendix). In this patent the inventor describes a process for molding gutta-percha, a natural latex with thermoplastic properties. An extruded, hot, tubular parison was formed. Water pressure was then used to expand the parison against the contours in the mold. Armstrong did earlier work with gutta-percha as a coating or insulation for telegraph wires [16].

A few years later, in 1881, W. B. Carpenter was granted a patent for blow molding celluloid, or cellulose nitrate. Either two sheets or a tubular parison of celluloid is clamped between the two halves of the mold cavity. Steam is then used to soften and expand the material into its final form. Celluloid was a new material commercialized by John W. Hyatt in 1869 [5]. The resin is considered the first true thermoplastic.

Cellulose nitrate is highly flammable, and working with it was very difficult. A more practical material, cellulose acetate, was developed and eventually commercialized by 1919 [9].

In 1930 E. T. Ferngren, working with cellulose acetate, developed the first plastic squeeze bottle. Until that time blow molded plastics were used only for novelties and toys. This bottle was never commercialized. The patent, when issued several years later, was assigned to PLAX Corporation.

At about the same time that Ferngren developed his bottle, the chairman of the board of Hartford Empire Co., Goodwin Smith, saw the potential of plastics. Hartford Empire Co. was the developer and manufacturer of automatic glass blowing machinery [16]. The equipment was used by every glass bottle manufacturer except Owens-Illinois Glass Co., who built equipment for their own use.

During the mid-1930s Hartford Empire Co. bought the technology developed by E. T. Ferngren and formed the PLAX Corporation subsidiary. PLAX stood for Plastics Experimental. Its main purpose was to study and develop the feasibility of plastics blow molding. This was a broad charter that led to work in profile extrusion, sheet, thermoforming, and film as well as bottle blow molding. Figure 1-5 is one of the first automatic blow molders developed during this time.

By 1939 PLAX Corporation had developed machinery that could produce about 25,000 bottles per day from cellulose acetate. This machine used an intermittent extrusion process but with the end of the parison preclosed as it extruded downward. The technique was related to methods used in glass bottle manufacture. A limited quantity of plastic bottles was being used by Elizabeth Arden and Helena Rubinstein for sachet. The opportunity to produce Christmas tree balls (see Fig. 1-6), the first significant product for the firm, arose in the early 1940s when the traditional glass supplies from Europe were halted.

In 1933 researchers at ICI (Imperial Chemical Industries), England, discovered low-density polyethylene (LDPE) [5]. By 1939 the material was commercialized; however, most was soon being used for military and defense

Figure 1-5 One of the first automatic blow molders developed by PLAX from work by Ferngren. Work began in the early 1930s; this photograph was taken in 1938. Courtesy of Innopak Corp.

Figure 1-6 Two Christmas tree balls produced in 1941 from cellulose acetate. Center, a catsup dispenser molded in LDPE about 1947. All by PLAX Corp. Courtesy of Henry E. Griffith.

applications. PLAX Corporation used LDPE to produce spacers for a radar coaxial cable. The material was so important to the war effort that it arrived in fifty-pound diplomatic pouches handcuffed to the courier's wrist. Scrap from this operation was saved. By December, 1942, enough scrap was accumulated to attempt a trial blow molding run. The first bottle was not perfect, but it laid the groundwork for the first commercial bottles—the Stopette underarm deodorant squeeze bottle—in 1945]16] (see Fig. 1-7).

During the 1940s PLAX Corporation not only perfected their extrusion-based machine but also developed an injection blow molder. Most of the designs from this period were created by E. T. Ferngren, W. H. Kopitke, J. R. Hobson, and in the early 1950s, R. W. Canfield; but the real genius at PLAX Corporation was J. Bailey, Director of Research.

V. E. Hofmann and S. T. Moreland of Owens-Illinois Glass Co. developed an injection blow molder based on an eight-ounce Lester injection molding machine [16]. A few bottles were made during World War II for the Army Medical Corps, but the firm, principally a glass bottle producer, had little interest in plastics. Figure 1-8 is an illustration taken from the patent filed in 1939. Many years later Moslo sold a commercial machine that used the same principle.

By the late 1940s the PLAX Corporation had the makings of a very nice business in blow molded plastic containers. Owens-Illinois Glass Co., now realizing the opportunity, began negotiations for a license. In 1953 Owens-

Figure 1-7 Stopette underarm deodorant squeeze bottle, molded by PLAX Corp., the first high-volume LDPE commercial blow molding application. [16]

Illinois purchased a half interest in PLAX Corporation from the Emhart Corporation. The Emhart Corporation had been formed in the late 1940s by reorganization of Hartford Empire Co., in consequence of a restraint of a trade suit initiated by the government. Interestingly, Monsanto purchased this half interest from Owens-Illinois in 1957 and purchased the remaining half of PLAX Corporation from Emhart in 1962. The PLAX name, however, was soon dropped from use.

Developments from Owens-Illinois began to flourish in the mid-1950s. It was in this period that O. B. Sherman and G. V. Mumford developed the neck ring process, a clever technique, still in use today, in which the finish of the bottle is injection molded to provide quality, and the body of the bottle is extrusion molded to provide process flexibility. Figure 1-9 illustrates the basic sequence of operation.

In the early 1950s a number of other firms began to acquire and develop

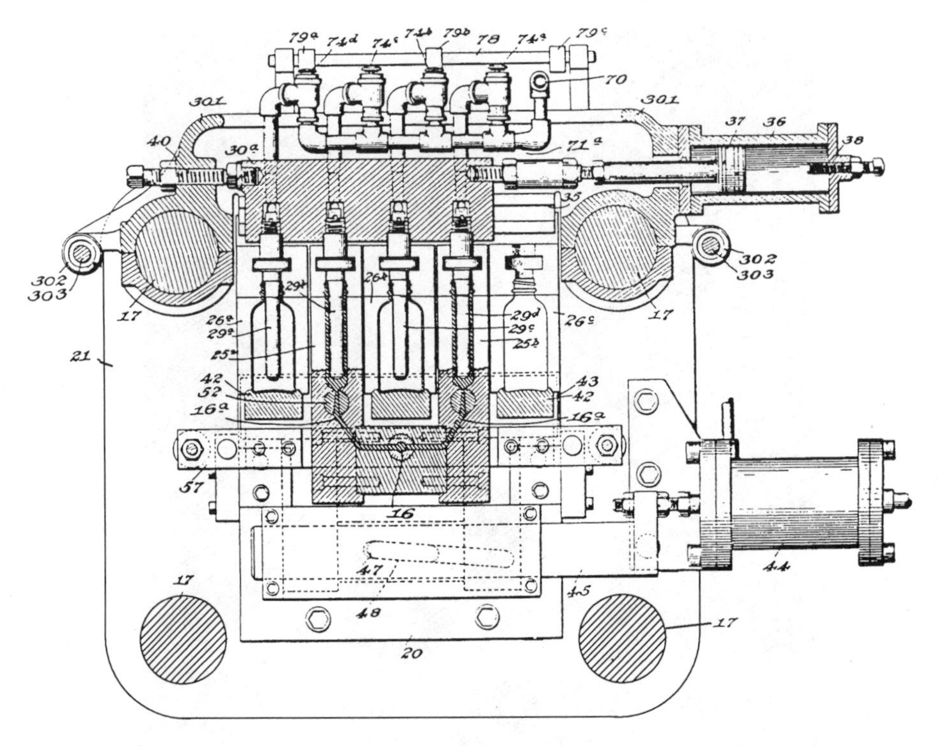

Figure 1-8 An illustration from the Moreland Injection Blow Molding patent filed in 1939.

blow molding processes. Wheaton Glass Co. further developed injection blow molding using technology from PLAX Corporation, A. Gussoni (an Italian inventor), and A. Borer (a Swiss inventor).

Continental Can Company purchased a continuous extrusion horizontal wheel technology from E. E. Mills. This was to supplement their vertical design by H. S. Ruekberg; however, this was not the only vertical design. C. C. Coates of the Royal Manufacturing Co. also developed a vertical wheel that was ultimately sold to American Can Company [16].

IMCO developed the in-place process, an extrusion blow technique. In this process a manifold is used with several parison extrusion heads. A parison is extruded from the first head. When it reaches the proper length a pair of molds close. At the same time a valve reroutes plastic flow to the second head to extrude the next parison. The sequence continues for each head, eventually returning to the first. The approach is no longer in use.

In the early 1950s Europeans, particularly Germans, also began to take interest in plastic blow molding opportunities. However, the technology developed in the United States was not available. The manufacturing of machinery and tooling was done by the bottle producers themselves, most of whom were in the glass business. Work was kept confidential and often was limited in scope. Developments in Europe therefore proceeded in isolation. The European glass industry had

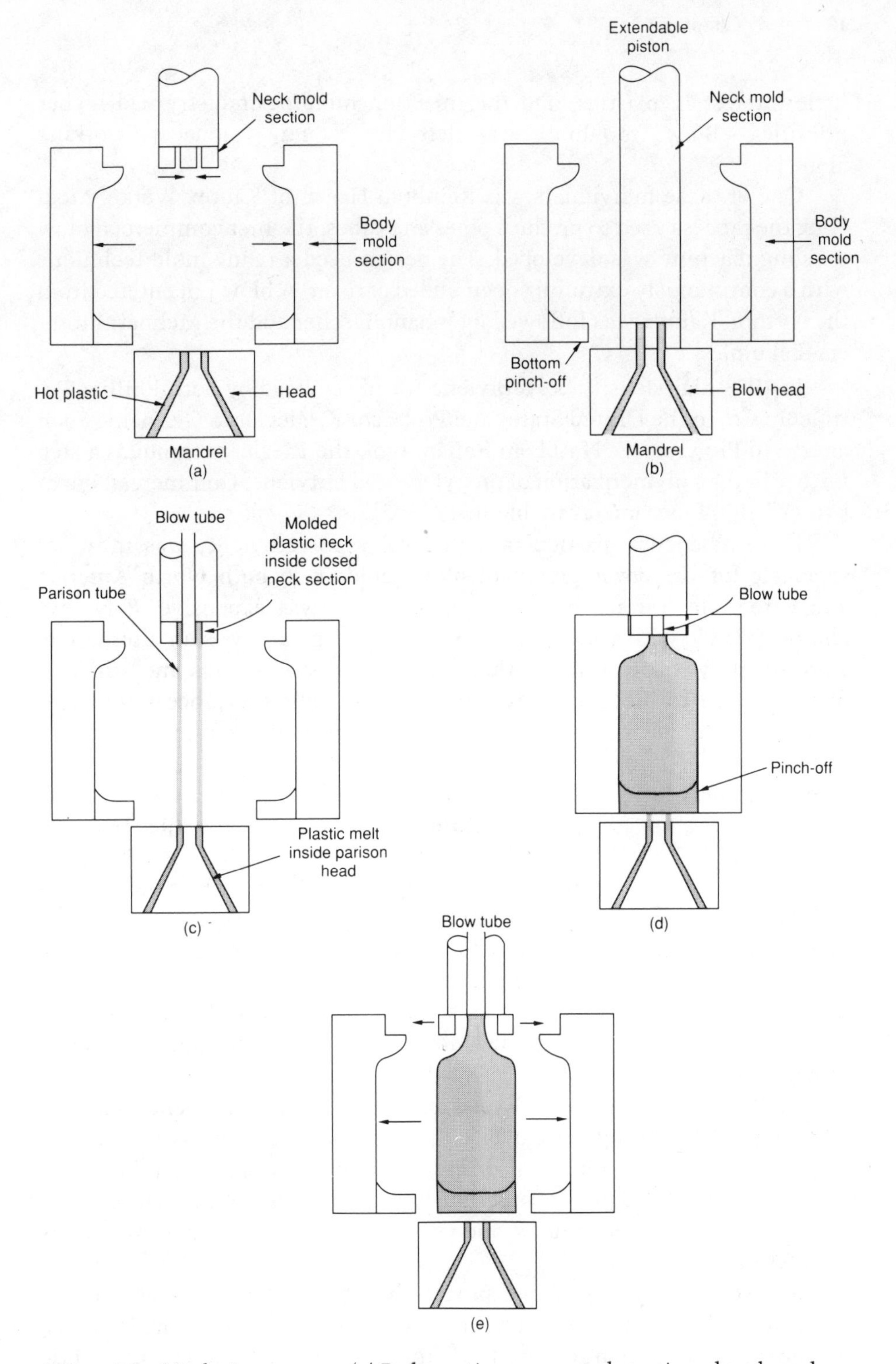

Figure 1-9 Neck ring process. (a) Body section open, neck section closed, neck section retracted. (b) Neck section extended to mate with parison nozzle (plastic fills neck section). (c) Neck section retracted with parison tube attached. (d) Body section closed, making pinch-off (parison blown to body sidewalls). (e) Body molds open, neck molds open, bottle about to be ejected. Courtesy of John Wiley and Sons.

little interest in plastics, and the injection molding industry had higher priorities. Blow molding was left to a few pioneers working alone [8].

One of these individuals was Reinhold Hagen of Kautex Werke. Modeling the process used to produce pipes and tubes, the first commercial blow molding machine was developed. The design used a rising mold technique with a continuously extruding open-ended parison. A blow pin entered from the bottom. Kautex was followed by Johann Fischer and the Mehnert brothers (Bekum).

In 1953 high-density polyethylene was discovered by both Phillips Petroleum Co. in the United States and Professor K. Ziegler in Germany. Soon afterward Professor G. Natta, an Italian, took the Ziegler techniques a step further in the polymerization of propylene and butylene. Commercial quantities of HDPE became available in 1956 [5].

Polyethylene, in particular high-density polyethylene, was most responsible for the development of blow molding in both North America and Europe. In Europe, however, polyethylene was expensive. Polyvinyl chloride (PVC), with visual clarity, was often considered as an alternative. Unfortunately, because of its thermal sensitivity PVC was and still is a difficult resin to process. Early blow molders split the process into two steps to minimize problems: the preform was manufactured separately and later reheated for blow molding. In the early 1960s continuous extrusion machinery originally developed for HDPE was improved to allow the processing of PVC, thus eliminating the need for the two-step approach [8].

Commercial blow molding machinery, mostly from Europe, became readily available in North America in the late 1950s. The explosive demand for HDPE detergent, cosmetic, and pharmaceutical bottles could not be satisfied by the original blow mold process developers. By 1960 over fifty-five companies were recognized as suppliers of the processing equipment and over one hundred firms in the United States alone were engaged in blow molding [11,12].

Packaging was not the only use for blow molding that benefited from the new resins. During 1959 over 40 percent of all the resins used in the process were used for technical or industrial applications. Most were toys and dolls, but the industry was seriously discussing fuel tanks, automotive arm rests, air conditioner ducts, and windshield cleaner solution containers as potential applications [11].

By 1960 the blow molding process had been established as a major element of the plastics industry. Beyond this point developments in the technology and the markets are too numerous to list. Most are just subtle refinements. Nevertheless three in particular deserve mention: the dairy milk bottle, the biaxially oriented beverage bottle, and the multilayer bottle. The technology of these developments and others has taken the blow molding industry to the 1990s.

IMPORTANT DATES IN THE HISTORY OF BLOW MOLDING

1800 B.C. Reliefs on the walls of Egyptian royal tombs record the art of glass blow molding [8].

1700–1600 B.C. Egyptians further develop the art, which includes blow molds requiring the parison to be flashed outside the cavity. The excess is ground smooth [8].

1851 S. T. Armstrong, U.S. patent reference to blow molding a plastic material other than glass: a tubular parison of gutta-percha, a natural latex with thermoplastic properties. Water is used as the pressuring medium.

1869 John Hyatt (U.S.A.) commences production of cellulose nitrate, which was discovered by chance by Schönbein (Switzerland, 1846) [5].

1872 "Celluloid" trademark registered by Hyatt [5].

1881 W. B. Carpenter, U.S. patent reference to blow molding of celluloid from sheet stock or tubular form. Steam is used to soften and expand the preform.

1899 Lederer (Germany) improves the method discovered by Schutzenberger (France, 1865) for the manufacture of cellulose acetate, a more practical material than cellulose nitrate [5].

1919 Cellulose acetate commercially available [9].

1928 Production begins of polyvinyl chloride (PVC), a material first discovered by Regnault (France, 1838) [5].

1933 Researchers at ICI (England) discover the high-pressure polymerization of ethylene, producing low-density polyethylene (LDPE) [5].

1936 PLAX Corporation (U.S.A.) granted a patent for the first plastic "flexible bottle" made from cellulosic materials. Work had begun on the project in 1930.

1938 The first of several patents issued to PLAX Corporation (Hartford Empire Co.) describing methods and machinery for automatically blow molding plastic articles. The first, an extrusion technique, but with a closed end, is based on earlier work in automatic glass blowing. Later, injection blow molding techniques are also used. All equipment is for private use.

1939 Commercial production, by ICI (England), of LDPE begins [5].

1941 Production of the first significant blow mold article—cellulose acetate Christmas tree balls molded by PLAX Corporation [16].

1942 An injection blow molding technique patented by Owens-Illinois (U.S.A.); as with PLAX Corporation, equipment is for private use.

1942 The first LDPE bottle molded by PLAX Corporation, from scrap polyethylene reclaimed from a war project [10,16].

1945 Stopette squeeze underarm deodorant, first commercial product packaged in LDPE bottle. Bottle molded by PLAX Corporation [16].

1947 Polyethylene tomato catsup dispenser to be filled by consumer from the glass package [16].

1950 E. E. Mills (U.S.A.) granted a patent for high output continuous extrusion rotary blow molder. The design is used privately by Continental Can Co. Other wheel designs, by H. S. Ruekberg and by C. C. Coats, follow quickly.

1950 Kautex Werke (Reinhold Hagen, Germany) develops and soon offers the first commercially available blow molding equipment. The design uses a rising mold technique with continuously extruded open ended parison [8].

1953 Discovery of high-density polyethylene (HDPE) by Phillips Petroleum Co. (U.S.A.) and Professor Ziegler (Germany) [5].

1953 Polymerization of propylene and butylene by Professor Natta (Italy) using Ziegler techniques [5].

1955 Kautex (Germany) ships automatic blow molder to U.S.A. [8].

1956 Commercial production, by Hoechst (Germany) and Phillips (U.S.A.), of HDPE begins [5,4].

1957 First HDPE bottle produced—an 8 fl oz baby nurser bottle blow molded by PLAX Corporation using Phillips material [15].

1958 First HDPE blow molded commercial package—a 16 fl oz cylindrical bottle for Breck Shampoo (U.S.A.). Liquid detergent bottles follow quickly [2].

1958 First PVC blow-molded commercial package for a liquid aromatic (Europe) [1].

1958 Commercial blow molding equipment displayed for the first time at National Plastics Exhibition (U.S.A.). Two machines, both European, utilize the continuous extrusion process. HDPE is the resin of choice [3].

1960 At the 16th Annual Technical Conference of the Society of Plastics Engineers, for the first time blow molding becomes a major topic of discussion (U.S.A.) [11].

1960 Over fifty-five companies, the majority European, are recognized as suppliers and manufacturers of blow molding equipment. Many of these firms were unknown a year earlier. Thirty years later only six of the original firms exist in a recognizable form [12].

1962 Shell Chemical Company (U.S.A.) begins vast program promoting an HDPE gallon bottle for milk [13].

1964 Melville Dairy, Burlington, North Carolina, becomes the first to blow mold bottles in-house for packaging fluid milk. Blow molder and resin are supplied by W. R. Grace & Co [14]. By the mid-1980s virtually all plastic milk bottles are made in-house using Uniloy equipment.

1968 Owens-Illinois (U.S.A.) produces a hollow-handled PVC liquor bottle for American Distilling Co. In Europe PVC is a far more popular blow molding resin, having been used for several years for water and table wine applications [10,8]. Because of governmental concerns the plastic bottle is later removed from the market. The plastic liquor bottle does not return until early 1980s, as a biaxially oriented polyethylene terephthalate (PET) container.

1969 Monsanto (U.S.A.) produces an experimental Coca-Cola bottle from methacrylonitrile/styrene.

1970 DuPont (U.S.A.) produces an experimental beverage bottle from biaxially oriented PET resin. DuPont receives patent in 1973 [10].

1970 Coca-Cola test markets the world's first plastic carbonated beverage bottle, a methacrylonitrile/styrene bottle by Monsanto.

1972 Toyo Seikan (Japan) develops basic multilayer bottle from polypropylene and ethylenevinyl alcohol resins for food product applications.

1974 Monsanto makes additional improvements by introducing a biaxially oriented acrylonitrile/styrene beverage bottle; however, health concerns and legislation banning the use of this material for this application soon force it from the market [10]. The material now has full clearance when a patented post mold treatment is used to bind all residual monomer.

1975 Cincinnati Milacron (U.S.A.) and Gildamister Corpoplast (Krupp, Germany) begin to offer commercial blow molding equipment for biaxial orientation of PET. The equipment is based on the two-step process, in which the preform and bottle are produced on separate machines in separate operations.

1976 First PET bottles for a commercial filling application, for Pepsi-Cola, molded by Amoco, Inc. (U.S.A.).

1977 Nissei, ASB (Japan) begins to offer biaxial orientation PET blow molding equipment based on the one-step process, in which the preform and bottle are produced on the same machine.

1983 Multilayer blow molding comes to the United States with the introduction of the Heinz catsup squeeze package.

APPENDIX: EARLY U.S. PATENT REFERENCES

Patent No.	*Inventor*	*Date*
8,180	Armstrong	Jun. 24, 1851
237,168	Carpenter	Feb. 1, 1881
1,263,141	Strauss	Apr. 16, 1918
1,592,299	Howard	Jul. 6, 1921
1,654,647	Heist	Jan. 3, 1928
1,863,339	Humphrey	Jun. 14, 1932
2,030,059	Ferngren	Feb. 11, 1936
2,099,055	Ferngren	Nov. 16, 1937
2,128,239	Ferngren	Aug. 30, 1938
2,175,054	Ferngren	Oct. 3, 1939
2,222,461	DeWitt	Nov. 19, 1940
2,230,188	Ferngren	Jan. 28, 1941
2,230,189	Ferngren	Jan. 28, 1941
2,230,190	Ferngren	Jan. 28, 1941

APPENDIX *(continued)*

Patent No.	*Inventor*	*Date*
2,260,750	Kopitke	Oct. 28, 1941
2,262,612	Kopitke	Nov. 11, 1941
2,283,751	Ferngren	May 19, 1942
2,285,150	Ferngren	Jun. 2, 1942
2,288,454	Hobson	Jun. 30, 1942
2,290,129	Moreland	Jul. 14, 1942
2,298,716	Moreland	Oct. 13, 1942
2,315,478	Parkhurst	Mar. 30, 1943
2,331,687	Hobson	Oct. 12, 1943
2,331,688	Hobson	Oct. 12, 1943
2,331,702	Kopitke	Oct. 12, 1943
2,348,738	Hofmann	May 16, 1944
2,349,176	Kopitke	May 16, 1944
2,349,177	Kopitke	May 16, 1944
2,349,178	Kopitke	May 16, 1944
2,353,825	Hofmann	Jul. 18, 1944
2,401,564	Hofmann	Jun. 4, 1946
2,452,080	Stephenson	Oct. 26, 1948
2,503,171	Posner	Apr. 4, 1950
2,515,093	Mills	Jul. 11, 1950
2,579,390	Mills	Dec. 18, 1950
2,579,399	Ruekberg	Dec. 18, 1950
2,632,202	Haines	Mar. 24, 1953
2,674,006	Bailey	Apr. 6, 1954
2,710,987	Sherman	Jun. 21, 1955
2,715,751	Weber	Aug. 23, 1955
2,750,024	Coates	Jun. 19, 1956
2,750,625	Colombo	Jun. 19, 1956
2,783,503	Sherman	Mar. 5, 1957
2,784,452	Ruekberg	Mar. 12, 1957
2,787,023	Hagen	Apr. 2, 1957
2,789,312	Borer	Apr. 23, 1957
2,789,313	Knowles	Apr. 23, 1957
2,790,994	Cardot	May 7, 1957
2,792,593	Hardgrove	May 21, 1957
2,804,654	Sherman	Sept. 3, 1957
2,805,787	Sherman	Sept. 10, 1957
2,810,934	Bailey	Oct. 29, 1957
2,853,736	Gussoni	Sept. 30, 1958
2,854,691	Strong	Oct. 7, 1958
2,858,564	Sherman	Oct. 14, 1958
2,878,520	Mumford	Mar. 24, 1959
2,896,251	Sherman	Jul. 28, 1959
2,901,769	Sherman	Sept. 1, 1959
2,903,740	Parfrey	Sept. 15, 1959

APPENDIX *(continued)*

Patent No.	*Inventor*	*Date*
2,913,762	Knowles	Nov. 24, 1959
2,914,799	Canfield	Dec. 1, 1959
2,918,698	Hagen	Dec. 29, 1959
2,919,462	Friden	Jan. 5, 1960
2,952,034	Fortner	Sept. 13, 1960
2,964,795	Schaich	Dec. 20, 1960
2,975,472	Colombo	Mar. 21. 1961
2,975,473	Hagen	Mar. 21, 1961
3,005,231	Pechthold	Oct. 24, 1961
3,011,216	Gussoni	Dec. 5, 1961

REFERENCES

1. Brown, G. S. "The Mechanical Processes of Blowmolding." *Plastics World* (May 1959): 16, 17, 20, 21.
2. "High Density Polyethylene Shampoo Containers." *Modern Plastics* 35 (July 1958): 156.
3. "Report on the Show." *Modern Plastics* 36 (January 1959): 100.
4. Jones, R. V. "High Density PE—Polymer Success Story." *Plastics World* (Apr. 1969): 32–39.
5. Domininghaus, H. *Introduction to the Technology of Plastics.* Hoechst Aktiengesellschaft, Frankfurt, 1975, pp. 7–9.
6. "Markets for Materials 1958." *Modern Plastics* 36 (January 1959): 79.
7. "The Plastiscope." *Modern Plastics* 36 (February 1959): 43.
8. Holzmann, R. "The Development of Blow Moulding from the Beginnings to the Present Day." *Kunststoffe* 69 (1979): 704–711.
9. DuBois, J. H., and F. W. John. *Plastics.* New York: Reinhold Publishing Corp., 1967, p. 12.
10. *50 Years of Progress in Plastics.* Resin Publications, Denver, 1987, pp. 22–41.
11. Bracken, W. O. "Blow Molding." *Technical Papers,* Vol. VI. 16th ANTEC, Soc. Plastics Eng., 1960.
12. Bracken, W. O. "Blow Molding Developments." *Technical Papers,* Vol. VII. 17th ANTEC, Soc. Plastics Eng., 1961.
13. Anastos, C. "Plastic Gallon Milk Container Is Here!" *The Milk Dealer* (January 1963): 48–49.
14. Scott, R. H. "Two Views of Plastic Milk." *The Milk Dealer* (December 1964): 54–58.
15. "Plastic Products." *Modern Plastics* 34 (December 1957): 128–129.
16. DuBois, J. H. *Plastics History U.S.A.* Cahners Pub. Co., Boston, 1972, pp. 14, 16, 346–365.
17. Conversation with Henry E. Griffith, VP and Gen. Mgr., PLAX Corp., during the early years.

SELECTED READINGS

Bailey, J., "Blow Molding," *Modern Plastics* 22 (Apr. 1945): 127–133, 198, 200.
Brown, G. S. "Blow Molding." *Modern Plastics Encyclopedia,* New York, NY, 1960, pp. 721–723.
Kovach, G. P. "Forming of Hollow Articles." In E. C. Bernhardt, *Processing of Thermoplastic Materials.* Robert E. Krieger Pub. Co., Huntington, NY, 1959 (repr. 1974), pp. 511–522.

SECTION 1

PROCESSES AND EQUIPMENT

CHAPTER

2

Extrusion Blow Molding

DON PETERS

Blow molding is the forming of a hollow object by "blowing" a thermoplastic molten tube called a parison in the shape of a mold cavity. Extrusion blow molding is the most widely used of many blow molding methods.

There are probably more differences in equipment for blow molding than for any other plastics fabrication technique. A blow molding machine may be the size of an office desk or may occupy a large room, making hollow objects as small as a pencil or as large as 5,000 gallons capacity or greater. There are also a great many operating variables in a blow molding process, which make it one of the more complex processes. This chapter covers five topics:

- Blow molding markets, summarized briefly.
- Processes and Equipment
- Controlling wall distribution, the heart of blow molding. Unlike injection molding, in which the wall thickness is automatically set by the dimensions of the mold core and cavity, many variables in blow molding affect wall thickness.
- Special blow molding techniques using moving sections mold with which the Phillips Plastics Technical Center has done much work.
- Blow molding applications and resins

BLOW MOLDING MARKETS

Blow molding grade high-density polyethylene (HDPE) comprises about 69 percent of all materials that are blow molded. The use of polyethylene terephthelate (PET) has increased rapidly in recent years and now claims about 21 percent of the blow molding market. PET use will probably continue to grow at a faster rate than that of high-density polyethylene, however, HDPE is still the major material of the blow molding industry.

Of the 8.2 billion pounds of HDPE processed domestically in 1988 roughly 37 percent (3 billion pounds) went into blow molding. About 8.1 billion pounds of HDPE were consumed in 1987, and of that about 36 percent was blow molded.

About 25 percent of the total blow molding poundage goes into milk bottles. That is a tremendous market, but consumption for household chemical containers (bleach bottles, detergent bottles, etc.) is even larger: 33 percent. About 76 percent of the total resin usage goes into bottles, and the other 24 percent is in industrial items—gas tanks, bulk-goods drums, seats, toys, and other large items. The rate of growth is faster in the industrial products area, but total consumption is lower.

PROCESSES AND EQUIPMENT

As discussed in the previous chapter, blow molding can be classified in two major categories, extrusion blow molding and injection blow molding, with many subdivisions. There are two extrusion blow processes, continuous and intermittent. Intermittent extrusion is subdivided into reciprocating screw, ram, and accumulator parison extrusion. Injection blow molding "injects" the molten parison onto a metal core rod. This process is covered in detail in chapter 3.

Continuous Extrusion Blow Molding

A stationary extruder plasticizes and pushes molten polymer through the head to form a continuous parison (Fig. 2-1). Because the parison does not stop moving, it is necessary to transfer the parison from the die head to the mold(s) either by means of arms or by moving the mold(s) to the parison(s). An example is a model B-2 Kautex machine once used in the Phillips 66 Company Plastics Technical Center (PTC) (Fig. 2-2). This machine transfers the parison away from the die head, a nearly obsolete process. The more common procedure is to move the molds to the parison. Bekum, (Fig. 2-3), Kautex, Fischer, Hayssen, Johnson Controls, and others build commercial continuous extrusion equipment that uses reciprocating action of molds or parisons.

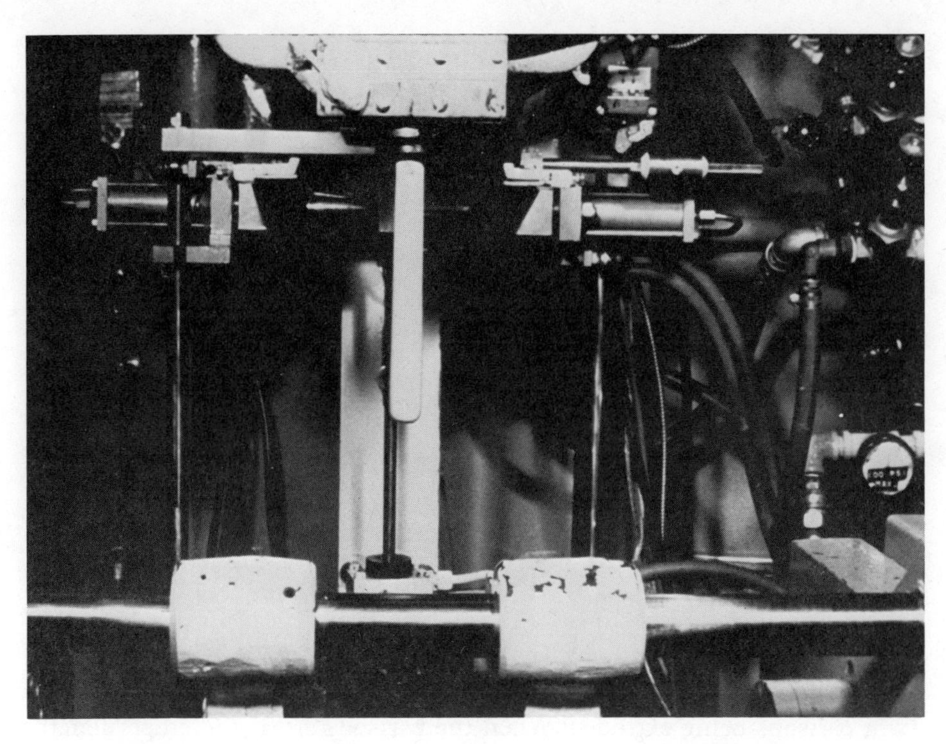

Figure 2-1 Model B-2 Kautex extrusion blow molding machine.

Figure 2-2 Parison transfer on a B-2 Kautex machine.

Figure 2-3 Bekum reciprocating action continuous extrusion machine.

In the Bekum machine (Fig. 2-3) the molds move diagonally up to the twin parisons being extruded. When the parisons are at the proper length, the mold halves close on the parisons and a hot knife cuts the parison between the top of the mold and die head. The press and molds then move back down to the molding station (shown by the alternate press in molding position) (Fig. 2-3). Two calibrating blow pins move down inside the top of the parisons to compression-mold a neck seal, introduce air inside the parisons, and blow the parisons out against the cavity walls. The two presses with double cavity molds move alternately back and forth, giving four-mold production. In other machines, the action may be a straight horizontal movement of the molds or parison.

Three of the equipment manufacturers mentioned above are German; through the years they have been the leaders in blow molding technology, although American-made equipment is superior in some respects. Some of the biggest blow molders in the United States use continuous extrusion, including Continental Can (also called Continental Plastic Industries). The equipment uses large extruders and wheels that may hold up to twenty-five molds (Fig. 2-4). As the wheel rotates, molds index and close on the parison. The machines use a needle to puncture the parison in order to introduce air inside. Rotary equipment is expensive. It is high production rate equipment and that must produce million of units to be economical. Innopak (formerly Monsanto and PLAX) also has this type of equipment. Graham Engineering has a rotary machine that is usually equipped with either nine or twelve double molds on a vertical wheel (Fig. 2-5). The parison is extruded upward, actually being pulled up by the preceding mold. The molds on this machine also use a needle blow.

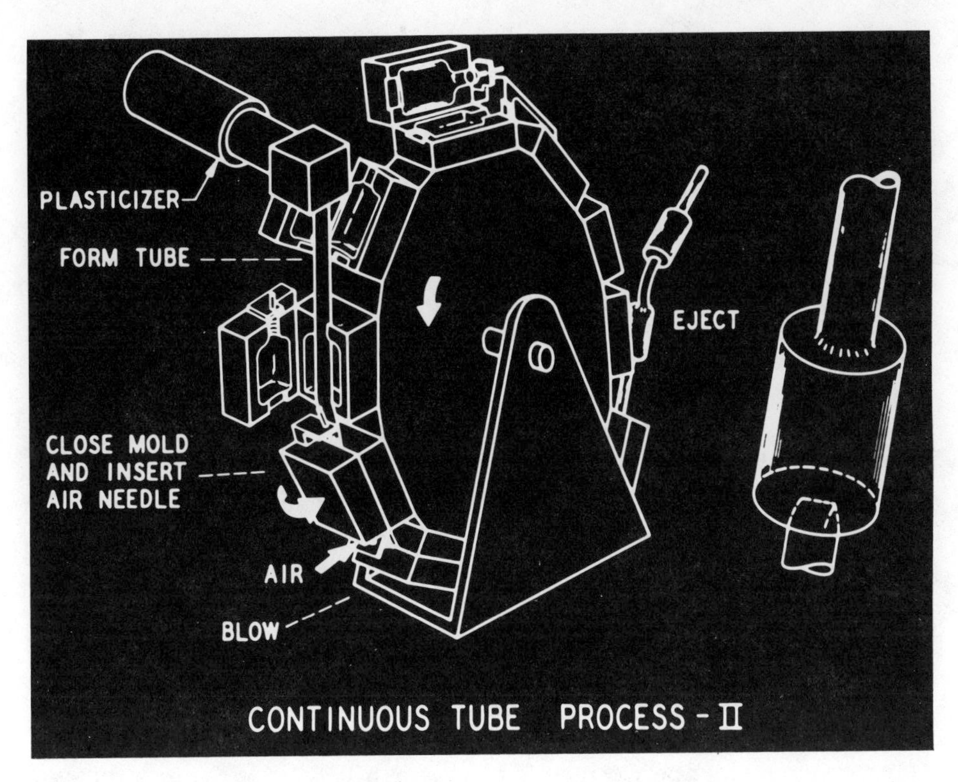

Figure 2-4 Continental Can Co. mold wheel configuration.

Figure 2-5 Graham Engineering wheel.

Intermittent Extrusion Blow Molding

Reciprocating Screw Machining

Intermittent extrusion may be also called shot extrusion. Parison shot extrusion is accomplished by means of a reciprocating screw almost identical to those used in injection molding machines. The resin is plasticized or melted; as it melts, the screw rotates and retracts, "charging" the shot in front of the screw. The screw is rammed forward by hydraulic means, pushing the plastic through the head and out the die head tooling as a parison (Fig. 2-6). At this point, the mold(s) close. Parison shot extrusion is rapid compared to continuous extrusion. For a typical blow molding machine producing milk bottles, the parison extrusion time is 1.5 seconds. A two-head Uniloy dairy machine used by Phillips PTC for studies, customer problems, and operator training is shown in Fig. 2-7. It has the same controls as commercial Uniloy four- and six-head machines (Fig. 2-8) used to mold more than 90 percent of the milk bottles in the United States.

Figure 2-9 shows the cooling conveyor table of a commercial four-head machine. In a normal milk bottle production line there are either four or six one-gallon bottles on the cooling table. The cooling table delivers the bottles with "flash" (the part of the parison exterior to the cavity that is compression molded between mold halves) to the trimming conveyor. These bottles are automatically indexed into the trimmer, which cuts or knocks off the neck, handle, and flash. The bottles are automatically leak tested on-line and are conveyed to a silk screen printer where the appropriate dairy

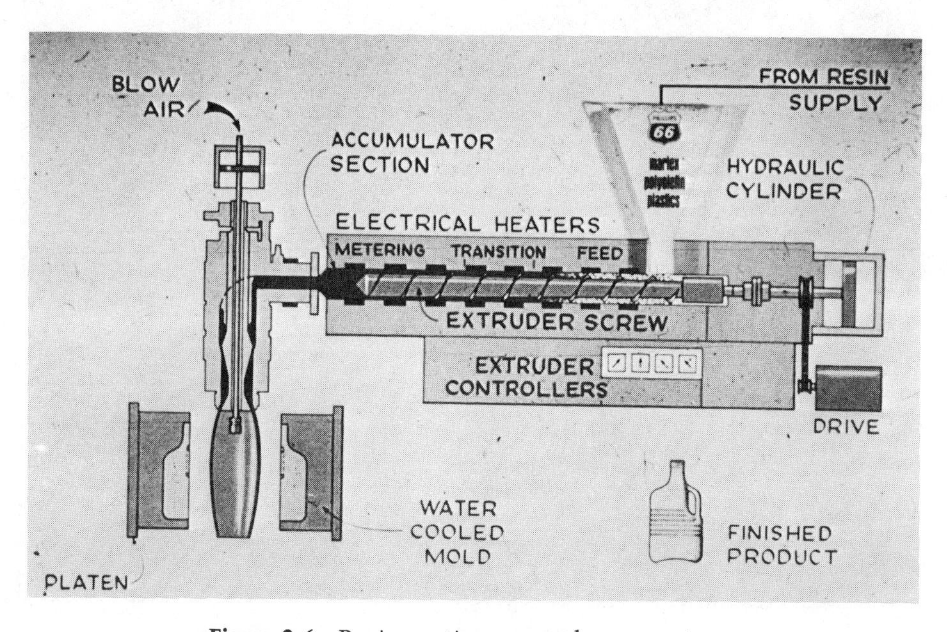

Figure 2-6 Reciprocating screw shot extrusion.

Figure 2-7 PTC dairy machine.

Figure 2-8 Four-head dairy machine.

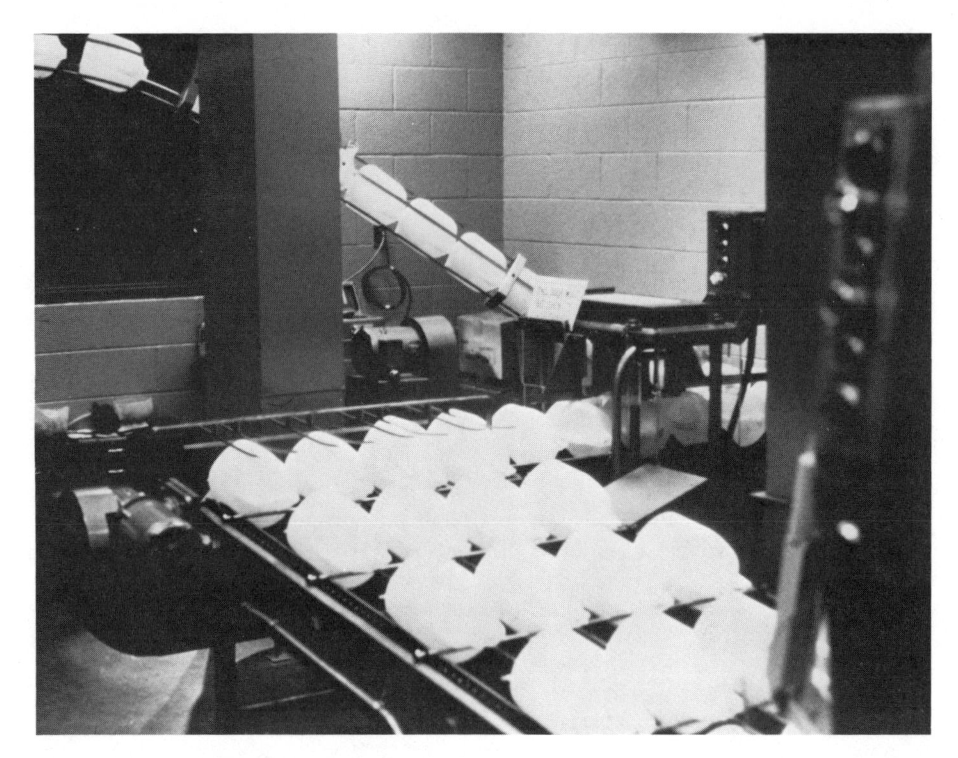

Figure 2-9 Uniloy commercial dairy bottle machine.

Figure 2-10 IMPCO model B-30, five-pound parison shot extrusion machine.

logo is printed on each milk bottle. From the decorating operation the bottles are moved on a combination conveyor/storage line, eventually going to the milk filling equipment. The conveyor system has a large amount of "surge" (backlog inventory) to keep the filling lines going in case the blow molding machine is shut down for short periods of time. In a normal milk bottle molding operation the plastic bottle is not touched from the time it is blow molded until it has been filled, capped, and placed in a carton that normally contains four one-gallon bottles.

The largest reciprocating screw blow molding machines sold commercially in the United States are the Improved Machinery Co. (IMPCO) models B-24, B-30, and B-50, which respectively have four-, five-, and twelve-pound parison shot extrusion capacities (Fig. 2-10.) These machines are used mostly for small industrial parts.

Ram Extrusion

Figure 2-11 is a schematic diagram of an intermittent extrusion blow molding machine with a ram (piston) accumulator remote from the die head. The ram pushes the accumulated material through the die head to form the parison as a shot. Although this is an obsolete process, this type of equipment is still operating in the United States. The extruder is stationary and con-

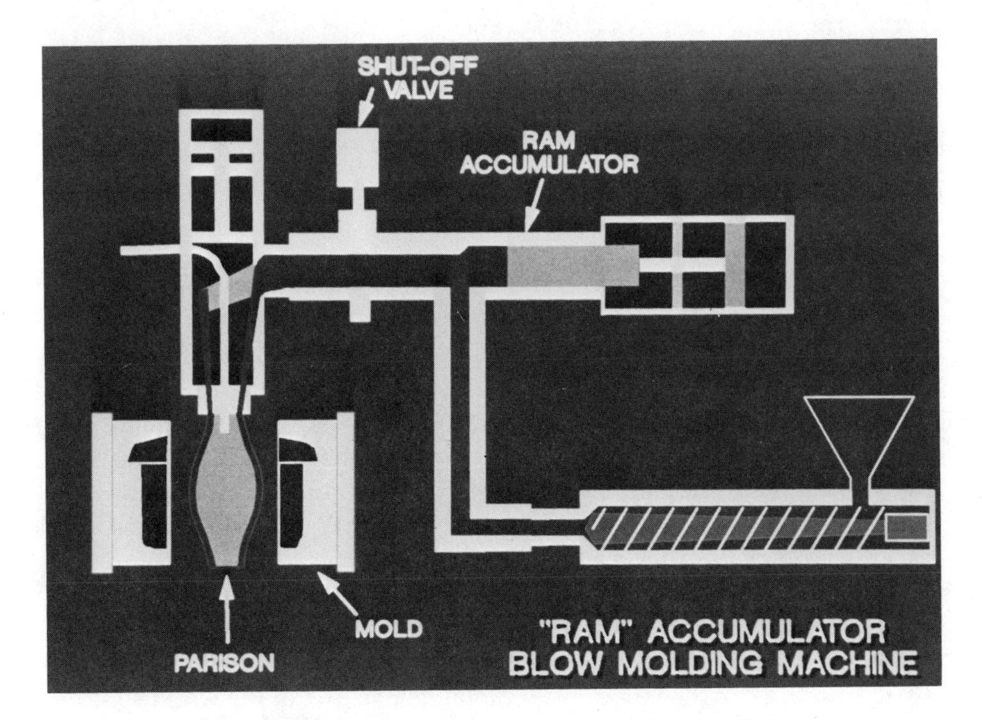

Figure 2-11 Ram accumulator extrusion blow molding.

Figure 2-12 Producto ram extrusion machine.

tinuously feeds molten polymer into an accumulator. The extruder speed is adjusted to fill the accumulator at the appropriate rate so that the parison is extruded immediately after the mold has opened and the part has been removed. An example of this kind of equipment is the Producto machine shown in Figure 2-12. Black, Clawson and Hartig have a great deal of this kind of equipment still in operation.

Accumulator Head

Parison ram extrusion has generally been replaced by equipment using accumulator heads. A diagram of a Kautex accumulator head is shown in Figure 2-13. The accumulator head contains a complete parison shot in annular form. This results in a parison of much more uniform circumferential wall thickness. The parison is extruded by moving an annular ring downward to displace the material through a die bushing and mandrel. The Hartig machine in the PTC has a fifty-pound accumulator head, shown in Figure 2-14. This machine is used to make fuel tanks, 30-gallon drums, and other large items.

Hartig Division of Somerset Technologies, Johnson Controls, Hayssen, and APV manufacture blow molding machines with accumulator heads that are similar in concept to the Kautex head, although all have significant differences. In 1977 the largest blow molding machine in the world was the Kautex shown in Figure 2-15, in which four extruders feed an accumulator

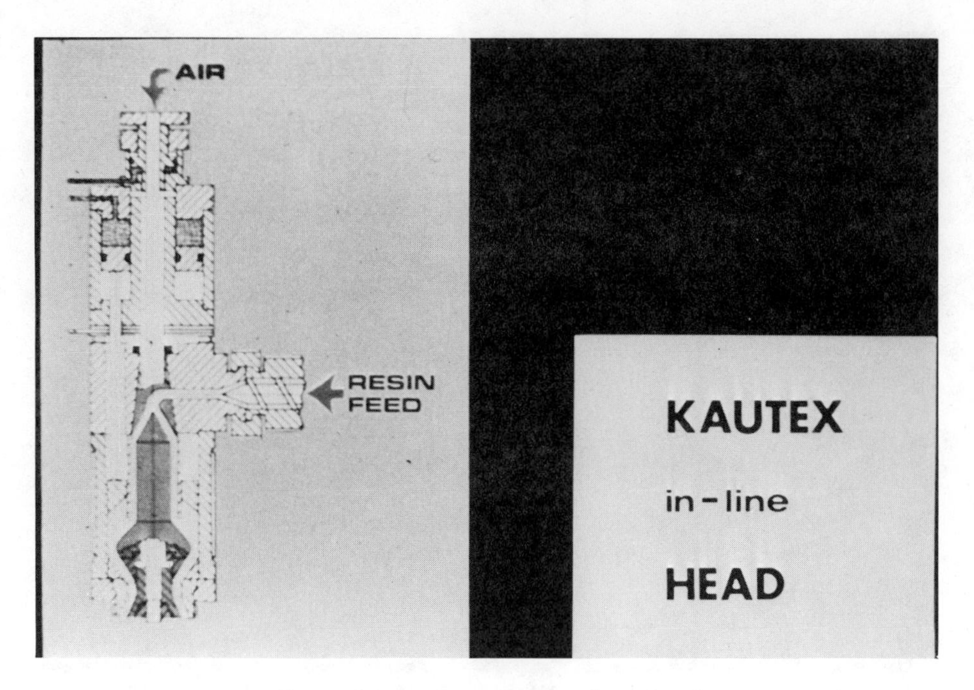

Figure 2-13 Kautex accumulator head.

Figure 2-14 PTC Hartig machine with fifty-pound accumulator head.

Figure 2-15 Large Kautex machine with four extruders.

head with 300 pounds shot capacity. The machine, located in Germany, is shown blow molding a 1,320-gallon heating oil tank. Larger machines have been built since that time and reportedly are located in either East Germany or Russia. Machines in this size range normally use high molecular weight resins so that the parison will have enough melt strength to hang on the die head during mold closure.

Oriented Process

There are several different concepts in oriented parison blow molding, one of which is the Orbet process developed by Phillips. By orienting a parison at a temperature slightly below the melting point of the polymer, significantly improved impact performance, stiffness modulus, and clarity can be achieved. An Orbet orientation blow molding machine built by Beloit is shown in Figure 2-16. It is a two-process operation. The parisons are made in an extrusion operation much like pipe. The cut-to-length, open-ended parisons are reheated in an oven (Fig. 2-17) to just below the melting point, then are transferred to the molding station (Fig. 2-18) where they are stretched longitudinally by mechanical means (right, Fig. 2-19). Axial orientation occurs when the parison is blow molded (left, Fig 2-19). The PET beverage bottle business is oriented parison blow molding using an injection molded parison.

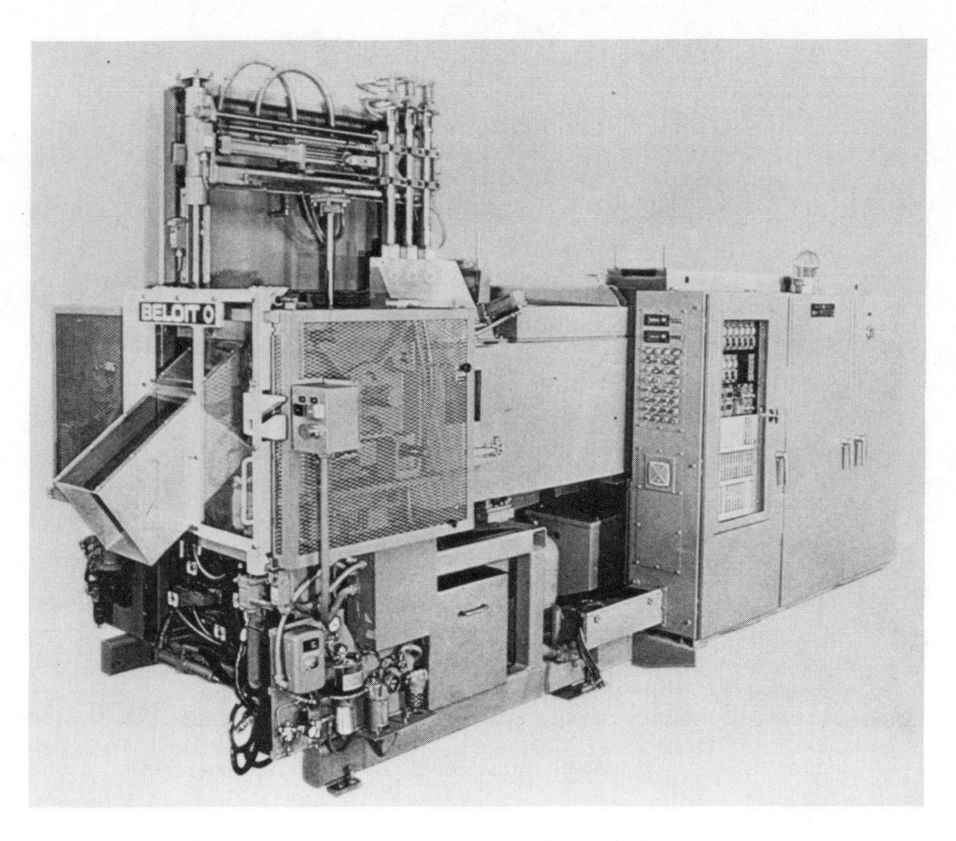

Figure 2-16 Beloit Mark IV oriented blow molding machine.

Figure 2-17 Parisons being reheated in Mark IV oven.

Figure 2-18 Parisons being transferred to Mark IV molding station.

Figure 2-19 Parisons being stretched in Mark IV machine.

CONTROLLING WALL DISTRIBUTION

Control of wall distribution is the heart of blow molding. There are two primary techniques in extrusion blow molding for controlling wall distribution: programming and die shaping. Programming is the control of the wall thickness, from top to bottom, of the parison as it emerges from the die head tooling during extrusion. In die shaping, sectors of the die bushing or mandrel are machined to thicken the parison longitudinally in those areas where the part being formed requires greater thickness. The diameter of the die tooling is very important, for it determines the parison diameter. Too small a parison will rupture or "blow out" because of too much stretch. Too large a parison will result in too much flash, and cause trimming problems. The following discussion of controlling wall distribution assumes a parison of correct size.

Parison Programming, Extrusion Blow

Figure 2-20 shows a programming die head. The center rod or drawbar is movable, by hydraulic means. The drawbar is normally attached to a mandrel that forms the inside of the parison. A bushing shapes the outside

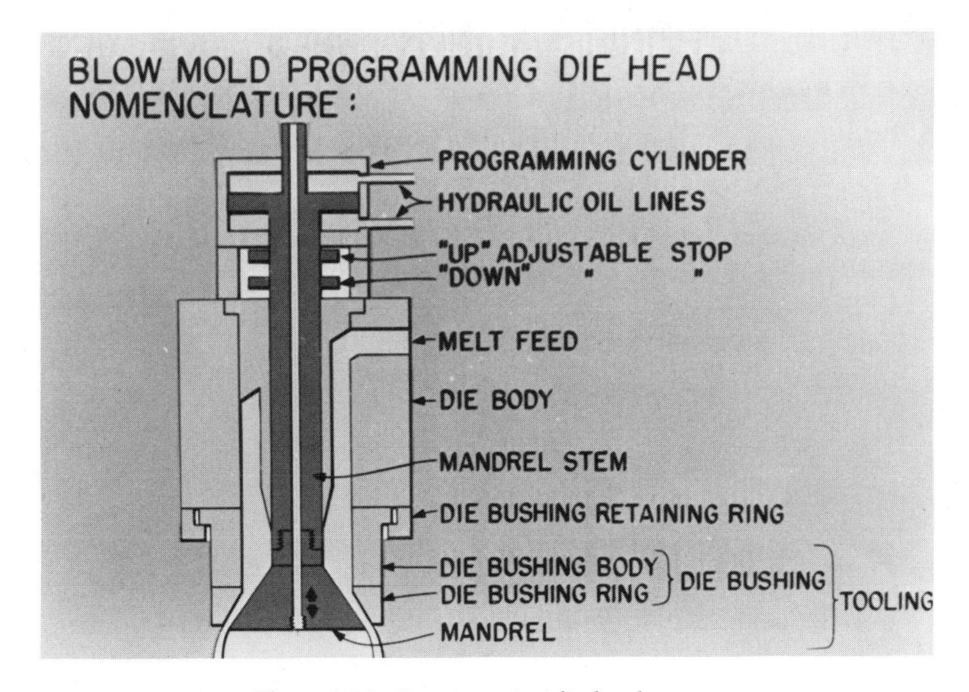

Figure 2-20 Programming die head.

of the parison. The die head bushing and mandrel have either diverging (Fig. 2-21) or converging lands. The angular lands allow die gap and resultant parison thickness changes with vertical movement of the mandrel; the sequence of mandrel movements constitutes the parison programming.

Control of the programming is normally accomplished electronically, either with a patch panel or by microprocessor (Fig. 2-22). A patch panel is programmed with pins or slides for the amount of mandrel movement. Each point on the patch panel feeds a signal to a servo valve, which admits oil to a hydraulic cylinder at the top, resulting in drawbar and mandrel movement that is sensed by a linear variable differential transformer (LVDT). The transformer feeds back a signal to the patch panel to produce electrical null, at which point the servo valve shuts off oil, stopping movement. The programming proceeds to the next point on the patch panel. There may be ten, twenty, thirty, or more points on a programming panel; each point can control one change in wall thickness during a single parison extrusion. Most of the electronic patch panel programmers used today have thirty-four points. The two manufacturers of such programmers in the United States are Hunkar and Moog. Figure 2-23 is a photograph of a Hunkar programmer on the large machine in the PTC.

There is also a mechanical means of programming by use of a rotary cam (shown in Fig. 2-24) to admit oil in and out of a cylinder. In this case, the external body or the bushing is moved up and down for programming.

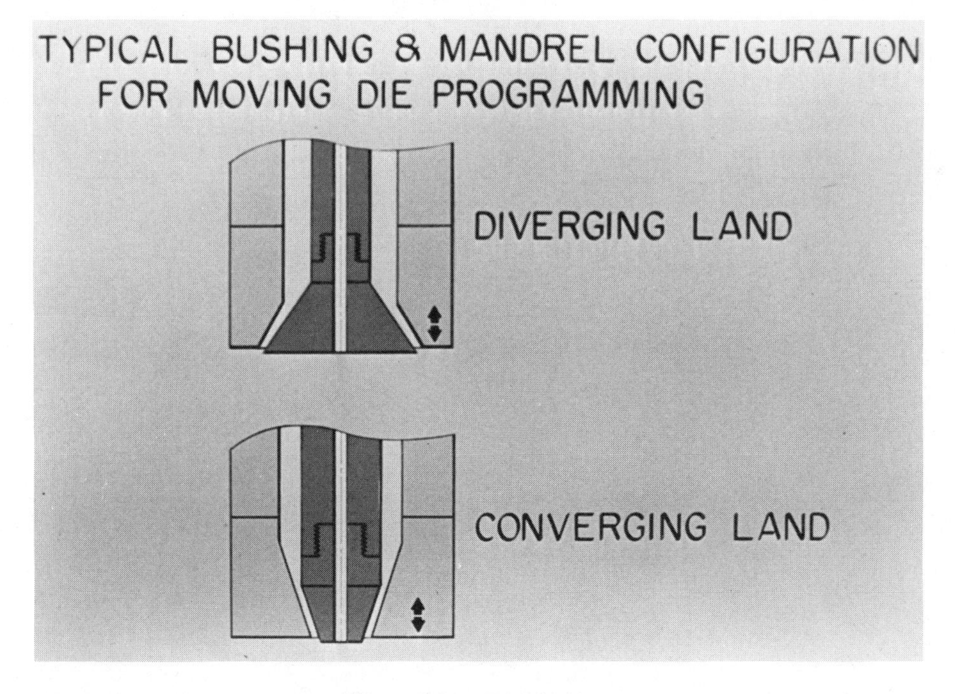

Figure 2-21 Die tooling.

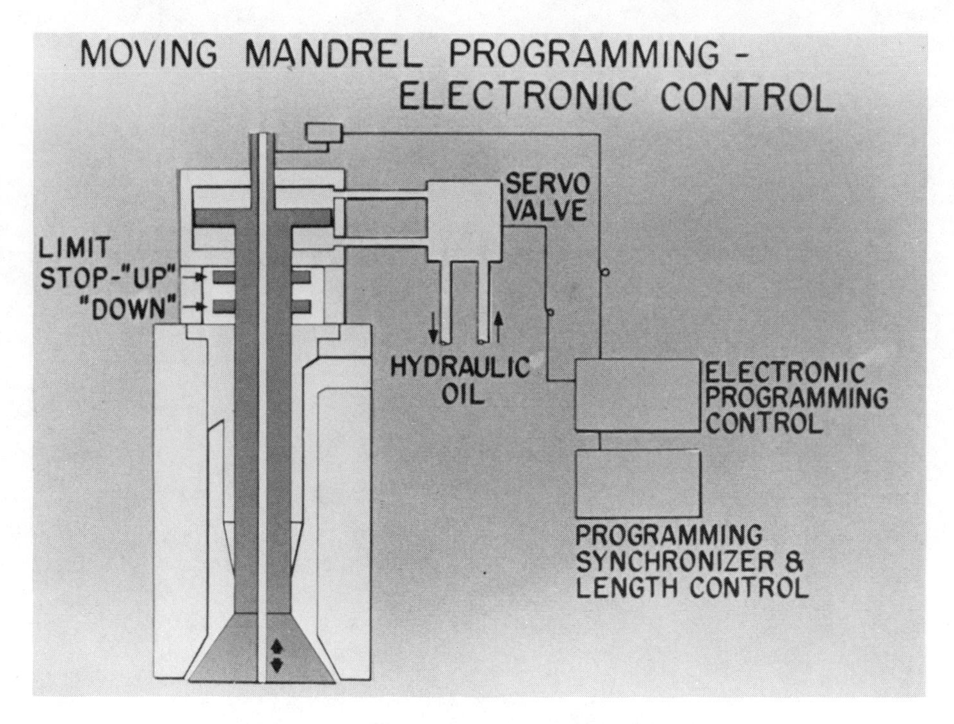

Figure 2-22 Moving mandrel programming.

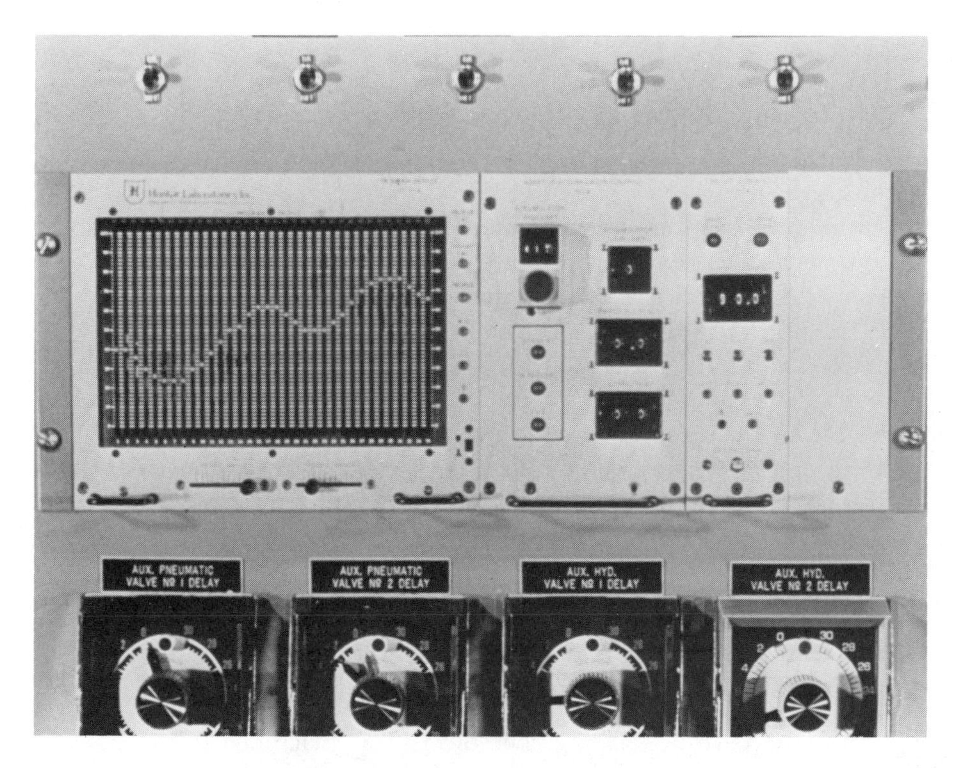

Figure 2-23 Hunkar programmer.

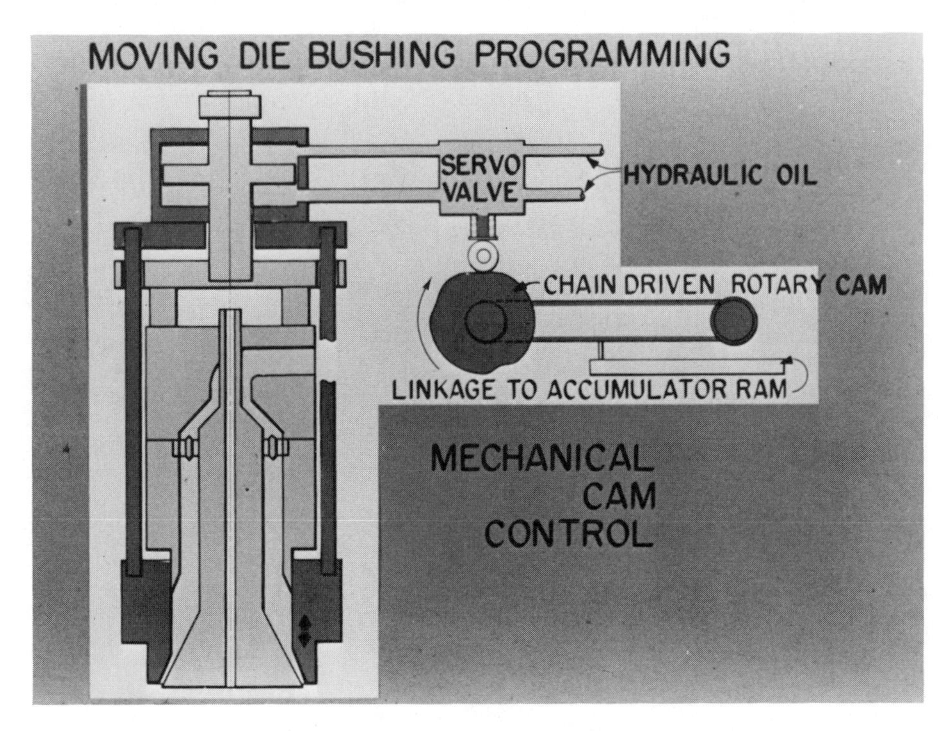

Figure 2-24 Cam-controlled programming.

It is an effective programming system, but is a much more massive head because drawbars and cylinders must be located externally and hydraulically balanced. The cam has been made largely obsolete by electronic and computer systems.

Microprocessors are normally used for parison programming control in today's new equipment. Barber-Coleman is the major supplier of microprocessors in the United States. The Sterling Company (now APV) first introduced microprocessor control for parison programming on their blow molding equipment. They also control most of the machine sequencing by proportional hydraulics and microprocessors. Other equipment manufacturers are now using microprocessor controls, either as standard equipment or as an option. It is expected that this type of control will eventually be used in all blow molding equipment, for nearly all functions.

Die Shaping, Extrusion Blow

The other primary technique often used to improve wall distribution on a blow molded part is die shaping. Irregularly shaped parts such as a gas tank usually need die shaping. The die land is machined to increase the annular die gap in the sector or sectors corresponding to areas of the part

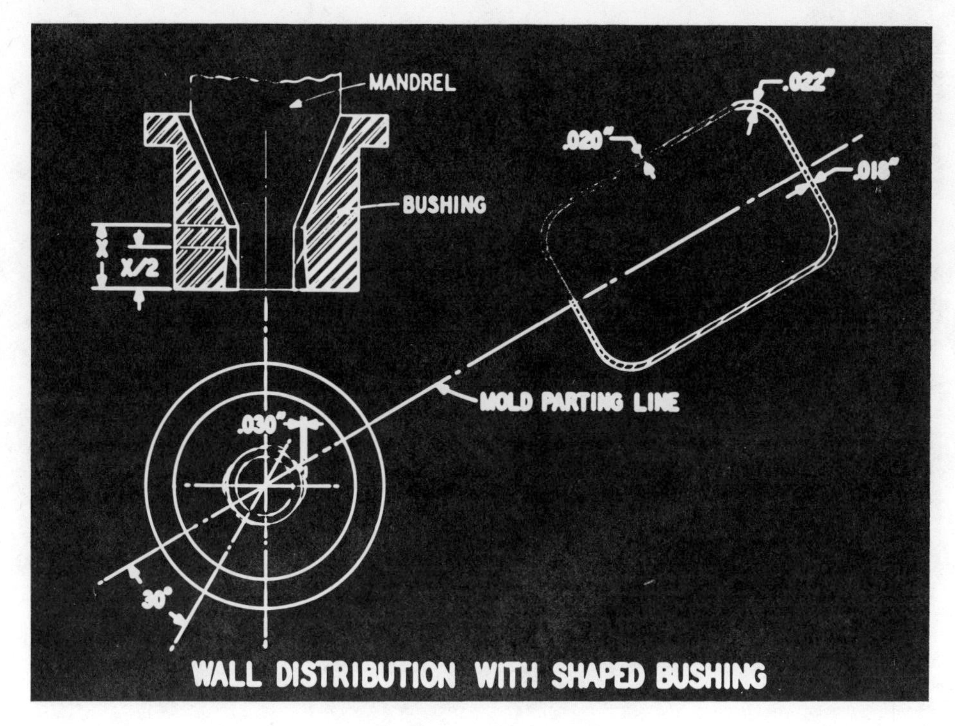

Figure 2-25 Die shaping.

that need to be thickened. Figure 2-25 shows a bushing with four angular shapes in the die land to increase parison and wall thickness in a square item. Shaping can also be done on the mandrel. The choice of whether to shape a bushing or a mandrel is normally dictated by whichever is the easiest piece to machine—usually the bushing with diverging tooling, and the mandrel with converging tooling.

Choosing the areas to be shaped is generally done by marking the parison with a grease pencil as it is extruding, although computer calculated methods are being developed. By trial and error, the area or areas of the parison that require shaping are marked (Fig. 2-26). The corresponding positions are then marked on the die tooling. Each position is marked because flow patterns through the heads are not easily calculable and shaping positions will not precisely correspond to the symmetry of thin areas in the part. A properly marked parison for a square battery box is shown in Figure 2-27.

The depth, width, and land length of the shaping cut in the die are currently based on experience. Thin-wall gallon bottle tooling probably would be shaped 0.022–0.003 in. deep. That is very little, and requires a precise setup by the machinist (Fig. 2-28). Intermediate thickness shaping may be 0.060–0.080 in. deep and thick wall shaping 0.150–0.180 in. deep. Figure 2-29 shows two intersection shapes being cut in a 12 in. diameter bushing for a fuel tank. The cuts are 0.080 and 0.100 in. deep, which is typical for this type of part.

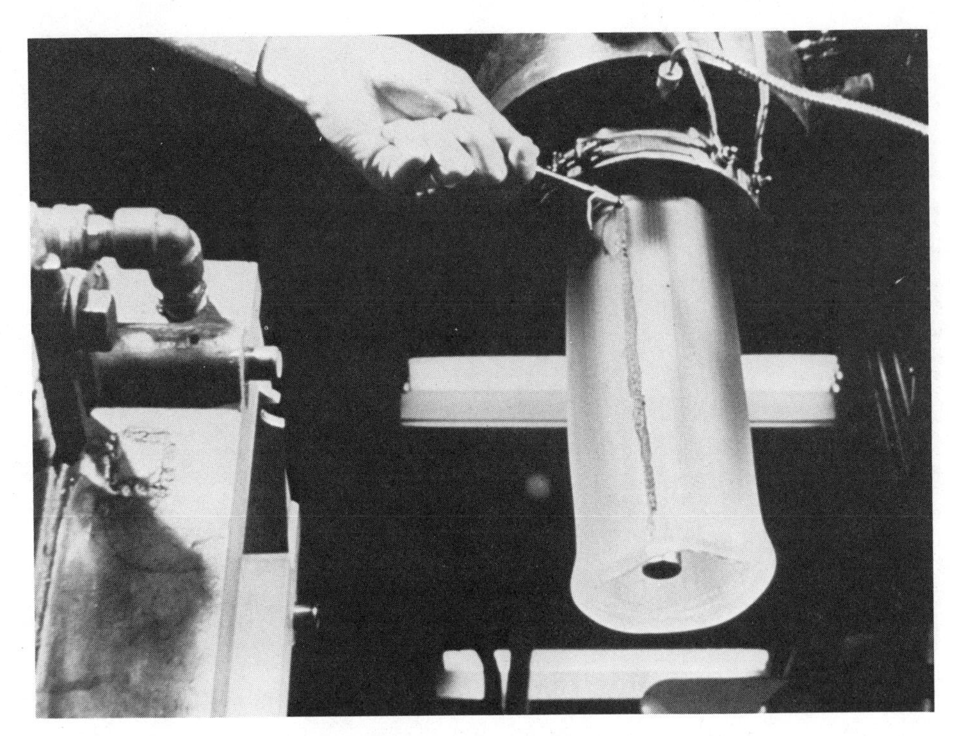

Figure 2-26 Marking a parison.

Figure 2-27 Marked part.

Figure 2-28 Shaping a bushing for a thin-wall product.

Figure 2-29 Shaping a bushing for two part thicknesses.

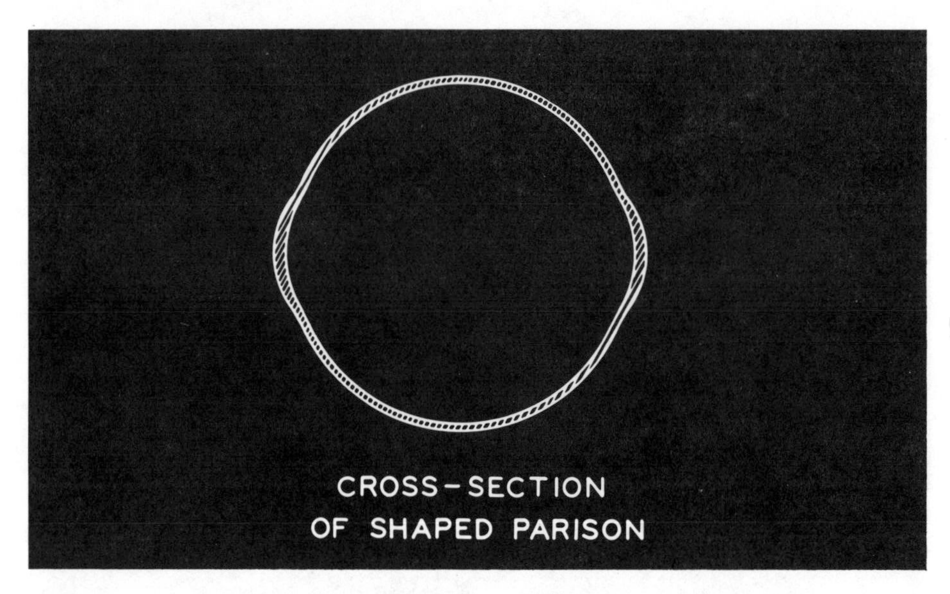

Figure 2-30 Correct shaping.

Some round symmetrical parts, such as a drum, also require die shaping because the areas or parison forming the top and bottom corners have to blow (stretch) the radius of the part 90 degrees from the parting line. Correct shaping results in a smooth, thicker bulge, which will cause few problems in blow molding (Fig. 2-30). Shaping too deep will result in the parison extruding more rapidly and a longer parison in those areas, but the shaped areas are still a part of the slower moving, unshaped parison. The convoluted, unstable parison in Figure 2-31 is an example of too much shaping. The solution is to machine out part of the shaping. This is more easily done than to shape deeper, because the machining setup is concentric rather than eccentric (as required for shaping).

Usually a combination of programming and die shaping is desirable. This is the case in a conical-shape part where shaping is cut on a converging mandrel (Fig. 2-32). The die mandrel moves up to the maximum die gap to extrude a thick parison for the corners. At this position, the shaping is all within the die and maximum benefit from the shaping is achieved. As the die gap is reduced to thin the parison for the middle area of the part, the shaped mandrel extends out of the bushing, which reduces the useful shaped land length. This is effective with a cylindrical part because a small amount of shaping is needed even in the center of a round part. When mold halves close, both ends of the parison are flattened, resulting in an elliptical or oval parison. The areas near the end pinch-offs will blow to the wall first (since they are very close) and as a result be relatively thick. The corner areas 90 degrees from the pinch-off are formed from parison that blows the radius of the part, requiring maximum programming and shaping. The areas 90 degrees from pinch-offs blow more uniformly.

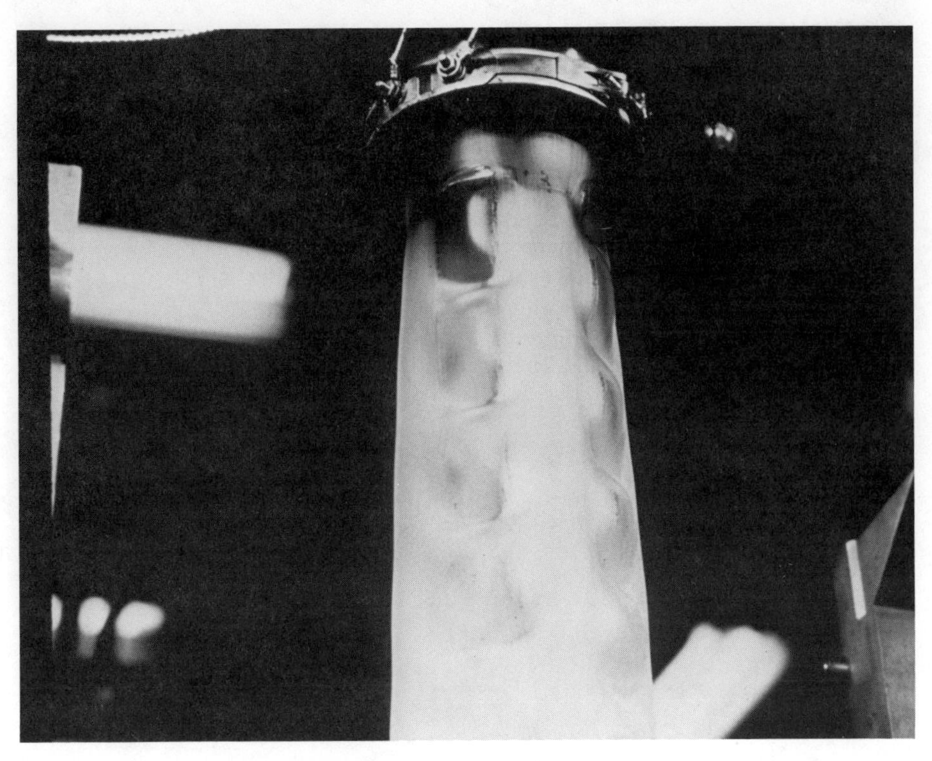

Figure 2-31 Overshaping.

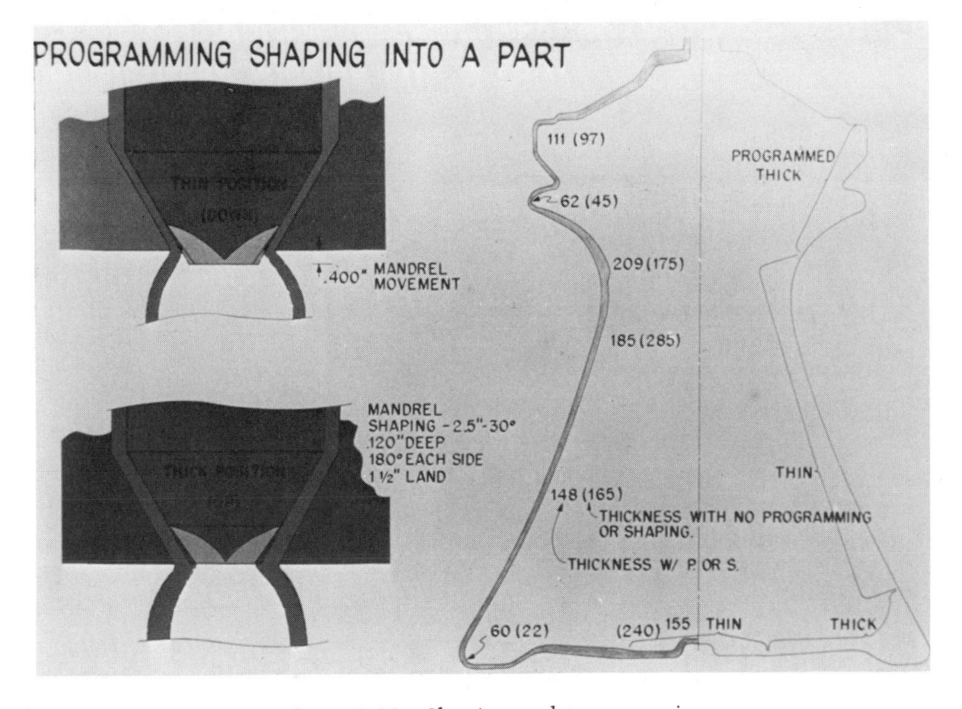

Figure 2-32 Shaping and programming.

MOVING SECTION MOLDS

The Phillips Plastics Technical Center has done extensive work with moving section molds. One technique is being used commercially to make a threaded neck off the parting line of a mold. This is done with a reciprocating plug. The sequence is:

1. Blow the parison against the extended plug (Fig. 2-33)
2. Retract the plug during the blow operation (Fig. 2-34)

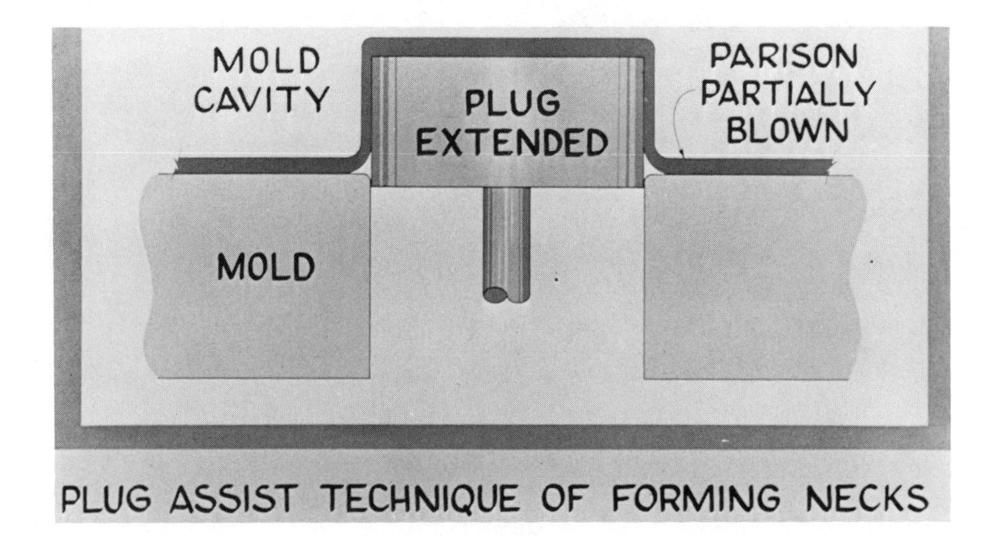

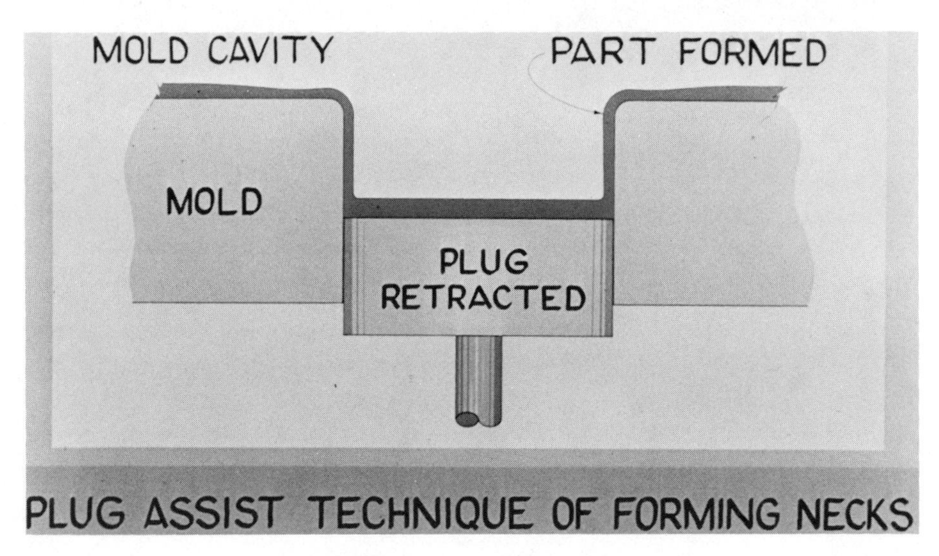

Figure 2-34 Plug assist: plug retracted.

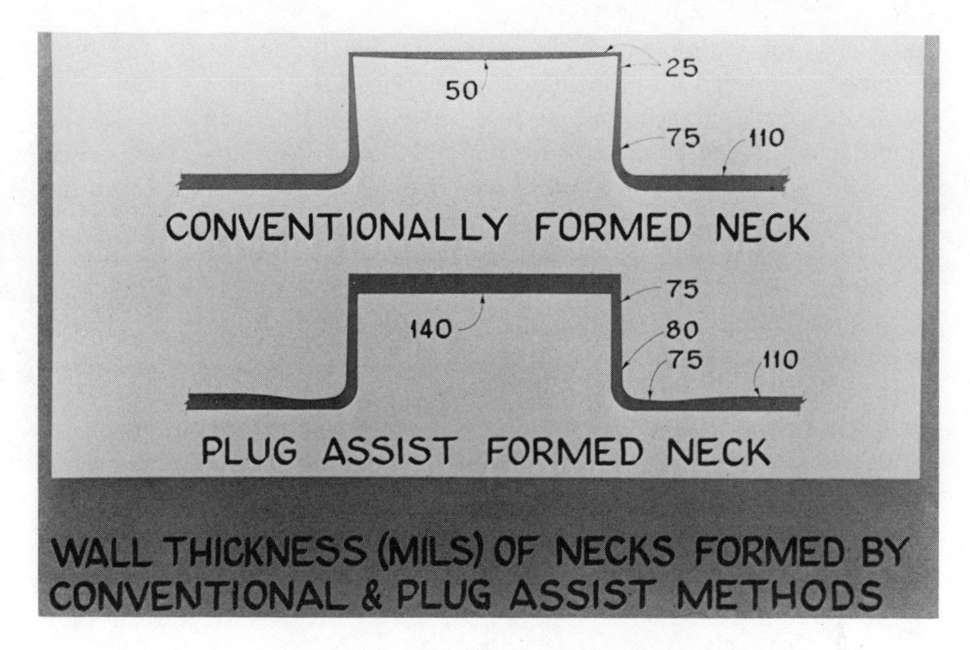

Figure 2-35 Plus assist wall distribution.

Figure 2-36 Scorpion snowmobile fuel tank.

Wall distribution is significantly improved from conventionally molded necks, as shown in Figure 2-35. Blow time and plug movement times must be precise. This type of neck was made commercially on a snowmobile fuel tank (Fig. 2-36) by Crosby Manufacturing, using high molecular weight HDPE. It saves either a postmolding operation of a fusion-welded neck, or a molded-in encapsulated anchor ring for postmold neck attachments.

Moving section molds are used to make water cooler lids having integral handles. Many millions of cooler lids have been made by both Igloo and Gott. An integral handle lid has been in production at Igloo for twenty years and Gott Corporation has made an integral handle, internally threaded, double-wall cooler lid for seventeen years (Fig. 2-37). Gott's screw-on lid carries the cooler body. The one-piece handle avoids having to install a handle in a subsequent operation. One mold half consists of a split cavity (Fig. 2-38) powered by air cylinders to form the integral handle. The other half has a threaded, rotatable core to form the internal threads (Fig. 2-39). The molding sequence of this part is as follows. The extruding parison is prepinched (sealed at the open end). Very low pressure is introduced into the parison before mold closure (the slight positive pressure results in the parison covering the handle area between the split cavities, and at the same time insures that the parison will be wide enough to flash the core). The mold halves and split cavity mold sections close almost simultaneously, resulting in a two-plane pinch through the handle and around the lid skirt. Blow air is introduced, forming the lid. The part is cooled. The thread core rotates, unscrewing from the internally threaded lid. The split cavities and mold halves all open at about the same time (Fig. 2-40), ejecting a part with flash as shown in Figure 2-41, which trims to the lid shown previously (Fig. 2-37).

Mauser drums, made world wide (Fig. 2-42), use an allied technique to make the L-ring that makes it possible to handle a plastic drum similarly to a metal drum. The PTC built a moving section mold that compression-molds a handling ring on a 30-gallon drum (Fig. 2-43) in an effort to improve the handling characteristics and therefore the market for such a drum. By using this technique, a solid chime, or ring (Fig. 2-44), can be molded at the head of the drum on which standard drum handling equipment can be used.

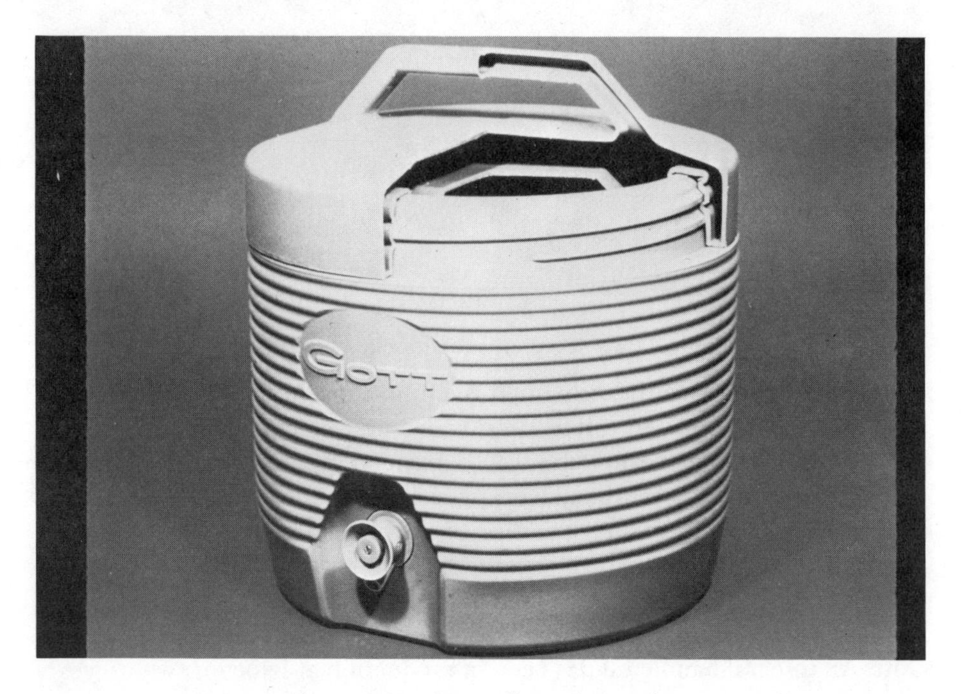

Figure 2-37 One-gallon Gott cooler.

Figure 2-38 Prototype mold, split cavity.

Figure 2-39 Prototype mold, core.

Figure 2-40 Prototype mold, mold opening.

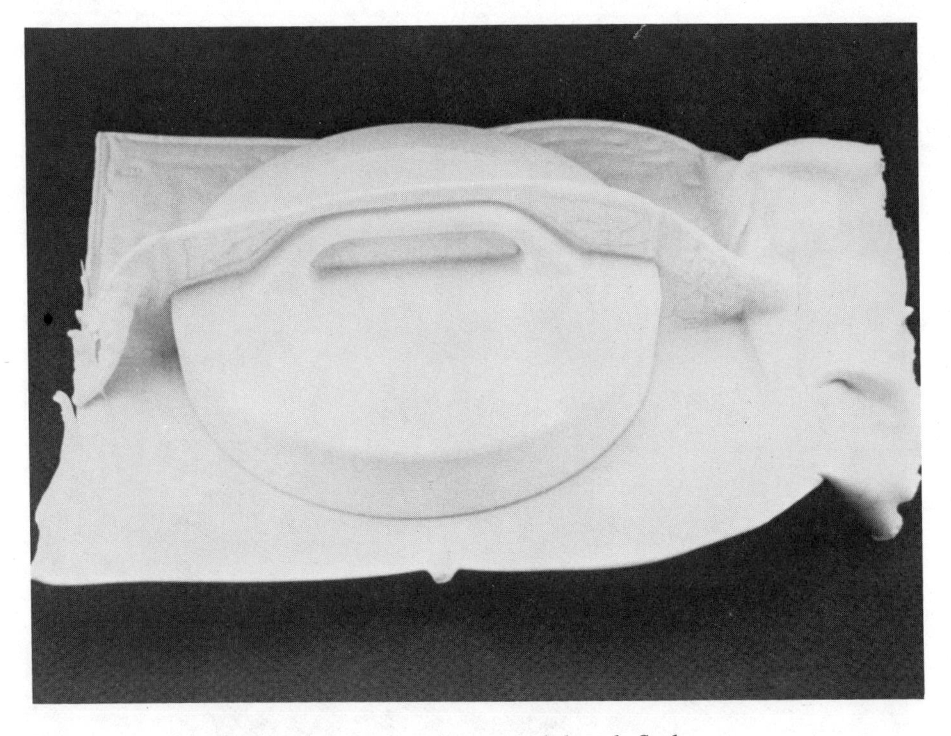

Figure 2-41 One-gallon Gott lid with flash.

Figure 2-42 Mauser drum.

Figure 2-43 Thirty-gallon drum.

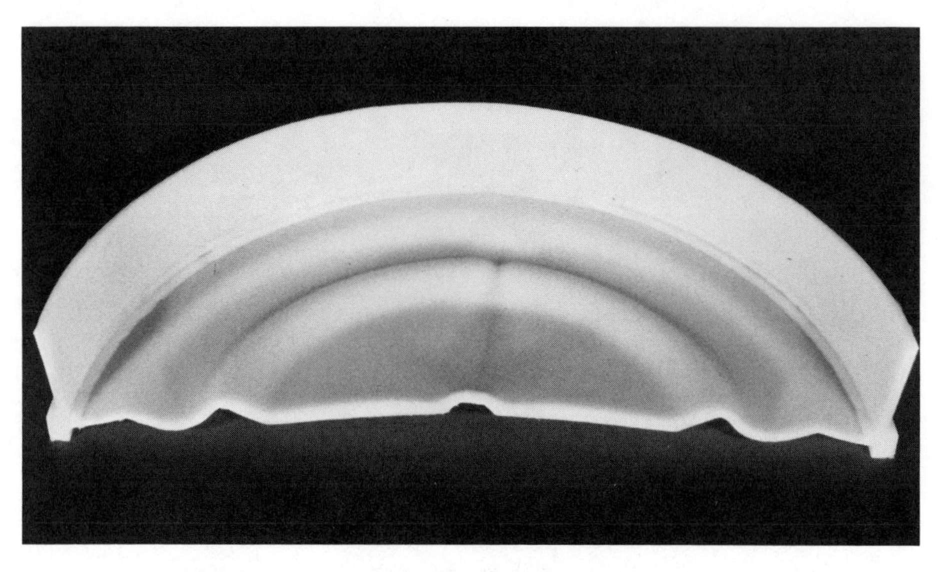

Figure 2-44 Section of a drum.

Sections of the chime are shown in Figure 2-45. Although these drums can be molded with excellent control of wall thickness, and produced consistently, poor impact resistence can become a problem with this design. The reduced impact tolerance is due to three factors:

1. A thick, solid, relatively nonflexible ring is adjacent to a much thinner flexible side wall
2. Normally a notch is created by the moving mold section at the juncture of the solid ring and side wall
3. Built-in stresses result from the greatly different cooling rates of the heavy solid ring and thinner adjacent side wall

For these reasons, it was necessary to use high molecular weight (MW) resins (in the 2 HLMI [high loss melt index] range) to pass a six-foot, 0° F drop impact test. The standard L-ring drum made by Mauser also requires a high MW resin, usually about 1.7 to 2.3 HLMI. Recent improvements in L-ring drum design now allow successful use of 5–10 HLMI resins in these drums.

With this molding technique, a hollow-wall, semihollow, or completely solid chime can be molded by changing the timing when the mold sections move. The three chimes are shown in Fig. 2-45. The key to the success of this technique is precise control of the process so the part will be reproduced identically time after time. A solid chime can be molded with good consistency, but hollow and semihollow chimes are more difficult. Advances in resins and equipment technology have resulted in this being a practical molding method.

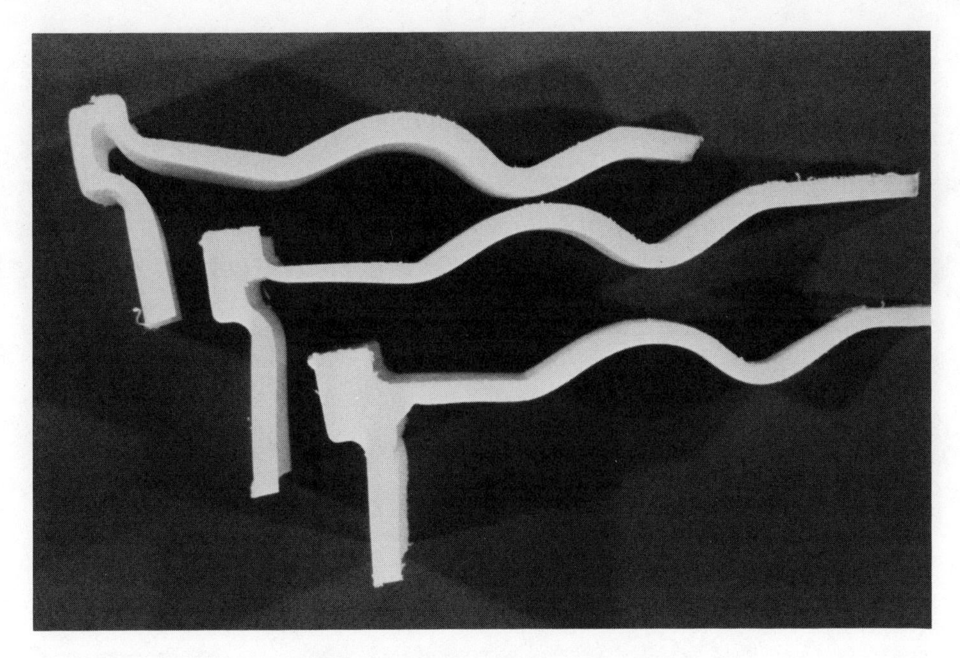

Figure 2-45 L-ring cross sections.

BLOW MOLDING APPLICATIONS AND RESINS

The following material illustrates eighteen applications of blow molding to manufacture plastics products and discusses the resin requirements. A general relationship of melt index (MI) [which is an inverse measure of molecular weight (MW)], density and stress cracking resistance (ESCR) exists in part properties. Parts made from low MI resins exhibit greater toughness and ESCR. Density affects stiffness—the higher the density, the greater the stiffness and the lower the ESCR at a given MW. Processability is usually more difficult with the high MW resins.

One-gallon milk bottles (Fig. 2-46) are commonly made using 55 to 72 grams of HDPE. A resin with high stiffness (0.966 to 0.964 density) and good processability is required, but no significant degree of environmental stress crack resistance (ESCR) is necessary. A 0.7 Melt Index homopolymer HDPE is the preferred resin. About 750 million pounds of HDPE is used each year in milk bottles.

In HDPE pharmaceutical bottles (Fig. 2-47) there also is essentially need for stress cracking resistance. A very clean resin with good processability is needed. Both 0.7 MI homopolymers and 0.3 MI copolymer resins are widely used for pharmaceutical bottles.

Antifreeze bottles (Fig. 2-48) are typically made in the one-gallon size. Because they are filled, placed in cartons, and palletized for warehousing, top loads are present, particularly on bottom layers. For this reason an intermediate resin with reduced density and stiffness but having high ESCR resistance is required. Resins with 0.3 MI, 0.955 density are normally used.

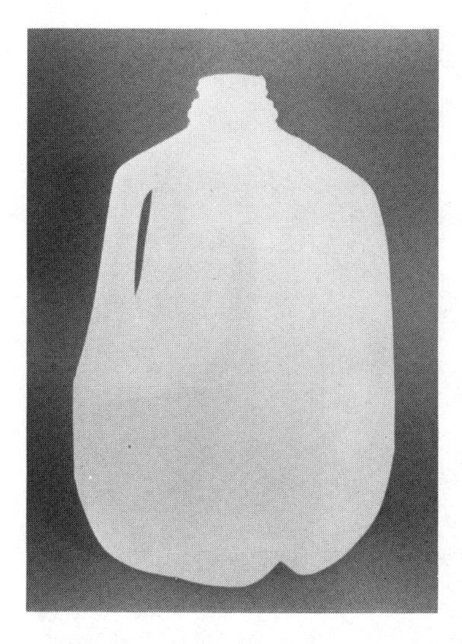

Figure 2-46 One-gallon milk bottle.

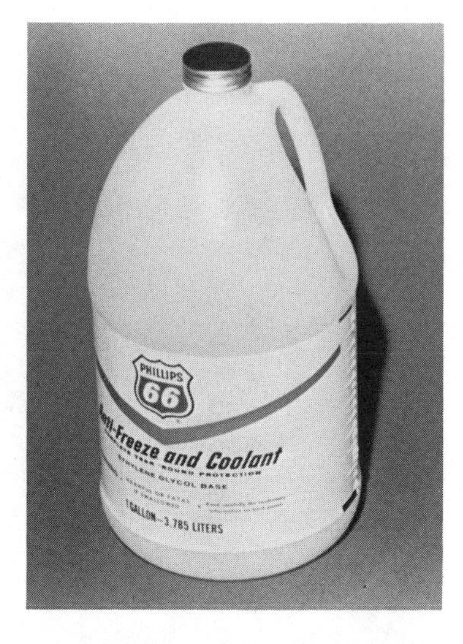

Figure 2-48 Antifreeze bottle.

Figure 2-47 Pharmaceutical bottles.

Polypropylene copolymer is used where for bottles see-through clarity and improved hot fill characteristics are needed (Fig. 2-49). Polypropylene is often used in such items as syrup bottles.

Butadiene/styrene copolymer KR03 has excellent clarity and good toughness. It is often used where styrene does not have enough impact strength. A typical application is shown in Fig. 2-50, canisters that display wrapped candy.

Figure 2-49 Polypropylene bottles.

Figure 2-50 K-Resin® canisters.

Intermediate density 0.3 MI HDPE is used for a wide range of water coolers and ice chests such as those by Rubbermaid/Gott in Fig. 2-51. The resins are a compromise between stiffness and stress cracking resistance. Because the cooler and chests are foamed, material stiffness is needed mostly for easy assembly and in the nonfoamed lids.

In the one-piece HDPE chair seat and back shown in Fig. 2-52, the center section seat is compression-molded into a single wall and then bent on

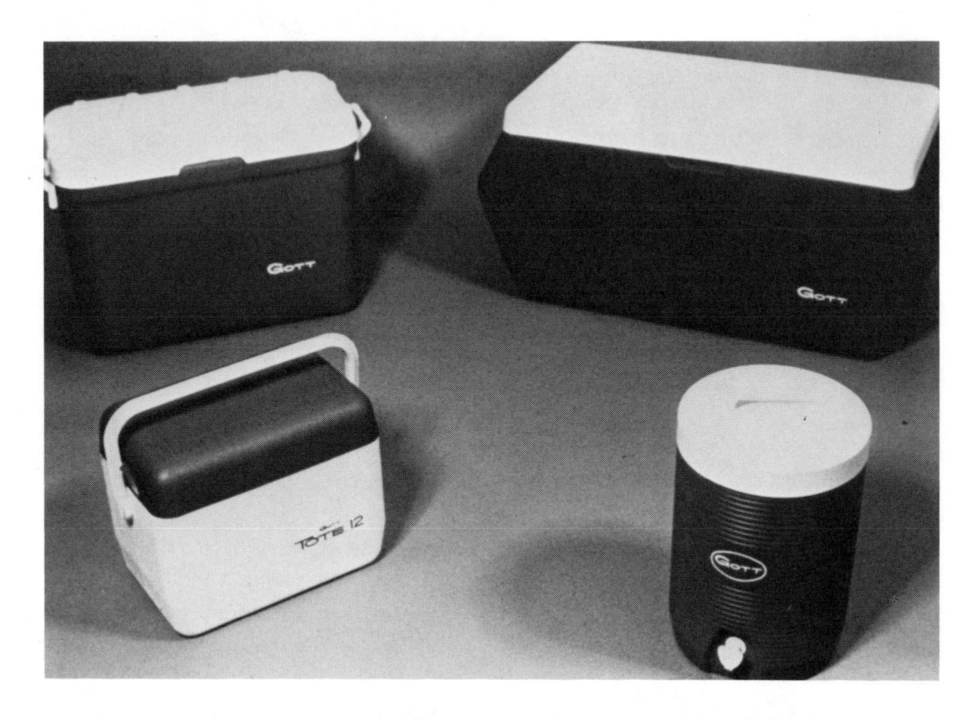

Figure 2-51 Ice chests and coolers.

Figure 2-52 One-piece chair.

installation in the chair frame. For this application good stress cracking resistance is required. High molecular weight HDPE (low MI or HLMI resins) with about 0.950 density are generally used.

Coliseum seats made of HDPE (Fig. 2-53) are subject to outdoor weathering and hardware attachments. Resins with excellent ESCR and good toughness are required.

The lid of the Monoaural record player case (Fig. 2-54) was made with a vacuum assist on the cavity side to keep the parisons from sticking together during mold closure. The bottom of the case was made conventionally. HDPE with a 0.3 MI, 0.955 density was used in this part.

The Canadian Coleman Sportable HDPE cooler (Fig. 2-55) was a very difficult part to blow mold but was made commercially for eight years. It is actually three parts: a half-lid, the body, and another half-lid made in tandem in one mold with needle puncture blow for each part. Vacuum assist and core air billow techniques were used to blow mold the coolers. Polyurethane foaming filled the double walls (Fig. 2-56). A 0.3 MI 0.955 density HDPE was used for this part.

Figure 2-53 Stadium seats.

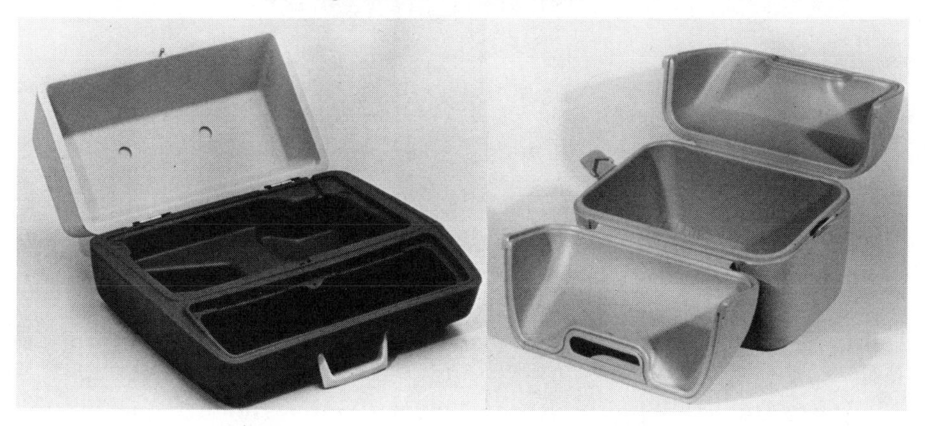

Figure 2-54 Double-wall player case. **Figure 2-55** Sportable cooler.

Figure 2-56 Sportable cooler, sectioned

Figure 2-57 Garbage cans.

The only real requirement for a garbage can (Fig. 2-57) is toughness and reasonably good stiffness. High MW HDPEs perform well in this application.

Many different chemicals are packaged in tight-head drums (Fig. 2-58). This application requires a HDPE material with excellent ESCR and toughness.

Because the small fuel tank shown in Fig. 2-59 does not have any inserts, many HDPEs would perform well in this application. However, because of it is a fuel tank, high MW, excellent ESCR-rated HDPEs are used for toughness and durability.

The snowmobile fuel tank by AMF in Fig. 2-60 has a metal insert and therefore requires a resin with good ESCR.

Figure 2-61 shows an approximately 25-gallon fuel tank that has been standard on the Dodge pickup trucks since 1972. High MW 10 HLMI HDPE resins are used in these fuel tanks.

A prototype fuel tank made for the U.S. Army Truck and Automotive Command (Fig. 2-62) contained a large number of metal inserts and, therefore, required a resin with excellent stress cracking resistance. A 6 HLMI 0.943 density HDPE was used.

Figure 2-58 Tight-head drums.

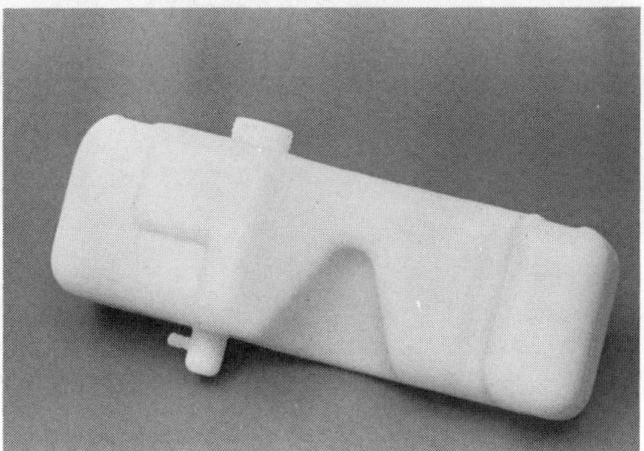

Figure 2-59 Small fuel tank.

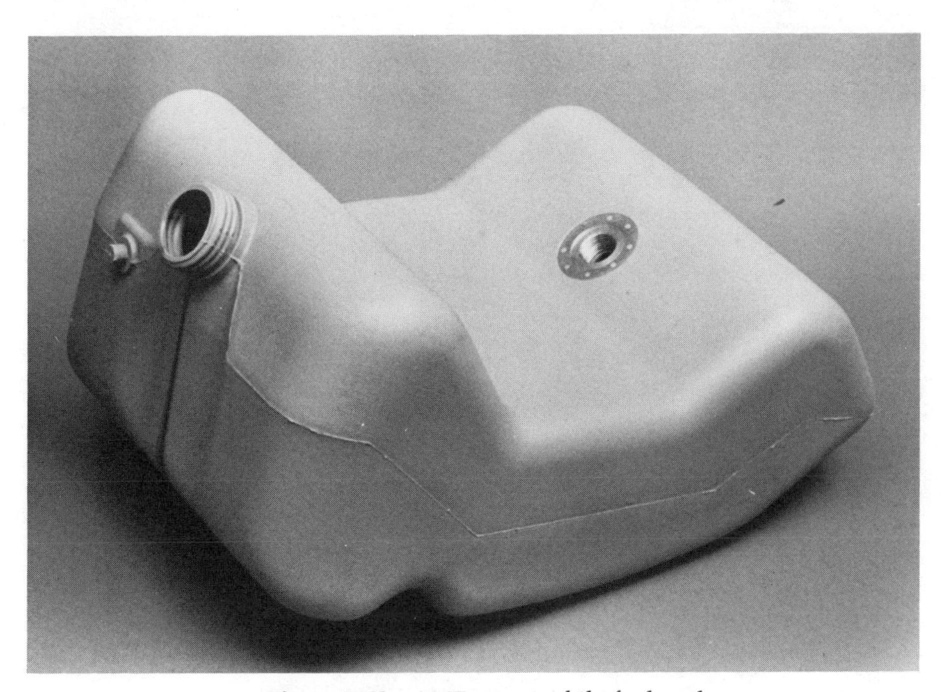

Figure 2-60 AMF snowmobile fuel tank.

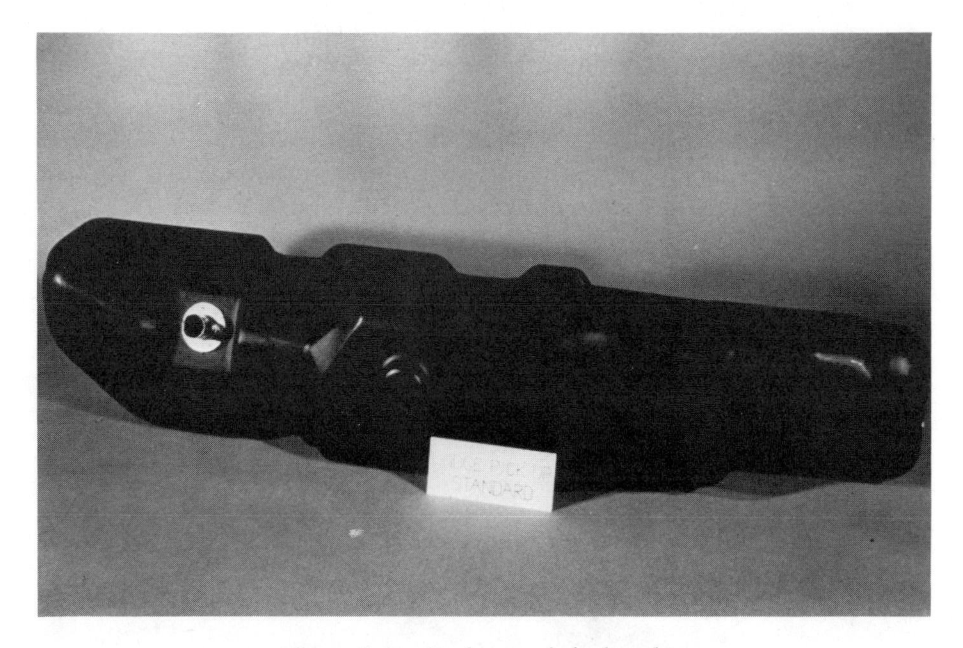

Figure 2-61 Dodge truck fuel tank.

Figure 2-62 M151 fuel tank.

Figure 2-63 Blow molded drums.

Standard 10 HLMI materials perform very well in conventional bulk-goods drums like those shown in Fig. 2-63. The drum second from the left is an exception; it is a Mauser drum with compression molded L-ring and requires a resin with much greater inherent toughness. Currently 2–5 HLMI resins are used for this type of drum.

There are hundreds of other blow molding applications and products. The use of moving section mold technology is expected to increase blow molded part capability in future markets.

CHAPTER

3

Injection Blow Molding

SAMUEL L. BELCHER

Injection blow molding combines injection molding with blow molding. The injection molding phase consists of injection molding a thermoplastic material into a hollow, tube-shaped article called a preform. The preform is transferred on a metal shank, called the core rod, into a blow mold. In the blow mold, clamp tonnage is established to withstand the gas pressure that is used to blow the preform from the core rod to its predetermined shape. The air or other gas enters through the specially designed core rod to force the core rod to open and to blow the thermoplastic material, which is below its melt temperature, into the shape defined by the mold. Thus the term injection blow molding.

The earliest injection blow molding is normally credited to W. H. Kopitke, who was awarded patents in 1943 while he was with the Fernplas Company. The early injection blow molding systems were actually modified injection molding machines with specialized tooling mounted between the platens of the injection molding press. The most referenced process was the system developed by Piotrowsky; others, such as Moslo and Farkus, offered variations and improvements (Fig. 3-1).

The injection blow molding systems of today basically employ the Gussoni horizontal indexing system that was patented in Italy in 1961. Wheaton also is credited with developing this system, but did not file any patents at that time. They had acquired the rights to a Swiss process called Novaplast

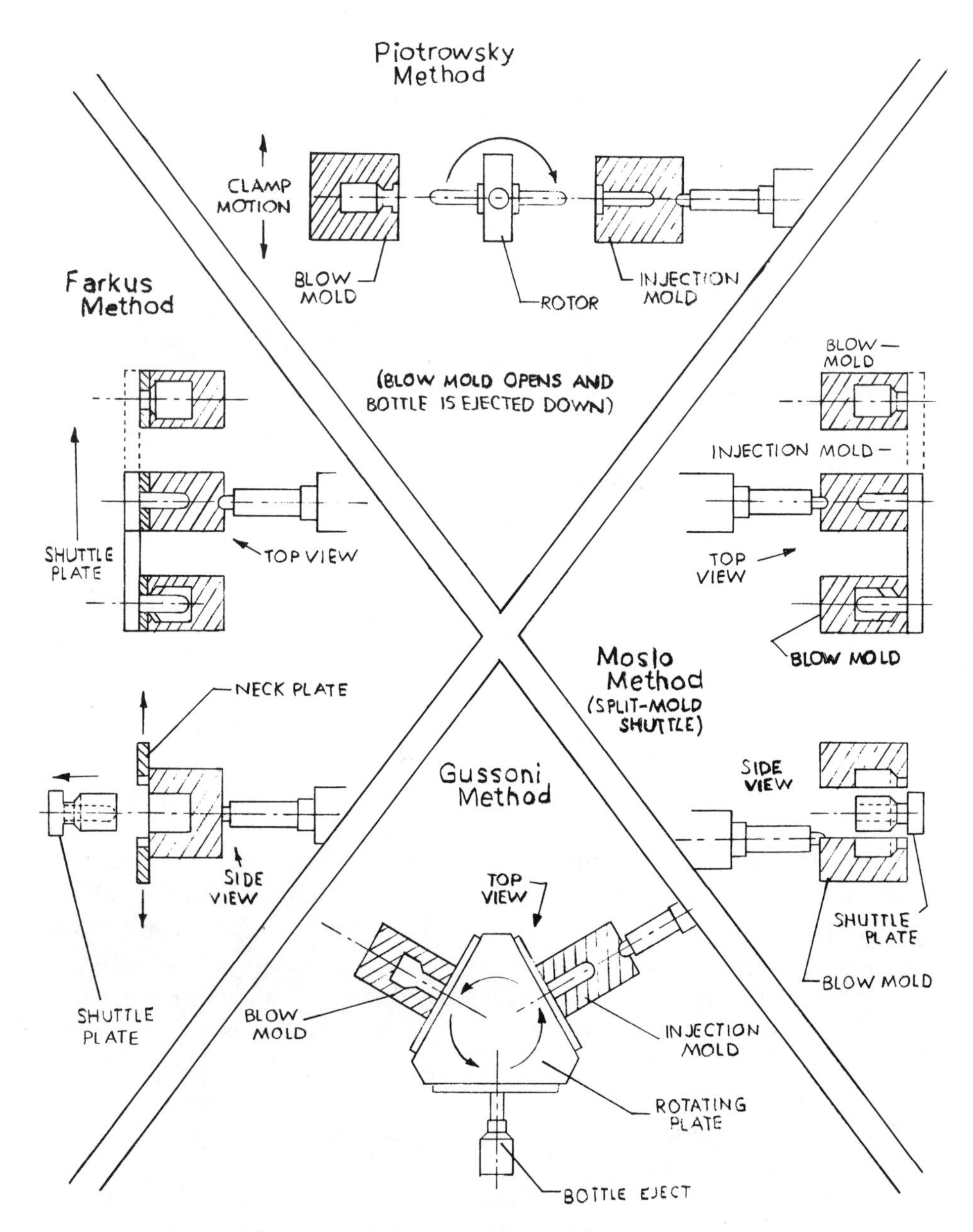

Figure 3-1 Injection blow molding methods.

in the early 1950s and developed their injection blow molding process and machines, including tooling.

PROCESS

Figure 3-2 depicts the injection blow molding process used in the plastics industry today on standard three-station injection blow molding systems. The trend in the industry is to four stations. Figures 3-3 and 3-4 respectively

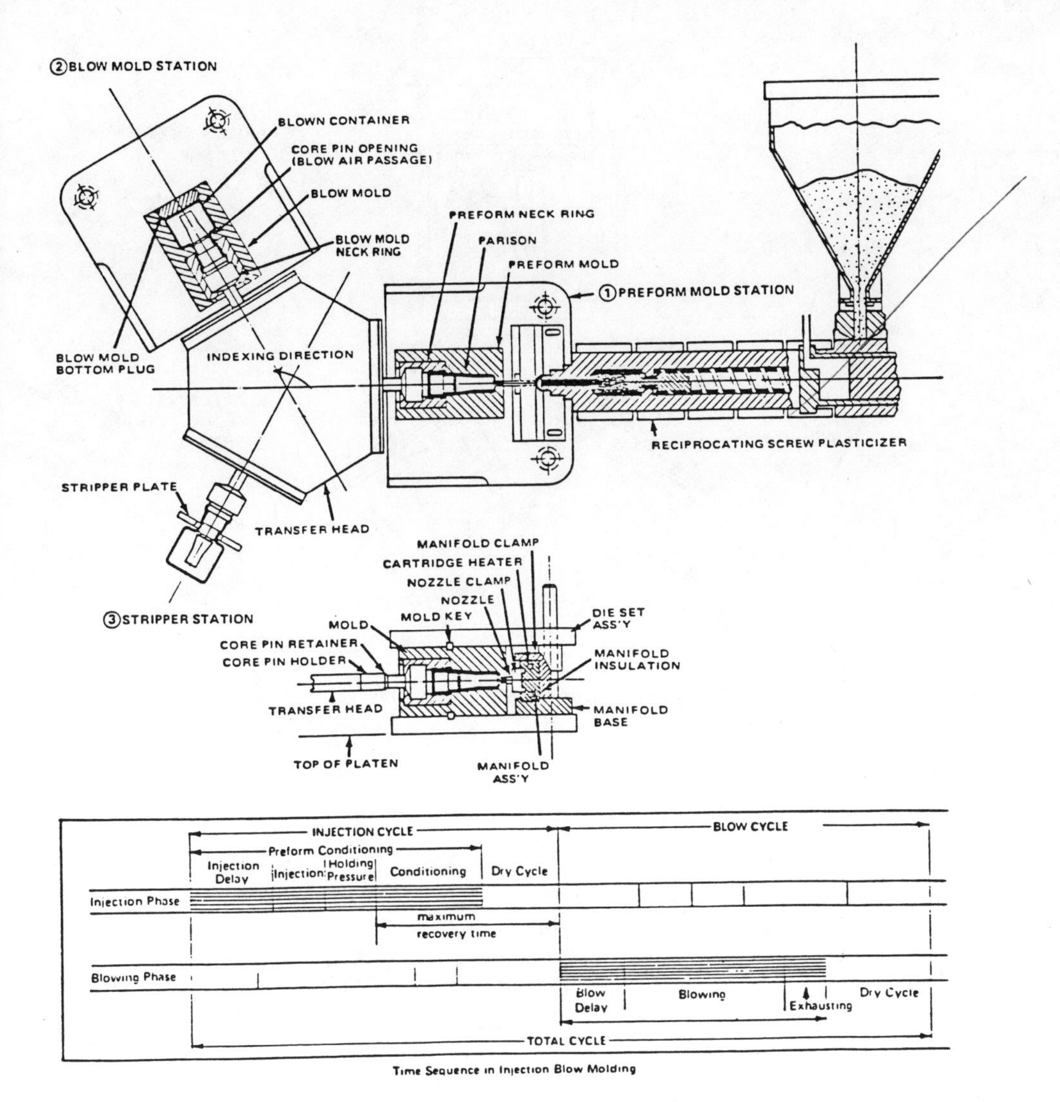

Figure 3-2 Three-station injection blow molding process. Courtesy of Rainville Div., Johnson Controls, Inc.

depict the three-station and four-station clamp and processing portion of the standard injection blow molding machines now available to the blow molding industry.

An important difference between the three-station and four-station indexing heads is the dry cycle time each achieves. Dry cycle time comprises the opening of the molds or clamp, the index time (to accomplish 120 vs. 90 degree movement), and the actual close time. Because no processing takes place during the dry cycle time, it should be as short as possible while permitting the machine to function smoothly and consistently. The open, index, and close time should be as close to 1.5 seconds as possible. Several companies, including Ferguson and CAMCO, offer standard indexing heads

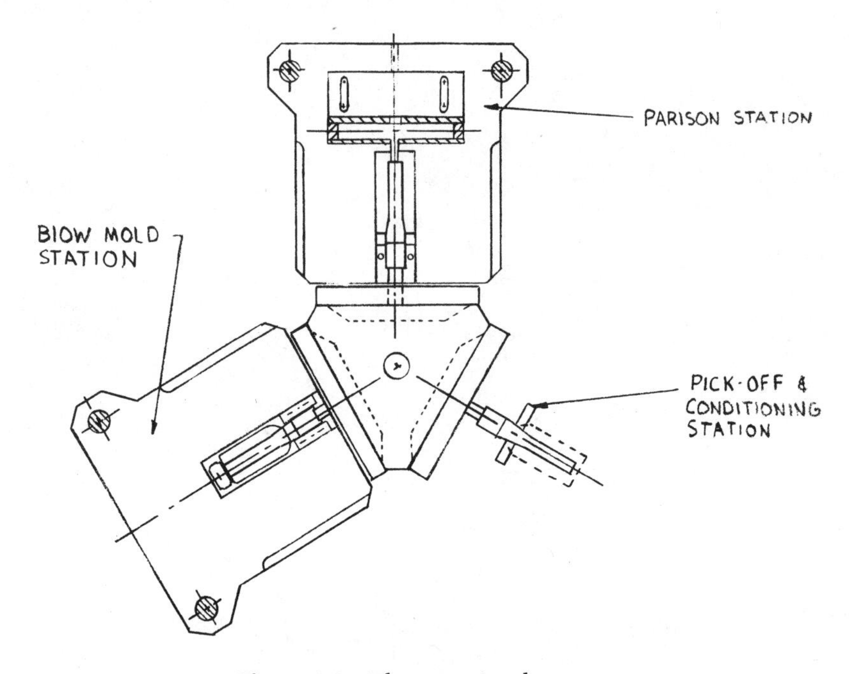

Figure 3-3 Three-station layout.

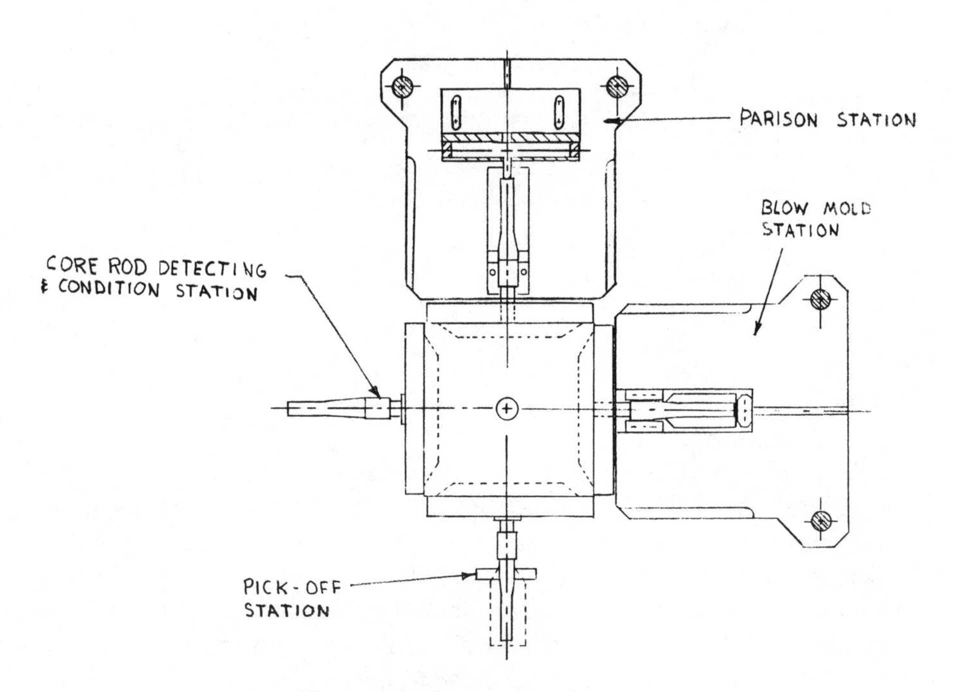

Figure 3-4 Four-station layout.

that achieve this speed. The use of proportional hydraulics reduces shock to the machine from clamp closing and opening, thus reducing tool wear and damage and the need for machine maintenance.

The fourth station has the advantage of requiring movement over shorter distances. It acts as a safety station where the core rods can be checked for preforms or bottles not stripped off, prior to indexing to the injection station to once again pick up the plastic material. This station can also be used to condition the outside and inside of the core rods to insure they are at the correct temperature prior to entering the injection station. This is necessary with some of the latest engineering thermoplastic materials, including polyethylene terephthalate (PET).

Figure 3-5 displays the components of a three-station injection blow molding machine. The machine has an injection unit or plastifier (a reciprocating screw extruder in Fig. 3-5) at the injection and preform station, a blow station, and a stripping or ejection station. A four-station machine also has a safety or core rod conditioning station.

As in any injection molding machine, the function of the plastifier is to convey, mix, melt, and inject the homogeneous melt into the injection mold. Normally, the plastifier is of the horizontal reciprocating screw type. However, Jomar predominantly utilizes a nonreciprocating vertical screw type. The vertical screw can be used with most materials, especially where low shear and low melt temperatures are required. However, the vertical screw is normally not preferred for high torque materials. It also has definite disadvantages with materials that must be dried. The mounting of the hopper drier poses structural and safety problems, and the drier mounting arrangement may result in poor drying. Shot control is not as consistent as on a horizontal reciprocating plastifier, and nozzle leakage has been an industry problem. Precise shot control on either type of plastifier continues

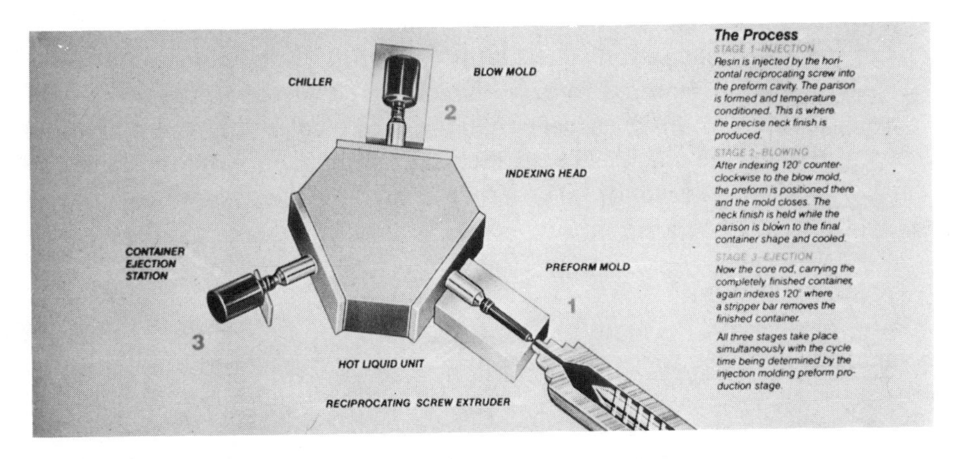

Figure 3-5 Three-station injection blow molding machine. (1) Injection. (2) Blowing. (3) Ejection. Courtesy of Rainville Div., Johnson Controls, Inc.

to be an area for improvement. Future machines may well have all-electric drives for the plastifier units.

The injection station is the critical station in the injection blow molding process. The axiom "If you make a good preform, you will blow a good bottle" is quite true. The preform design is the first critical stage, and proper preform design is both an art and a science. No textbook can cover all the factors for preform design at this time. Computer-aided design (CAD) is being used, however it will take some time to incorporate the "hands-on" aspect of preform operations into computer preform programs.

At the injection station, the molten homogeneous plastic material enters the injection mold by a gate. It fills the void between the female portion of the injection mold and the male core rod. Once the skin of the preform or parison, which is in contact with the female portion of the mold, has solidified sufficiently to allow the core rod to lift the parison out without fracturing the skin or causing sag, the injection clamp opens and the dry cycle portion of the total cycle begins. Normal injection pressure is 1500 to 6000 psi. Thus, the injection station mold clamp must have sufficient hydraulic clamp tonnage to prevent the molds from opening due to the injection pressure necessary to fill the void.

The injection clamp tonnage required to keep the injection molds closed is calculated as follows:

$$\frac{\text{Projected flat area of preform, sq in.} \times \text{Injection pressure, psi} \times \text{Total no. of cavities}}{2000\ \text{psi}} = \text{Minimum injection clamp force, tons}$$

Normally, a minimum 10 percent safety factor is added to the result.

At the blow mold station, the core rod lays the preinjected parison in the bottom half of the blow mold; this action is part of the dry cycle. The blow clamp then brings the top half of the blow mold together with the bottom half and clamp tonnage is built up, normally by hydraulic pressure. The blow clamp required tonnage is calculated in the same way as the injection clamp tonnage, except that the projected area is the maximum area of the desired shape. The air pressure in injection blow molding normally does not exceed 180 psi, for PET, and can be as low as 8 psi for polypropylene. The formula is:

$$\frac{\text{Projected flat area of final molded shape, sq in.} \times \text{Blow pressure, psi} \times \text{Total no. of cavities}}{2000\ \text{psi}} = \text{Minimum blow clamp force, tons}$$

Once again, a minimum safety factor of 10 percent is added to the result.

At the stripper station the blown product is removed from the core rod. This is normally accomplished by a hydraulic cylinder with a Tempsonic control to activate the stripper bar. Some companies prefer an air cylinder stripper. The stripping is accomplished by a plate with U-shaped cutouts that just clear the threads or finish of the blown item. As the cylinder activates, the U-shaped bar pulls against the shoulder of the blown item to strip it from the core rod. In some machines, prior to stripping a gas flame is passed over and beneath the container to oxidize the surface for secondary decorating. This feature is usually custom designed by the using company rather than supplied by the machine builder.

In the four-station machine offered by Bekum and built by Wheaton for their in-house use, it is possible to have a chamber that conditions the outside skin of the core rod by blowing air against its surface while internal air is blowing thru the core pin. The conditioning chamber is a replica of the injection female mold, with the following exceptions:

1. Clearance between cavity and core rod is not greater than 0.030 in. (0.75 mm)
2. Many small holes allow compressed air to impinge on the core rod tip and shank. The holes are 0.015 in. diameter, maximum. The air exiting these small holes expands, giving off heat, thus acting as refrigerated air. This air is vented in the same way as in the blow molds.
3. The tooling is usually produced from aluminum for lower cost machining, less weight, and better conductivity than steel (10 to 1)

With a four-station machine it is possible to configure the machine to also facilitate in-mold labeling. In this instance, the third station is used to place

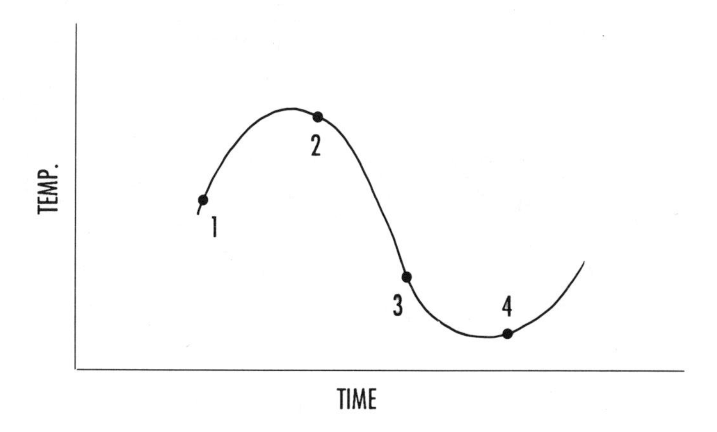

Figure 3-6 Core rod heat cycle. (1) Core rod picks up material. (2) Container is blown. (3) Container is stripped. (4) Core rod is conditioned.

the labels on the blown containers. Stripping and conditioning take place in the fourth station.

Correct conditioning of the core rod is one of the main factors in achieving highly efficient, blemish-free container production.

In any injection blow production cycle the core rod passes through a heating and a cooling cycle (Fig. 3-6); maintaining this cycle within limits allows for efficient quality production. Temperatures can be monitored by thermocouples placed at specific points on the core rod to arrive at the temperature limits for each material.

TOOLING

Preform design and production is the heart of the injection blow molding system. Each container shape has its own preform design and therefore its own core rod design. If the initial design of the preform and preform tooling are not suited for the specific container to be produced, very little can be done in the process to insure that good containers will be produced. Instead, redesign and tool rework or replacement become a very expensive necessity.

The first step in a tool design is evaluation of the container shape. The container must not be too tall in relation to its neck finish diameter. This is checked by determining the ratio of the core rod length (O.A. HG'T in Fig. 3-7) divided by the finish *E* diameter. In general, the *L*/*D* ratio should not exceed 10:1. Below this ratio core rod deflection will be minimized and uniform material wall distribution will be maintained. The use of programmed injection speed for preform cavity fill can extend the *L*/*D* maximum guidelines.

The container should have a blowup ratio of 2.5:1 or less for optimum

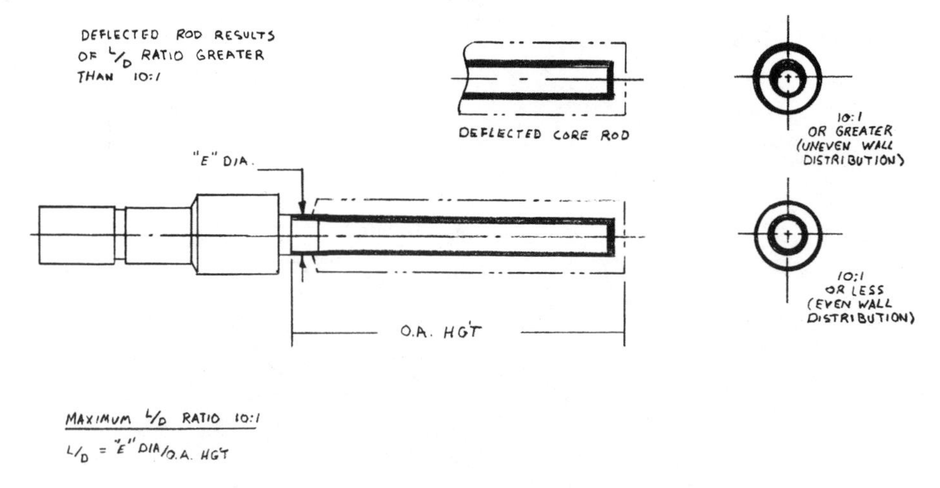

Figure 3-7 Core rod calculations for deflection.

processing. The blowup ratio is container body diameter divided by finish *E* diameter. Larger blowup ratios are possible; however, the greater the ratio, the greater the chance for nonuniform wall distribution. In exceptional cases, blowup ratios up to 3.5:1 can be achieved. When the recommended blow ratio is exceeded, it is usually in only a portion of the container. Each material possesses its own blowup property. In injection blow molding, the hoop stretch ratio is normally the only stretch ratio with which the designer is concerned, since there is negligible axial stretch in true injection blow molding.

Oval containers must be checked to see if they fit within acceptable ovality ratios, the ratio of container width to depth. Up to an ovality ratio of 1.5:1 satisfactory containers can be made with preforms of circular cross section. At a 2:1 ratio, a round core rod and oval preform mold can be used. Ovality ratios greater than 2:1 usually require oval core rods and oval preform molds to minimize weld lines. Increased tooling cost, development time, and development cost must be recognized as the ovality ratio increases. The upper limit is somewhere near 3:1, depending on the material selection.

Preform Design

Once it has been determined that the container shape fits within acceptable guidelines, a preform design can be developed. The normal minimum wall thickness is 0.078 in. in the annular area anywhere along the profile, except in the neck finish. There may have to be some compromise between parison wall thickness and blowup ratio. If a given bottle weight is to be maintained and the minimum wall thickness observed, the preform will have to be smaller in diameter, which increases the blowup ratio.

When designing preforms for an oval container, the parison is usually ovalized in the direction of the container depth. The ratio of the maximum wall thickness to minimum wall thickness across a preform cross section should be less than 1.5 to prevent weld lines.

Generally, preform thickness greater than 0.250 in. is unstable during blowing because the thick section cannot be adequately conditioned in the preform cavity.

Core Rod Design

The preform design usually establishes the core rod length and diameter. In all cases the minor diameter of the core rod (preform inside diameter) must be smaller than the smallest diameter in the neck finish of the preform to facilitate removal of the container in the ejection stage or stripper station. Further core rod criteria are discussed under core rod assembly.

Injection Blow Mold Tools

Injection blow mold tools must be machined to precise tolerances due to their critical function in the overall molding cycle. The preform molds are more critical than the blow molds since they are subjected to direct injection molding pressures. The core rods are subjected to similar conditions as the preform molds and must be concentric within 0.002–0.003 in. (0.05–0.08 mm) to initiate good blowability. Due to the precise machining and interrelations of the number of component parts making up the complete injection blow tool, excessive dimensional variations between components cannot be tolerated. The head support design of the core rod is critical to maintaining good core rod concentricity in the injection mold station. This is part of the parison neck ring design.

Parison Neck Ring (Fig. 3-8a)

The parison neck ring forms the finished shape of the threaded or neck section of the container. It also centers and securely retains the core rod inside the preform cavity to prevent core rod deflection during the injection part of the blow molding cycle.

The parison neck ring details are determined by multiplying the desired dimensions by the shrinkage factor appropriate for the polymer being used. The parison neck ring is most commonly made from nondeforming hardened A2 or H13 tool steel.

Parison Cavity (Fig. 3-8a)

The preform mold consists of stationary and movable halves, with the cavity dimensions determined by the core rod layout. The width of the preform mold is governed by the number of blow mold cavities that will fit in the specific machine, limited by the length of the trigger bar.

Temperature control (material conditioning) is an essential part of the cavity design. Temperature zones are created by channels drilled in the mold body for the circulation of a liquid cooling medium. The channels are connected to form separate individually controlled temperature zones. The number and location of the zones is a critical factor in the design since it affects the efficiency that can be obtained in production. Each temperature control channel is positioned along the parison cavity profile so that the temperature can be varied along the preform to effect uniform blowing of the preform wall.

Individual parison mold blocks are normally fabricated from 1017 hot

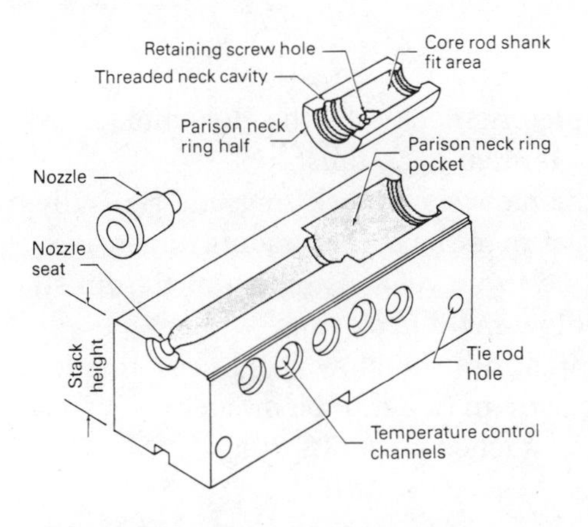

Figure 3-8a One half of a parison mold cavity, with nozzle and neck ring details.

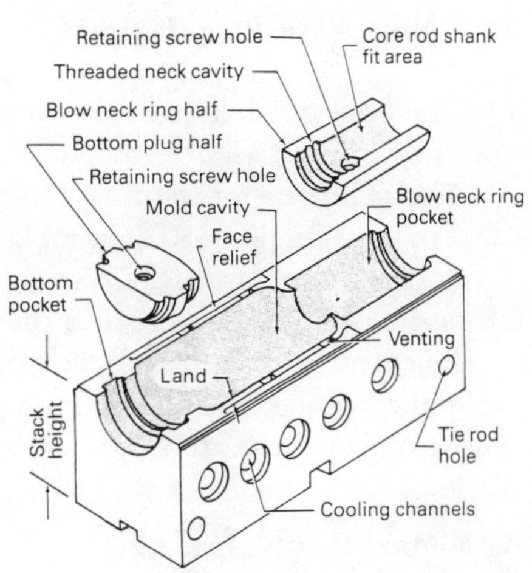

Figure 3-8b One half of an injection blow mold, with bottom plug and neck ring details.

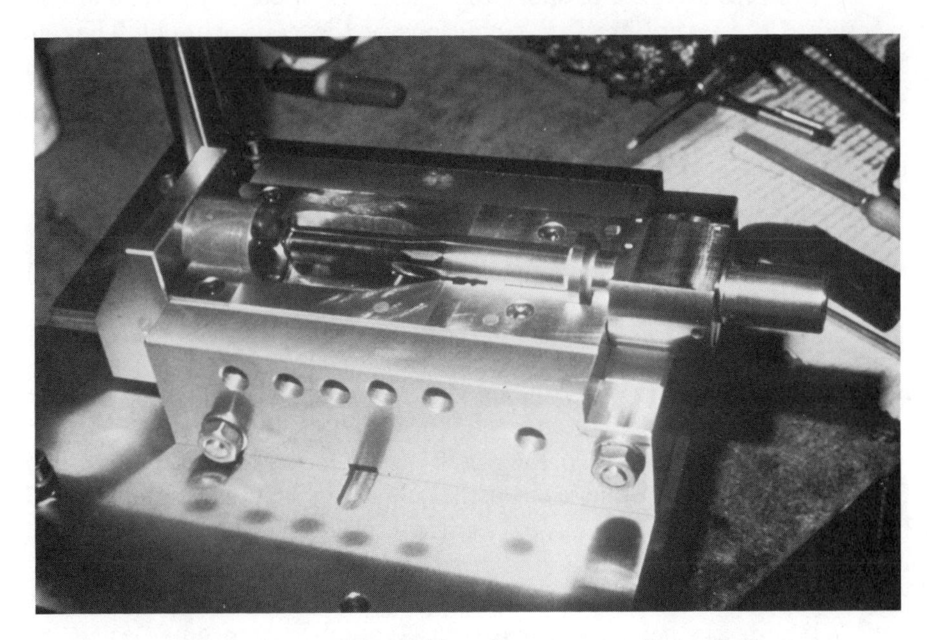

Figure 3-8c Blow mold tooling setup on die set half.

rolled steel or P20 prehardened steel for use with polyolefin resins. Molds for rigid resins, such as styrenics, nitriles, and carbonates, are usually made of an oil-hardening steel (A2 or equivalent) rough machined and hardened to RC 40–45. In both cases, final polishing is usually followed by hard chroming of the molding surfaces, although chrome is not required for olefins.

Blow Mold Neck Ring (Fig. 3-8b)

The neck finish is formed in the preform neck ring; the blow mold neck ring contains and secures the already formed neck finish.

The diameters of the finish in the blow mold neck rings are generally 0.002 to 0.010 in. (0.05–0.25 mm) larger than the corresponding dimensions of the preform neck ring. This is done to prevent distortion of the finish during transfer and seating into the blow mold neck ring.

Independent cooling is incorporated into the blow mold neck ring for thermal control. Depending on the plastic material to be molded, the neck rings are made of A2 tool steel or prehardened steel (RC 30).

Blow Mold Cavity (Fig. 3-8c)

The blow mold cavity consists of stationary and movable halves, similar to the parison mold cavity.

Blow molds are used to shape the final form of containers and are not exposed to the severe pressure conditions of the parison mold. They are subjected to the minimal mold clamp pressures needed to withstand parison-blowing air pressure of 100 to 150 psi.

The most important consideration in the blow mold cavity design is the provision of maximum cooling. Here, as in the parison mold cavity, peripheral channeling produces maximum cooling.

Depending on the type of plastic material to be molded, these cavities are made from A2 tool steel, prehardened steel (RC 30), aluminum, or beryllium copper.

Bottom Plug (Fig. 3-9)

The bottom plug of the blow mold cavity forms the bottom configuration of the final container. It is usually made in halves, retained in each half of the mold, and can be water cooled. For polyolefins, container push-up heights of up to 0.0625 in. are stripped from the molds. With polystyrenes, polycarbonates and other rigid materials, a maximum push-up height of only 0.031 in. can be stripped with this type of bottom plug design. For containers with rigid materials that require a deep bottom push-up height, a retractable bottom mechanism is necessary.

By using retractable bottom plugs, it is possible to manufacture rigid containers with up to a 0.375 in. push-up height. This design requires an air cylinder, cam, or spring-actuated mechanism that retracts the one-piece bottom plug prior to, or simultaneously with, mold opening. The latter requires careful design, especially when maximum-depth push-ups are required, because of the simultaneous movement of the bottom plug and finished container.

Figure 3-9 Cam-operated bottom plug. Courtesy of Bekum, West Germany.

Core Rod Assembly (Fig. 3-10a)

The core rod forms the inside diameter of the neck finish and the inside shape of the parison design. It incorporates the air channels and valving used to blow the plastic melt to the final shape of the bottle.

The type of molding material and the length of the core rod molding area generally determine the location of the air outlet to use. The top opening core rod is used when more heat is needed in the body of the parison to blow the resin, and when it becomes necessary to blow a difficult shoulder of a container. The bottom opening core rod is used when the *L/D* ratio of the molding area becomes greater and core rod ridigity is necessary to alleviate core rod deflection.

A cam nut, spring, and star nut assembly located at the back end of the core rod is the mechanism that opens/closes the air passage during blowing (open position) and parison injection (closed position). (See Fig. 3-10b.)

The core rods are generally made from L6, an oil-hardened tool steel, polished and hard chromed. Beryllium-copper is used for the core rod shank or body in certain instances.

Injection Nozzles (Fig. 3-11)

Injection nozzles allow passage of the melt from a manifold to the parison molds. Nozzles orifices can be from 0.040 to 0.187 in. (1.0–4.75 mm)

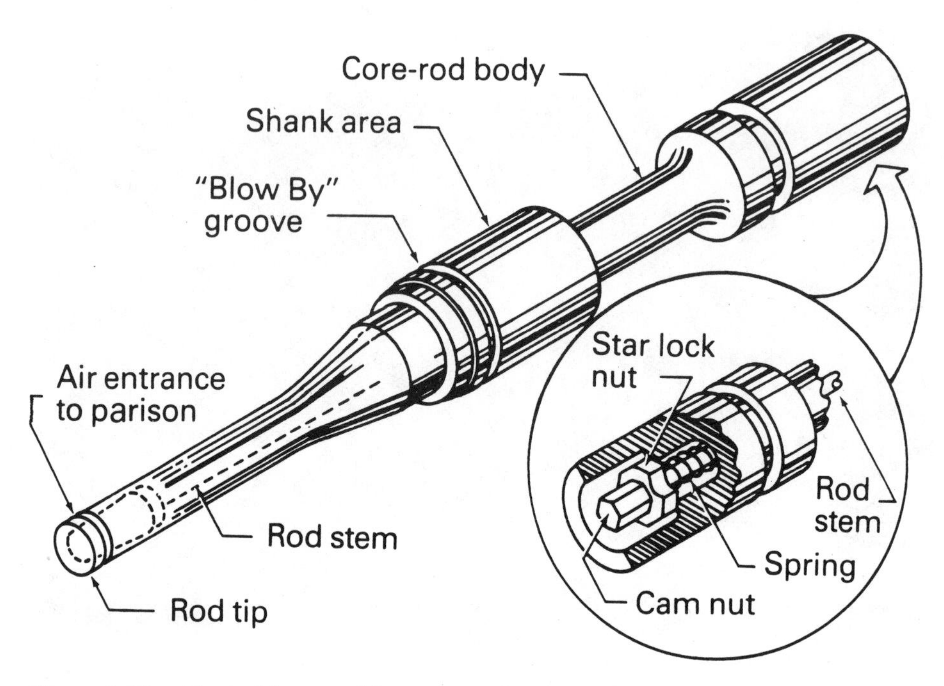

Figure 3-10a Typical bottom-blow core rod, principal elements. Detail: Rod tip mechanism that closes the air passage during the parison injection cycle.

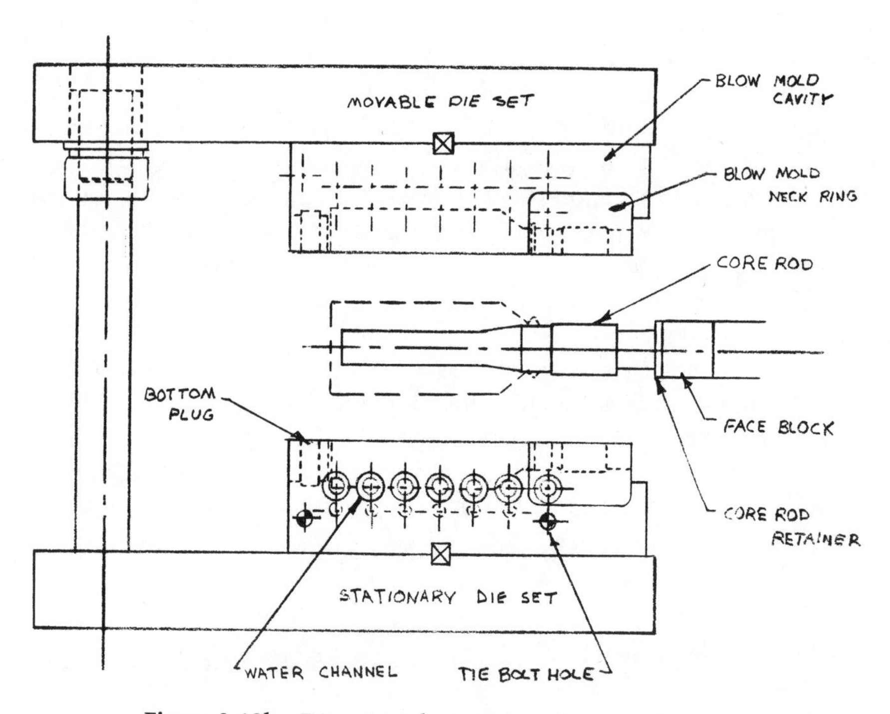

Figure 3-10b Die set with parison and core rod setup.

Figure 3-11 Core rod ready to drop into parison mold, showing manifold and nozzles. Courtesy of Bekum, West Germany.

diameter and can vary from mold to mold in a given mold set by approximately 0.010 in. (0.25 mm) to effect uniform fill of all parison molds. The nozzles are generally heated by surface contact with the heated manifold and parison mold or they can be individually electrically heated.

Manifold (Fig. 3-12)

The hot melt is injected from the machine nozzle into a manifold that performs the same function as a hot-runner manifold in regular injection molding. The manifold assembly used for injection blow molding is far less sophisticated than the type used in hot-runner or insulated-runner injection molds. Simple manifold designs have been adequate because most resins used for injection blow molding are easily processed. With the advent of heat-sensitive materials more sophisticated special manifold designs have become necessary.

The manifold assembly is fastened to the parison die set. The manifold assembly itself is made up of the manifold, a manifold base, manifold clamps, and nozzle clamps. The injection nozzles are held by the nozzle clamps with retaining screws accessible from the rear of the manifold.

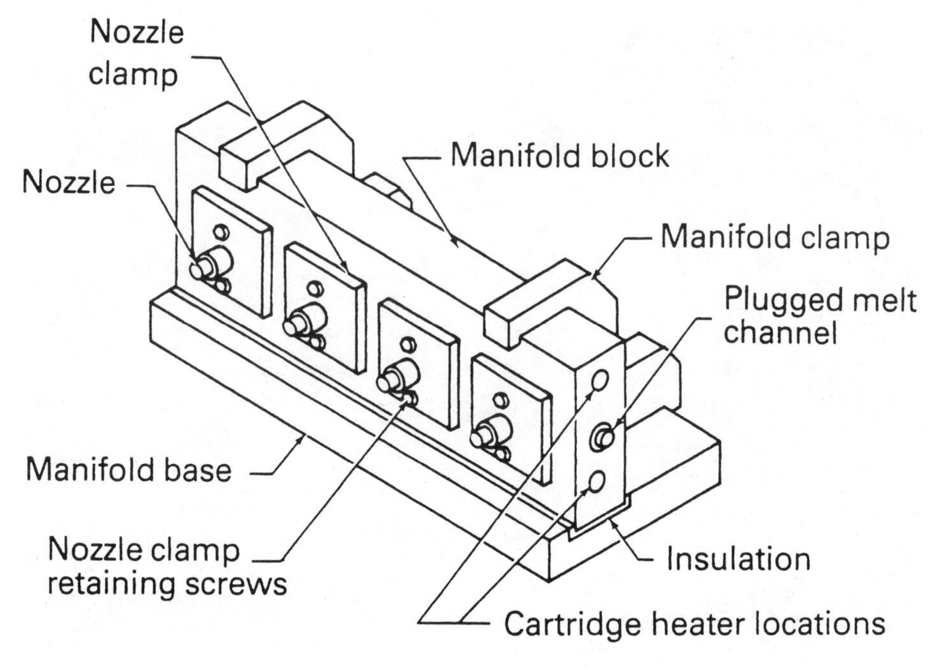

Figure 3-12 Manifold for injection molding of parisons. Individual nozzles are clamped to the manifold block, which houses a hot runner for the melt.

Die Sets (Figs. 3-13a, 3-13b)

Die sets hold the blow molds and parison molds in the machine platens. The molds are keyed in two directions and secured by screws from the top and bottom of the die set. This crosskeying precisely lines up the molds for cavity part-line alignment. Guideposts and bushings are used to maintain alignment between the upper and lower mold halves.

Pick-off Assembly or Stripper (Fig. 3-14)

The pick-off assembly is attached to the machine stripper mechanism and is located at the third station of the machine. This assembly serves two functions: one is to pick off the finished container after molded, the other is to externally cool the core rod in specific areas along the core rod profile.

Resins

Each thermoplastic resin has its own set of tooling design parameters. Hot melt density, shrink factors, stretch ratios, blow pressure, venting cri-

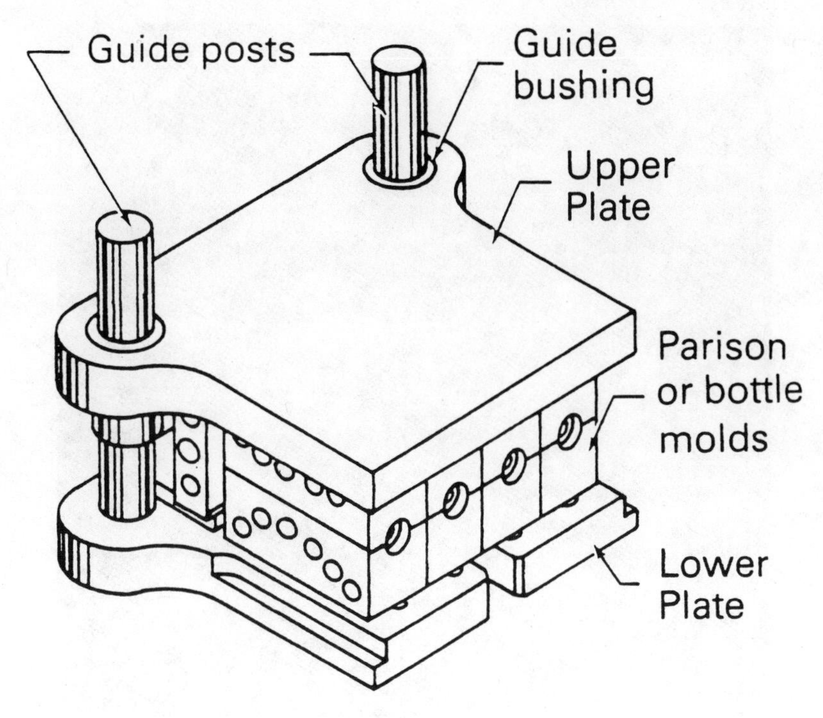

Figure 3-13a Die set for maintaining position and alignment of injection blow mold cavities.

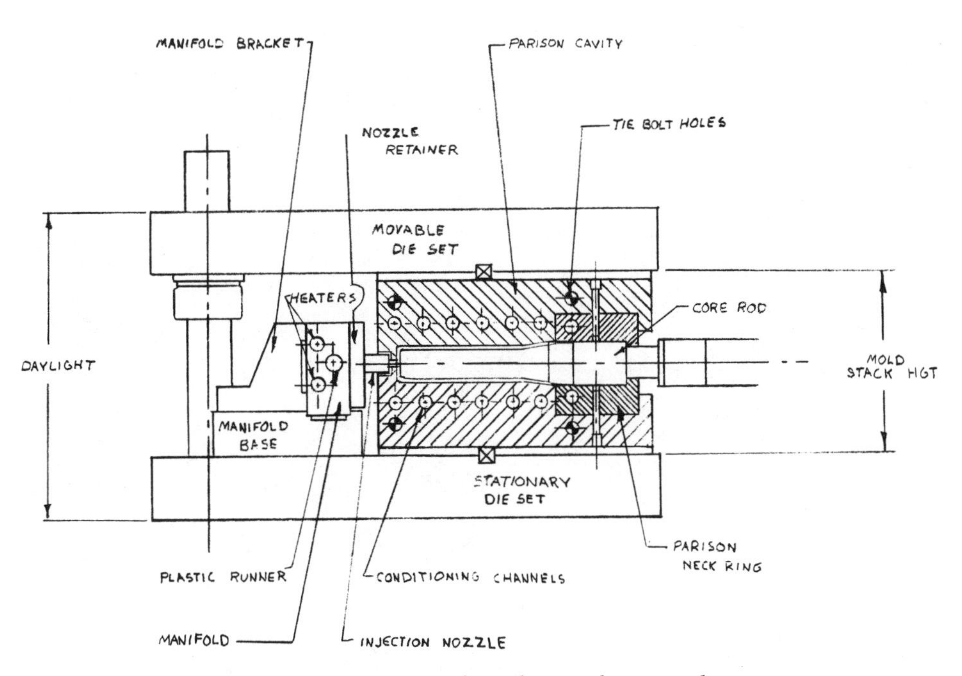

Figure 3-13b Die set with tooling and nomenclature.

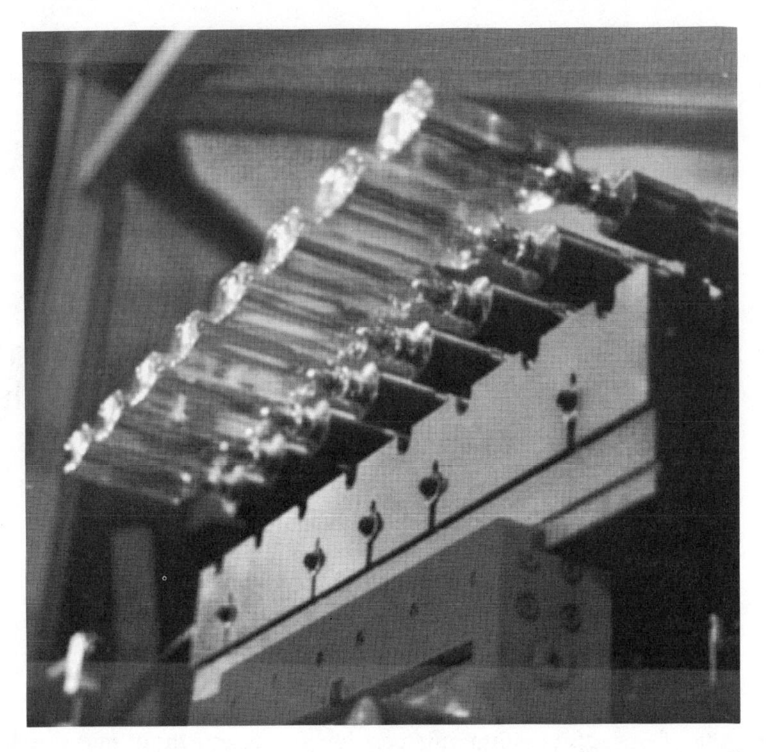

Figure 3-14 Stripper waiting for head to drop into position for stripping. Courtesy of Johnson Controls, Inc.

teria, and surface area of the tooling must all be known prior to designing any tooling.

Following is a list of materials and the hot density factors needed to convert cu/in. to grams:

Polyethylene (LDPE) (13.077)
Polyethylene (HDPE) (12.5)
Polystyrene (PS) (16.1)
Styrene Acrylonitrile (SAN) (16.5)
Acrylic (XT375) (17.3)
Polypropylene (12.4)
Polypropylene Alloy (12.5)
Kynar (28.0)
Ethylene Vinyl Acetate Copolymer (EVA) (16.35)
Celcon (19.5)
P.V.C. (20.1)
Barex (18.35)
Polycarbonate (Lexan) (18.75)
PET (21.35)

INJECTION BLOW VS. EXTRUSION BLOW

There are several instances where the container size or other factors will dictate whether injection or extrusion blow can best be utilized for efficient production. Many factors influence the selection, but ultimately the unit cost will determine the choice of the process.

In general, injection blow molding becomes very attractive at the level of one million up to six or seven million units, for containers of 10-ounce volume or smaller. The choice of method is less clearcut for production of containers between 10- and 24-ounce volume. Normally, investment costs will be greater for injection blow than for extrusion blow molding, however, the final quality of an injection blow molded container will be superior to an extrusion blow molded container.

The following list comparing extrusion blow molding to injection blow molding can be used in selecting the production method.

EXTRUSION BLOW	INJECTION BLOW
Scrap From 5 to 30% scrap must be ground, mixed with virgin, and recycled. That increases equipment investment and maintenance costs 10 to 15% and causes a variation in bulk factor in raw material, resulting in instability of process.	Normally no scrap with proper tooling and controlled shop efficiency. *All bottles are completely finished in mold.* Scrap comes from start-up and errors in color mix. Rerun scrap does not affect process material.
Orientation of Plastic Except in special Bekum two-stage extrusion blow molder, all extrusion parisons are blown at too high a temperature to orient the plastic.	Injection blow inherently gets some hoop orientation from the injection stage during cavity filling. Also, since plastic temperature drops about 150° styrene orients into a strong container. No effect on polyethylene.
Tool Cost Only blow mold and extrusion tooling required; cost is 30 to 40% of injection tooling cost. Superior for very low volume requirements, under 50,000 per year.	Must have three or four core rods for each cavity and an injection mold. Tooling is precise and expensive. However, high price is justified by higher efficiency and is sometimes indicated by preference for this method.
Clarity Extrusion die lines appear with some materials, or if tooling is improperly finished.	Containers are always clear when using transparent materials. Polypropylene is clearer because some hoop orientation occurs during blow up.

EXTRUSION BLOW	INJECTION BLOW
Bottom Pinch-off Scar Often creates objectionable appearance, or weak point where seal is not adequate.	No pinch-off.
Bottom Push-up Is difficult because of need to pinch off and seal parison.	Great design leeway for bottom shapes with retractable bottom plugs in blow mold.
Machine Cost For medium-size bottles cost is similar to injection blow for same production.	Cost per thousand bottles per hour improves greatly for higher number of cavities, such as 10 to 14 cavities on a 10 sec. cycle.
Widemouth Containers Must be post-finished, inspected, trimmings removed. Neck is poor in quality and generally weak. Very thin neck possible if desired. Tolerances very limited.	Injection-molded neck of any thickness approx. 0.040 in. Finish and tolerance excellent.
Inside Neck Tolerances Good on some equipment, but moving tool parts wear. Accurate undercut difficult.	Accuracy excellent for plug insert for squeeze bottle or cap seal for flexible materials (nose or deodorant spray).
Special Shapes for Safety Caps Post-forming limited in shape and tolerances.	Offers wide range of any form that can be injected molded.
Rigid Containers Neck finish poor because of difficulty with pinch-off and post finishing. Very large parts can be blown. Handleware can be produced.	Excellent results, with many molds in use and more being readied for production.

THE FUTURE

Injection blow molding as an industry will continue to grow approximately 5 to 7 percent a year through the early 1990s. However, tooling costs must be lowered, and this will take new insight by not only machinery designers and manufacturers but also tool designers and manufacturers.

Injection blow molding will also add stretch molding capability to the

Table 3-1 Three-station injection blow molding machines

Model No.	*Preform Clamp Tonnage, tons*	*Blow Clamp Tonnage, tons*	*Injection Unit Screw Dia., in.*	*Plasticizing Capacity, lb./hr.*
Rainville				
54-3	42	12	1.75	135
70-3	58	12.5	2.0	275
90J-3	77	12.5	2.5	325
122-3	103	19	2.5	325
187-3	158	25	2.5/3/5	325/425
Jomar				
15	12	3	1.0	30
30	25	6	1.38	72
40	35	6	1.38/2.0	72/142
65	50	17	2.0	142
85	68	17	2.0/2.5	142/285
125	110	25	2.5	285
160	150	25	2.5	285
Four-station injection blow molding machines.				
Bekum				
SBM 4100	100	17.4	2.4	300
Captive/Sabel				
504-180	180	25	60 mm two-stage	500

process, which will broaden the customer base, and the industry will use the growing parison knowledge to produce better and more diverse products.

With the advent of statistical process control and new control technology, creative thinking must enter the picture to allow the container manufacturer to produce more quality containers per investment dollar and labor dollar; this means new machinery designs. Perhaps the next generation of injection blow molding machines will more closely resemble high-speed glass container production machines.

With the decline in family size and the growth of safety in packaging, the market for smaller size containers will continue to grow. The demand will be met with new machines and materials.

SELECTED READINGS

Beckman, R. E. "Blow Mold Cooling Analysis Boosts Quality and Output." *Plastic Technology* 32(13) (December 1986):13/5.

Brochschmidt, A. "How Big Is Blow Molding? Results of First Ever U.S. Census." *Plastic Technology* 32(12) (November 1986):77–79.

Chung, Tai-Shung. "Principles of Preform Design for Stretch Blow Molding Process." *Polymer Plastic Technology Eng.* 20(2) (1983): 147–160.

Dreps, J. R. "Injection Blow Molding Bottles," *Plastics Engineering* 34 (Jan. 1975).

General Information about Hoechst—Thermoplastic Polyester Resin. Technical Bulletin 2, 1980. American Hoechst Corp., Hoechst Fibers Industries.

Glossary of Plastic Bottle Technology. Plastic Bottle Institute, Society of the Plastics Industry, 1980.

Glossary of Plastics Terms. Bartlesville, OK: Phillips Petroleum Co., Chemical Dept., Plastics Div., 1963.

Hunkar, D. B. "Role of Control Technology in High Performance Container Manufacturing." *High Performance Container Technology,* Second Blow Molding Conference, Blow Molding Div., Soc. Plastics Eng., 1985, pp. 241–246.

Johnson Controls. "Uniloy Puts Its Technology on Show." *Plastic Rubber Weekly* No. 1167 (Dec. 6, 1986): 18.

Jomar Industries. "Vertical Layout in Jomar Range." *Plastic Rubber Weekly* No. 1167 (Dec. 6, 1986): 18.

Kovack, G. P. In *Processing of Thermoplastic Materials.* E. C. Bernhardt, ed. Huntington, NY: R. E. Krieger, 1959 (repr. 1974), pp. 511–522.

Krantz, G., "Plastic Processing Machines." *Kunstoff Magazine* 3 (1985): 553.

Morgan, B. T., N. R. Wilson, and D. L. Peters. *Modern Plastics Encyclopedia,* Vol. 46, No. 10A. New York: McGraw-Hill, 1969–1970, p. 525.

Nancekwell, J., "Check Out All Options Before Picking a Process." *Canadian Plastics* 44(7) (July/August 1986): 36–40.

Nelson, J. *Automotive Boots Manufacturing.* Toronto: ABC Plastic Molding Co., 1983.

Operators Guide—Controlling Shrinkage of HDPE Bottles. Midland, MI: Dow Chemical Co., 1968.

Presswood, K. J. "Oriented PVC Bottles: Process Description and Influence of Biaxial Orientation on Selected Properties." *Technical Papers.* 39th ANTEC, Soc. Plastics Eng., 1981, pp. 718–721.

Robinson, J., ed., "Plastic Molding Equipment, Processes and Materials." *Polymers & Plastic* (1981): 299.

CHAPTER

4

Stretch Blow Molding

SAMUEL L. BELCHER

"The most significant new package since the introduction of the two-piece can" was the statement made in the packaging industry at the introduction of the stretch blow molded PET (polyethylene terephthalate) soft drink container in 1977. With the commercial introduction of this container, stretch blow molding became a common term in the blow molding industry.

Biaxial stretch blow molding is the method of producing a plastic container from a preform or parison that is stretched in both the hoop direction and the axial direction when the preform is blown into its desired container shape. Figure 4-1 shows a variety of containers produced in this way.

Stretch blow molding is possible for various thermoplastic materials such as acrylnitrile (AN), polystyrene (PS), polyvinyl chloride (PVC), nylon, polycarbonate (PC), polysulfone, acetal, polyarlyate, polypropylene (PP), surlyn, and polyethylene terephthalate (PET). Amphorous materials such as PET, which have a wide range of thermoplasticity, are easier to stretch blow than partially crystalline materials such as polypropylene.

Normally, when a container has been produced by biaxial stretch blow molding, the properties of the raw material are enhanced, as shown in Table 4-1.

Biaxial stretch blow molding of a thermoplastic stretchable material will normally increase the material's tensile strength, barrier properties, drop impact, clarity, and top load in a container such as a bottle. With these properties increased it is usually possible to reduce the overall weight in a small container by 10 to 15 percent less than that of an equivalent container produced in another way. In larger containers, 64-ounce up to one-gallon size, the weight reduction may be as much as 30 percent less than a container that was not stretch blow molded. Stretch blow molding produces a container from less raw material and with improved economics and bottle properties.

Figure 4-1 PET bottles.

Table 4-1 Gas and water vapor barrier properties of glass-clear resins

Polymer	O_2^*	CO_2^*	*Water***
Oriented PVC via extrusion blow molding (high impact)	9.0	16.5	0.7
Oriented PET	11.1	22.2	1.5
Oriented PVC via extrusion blow molding (normal impact)	9.6	5.2	1.1
Nonoriented PVC via extrusion blow molding (high impact)	12.2	35.7	1.5
Nitrile	1.1	1.3	4.4
Polycarbonate	215.0	400.0	9.0

*cc/mL/24 h/atm/100 sq in./73°F
**g/mL/24 h/atm/100 sq in./100°F, 90% RH
Test results provided by Occidental Chemical Corp.

Table 4-2 Stretch ratios

Material	*Stretch Ratio*	*Orientation Temperature Range °F*
PET	16/1	195–240
PVC	7/1	210–230
PAN or AN	12/1	240–260
PP	6/1	260–280
Polystrene (Crystal)	12/1	290–320

Approximate temperatures for the noted materials to yield maximum biaxially oriented bottle properties are listed below.

PP	260°F (128°C)
AN	257°F (125°C)
Polyacetal	318°F (160°C)
PVC	210°F (100°C)
PET	195°F (105°C)
Polystyrene	310°F (105°C)

Stretch blow molded containers of AN, PVC, and PET can have an average side wall thickness of 0.012 to 0.015 in. Normal blow molding would call for average wall thickness of 0.018 to 0.022 in., with no less than 0.008 to 0.010 in. in the thinnest area. The latter is true in standard blow molding as well as in stetch blow molding.

As noted above, each material has a temperature at which the heated parison or preform should be stretched for optimum orientation properties. Each material also exhibits its own natural stretch ratio. The stretch ratios for the most used materials are listed in Table 4-2.

PROCESSES AND MACHINES

One of the earliest materials that was stretch blow molded was polypropylene. The Beloit Co. produced a commercial stretch blow molding machine utilizing the cold tube process for polypropylene in the 1960s.

The cold tube process used an extrusion line to produce PP pipe. When cooled, the pipe was cut into predetermined lengths that were stored prior

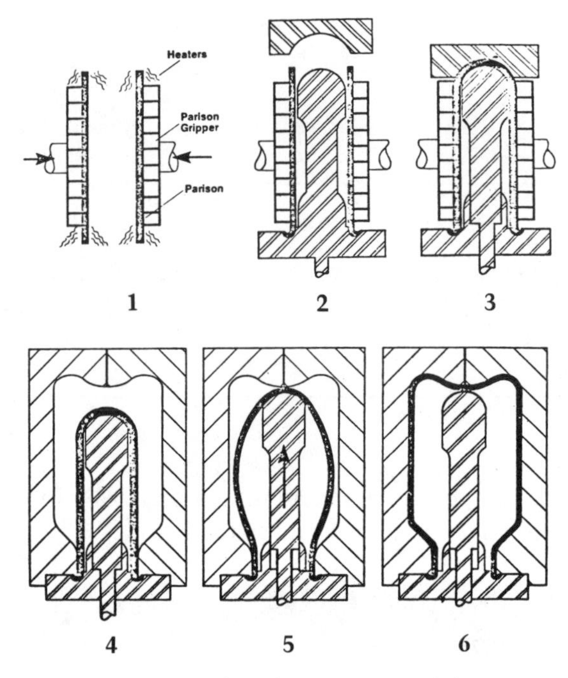

Figure 4-2 Cold tube process, product forming steps. (1) Precut cool pipe length is heated. (2) Neck finish is die formed. (3) Bottom end is closed. (4) Preform in stretch blow mold. (5) Forming mandrel produces axial stretching. (6) Blow air produces hoop stretching.

to being reheated and stretch blow molded into finished containers. Figure 4-2 depicts the process from the point at which a precut length of pipe, the parison, is passed into an oven to be heated by metal sheath heaters. Heating was by convection or heated air, as in a baking oven. In part 1 of Figure 4-2 the pipe length is held on a pin and heated so that the wall thickness of the precut pipe length is uniform. Part 2 of the figure shows the lower end being formed by matching dies to provide the neck finish of the preform. At the same time, part 3, matched dies (inside-outside) close the preform end to form a bottom (top of diagram) prior to transfer of the formed preform to the blow mold. Part 4 shows the preform inside the closed blow mold. The forming mandrel acts as a stretch rod. It moves up to the top of the blow mold, part 5, stretching the preform axially. Once the preform has been stretched to the top of the closed blow mold, blow air enters and blows the axially stretched preform in the hoop direction until the side wall of the desired container is produced, part 6. Provided the material was at the correct temperature, it is now biaxially oriented.

Baxter Laboratories utilizes the cold tube process to produce intravenous solution bags for the medical field. It was also used for producing the first biaxially oriented PET soft drink beverage container.

In the late 1960s, companies such as Owens-Illinois, Continental Can, DuPont, and Cincinnati Milacron worked with 1.04 IV PET (IV: intrinsic viscosity, a measure of materials molecular weight) to extrude pipe approximately one inch in diameter. The PET pipe was cut into six-inch lengths by band saw, tube cutters, hot wire, or radial saw. The pipe was heated to its forming temperatures and placed in a male/female die to either form the threads or close the end as a test tube. The formed tubes were then reheated by metal sheath heaters and stretch blown in a unit cavity mold station. At this time it was not known that PET could be injection molded into a preform shape.

In late 1972 and early 1973, Broadway Company in Dayton, Ohio (William Gaiser), was working with Cincinnati Milacron, DuPont, and the Pepsi-Cola Co. to make the first injection-molded PET preform (Fig. 4-3).

In 1976 Cincinnati Milacron introduced the first commercial reheat stretch blowing machine, the RHB-V (Fig. 4-4). It was sold to Amoco, Atlanta, Georgia.

The RHB-V was a four-lane, four-cavity machine rated at 2800 bottles per hour. The oven utilized metal sheath heaters, with the preforms rotating at approximately 60 rpm in front of horizontally mounted heating rods (Calrods) rated at 1500°F each. Figure 4-5 depicts the operation of the RHB-V.

Figure 4-6 shows stretch blow molding of a PET injection-molded preform in a two-stage machine that heats the preform with the finish or threaded area at the top. Such machines are made by Cincinnati Milacron, Van Dorn, and Bekum.

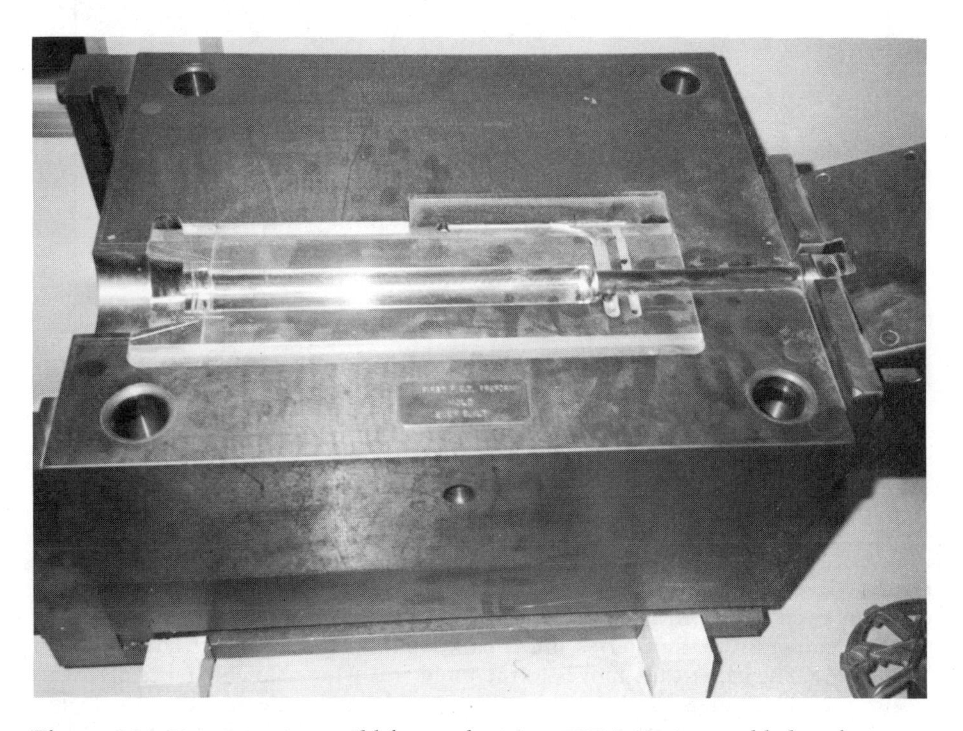

Figure 4-3 First injection mold for producing a PET injection molded preform. Courtesy of Broadway Company, Dayton, Ohio.

Figure 4-4 The RHB-V, first commercial reheat stretch blow molding machine, 1976. Courtesy of Cincinnati Milacron, U.S.A.

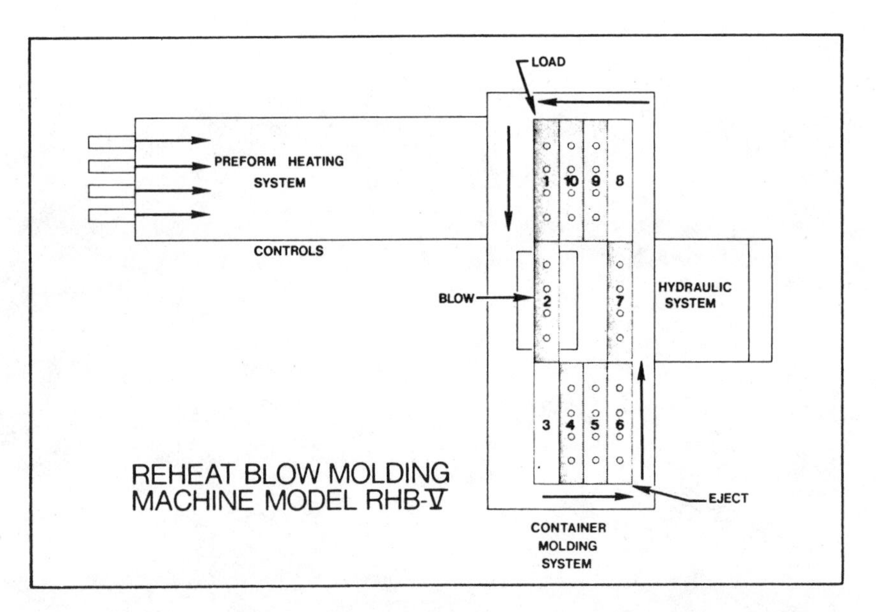

Figure 4-5 Oven, pallet shuttle system, and blowing station of RHB-V. Preforms enter the oven in four rows at the left of the machine. After reaching blow temperature, preforms are transferred into a pallet four at a time at the load station (position 1). The pallet then moves to the mold area where blowing takes place (position 2). The pallets index one position each blowing cycle, in a counterclockwise direction. Bottles eject at position 6. Pallets then continue around to position 1 to pick up another set of preforms. Courtesy of Cincinnati Milacron, U.S.A.

The Cincinnati Milacron RHB-V has been improved to a present rating of 5500 bottles per hour for a two-liter base cup design, using horizontally mounted quartz lamp heaters rated at 4000°F (2204°C) each. The RHB-VI for one-liter beverage bottles is rated at 3500 bottles per hour, and the RHB-VII for 16-ounce beverage bottles is rated at 7200 bottles per hour. The RHB-VI utilizes metal sheath heaters and blows six one-liter bottles per cycle. The RBH-VII utilizes quartz heaters and blows nine 16-ounce soft drink bottles per cycle. All the RHBs are indexing style machines, based on the original RHB-V design.

At the same time that DuPont, Cincinnati Milacron, and Pepsi-Cola were working on the PET soft drink bottle, Monsanto was working with the Coca-Cola Co. to use LOPAC (low-oxygen polyacrylnitrile) material for a beverage container. Coca-Cola and Monsanto were actually first in the market with the one-liter base cup design soft drink bottle, and they test marketed a 16-ounce soft drink beverage bottle prior to the first commercial PET soft drink bottle. Monsanto used proprietary equipment to produce these LOPAC containers. In 1976 LOPAC came under scrutiny by the U.S. Food and Drug Administration, and PET became the beverage bottle material.

Nissei of Japan introduced its first single-stage machine, the ASB 650, for production of PET beverage bottles in 1978. This machine allowed the customer to have all the bottle processing in one machine. Producing six at a time, it was rated at 900 bottles per hour for a two-liter soft drink bottle,

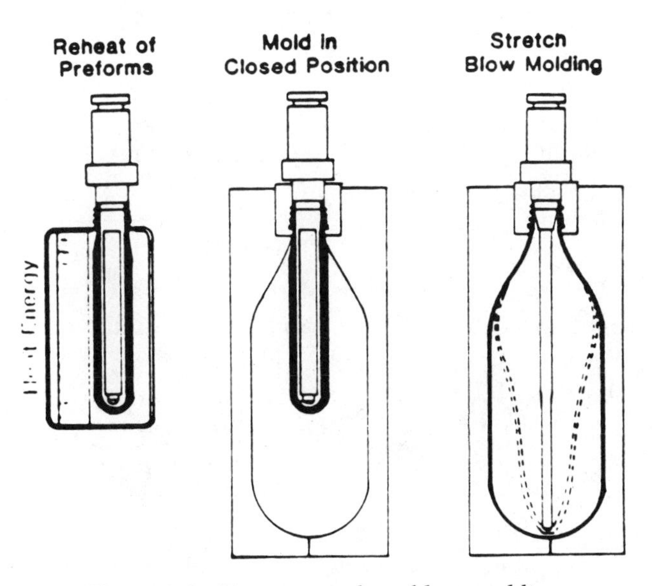

Figure 4-6 Two-stage reheat blow molding.

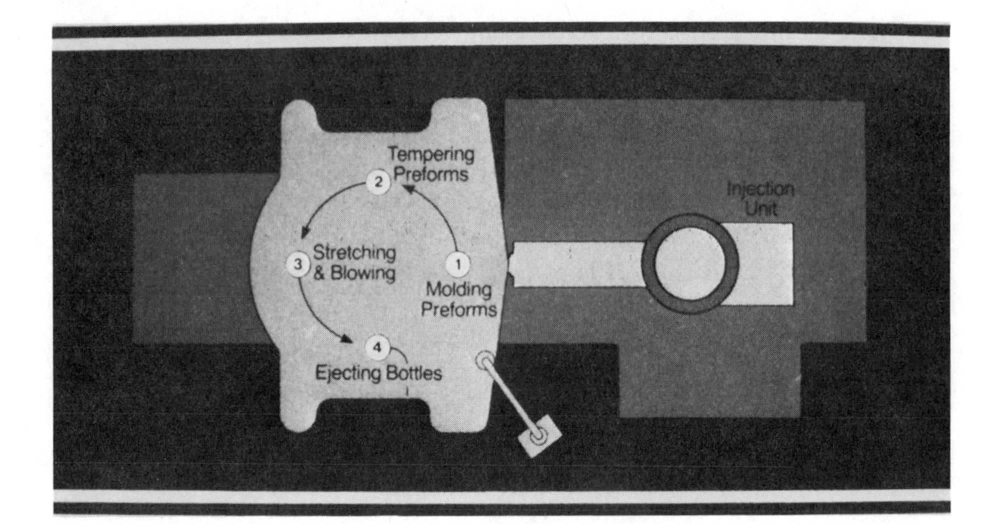

Figure 4-7 Single-stage process.

and 1200 bottles per hour for a one-quart size container. The process is depicted in Fig. 4-7.

The single-stage machine was the original machine of choice because it carried the lowest investment cost and no one foresaw the overnight success of the PET soft drink beverage bottle. Figure 4-8 describes the process. Figures 4-9 and 4-10 show single-stage machines offered by Cincinnati Milacron and Nissei of Japan, respectively.

Owens-Illinois, through an agreement with Krupp Corpoplast of West Germany, entered the PET soft drink market with the first rotary two-stage machine in 1978. Known as the B-40, the machine was rated at 2800 64-ounce soft drink bottles per hour, later changed to a two-liter bottle. The B-40 has been improved until it now can produce over 4000 bottles per hour. Figure 4-11 shows a present-day B-40. Figure 4-12 depicts the two stage heating and blow molding of PET soft drink bottles.

Corpoplast and Owens-Illinois ended their mutual agreement in 1985, freeing Krupp to sell their machines and technology to any PET bottle producer. Since then Corpoplast has added extensively to their machine offerings, as shown in Fig. 4-13, 4-14, and 4-15.

Figure 4-16 is a layout diagram of the Corpoplast B-40. It shows the basic operation of wheel-type PET stretch blow molding machines manufactured by Krupp, Corpoplast and by Sidel SMTP of France. The greater the output, the more heating stations, equilibration areas, and number of blow cavities. Figure 4-17 shows a Sidel wheel machine capable of 8000 to 10,000 PET stretch blow molded soft drink bottles per hour.

A machine that has been successful in Europe and countries with smaller markets is the two-stage SSB-20 machine offered by MAG Plastics, Switzerland, capable of producing approximately 2000 two-liter bottles per hour. It is similar to the Cincinnati Milacron RHB 2000 except that it heats with the finish or thread area down. It is pictured in Figure 4-18.

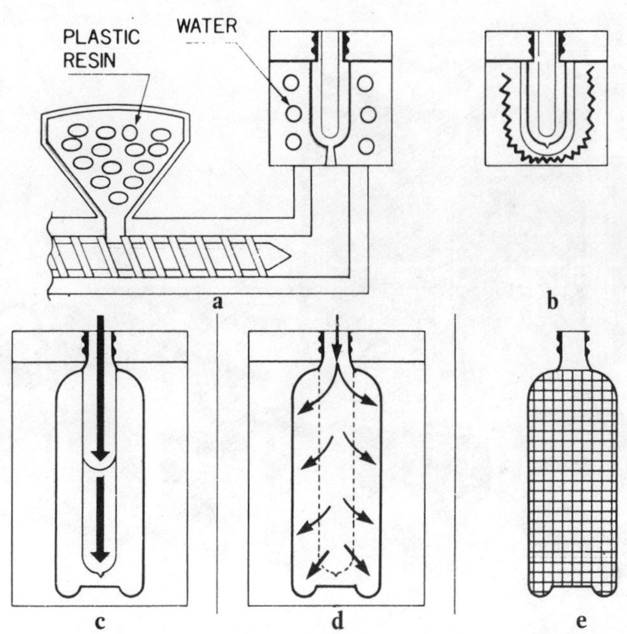

Figure 4-8 Single-stage PET process. (a) Milk-white material (crystallized PET chips) is melted by heating, then injected into the mold and rapidly cooled to form a transparent parison. (b) The cooled parison is reheated with an electric heater to soften it. (c) The softened parison is stretched to about twice its original length in the bottle mold. (d) Compressed air is blown into the stretched parison to expand it into the bottle mold. (e) The process has arranged the molecules of the material in both lengthwise and crosswise orientations, making a finished product that is stronger and more attractive than those produced by older processes.

Figure 4-9 Single-stage PET machine. Courtesy of Cincinnati Milacron, U.S.A.

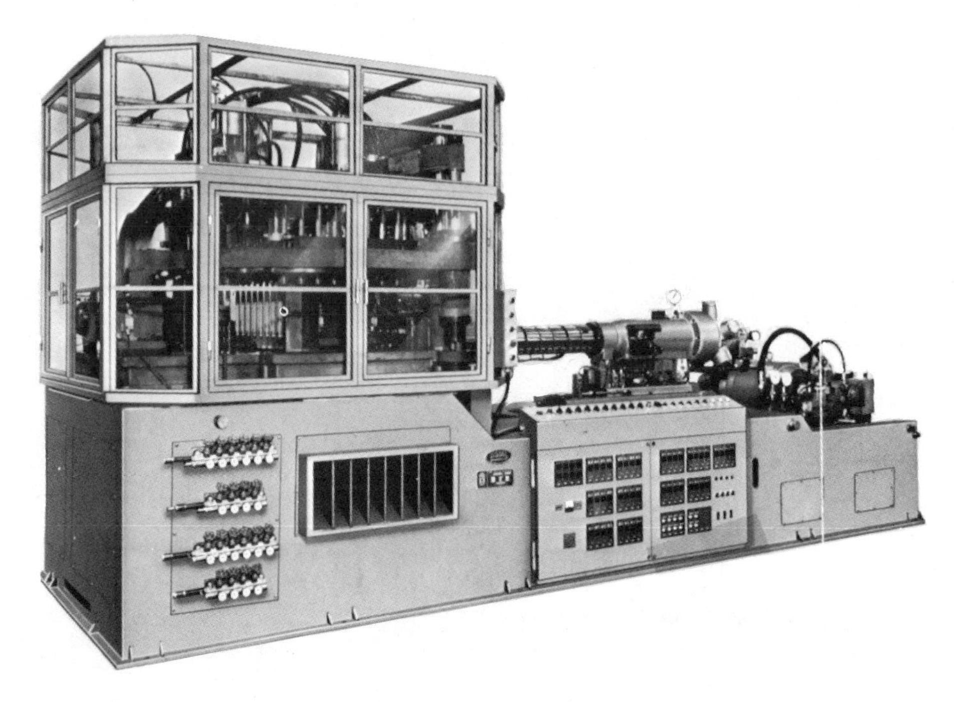

Figure 4-10 ASB 650 single-stage machine. Courtesy of Nissei ASB Machine Co., Ltd., Japan.

Figure 4-11 Krupp Corpoplast B-40. Courtesy of Krupp Corpoplast, West Germany.

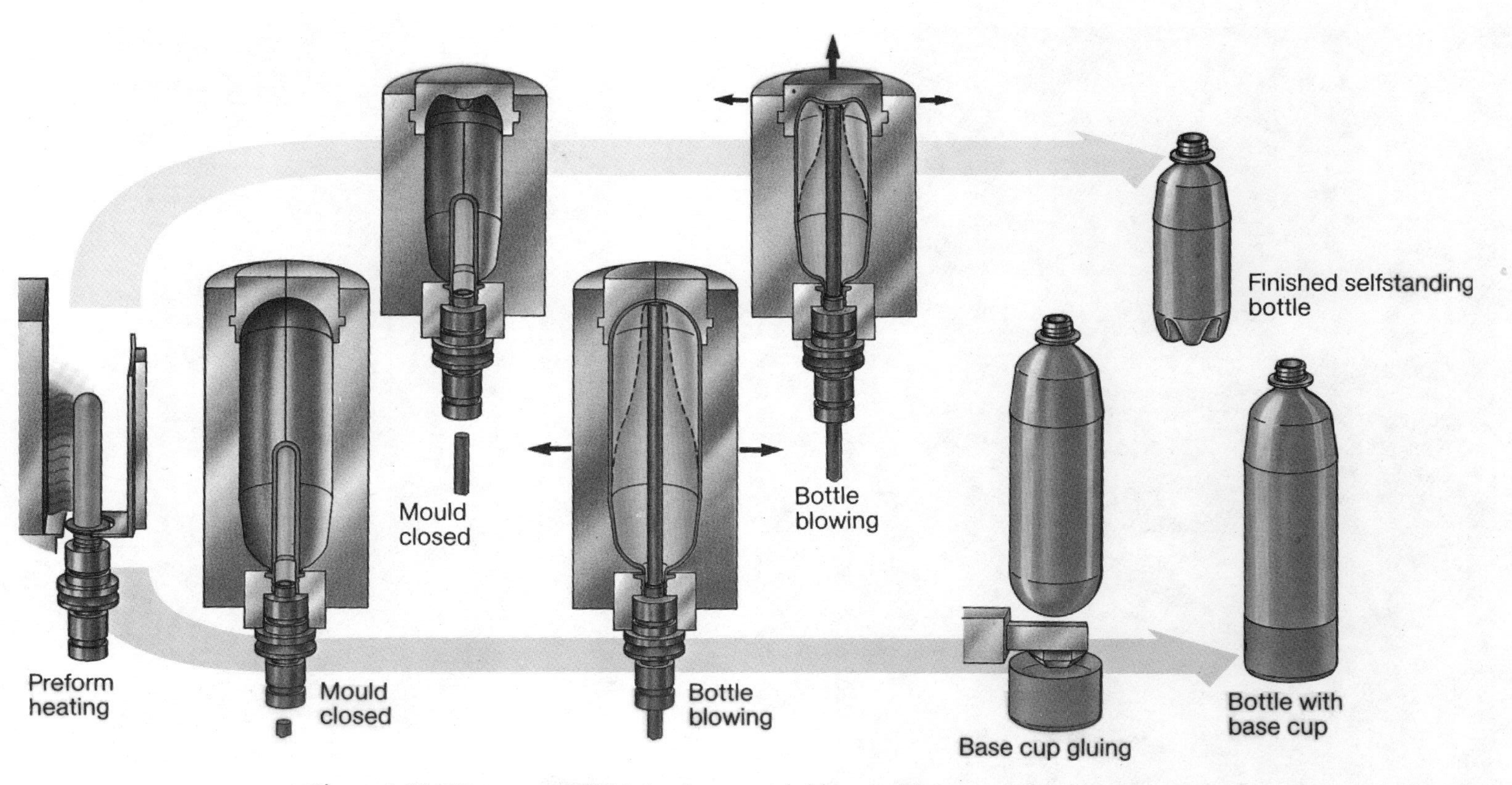

Figure 4-12 Two-stage PET injection stretch blow molding machine. First station: Condition the preform. Second station: Condition the preform. Third station: Stretch blow PET bottles. Fourth station: Eject bottles. Courtesy of Krupp Corpoplast, West Germany.

Figure 4-13 Blow molding line, Corpoplast B-25. Courtesy of Krupp Corpoplast, West Germany.

Figure 4-14 Blow molding line, Corpoplast B-80. Courtesy of Krupp Corpoplast, West Germany.

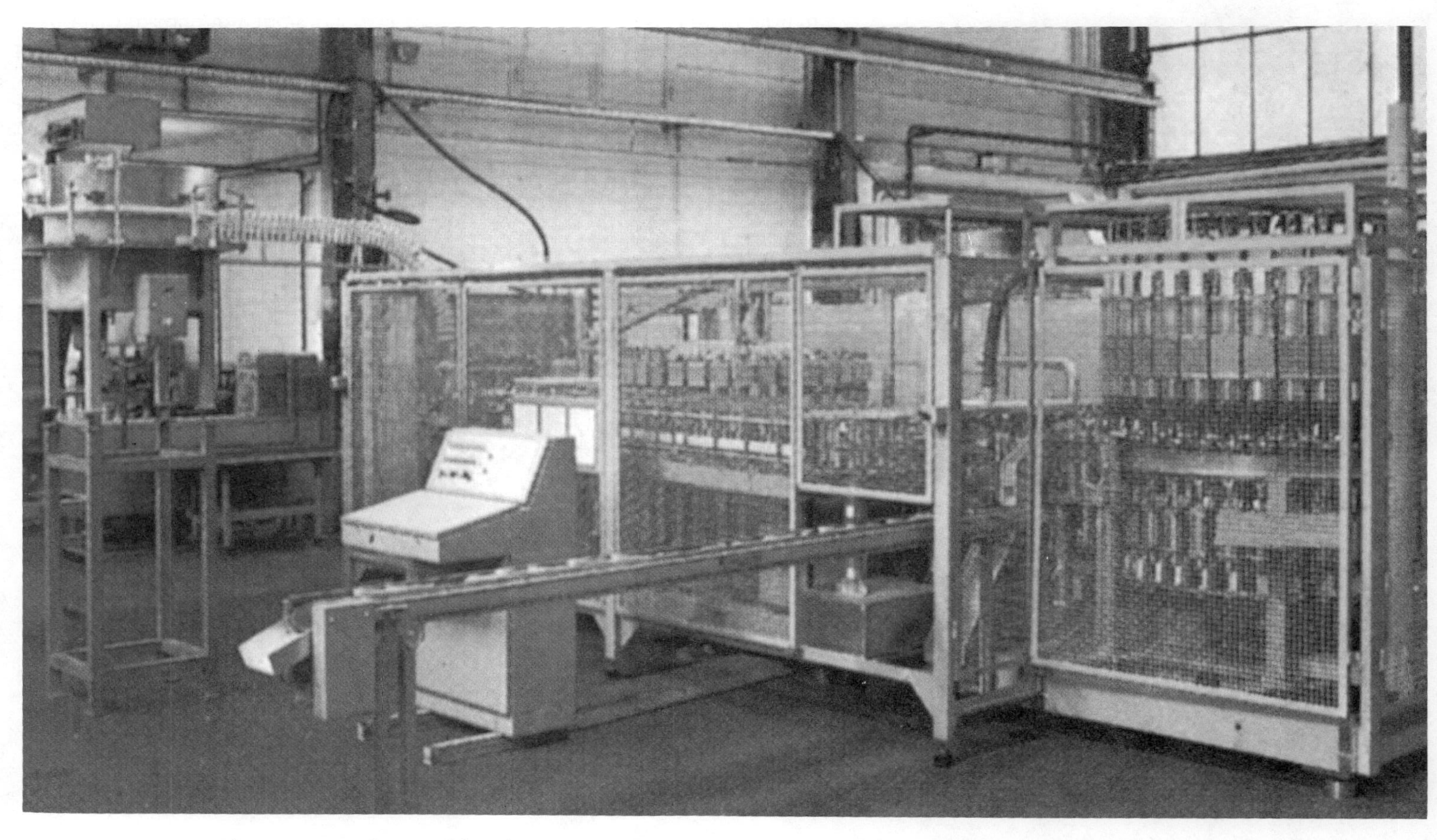

Figure 4-15 Blow molding line B-160. Courtesy of Krupp Corpoplast, West Germany.

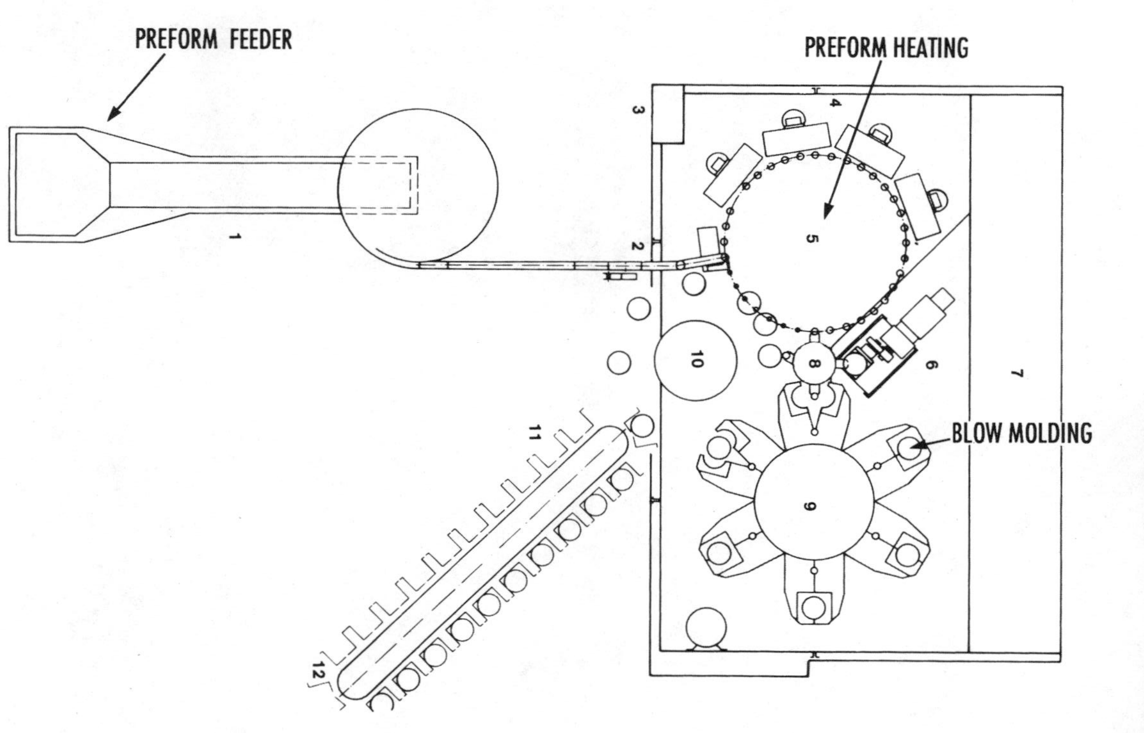

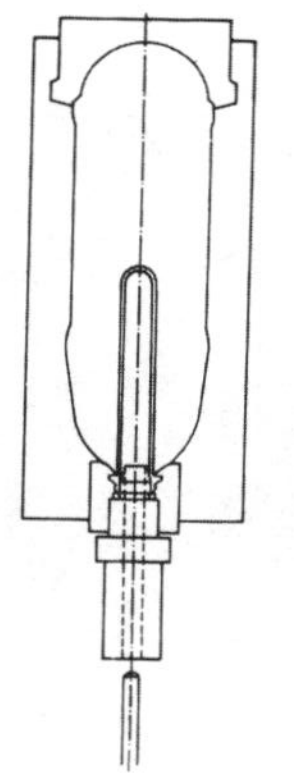

Figure 4-16 Two-stage PET wheel machine B-40 showing preform feeding, preform heating and blow molding. Courtesy of Krupp Corpoplast, West Germany.

Figure 4-17 Sidel SMTP wheel-type stretch blow molding machine. Courtesy of Sidel of America, Atlanta, Georgia.

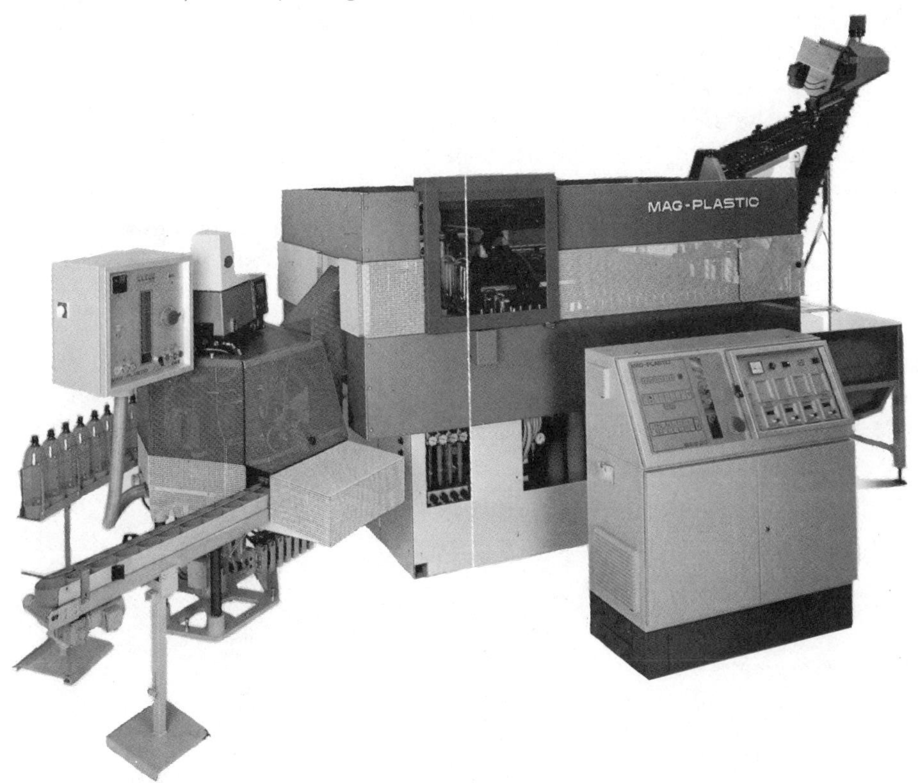

Figure 4-18 MAG-PLASTIC SSB-20. Courtesy of MAG Plastics, Switzerland.

The main difference between the Cincinnati Milacron and MAG-Plastic machines is that RHB 2000 is capable of producing PET bottles with finish sizes ranging from 28 to 89 mm while finish size in the SSB-20 ranges from 12 to 50 mm.

In two-stage PET stretch blow molding, injection-molded PET preforms are fed into the blow molding machine. They are transported through an oven where they rotate continuously to achieve even heating throughout. The finish or threaded area of a preform is protected by heat shields so that it is not heated. Once heated, the preforms exit the oven and are allowed to equilibrate.

Equilibration is one of the most important steps in producing a quality PET stretch blow molded container. Since the preforms are heated by infrared energy, the heat to the outside skin of the preform is higher than to the inside skin. Upon exiting the oven, the outside skin can be as much as 35°F (19°C) hotter than the inside skin. In equilibration, the outside skin temperature drops while the inside skin temperature rises due to the residual heat in the material between the two skins. Although the outside skin also receives residual heat from this mass, it is in contact with the atmosphere and therefore can cool quicker than the inside skin. The inside skin is in contact with the heated mass on one side and with interior heated air and the surface of the supporting metal or plastic collet or mandrel on the other side, both of which are warmer than the outside skin. Figures 4-19 and 4-20 depict the temperature difference, Δt, of the preform skins during heating and at equilibration.

Once the preform has equilibrated, it is transferred to the waiting blow mold. The blow mold closes just beneath (Cincinnati Milacron; Fig. 4-6) or above (Corpoplast, Sidel; Fig. 4-12) the threaded area and the capping ring (the bumper role in glass industry terminology). Once the blow mold has closed and clamp tonnage is achieved, a center rod or stretch rod enters the

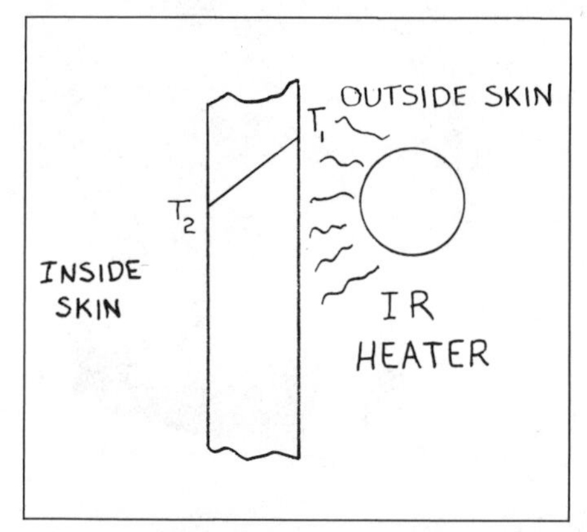

Figure 4-19 Temperature difference, Δt, during heating of injection-molded PET preform.

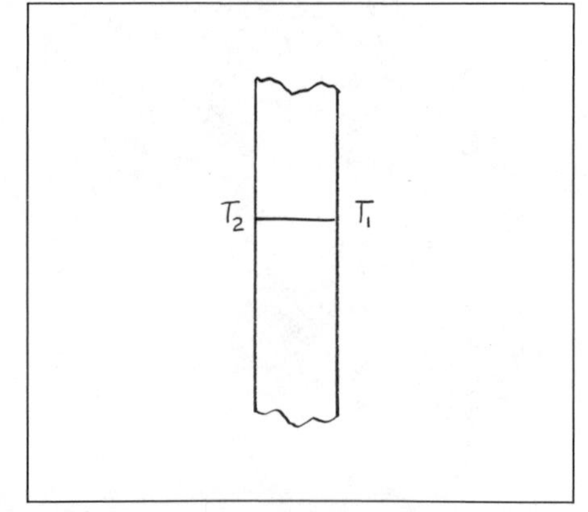

Figure 4-20 Desired Δt of outside skin and inside skin after equilibration.

preform through the stuffer and axially stretches the preform until it bottoms against the interior of the blow mold opposite the stuffer. (A stuffer is a seal through which the blow air enters, and can be either a topseal or an inside seal.) At this time, the first-stage oil-free air enters and blows the heated, axially stretched preform until the material fills the female cavity of the mold. This low-pressure blow air is usually 150 to 220 psi. Once the bottle is formed, high-pressure oil-free blow air enters and fills the bottle up to as much as 600 psi. The bottle's body cools by contact with the cold blow mold body [normally 38° to 50°F (3°–10°C)]. The blow air is then exhausted, the center rod retracts, and the blow mold opens to allow the stretch blown bottle to be ejected.

In the single-stage process (Fig. 4-8) the total process takes place in one machine. The PET preform is injected molded in the first station and upon proper cooling, the cavity mold half drops down and the core rod mold half rises, leaving the PET injection-molded preform supported by the thread splits in the horizontally rotatable table. The table then indexes 90 degrees to the conditioning station, where each preform is equilibrated or conditioned via internal and external Cal-rod heaters to the proper temperature for stretch blow molding. The table again indexes 90 degrees to the third station, where the preform is stretch blow molded. Two differences exist here from the two-stage process, and they can be critical. First, because the preforms are held in the thread splits there is no need for a capping or bumper roll to take the pressure from the stuffer sealing the preform prior to entry of the center rod and the blow air. Second, the single-stage process normally uses only one air pressure, approximately 300 psi. However, some companies have found it necessary to use a two-stage blow (low pressure, high pressure) for containers such as the free-standing beverage bottle design patented by Continental Can, or for bottles that require high definition, such as a multi-ribbed design.

TWO-STAGE AND SINGLE-STAGE PROCESSES COMPARED

There are distinct advantages and disadvantages to each of the two PET stretch blow molding processes.

Single-stage Process Advantages

- Lowest cost investment
- Requires no capping ring or bumper roll
- Produces the most pristine bottle, with fewer blemishes
- All processes are in one machine
- Relatively easy to change over if the same finish is used on different size (volume) containers

Moderate heat setting of PET bottles can be achieved by use of hot blow molds since the injection molding of the preform controls the length of the total cycle time, thus allowing for extra time in the blow molds
Relatively low volume production

Single-stage Process Disadvantages

Produces a heavier bottle
Lower barrier properties in the bottle because produced at a higher temperature than the two-stage process
Large storage capacity required to store bottles, because when the preform is produced it must be blown
Cannot separately optimize either the injection molding or the blow molding process because each depends on the other
When the system is down, total production is down
Bottles exhibit lower static attraction than two-stage bottles
Overall lower efficiency than two-stage system
Requires more skilled operator who must know the total process—injection molding and blow molding
Cannot produce a preform without blowing a bottle

Two-stage Process Advantages

Lowest cost bottle
Lowest weight bottle
Best barrier bottle
Can optimize preform design and injection molding
Can optimize blow molding of the bottle
Can have centralized preform production feeding to different blow molding sites
Can either make or buy preforms
Can plan for cyclical production
High productivity
Bottles are more resistant to static and dust
Quick changeover if the same finish is used on different size (volume) bottles
Easier to troubleshoot because each process is separate

Two-stage Process Disadvantages

Requires the preform to have a capping ring, bumper roll, or some other feature to aid in feeding the preform to the blow mold and to resist stuffer pressure in the blow mold

More blemishes than single-stage bottle
High cost of investment
Restricts some finish designs and bottle designs due to preform handling and preform rotation in the heating areas (e.g., ovalized thickness in preforms would not heat evenly)

Volume versus Economy

Usually if the volume of a single design bottle exceeds seven to nine million units, it pays to install the two-stage system due to lower gram weight per bottle, lower labor cost per bottle produced, and overall lower cost per bottle produced in total cost analysis.

ORIENTATION

Biaxial stretch blow molding was defined at the beginning of this chapter, and material property enhancements are given in Table 1-1. To further understand biaxial orientation and stretch blow molding it is necessary to look at the basics of stretch blow molding, where production begins with the preform, whether it be injection molded or formed as an extruded tube. Figure 4-21 shows a two-liter PET soft drink bottle alongside the injection molded preform from which it is produced.

In stretch blow molding two preform-to-product ratios are used. The hoop ratio, H, is defined as the ratio of the largest inside diameter of the blown article, D_1, divided by the inside diameter, D_2, of the main preform or parison body before blowing:

$$\text{Hoop ratio, } H = \frac{D_1}{D_2}$$

The axial ratio, A, is defined as the length, L_1, measured from where axial stretch is initiated in the container to the inside bottom of the container, divided by the length, L_2, in the preform from where stretch will be initiated to the inside bottom of the unstretched preform:

$$\text{Axial ratio, } A = \frac{L_1}{L_2}$$

The total blowup ratio, BUR, is equal to the hoop ratio times the axial ratio:

$$\text{Blowup ratio, } BUR = H \times A = \frac{D_1}{D_2} \times \frac{L_1}{L_2}$$

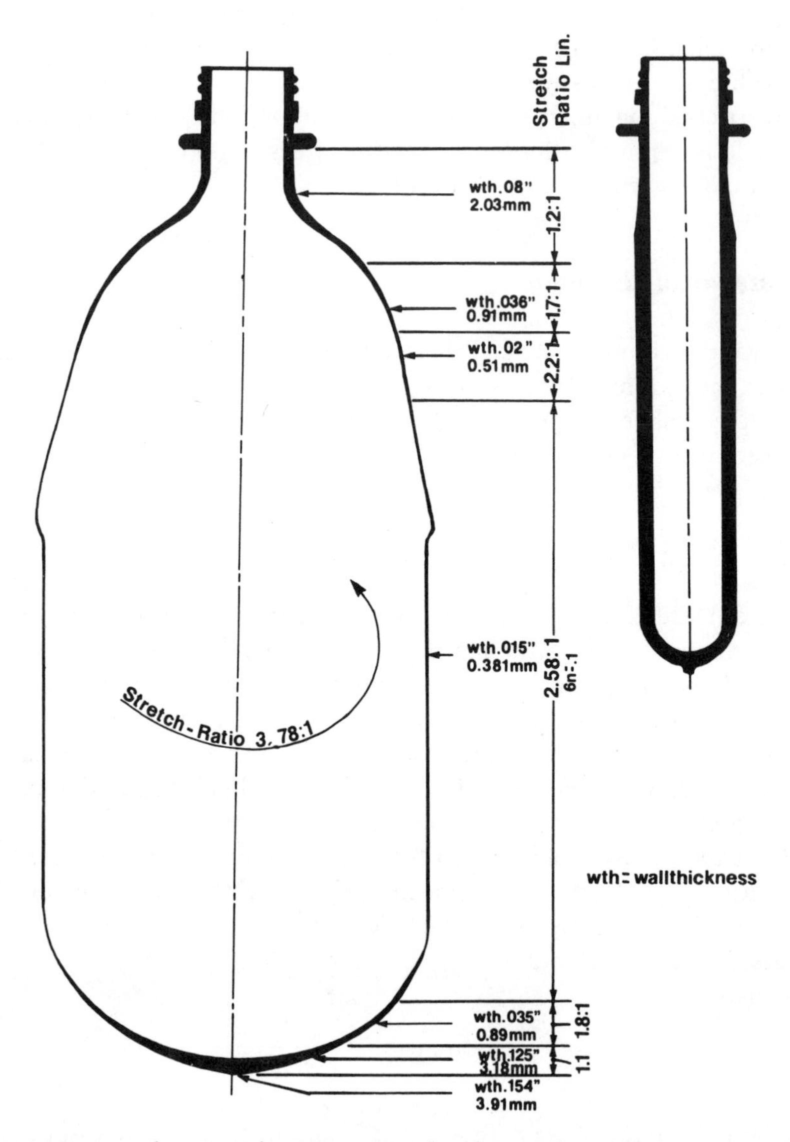

Figure 4-21 Stretch ratios of a generic type bottle, two-liter size. Courtesy of American Hoechst, U.S.A.

In designing a container to hold pressure, such as a soft drink beverage bottle, the BUR should be 10 or greater. The two-liter beverage base cup design PET bottle has a BUR of 10.484. Of the two values that determine the blowup ratio, the hoop ratio is the most important. Hoop ratios usually range from 4 to 7 while the axial ratio can be from 1.4 to 2.6.

It is easy to understand that the orientation ratios will vary throughout the stretch blown container. Figure 4-21 shows the stretch ratios actually produced in a standard two-liter PET stretch blow molded base cup type soft drink beverage container.

Each material noted has its own natural stretch ratio. This is the ratio between the material dimension at no stretch and the corresponding dimension at the limit where the material fractures or simply stops stretching. The natural stretch ratios of several materials are:

PET	16
Polpropylene	6
Polyvinyl chloride	7
Polycarbonate	6
Polystyrene	12
Acrylnitrile	12

Polyester, which is PET, sold by DuPont as Mylar, is totally biaxially oriented. When produced as a film, it is stretched four times in each direction. Thus for PET:

$$BUR = H \times A = 4 \times 4 = 16$$

In designing for stretch blow molding and using the known stretch ratios for each material, the results can be checked in the laboratory.

If a plaque is compression molded from the original material—for example, 0.72 IV (intrinsic viscosity) PET—the tensile strength of an ASTM dog bone cut from the molded plaque can be checked on an Instron or similar machine. The tensile property should approximate 6700 psi, which would closely match the raw material specification data supplied by the resin producer. This is basic data that can be used to determine how much orientation is achieved in the desired stretch blown container.

If strips are cut from a biaxially stretch blow molded container, one in the hoop direction and one in the axial direction, and each is tested for tensile strength, new results will be recorded. (See Fig. 4-22) These results should equal the appropriate ratio (hoop or axial) multiplied by the raw material tensile property value. The hoop ratio is the inside diameter at the largest point in the bottle divided by the inside diameter of the preform or parison. It is used to check the orientation in hoop direction and the overall blowup ratio (BUR) of the final bottle. The axial ratio is the inside diameter of the blown bottle divided by the inside diameter of the preform or parison. It allows for the orientation to be calculated and for the BUR.

For example, with 0.72 IV PET, if the hoop ratio D_1/D_2 is by design 5, then $5 \times 6700 = 38{,}500$ psi. Thus, the tensile test result with the strip cut in the hoop direction should be close to 38,500 psi. If the test result is not within 2000 psi of the calculated value, the parison or preform was either too hot (low tensile value) or too cold (fractured before reaching the desired orientation temperature).

The BUR is also used to determine the thickness of the preform or the bottle, depending on where the design begins (BUR = hoop ratio × axial ratio). For instance, if the average desired wall thickness in a production

Figure 4-22 Hoop and axial ratios in blow molded bottle.

container is 0.020 in. and the BUR is approximated to be 10, then

$$\begin{aligned}\text{Desired wall thickness} \times \text{BUR} &= 0.020 \times 0.10 \\ &= 0.200 \text{ in. wall thickness in main body of preform}\end{aligned}$$

The original two-liter PET bottle had a blow up ratio of 10.484 and the preform main body was 0.157 in. thick.

The closer the parison or preform is to the ideal orientation temperature when blown, the better the theoretical and actual tensile results will correlate.

The axial ratio L_1/L_2 can be checked in the same manner, using a strip cut from the axial dimension of a container.

A quick and easy check for orientation in a PET biaxially oriented stretch blow molded container can be made on the production floor, without going to the laboratory. The production container is cut so that a sidewall piece can be torn by hand in the axial direction and also in the hoop direction. If the tear is smooth, as if cut by a knife, there is little or low orientation in that direction. Biaxial orientation is revealed by a tear that is very ragged, with many fractures, and at the ragged edges layers that can actually be separated.

PET has one property that no other thermoplastic blow moldable material has: self-leveling. Other blow moldable thermoplastics stretch like bubble gum, while PET stretches like a rubber balloon. In resins such as HDPE, LDPE, PVC, AN, and others, the weakest or hottest area will respond to blow pressure first. If the other material in the preform is not close in temperature, that area will stretch as a bubble and burst. In PET the weakest or hottest section will start to expand, but in doing so its tensile value increases. Thus it is stronger than material adjacent to it and will remain stable until the adjacent material expands and achieves the same tensile strength; then the entire area will continue to expand. This produces even wall distribution within the stretch blown container.

Another unique quality of PET is its ability to be "free blown." It was learned in 1980–1981 that if a preform was heated to 205°F (96°C), preform designs could be checked in the laboratory without using a blow mold to restrict the preform shape when it was stretch blown. 205°F is the optimum temperature for optimum orientation.

A PET injection molded preform of 0.72 IV material can be heated in an oven, by infrared or radio-frequency energy, to the proper orientation temperature. Once this temperature is even throughout the preform, a closure can be applied through which a gas, say air at 20 to 25 psi, can enter and the preform will expand to a shape. If the shape of the desired bottle is to be long and round like a piece of sausage, a properly designed preform will blow to a close approximation of that shape. If the shape of the desired blown container is to be round, like a sphere, then the preform should stretch and blow to a small sphere. If the opposite happens, the preform is not designed correctly to stretch blow the desired container shape. Free blowing of PET has allowed most beverage bottle producers to stretch blow mold two-piece carbonated PET soft drink containers without the use of center rods (stretch rods), which in turn has allowed faster cycles, producing more bottles per hour per labor dollar.

As we said earlier in the chapter, preform design for biaxial stretch blow molding is both an art and a science, but through the use of computer programs, it is slowly becoming more of a science.

PREFORM DESIGNING

In designing the preform, there are two critical areas to consider. The first is the area below the finish where the transition occurs to the main body of the preform; it is referred to as the translation area. It is a critical area because translation usually occurs over a very short distance and the wall thickness is constantly changing as it blends from the main thickness in the preform body to the desired thickness in the threaded or finish area. (See Fig. 4-23.)

The second critical area in preform design is the tip or bottom of the preform (Fig. 4-24). The tip of the preform will not only produce the bottom of the container, but also the heel area. The tip must be designed to provide enough material to stretch out to form an adequate heel area, yet not produce a heavy bottom section, which is material that is not really usable. The tip must also allow an easy flow of the hot plastic material to fill the preform without causing the core rod to deflect, which would produce uneven wall distribution in the preform. For this reason the injection pressure should be as low as possible for the given material.

Containers for Scope mouthwash, produced by Amoco, were the first commercially manufactured nonbeverage biaxially oriented PET containers. The preform was ovalized to get the material into the oval shape of the container. Later, as preform technology increased, round preforms were used to take full advantage of PET's self-leveling ability.

Preforms in PET have been produced ranging in weight from as low as 10 grams, for an aerosol bag, up to 550 grams, for a 50-liter barrel. (See Fig. 4-25.)

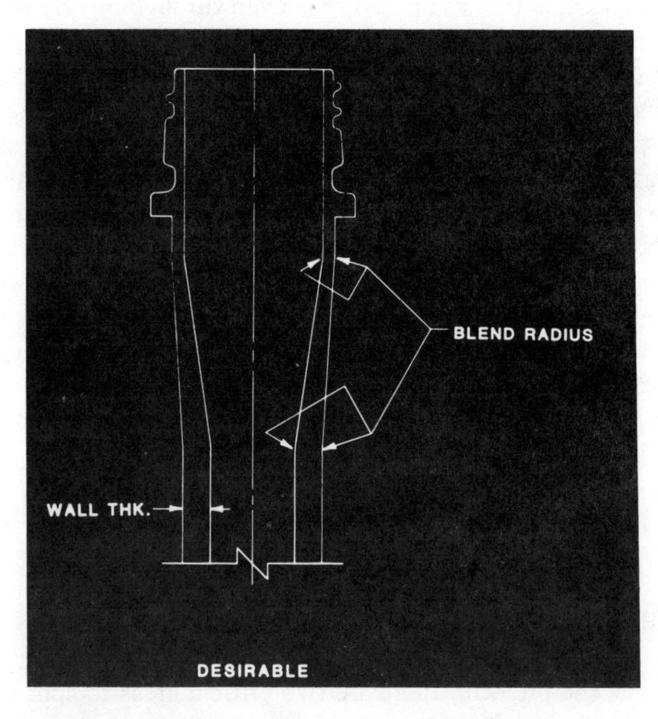

Figure 4-23 Critical areas in preform design.

Actual injection molds are shown in Figures 4-26, 4-27, and 4-28. PET preform hot runner injection molds began as eight-cavity molds in initial production; today Husky offers a seventy-two cavity production mold. Costs can range from $6,000 for a single cavity sampling mold to over $500,000 for a seventy-two cavity production mold.

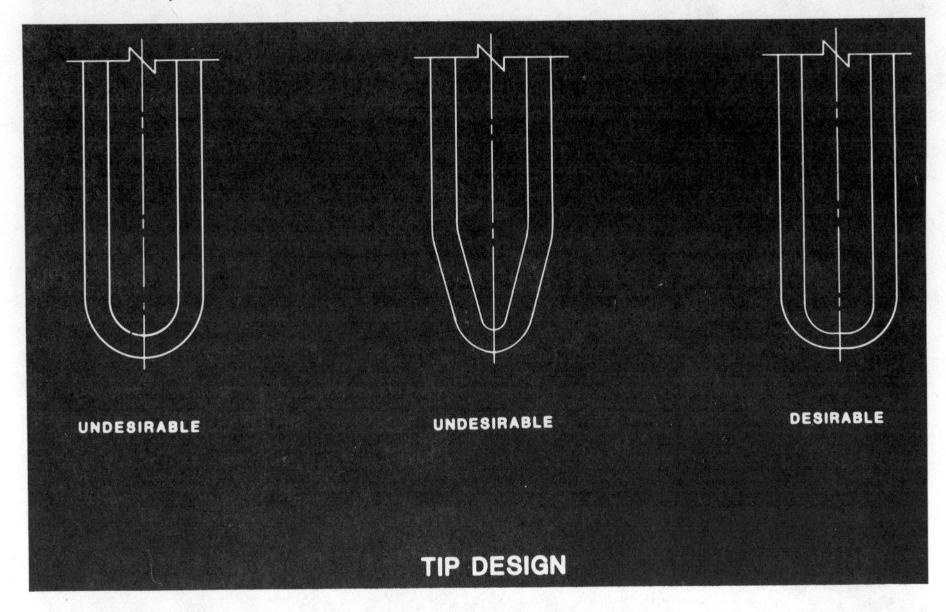

Figure 4-24 Tip design.

Figure 4-25 Injection-molded PET preforms.

Figure 4-26 Sixteen-cavity PET preform injection mold. Courtesy of Broadway Company, Dayton, Ohio.

Figure 4-27 Twenty-four cavity PET preform injection mold. Courtesy of Electra-Form, Inc., U.S.A.

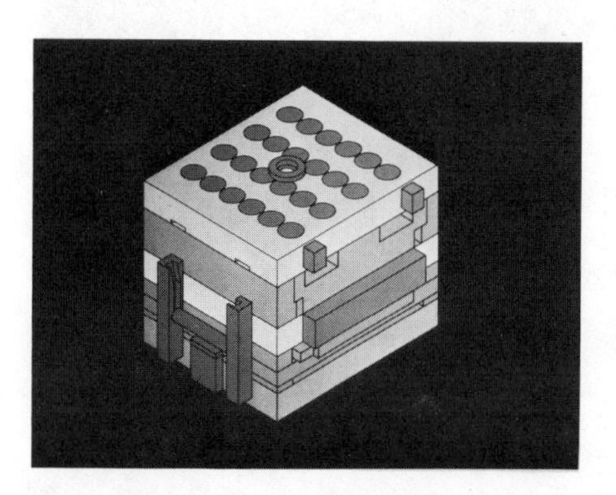

Figure 4-28 Twenty-four cavity PET preform injection mold graphic presentation. Courtesy of Electra-Form, Inc., U.S.A.

EXTRUSION STRETCH BLOW MOLDING

Polyvinyl chloride (PVC) has been biaxially stretch blow molded in Europe for many years. Companies such as Bekum, Krupp Kautex, and Battenfeld Fischer all offer machines for this method.

In extrusion biaxially oriented stretch blow molding, the free extruded hot parison is blown in a predetermined shape in a first station blow mold; this shape is then transferred to a second station blow mold where the final biaxially oriented container is produced. The system is entirely sequential, and containers can be produced two at a time either by shuttle blow molds (Kautex) or by inclined blow molds (Bekum). The process is shown in Fig. 4-29. Figure 4-30 shows the interior of a process machine, and Fig. 4-31 diagrams details of the equipment and the production cycle. Typical extrusion stretch blow molded bottles are shown in Fig. 4-32.

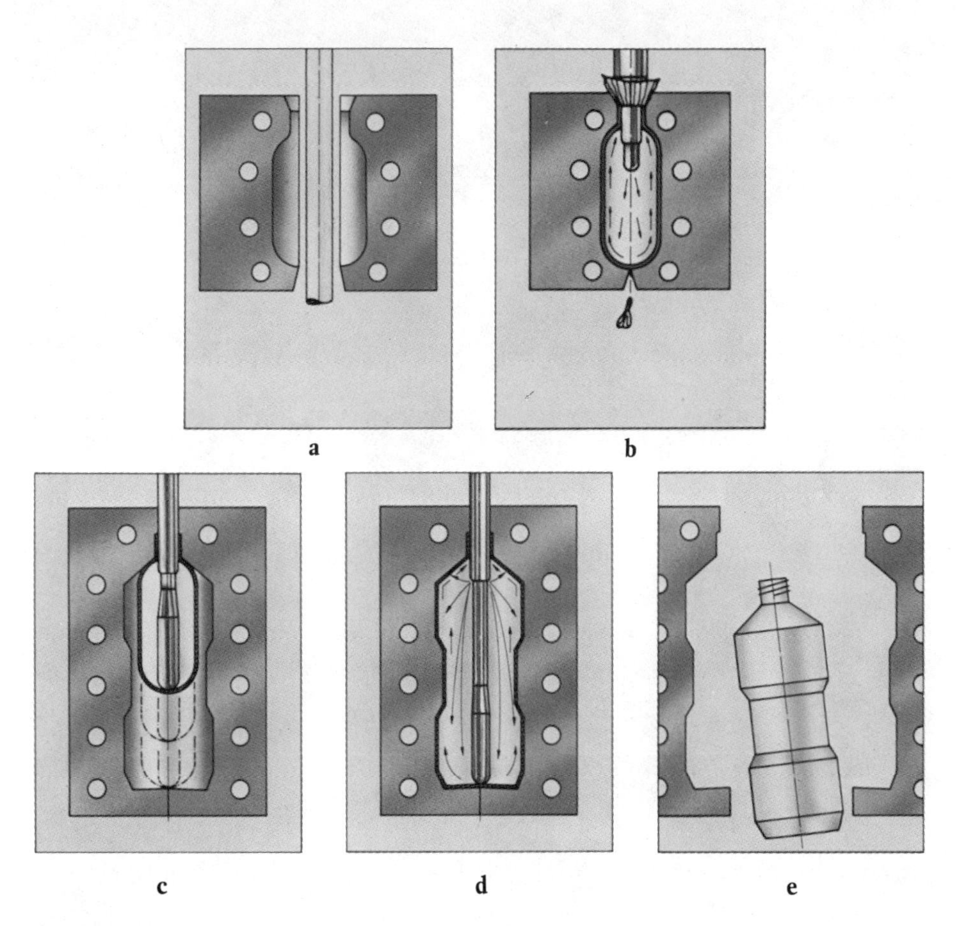

Figure 4-29 (a) Extrusion stretch blow molding: extrusion. (b) Extrusion stretch blow molding: blowing and conditioning. (c) Extrusion stretch blow molding: stretching. (d) Extrusion stretch blow molding: blowing. (e) Extrusion stretch blow molding: ejection. Courtesy of Battenfeld Fischer, West Germany.

Figure 4-30 Interior of extrusion stretch blow molding machine. Courtesy of Battenfeld Fischer, West Germany.

Figure 4-31 *(facing page)* Extrusion stretch blow molding cycle. (1) Preblow mold. (2) Preblow mandrel. (3) Completion blow mold. (4) Completion blow mandrel and stretch plug. (5) Bottom scrap removal. (6) Movable mold bottom. (7) Top scrap removal. Courtesy of Battenfeld Fischer, West Germany.

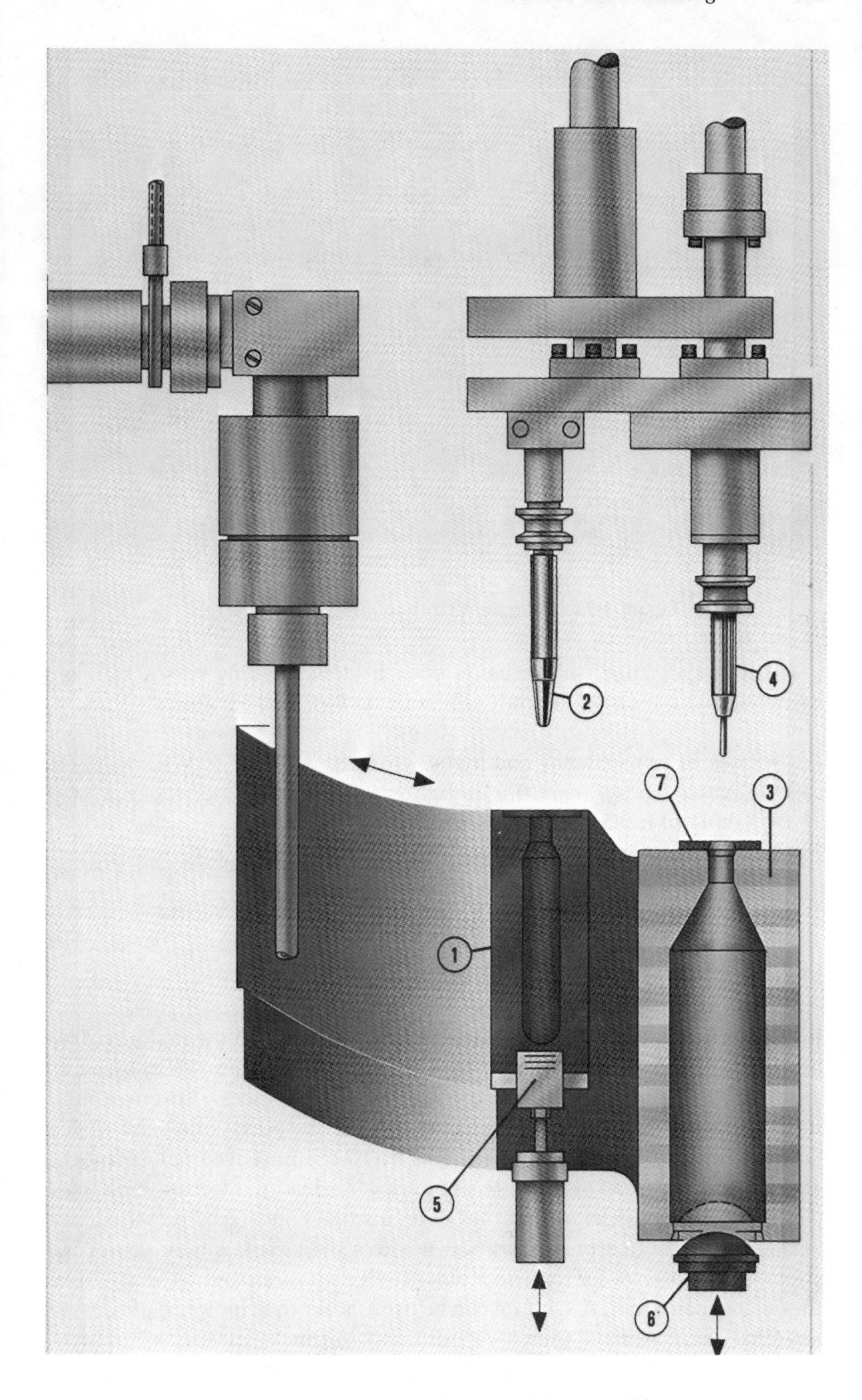

1
2
3
4
5
6
7

Figure 4-32 Extrusion stretch blow molded bottles.

Advantages cited for extrusion stretch blow molding versus standard extrusion blow molding of materials such as PVC and PP are:

- Greater transparency and higher gloss
- Greater rigidity, resulting in lighter bottles and material saved
- Saving of impact modifiers (PVC)
- Improved permeability
- Increased resistance to burst pressure
- Increased output capacity

INJECTION ORIENTATION BLOW MOLDING

In 1986 AOKI of Japan introduced injection orientation blow molding. This process originally was developed by Owens-Illinois in the 1960s, based on a patent by K. Allen, and was known as the "flair" process. Owens-Illinois used materials such as polypropylene and impact polystyrene. AOKI has taken advantage of new resins, such as PET, and improved polypropylene and polycarbonate resins. The basic process begins by injection molding a disk similar to a saucer, or in other cases a small cup-shaped preform. This is transferred to a secondary station where a plug assist and/or center rod stretches the disk or preform to a blow cavity's bottom and blow air forms the desired container. A vacuum can be used rather than blow air: plug assist stretches the materials, then a vacuum draw forms the desired container.

The AOKI method of injection orientation blow molding is depicted in Fig 4-33. Representative containers are shown in Fig. 4-34.

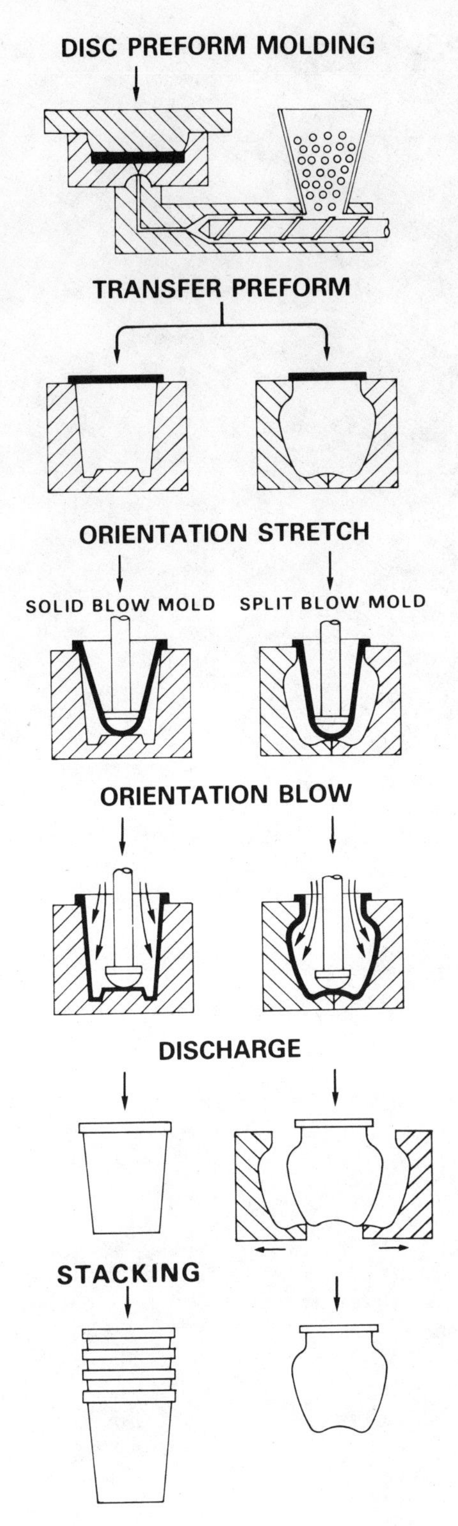

Figure 4-33 AOKI injection stretch blow molding process. Courtesy of AOKI, Japan.

Figure 4-34 AOKI injection stretch blow molded containers. Courtesy of AOKI, Japan.

The applications of stretch blow molding will continue to increase and new materials will enter the field. Economic factors and improved container features indicate the advantage of using stretch blow molding wherever feasible.

ACKNOWLEDGMENTS

AOKI, Japan
American Hoechst, Spartanburg, S.C.
Battenfeld Fischer, West Germany
Bekum, West Germany
Broadway Companies, U.S.A.
Cincinnati Milacron, U.S.A.
Eastman Chemicals, Kingsport, Tennessee
Electra-Form Inc., U.S.A.
Goodyear, Akron, Ohio
ICI, Wilmington, Delaware
Krupp Corpoplast, West Germany
Krupp Kautex, West Germany
MAG Plastics, Switzerland
Nissei, Japan
Sidel SMTP, France
Wentworth Mould, Canada

CHAPTER

5

Coextrusion Blow Molding

DAN WEISSMANN

Coextrusion blow molding refers to products which contain several layers in their wall structures and to the technology to produce such articles. The layers can be of the same material, colored material and not colored, recycled material and virgin, or of different materials.

The development of coextrusion technology is another step in taking advantage of the unique properties of plastics. This process makes it possible to combine materials with various attributes to create a finished product most suitable for a particular application. Additionally, the various parts of the structure can be optimized for the best balance between properties and cost.

Although several years have passed since the first use of coextruded blow molded containers—in the early 1970s, in Japan—coextrusion really became part of the blow molding scene with the introduction of the Gamma bottles produced by American Can Co. in 1983. The bottles were used to package Heinz Ketchup and Lipton barbecue sauce [1].

Today the uses of coextruded blow molded article are numerous and diverse. Some typical multilayer bottles are shown in Fig. 5-1. Packaging of various types is the main end use of coextruded products, with barrier properties being the main reason for the multilayer structure. This application is a result of the overall balance of the properties and the cost of the barrier materials used.

The multilayered structure of coextruded products is created by combining layers in a die before their extrusion as a parison. The multilayer parison is extruded and blown into a product in much the same way as a monolithic blow molding parison, as described in chapter 2. The main difference between single-material extrusion blow molding and coextrusion blow molding is in the extrusion system. In coextrusion each material in-

Figure 5-1 Coextruded multilayer bottles for catsup and jelly. Courtesy: *Plastic World* (Feb. 1986), Cahners Publishing Co., Newton, Mass.

corporated into the structure is extruded from its own extruder. The extrusion die must be designed so that the different materials can be formed into the proper layered structure.

COEXTRUDED STRUCTURES

The simplest coextruded structure is a two-layer combination, however, there are few application in which such a structure is used. The best example is where only a thin layer of the wall is colored, rather than coloring the entire wall. Three-layer structures are used where a significant amount of regrind, from both manufacturing and postconsumer recycling, is utilized. The regrind is used for the center layer, between layers of virgin material on the inside and the outside. The virgin material provides the desired appearance on the outside and protection from any contamination on the inside, in contact with the packaged goods. Because many benefits of coextruded structures come from combinations of materials that do not adhere readily to one another, another layer between two such materials is needed—an adhesive layer, also called a tie layer. Where barrier materials are used, the common structures are typically composed of five, six, or seven layers.

A multilayered structure is created to achieve certain performance criteria. Such criteria may be based on a physical need, for example, barrier or use temperature. They may also be based on economic considerations, which may dictate, for example, that some of the virgin wall material be replaced

Figure 5-2 Typical structure of a five-layer coextrusion. Courtesy: Dow Chemical Company.

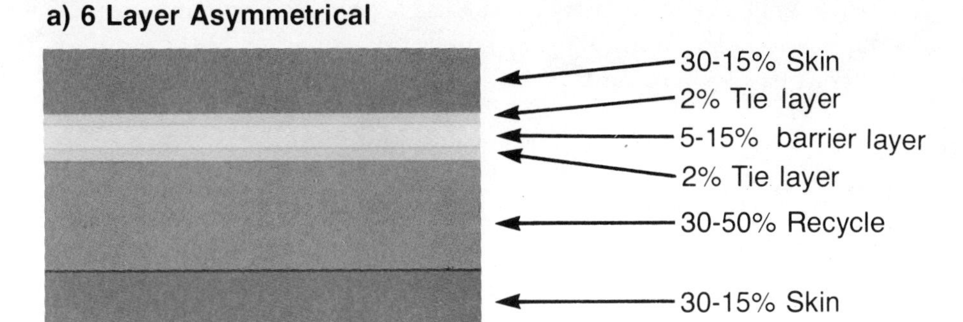

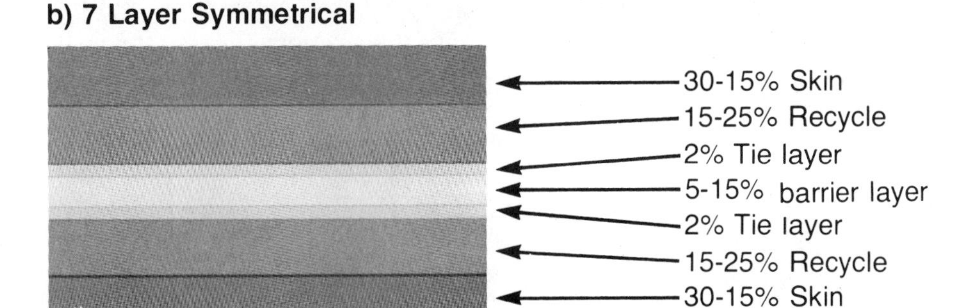

Figure 5-3 Typical coextruded structures containing recycled material in additional layers. (a) Six-layer assymetrical structure. (b) Seven-layer symmetrical structure. Courtesy: Dow Chemical Company.

by recycled material, or that colored material can be used in only a single thin layer of the structure.

A typical five-layer structure consists of an inside and an outside structural layer, a center functional layer, and an adhesive layer on each side of the functional layer (Fig. 5-2). When the sixth and seventh layers are used, they are the layers into which recycled material is returned (Fig. 5-3). These regrind layers replace some of the structural material, as indicated by the percentages listed in the figures, without affecting the functionality or the appearance of the final product. Any additional layers create higher levels of complexity both for the equipment and the manufacturing process.

Structural Layers

Structural layers can be of any thermoplastic material that can be blow molded. Because of economic considerations the basic blow molding resins, the polyolefins, are the most commonly used. Other resins are used where other properties are needed. For example, polycarbonate (PC) can be used as a structural layer where high temperature performance or high impact strength is needed; polyvinyl chloride (PVC), PC, or PETG (a form of polyethylene terephthalate, PET) where clarity is sought. Flexible PVC, polyuretane (PU), or etylene vinyle acetate (EVA) can be used where a flexible structure is desirable, as in a squeeze bottle.

Functional Layers

At present there are two main groups of functional layers: barrier layers, and layers that enhance physical properties. Physical property enhancements include higher use temperature for the product and better appearance.

Barrier layers are incorporated to enhance resistance to oxygen permeation. Barrier layers can also be used to resist the permeation of other gases, to reduce the transmission of water or other liquids through the wall, or to prevent the migration of a component of a packaged product through the wall—for example, hydrocarbons in agricultural chemicals. Another application of a barrier layer is to prevent the loss of flavor or fragrance, which are normally present only in minute quantities, from food or cosmetic products. Protection may also be needed to prevent transmission into the product from the outside [2].

The main difference between a functional layer and a structural layer is thickness. A functional layer only a few thousandths of an inch thick normally can provide sufficient performance, while structural layers are ten thousandths of an inch thick or more, depending mainly on the size and the physical requirements of the products (e.g., top load specifications for containers.)

Adhesive Layers

Adhesive layers are required to tie the functional and structural layers together when those materials do not otherwise bond sufficiently. The adhesive layers are normally very thin and therefore do not add to the physical strength or other properties of the product.

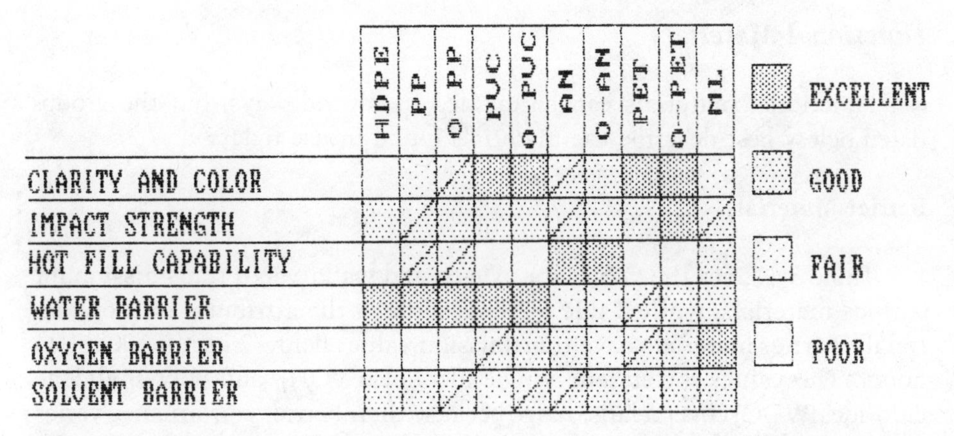

Figure 5-4 Properties of common plastic materials used in packaging applications [3].

MATERIALS

No single material can provide all the properties that may be desired for a particular performance in a product. The attributes of common blow molding materials especially suitable for packaging applications can be summarized as shown in Fig. 5.4 [3]. Certain types of materials are particularly well suited to supply the properties needed in each layer of the coextruded structure. Thus, materials can be divided into groups similar to the layer functions.

Structural Materials

Any material suitable for extrusion blow molding can be considered a structural material. Typical materials include both low-density and high-density polyethylene (LDPE, HDPE), polypropylene (PP), PETG, and PVC. Other materials—for example, engineering thermoplastics such as polycarbonate or polysulfone—may be used in special applications. Because polycarbonate and polysulfone provide certain unique performance characteristics, they may also be used as functional materials, as described later.

Typical characteristics of structural materials that make them suitable for blow molding are good melt strength and good strain hardening characteristics—the ability to propagate the deformation from the initial deformed area to undeformed parts of the parison when blown into a product. High melt strength makes it possible to extrude long and heavy parisons, which may be needed for a particular product. Good strain hardening during blowing makes it possible to control the stretching of the parison into the product so that sufficient wall thickness is reached everywhere. Typically, amorphous or partially crystalline materials will behave well due to their gradual softening on heating. Additional melt strength is gained by using extrusion grade materials, which are characterized by low melt flow or melt index.

Functional Materials

Functional materials can be divided in several ways, but the groups listed below best describe the materials found in use today.

Barrier Materials

Table 5-1 gives the values for oxygen and water vapor transmission for various materials, while Table 5-2 lists some of the attributes of such materials. Barrier materials have transmission values below 2 cc mil/100 in²/24 hours. The values for etylene vinyle alcohol (EVOH), and polyvinylidene chloride (PVDC) cover a large range because their barrier performance varies with the different formulations on the market. Additional variability for EVOH is due to the level of moisture in this material. The moisture adversely affects its barrier capability, although even at 100% RH the barrier is quite good. The difference between common blow molding materials and those defined as barrier materials is significant; and the relative barrier performance of various materials is shown in Fig. 5-5 [4]. As can be seen from the figure, EVOH and PVDC are by far the most effective barriers to oxygen and

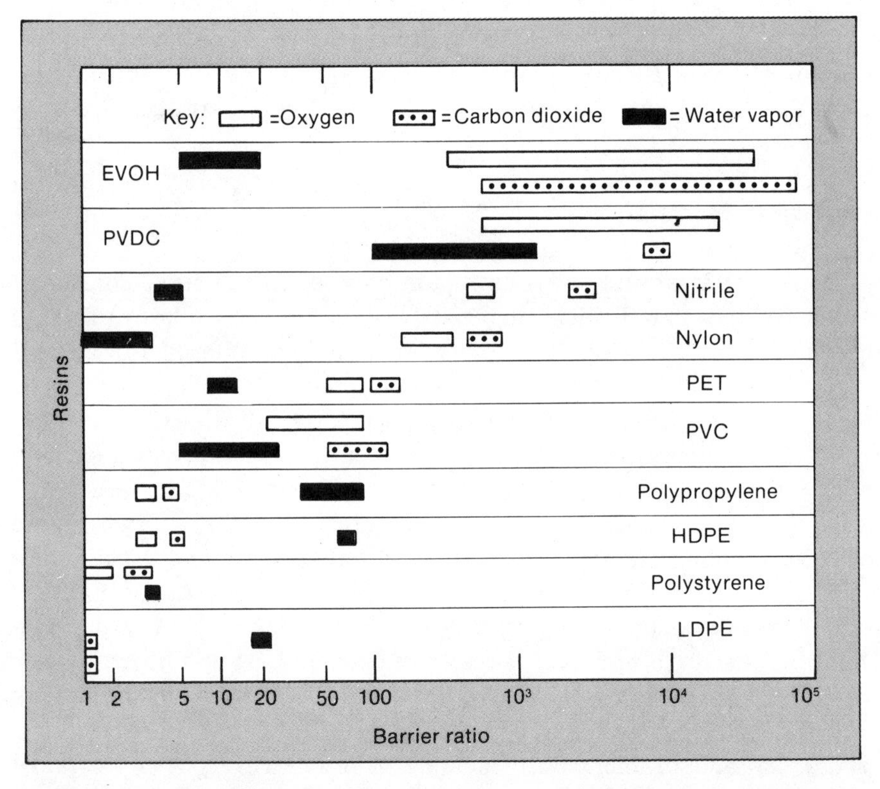

Figure 5-5 Comparison of barrier properties of various polymers used in packaging. Barrier ratio-thickness ratio to reach equivalent barrier. Courtesy: *Plastics Engineering* (May 1986).

Table 5-1 Barrier Properties of Commercial Polymers

Polymer	Transmission, cc mil/100 in.^{2}/24 h	
	Oxygen 25°C, 65% RH	*Moisture Vapor 40°C, 90% RH*
EVOH	0.05–0.18	1.4–5.4
PVDC	0.15–0.90	0.1–0.2
Acrylonitrile	0.80	5.0
Amorphous nylon	0.74–2.0	—
Oriented PET	2.60	1.2
Oriented nylon	2.10	9.0
Rigid PVC	14.0	3.0
LDPE	420	1.0–1.5
HDPE	150	0.4
PP	150	0.69
PS	350	7–10

Source: EVAL Co. of America. *Plastics Packaging* (July/August 1988): 19–21

Table 5-2 Features of Barrier Polymers Used in Food Packaging

Material	*Advantages*	*Disadvantages*
High-nitrile	Good O_2/CO_2 barrier Monolithic or coextrusion Scrap reuse Not moisture sensitive	Moderate moisture barrier Moderate impact resistance Limited grade offering
EVOH	Excellent O_2/CO_2 barrier Scrap reuse Extended grade offering	Coextrusion only Moisture sensitive
PVDC	Excellent $O_2/CO_2/H_2O$ barrier Coextrusion, lamination or coatings	Difficulty in scrap reuse No monolithic structures
Nylon (Selar PA)	Moderate $O_2/CO_2/H_2O$ barrier Monolithic or coextrusion	Moderate $O_2/CO_2/H_2O$ barrier High-cost O_2 barrier
Nylon (MXD6)	Excellent O_2/CO_2 barrier Potential low-cost O_2 barrier	Moderate moisture barrier No commercial monolithic containers

Source: BP Chemicals International. *Plastics Packaging* (July/August 1988): 19–21

carbon dioxide. Polystyrene and polyolefins, on the other hand, perform very poorly, and must usually be combined with EVOH or PVDC to effectively resist gas permeation.

EVOH currently is the main material used as an oxygen barrier in coextruded blow molding structures. The main reason is its relative ease of processing and its compatibility with other structural materials. PVDC, reportedly tested by several companies in a lab environment, is used mostly in thermoformed containers made from coextruded films. The main reason is the limited thermal stability of PVDC, which requires very close temperature control of the melt and fast encapsulation within the other layers of the structure. Those requirements cannot be achieved easily in the coextrusion blow molding die.

Tables 5-1 and 5-2 make it easy to see why certain combinations of material, for example, PP/EVOH/PP, have gained a dominant place in various applications, especially in food packaging, where both oxygen and water barrier properties are important. The tables also show very clearly the desirable combination of oxygen and water barrier characteristics in PET, which make it a very successful material for packaging.

A special case of barrier protection is the incorporation of a black layer within the structure to screen out ultraviolet rays.

High-temperature Materials

Figure 5-6 shows the thermal tolerance of various materials. The ability to withstand high temperature for any length of time is very important, especially in foods packaging where hot fill, retort, or pasteurization is applied to preserve the packaged product.

Typical materials that provide high-temperature protection are polypropylene, polycarbonate, nylon, and polyarylate. Such materials have a high heat distortion temperature or are partially crystalline. Crystalline regions within a material provide high heat resistance because their softening point is close to the melt temperature.

Appearance Materials

Several materials can be used to modify the appearance of the product. A thin, colored layer of the same material as the structural layer(s) can be used to achieve a desired color; this avoids the expense of coloring the entire material of the wall. Figure 5-7 shows such a product. Nylon or general-purpose polystyrene can be used to provide a glossy or printable surface, especially over PE or PP walls, which lack gloss and are very hard to print on.

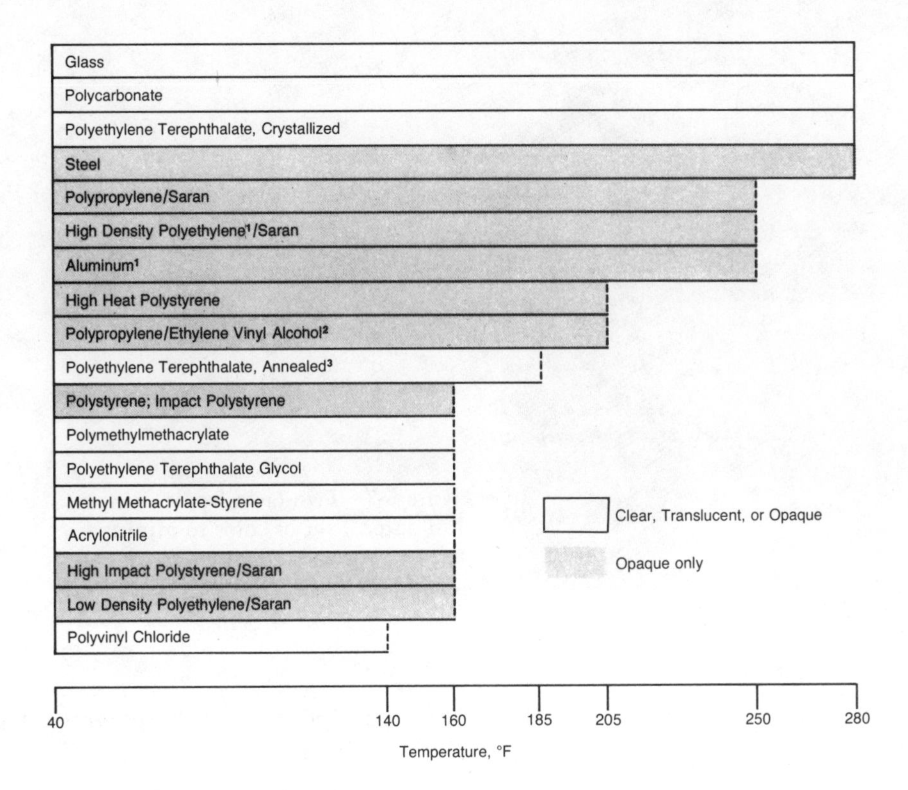

Figure 5-6 Thermal tolerance of typical packaging materials. Courtesy: Dow Chemical Company.

Miscellaneous Materials

Various engineering thermoplastics offer a variety of properties. Polysulfone provides good resistance to aggressive chemicals. Polycarbonate can be used where impact strength is desired. Clear materials must be used if the end product is to remain clear. They can be PVC, PETG, PC, or possibly PP. Flexible materials are required in packaging where squeeze capability is of interest; such materials include flexible PVC, polyurethane, or one of the many thermoplastic rubbers or elastomeric thermoplastics such as EPDM or Hytrel.

Adhesive Materials (Tie Materials)

Adhesive or tie materials are used in coextrusion to bond layers that are otherwise incompatible. Two types of adhesive resins are used. The first

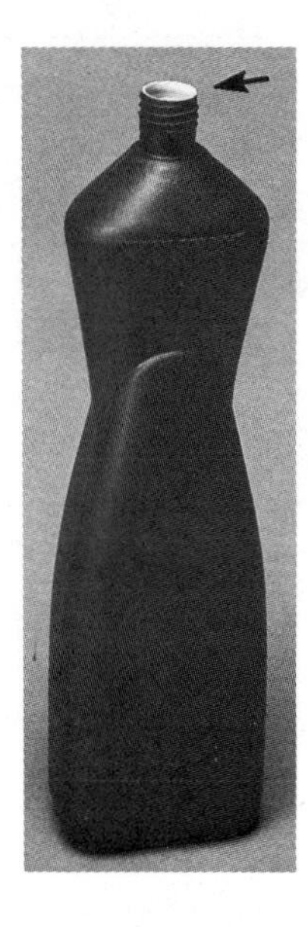

Figure 5-7 Two-layer bottle using a thin outer layer of color to minimize pigment cost. Courtesy: *Plastics Machinery & Equipment* (July 1989).

includes resins that are directly synthesized copolymer or therpolymer. The second are anhydride-grafted polyolefins. Through the chemical process, active sites are created in the material's molecules to increase its affinity for the materials of the layers being bonded in the coextruded structure [5]. In the first type of resins the bond between the adhesive layer and the joining layers is ionic, while anhydride-based adhesive materials create covalent bonds. See Fig. 5-8 [5, p. 35].

Adhesive materials are made by several manufacturers, under various trade names:

TRADE NAME	*MANUFACTURER*
Admer	Mitsui Petrochemical Industries Ltd.
Bynel	E. I. du Pont de Nemours & Co.
Modic	Mitsubishi Petrochemical Co. Ltd.
Plexar	Quantum Chemical Corporation, USI Div.
Pro-Fax	Hinmont U.S.A., Inc.

Most of the tie layer materials are polyolefins, and can be based on any of the typical polyolefins: LDPE, HDPE, PP, and EVA. Each supplier offers a large number of grades, each of which provides advantages for a particular application. The tie materials are used not only in blow molding, but also for film extrusion or blown films. Each process requires somewhat different melt characteristics.

A unique use of the adhesive layer as a moisture protector for an EVOH layer was developed by American Can Company [6]. To maintain the high barrier level of EVOH it has to be kept dry. A desiccant material incorporated into the adhesive layer absorbs moisture before it can reach the EVOH; thus the high EVOH barrier level is maintained.

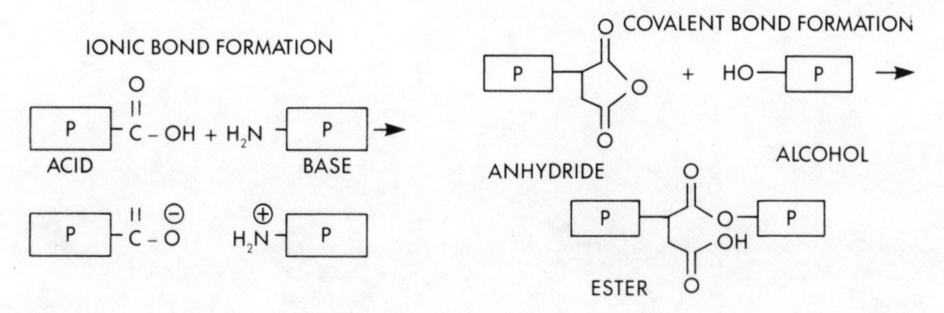

Figure 5-8 Chemical bonding of the layers is ionic with copolymers and terpolymers, or covalent with anhydride-grafted resins. Source: Du Pont. Courtesy: *Plastics Packaging* (Oct/Nov 1988).

COMMERCIAL COEXTRUDED STRUCTURES

Table 5-3 summarizes common commercial wall structures in coextruded products [7]. Also listed are typical applications and the particular properties needed for such applications. The use of recycled material in two symmetrical layers will make the number of layers seven.

Various asymmetrical multilayer structures have been successfully extruded and blown. They are asymmetrical in that the inside and outside layers are of different materials. In the following list of asymmetrical structures the middle layer was most frequently EVAL F, for a high oxygen barrier [8].

INNER LAYER	*OUTER LAYER*	*INNER LAYER*	*OUTER LAYER*
PP	Flexible PVC	PP	Nylon 11
PP	LDPE	PP	Nylon 12
PP	HDPE	PP	Polyether polyamide
PP	EVA	PP	Nitrile
PP	Polyurethane	PP	EVAL
PP	Polycarbonate	PP	Rigid PVC
PP	Nylon 6	Nylon	PP

Some of the above combinations were three-layer, while others required an additional two layers of adhesive. The structure of PP/tie/EVOH/tie/flexible PVC or polyurethane has the combined attributes of all three materials. The PP provides heat stability, the EVOH provides an oxygen barrier, and the PVC provides flexibility to make the product squeezable. If the flexible PVC is clear and the thickness of the PP is rather low, this structure can be made with good clarity. Clarity can be maintained without flexibility if rigid PVC is used for the outside layer. If nylon is used on the outside, printability and high gloss can be achieved, in addition to high use temperature and additional barrier protection.

Table 5-3 Common Commercial Coextruded Structures [7]

No. of Layers	Applications	Properties	*Combination of Layers*					
			1 inside	*2*	*3 middle*	*4*	*5*	*6 outside*
2	Detergents Cosmetics	Surface gloss	HDPE	—	—	—	—	LDPE
	Motor oil	Clear-Stripe	HDPE LDPE Stripe	—	—	—	—	—
3	Insecticides Pesticides Chemicals	Resistant to aggressive substances	PA	—	Bonding agent	—	—	PE Filler Regrind Color
	Milk	UV-protection	HDPE White	—	HDPE Black	—	—	HDPE White
	COSMETICS Hand lotions Hair cosmetics PHARMA-CEUTICALS Toothpaste Skin oils MEDICINE Infusions	Squeezable Printable Scratch-resistant, Aroma-barrier, 0_2 barrier	HDPE LDPE PP Regrind Color	—	Bonding agent	—	—	PA EVAL

	FOOD Soy sauce Ketchup	Short shelf life; hot fill	PP	—	Bonding agent	—	—	PA EVAL
	BEVERAGES Milk Liquor Handleware	Transparent Hot fill Gas barrier Rigid	PC	—	EVAL PET	—	—	PC
5 or 6	FOOD Ketchup Baby food Mayonnaise Edible oil Salad dressing Spaghetti sauce Fruit juice Soup concentrate	Squeezable Transparent (5) Translucent (6) Long shelf life Hot fill Gas barrier Moisture barrier	PP	Bonding agent	EVAL PET	Bonding agent	(Regrind)	PP
	Sterilized milk Carbonated drinks Liquor Tea Coffee Peanut butter Handleware	Rigid Returnable Gas barrier (N_2, O_2, CO_2) Moisture barrier Transparent	PC PP	Bonding agent	EVAL PET PETG PAN PVDC	Bonding agent	(Regrind)	PC

Courtesy: British Plastics and Rubber.

PRODUCT DESIGN

The Design Process

The design of blow molded articles proceeds much like the design of any new product. First the desired shape is generated or selected based on creative or functional criteria, or commonly a combination of the two. The shape is then adjusted to contain a certain volume or to fit a particular dimension or size. Once the shape and dimensions are known, the other physical and performance characteristics are considered. These lead to the selection of materials, followed by the determination of the thicknesses, so that together they meet the requirements for functional performance and structural design established for a particular product.

Coextrusion and Materials Selection

The selection of a coextruded structure is a part of the material selection process. Usually certain performance requirements lead to consideration of a particular material. If the requirements cannot be met by a monolithic wall structure of that material, a coextruded wall structure should be considered. The best example is where barrier properties are required. A monolithic AN/S polymer can provide an excellent barrier. However, this material is brittle unless biaxially oriented and even then its impact resistance is limited. Alternatives that will provide the barrier and the impact resistance are coextruded olefinic material or a coextruded PET-based structure. Additional considerations that may lead to using coextruded structures are clarity, flexibility, or the need for a material with Food and Drug Administration clearance for contact with the packaged goods.

After the materials for the coextruded structure are selected, the individual layer thicknesses and the total wall thickness can be determined. The functional layer thickness is based on the desired performance level. The desired physical integrity of the product will determine the thickness of the structural layers and the total thickness. Considerations for the structural layers include total stiffness, dimensional stability, internal pressure resistance, top load, and other physical characteristics of blown articles. Other demands, such as aesthetics or functionality should also be considered [9].

The combination of the functional and the structural materials will determine whether an adhesive must be used, and if so, what type. The adhesive must be compatible with the two materials to be bonded. Inclusion of a high melt temperature material in the structure forces the selection of an adhesive material with a similar high melt temperature and thermal stability. The size of the equipment being used and the typical residence time in the extrusion system may also require the use of a higher thermally stable material, to prevent degradation. Stability of the adhesive as well as the other materials is important for recycling the trim and start-up scrap, as discussed later.

Coextrusion product design must consider the location of the functional layer in the wall structure. Where migration from the product into the wall is a problem, the functional layer, in this case a barrier layer, should be located as close to the product as possible. Conversely, to prevent migration from outside, the barrier layer should be placed on the outside. A typical example of inside barrier placement is found in containers for garden chemicals. Bottles for this application are made of PP/tie/nylon, with the nylon on the inside to prevent migration of the active ingredients of the packaged chemicals. When the barrier layer is EVOH, protection of the EVOH from moisture is important. In order to maximize the barrier provided, the EVOH may be put in the middle of the wall structure or somewhat closer to the outside. The effect of the location of the layer on oxygen transmission through a PP/EVOH/PP structure is shown in Fig. 5-9 [10].

Practical processing considerations still make it advisable that the differences in viscosities of the various materials be minimized at the operating conditions of melt temperature and flow rate (shear rate.) This means that some matching of materials based on their processing characteristics is necessary. A symmetrical structure is desirable, because it is easier to extrude a controllable parison when the inside and the outside materials are the same. These materials have the longest path with die wall contact and therefore effectively control most of the flow through the die.

Special Design Areas

Some areas of the product need special consideration. One is the pinch area. Because of the possible problem with adhesion of the materials coming together in the pinch, reinforced pinch design may be appropriate. Such design will maximize the contact of the two sides of the inner layer, try to seal the outside layer from one side to the other, and entrap the middle layers. This will insure proper performance and reduce the chance of delamination due to adhesion of incompatible materials. The cross section of a pinched-off area is shown in Fig. 5-10. Special care in design of the pinch is required where migration in or out of the container causes the product to spoil or become hazardous. Because the barrier layer integrity across the pinch cannot easily be maintained, additional thickness or special designs are needed. Similarly, in a product such as a container, all the layers come to the surface at the opening, which creates a possible delamination area.

Recycled Material

The ability to recycle materials trimmed from the products during production, as well as start-up scrap, plays a major role in the economics of

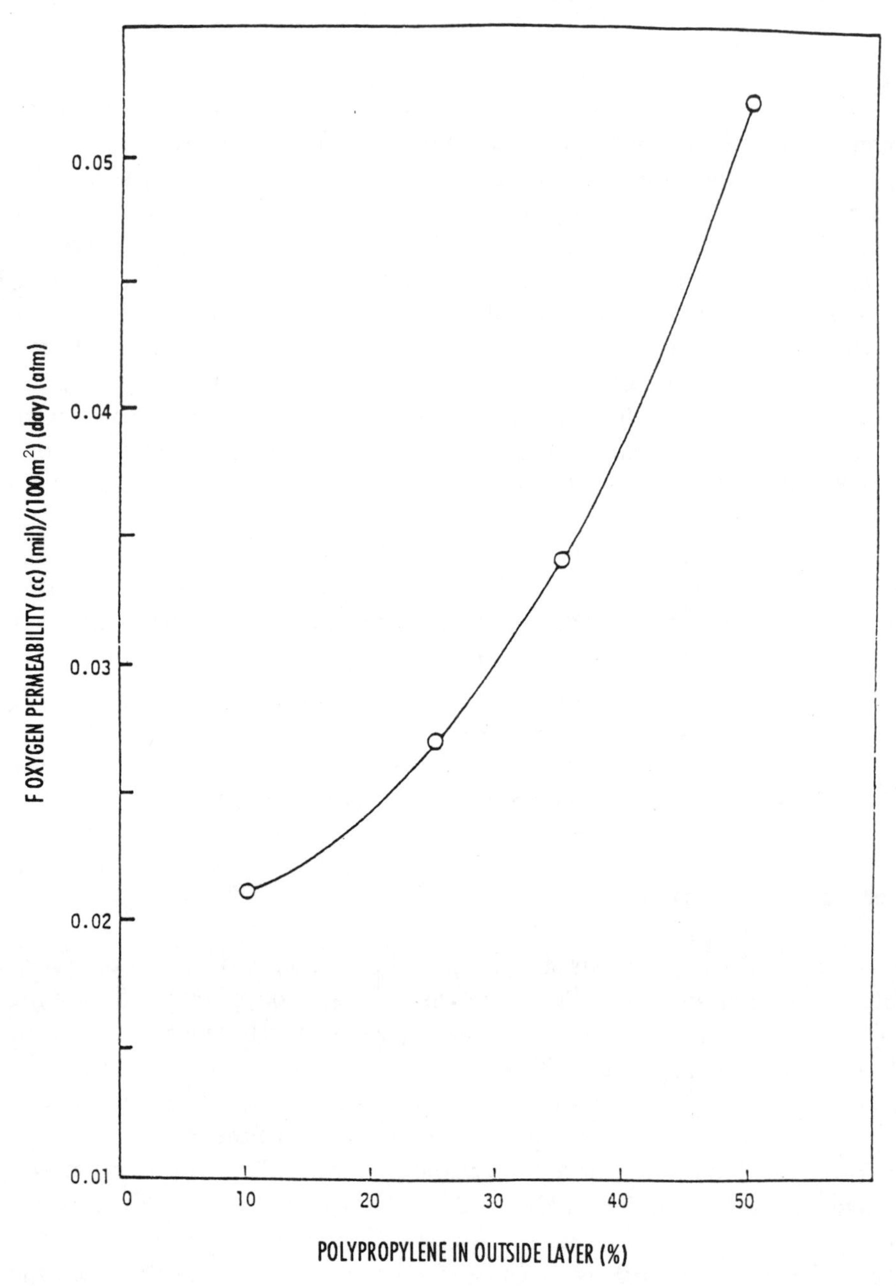

Figure 5-9 Effect of EVOH barrier layer location on the total permeation of the structure. Courtesy: Graumann, SPE RETEC Coextrusion III, 1985.

any blow molding operation. Returning the regrind to the product it comes from usually provides the highest value for this material.

In producing a single-material item it is very easy to return the material to the feed of the extruder after grinding it. The most important factors are to keep the regrind free of any contamination and add it as a predetermined proportion to the extruder feed. In coextrusion the use of recycled material

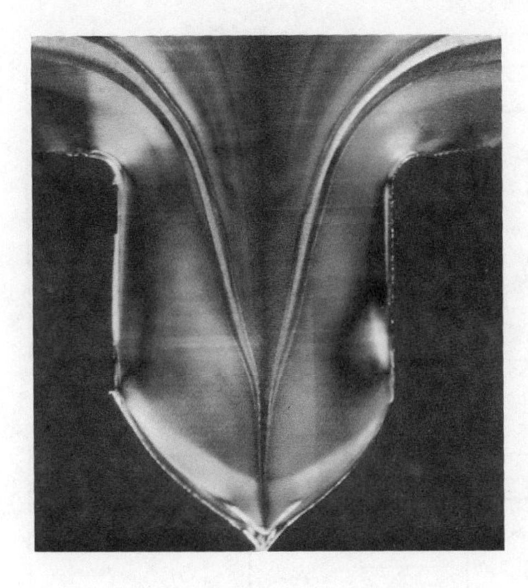

Figure 5-10 Microtome cutting of a pinched-off area of a coextruded article. Courtesy: Werksfoto Krupp Kautex. Krupp Plastics & Rubber Machinery, Inc., 330 Talmade Rd., Edison, NJ 08818.

is complicated by the fact that the regrind is really a mix of all the materials making up the wall structure: the structural material, the functional material, and the adhesive. This material can be returned to use either by direct extrusion of only recycled material, or by blending it with virgin material. It can be returned into specially added layers in the structure or, when blended, into the existing structural layers. Figure 5-11 illustrates the use of regrind in various structures [11]. The preferred way is to form an additional layer or layers inside the structure. Such layers are structural in nature and reduce proportionally the structural layer thicknesses of virgin material (Fig. 5-12).

Separating the regrind to a separate layer minimizes the effect of the regrind on the flow through the die and its effect on the appearance of the product. The extruder used to melt the mixed regrind can be designed especially to handle this type of mixed feed material. The disadvantages of forming separate regrind layers are the additional capital needed for the machine, the more complex die, and the cost of more complex operations. Under any circumstances, the total amount of EVOH that can be mixed is limited and should be kept under 15 percent [12].

When using regrind as part of the structural layer, the properties of the layer need to be taken into account. These properties depend on the ratio of the various materials in the mix. Furthermore, the original properties of the materials may have been changed due to their thermal history (exposure to high temperature, high pressure, shear rate, and the exposure time) in the previous extrusion. All of those conditions lead to material degradation, the rate of which depends on the material's thermal stability. Degradation effects can be as minor as slight color variation or as severe as complete breakup of the molecular structure with little or no retention of physical properties. The stability of various materials can be determined in different ways. Depending on the application, retention of tensile strength or impact

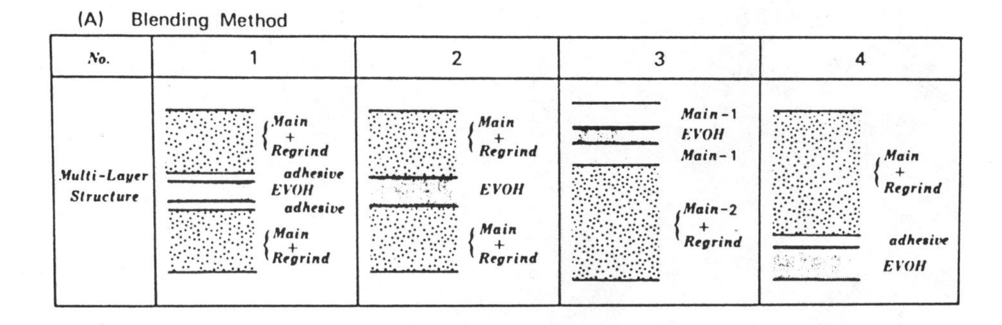

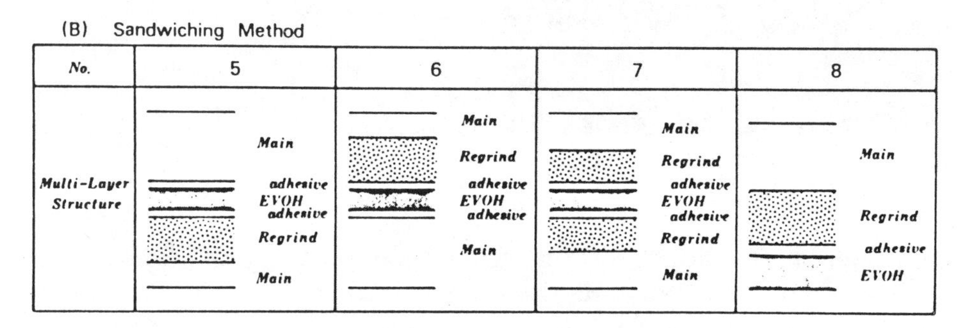

Figure 5-11 Typical structures of EVOH-based coextrusion composites using regrind. (A) Blending method. (B) Sandwiching method. *Proceedings, COEX '84,* p. 304.

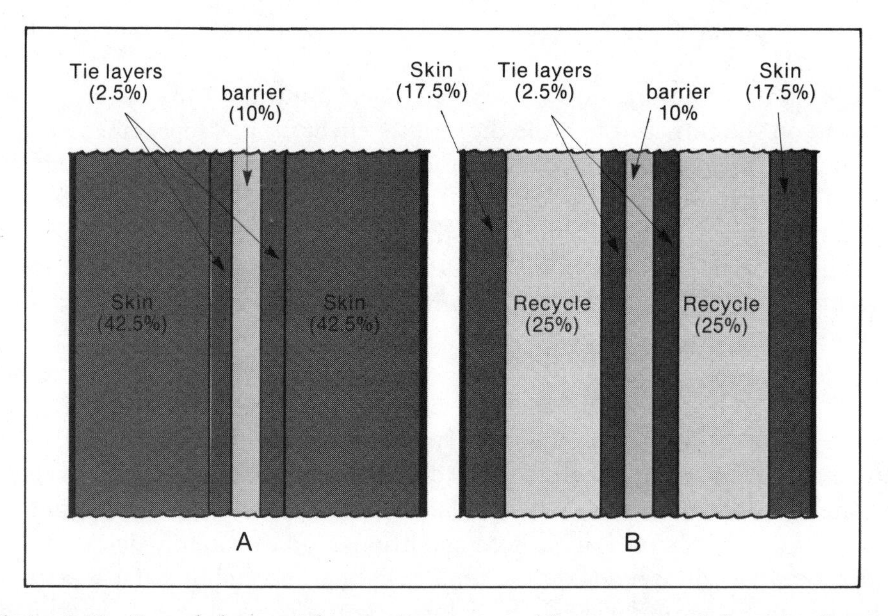

Figure 5-12 Recycled material replacing structural layers material for equivalent performance. (A) No recycled material, (B) with recycled material. Courtesy: *Plastics Engineering* (Feb. 1986).

may be evaluated. For barrier resin the molecular breakup can be evaluated through a measurement of its flow properties as a function of the thermal exposure. Typical thermal stability data are shown in Fig. 5-13 [11].

In continuous operation some successively smaller amounts will be extruded two, three, or more times. A simple calculation can establish the ultimate steady state mix (see Table 5.4). The properties of this mix need to be established and used for determination of the regrind layer's contribution to the performance of the product.

Table 5-4 Example of Calculation to Establish the Final Steady State Mix

Event	*1*	*2*	*3*	*4*	*5*	*6*
Virgin						
structural	85	35	35	35	35	35
functional	10	10	10	10	10	10
adhesive	5	5	5	5	5	5
	100					
2nd pass regrind						
structural		42.5	17.5	17.5	17.5	17.5
functional		5	5	5	5	5
adhesive		2.5	2.5	2.5	2.5	2.5
		100				
3rd pass regrind						
structural			21.25	8.75	8.75	8.75
functional			2.5	2.5	2.5	2.5
adhesive			1.25	1.25	1.25	1.25
			100			
4th pass regrind						
structural				10.625	4.375	4.375
functional				1.25	1.25	1.25
adhesive				0.625	0.625	0.625
				100		
5th pass regrind						
structural					5.3125	2.1875
functional					0.625	0.625
adhesive					0.3125	0.3125
					100	
6th pass extrusion						
functional						2.65625
structural						0.3125
adhesive						0.15625
						100

Note: Total recycled regrind is 50%. Each column is derived by taking the previous column and multiplying it by the regrind fraction, and shifting it down one level of extrusion pass. The numbers worked out are for the structure shown in Figure 5-12 where the regrind used is 50% of the structure. The example assumes starting without any regrind and adding the regrind from the second extrusion on. It is clear that the level of the virgin-only extrusion first pass is quickly diminishing and that the other levels are converging to a fixed value rather fast.

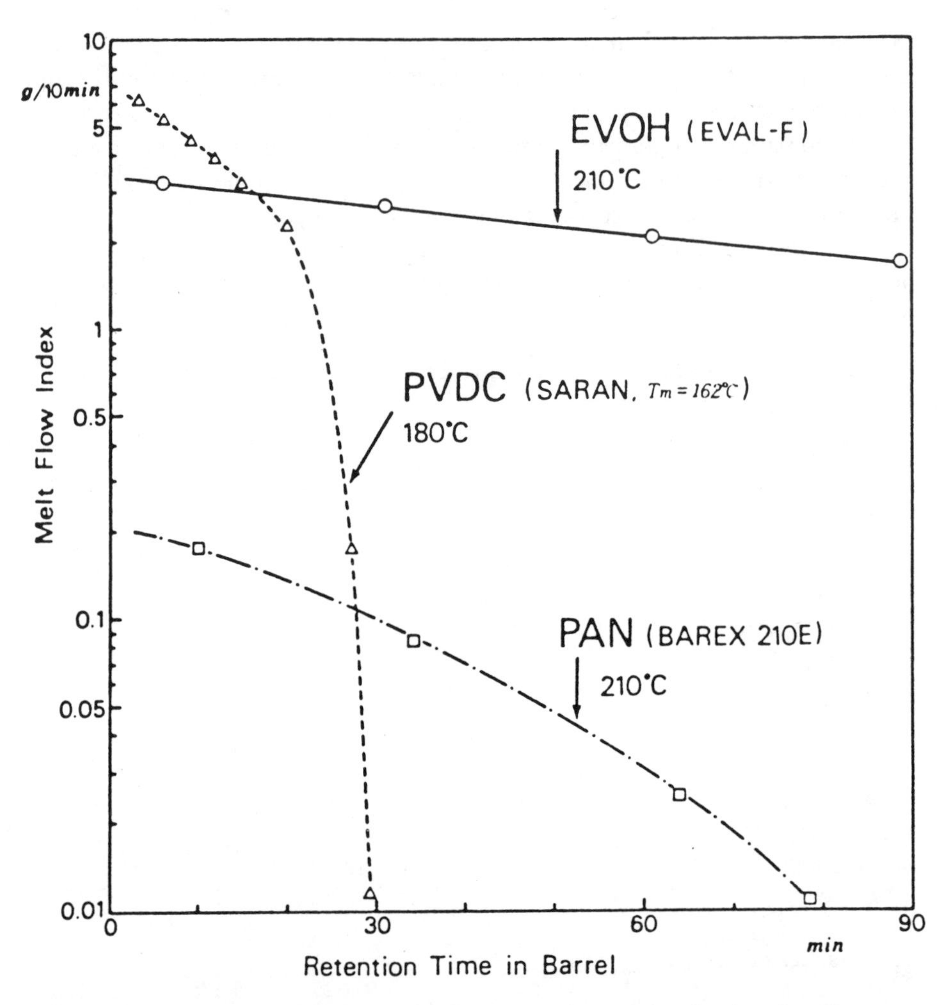

Figure 5-13 Thermal stability of three barrier resins. *Tm*, melting point. Courtesy: Schotland Business Research, Inc. *Proceedings, COEX '84* p. 292.

COEXTRUSION EQUIPMENT

Extrusion Systems

The extrusion system in coextrusion blow molding must supply several streams of melted material to the die simultaneously. Some of the streams are rather small in output compared to others. The required extrusion conditions also may differ from one material to another. However, of utmost importance is the stability and the uniformity of each stream of melt.

Extruders

As is typical of blow molding, the extruders used are the single-screw type. Well-designed extruders and auxiliary systems supporting them are essential for achieving a high-quality melt. The heart of the extruder is the screw, and the critical factor is the relation of the screw design to the flow properties of the material being processed.

Extrusion stability is achieved through proper screw design. Screw design must take into account each material's melt characteristics at the intended operating conditions, which are defined by flow rate, head pressure, and melt temperature. Extruder size is determined by the screw diameter and is selected based on the needed output. The length of each extruder, normally defined by the length to diameter ratio, *L/D*, is an important factor contributing to extruder stability. This ratio preferably should be 24/1 but at least 20/1. This length will insure complete melting of the pellets or feed and appropriate mixing of the melt to achieve a high level of thermal homogeneity and pressure stability. Good mixing is also very important where, in addition to the melting and thermal mixing, a dispersion of colorant, color concentrate, or any other additive is needed.

Extrusion stability is characterized by simultaneous uniformity of the pressure, as measured at the extruder outlet, and the melt temperature. The temperature, as function of time, needs to be uniform in all locations of the flow channel. Nonuniformity may be related to the operation of the extruder or to screw design. The cross-sectional uniformity of the temperature can be improved by the use of screw mixing devices or static mixers. These devices redistribute the melt in the flow channel and in doing so combine streams of various temperatures to create increased uniformity.

To improve on the stability of the output, extruders with grooved feed zones are used by some manufacturers. Grooved feed stabilizes the operation of the extruder by producing very high pressure in the feed area. The high pressure also generates high shear and shear heating of the material, and therefore this section of the extruder needs to be thermally insulated from the rest of the extruder and provided with a high level of cooling.

Changes in melt temperature will cause a change in the total resistance to flow in the die, resulting in a higher flow rate when temperature rises and vice versa. If the flow rate changes in any of the layers being extruded, the output ratio between the layers will change.

The temperature of the melt also affects the quality of the weld line being formed at each layer in the die. A low melt temperature reduces the mobility of the molecules in the melt, resulting in a lower level of reentanglement, compared to the level reached at a higher temperature.

Regrind from trimming operations or rejects is normally returned to a specially designated layer in the structure, as discussed previously. The extruder design for this material must take into account the composition of this material and its heat history in the operation. Such a mixture is prone to variation in itself, and if mixed with virgin material the ratio between the two can change due to hopper segregation or intentional changes in mix ratio. Such variations will affect the output of the extruder as well as the quality of the melt produced, which will affect the operation of the entire coextrusion die.

Extruders are manufactured by various companies as separate items or as parts of an extrusion blow molding machine. The size and number of extruders make their arrangement around the die quite important in order to maintain proper flow and short melt pipes, which connect the extruders to the die. One design, provided by Wilmington Plastic Machinery, combines all the extruders together in a single unit (Fig. 5-14). The extruders are arranged in such a way that all discharge openings are on the same plane and within a short distance of one another. This leads to a similar length and direct connection between all extruders and the die.

The exact size of each extruder in a coextrusion setup depends on the intended application, meaning the particular materials and the desired flow rates (based on product weight and the production rate of the machine). Typical extruders used in coextrusion blow molding machines are given in the following tables. Table 5-5, giving data for a Johnson Controls, Inc., extruder system is based on splitting the output of the extruder to the similar layers in the structure [13]. Table 5-6, drawn from data for the Battenfeld system, recommends the use of an individual extruder for each layer [13].

Table 5-5 Extruder Sizes Used for Various Coextruded Structures, Split Melt,* 35 Gram Container

Extruder No.; Screw Dia. Size	*Extrusion Output (lb/h)*						*Maximum Output; Resin(s)*
	3-Layer		*4-Layer*		*6-Layer*		
	SP	*DP*	*SP*	*DP*	*SP*	*DP*	
Ext. 1, 35 mm	0.41	0.83	0.41	0.83	0.83	1.6	12 ADH
Ext. 2, 35 mm	2.2	4.4	2.2	4.4	0.55	1.1	12 Nylon/EVOH
Ext. 3, 60 mm	25.1	50.2	13.8	27.7	19.4	38.8	110 PP/HDPE
Ext. 4, 60 mm	NA	NA	10.2	20.5	6.9	13.8	110 PP/HDPE

*Provides for splitting melt supply of adhesive and outside skin layers from the same extruder
SP: Single parison PP/HDPE, 6 bottles per min
DP: Dual parison PP/HDPE, 12 bottles per min
NA: Not applicable
Source: Johnson Controls, Inc. Courtesy: The Packaging Group Inc. (U.S.A.).

Table 5-6 Recommended Extruder Arrangement, Each Layer Supplied From a Separate Extruder

Container Size (liters)	*No. of Layers*	*Max. Die Diameter of Head (mm)*	*No. of Cavities*	*No. of Extruders × Diameter (mm)/(L/D)*
0.3–4	6	60	1	3 × 35/20 3 × 12/24
0.1–1	6	60	1	3 × 35/20 3 × 12/24
0.3–4	6	60	2	3 × 50/20 3 × 12/24
2–5	3	120	1	1 × 60/20 2 × 25/24
2–5	6	120	1	3 × 50/20 3 × 25/24
2–5	3	120	2	1 × 80/20 1 × 50/20 1 × 35/20
Up to 5	3–6	60/120	6–12	Depending on article

Source: Battenfeld-Fischer [21]. Courtesy: *Kunststoffe/German Plastics*, 77 (July 1987).

Figure 5-14 Wilmington coextrusion blow molding machine. Courtesy: Wilmington Plastics Machinery.

The actual decision is based on the value of the added flexibility and layer control gained by having individual extruders versus the cost they will add to the system.

Weight Feeders

Weight feeders, which control the input of feed into the extruder, can improve the stability of the extruder. When they are used, the extruder is operated in a starved feed mode and the screw flights are only partially filled. In this mode it is easier to maintain the internal balance between the various processes taking place inside the extruder, especially when the screw design is marginal. The amount of work done in melting and mixing the materials can now be independently controlled by variations of the screw speed, while the output remains the same, controlled by the feeder. Another advantage of using weight feeders is the ability to monitor easily the rate at which a material is being consumed in each extruder. This makes it easier to insure that a proper level, especially of the functional material, is put into each product. Additionally, individual control of each feeder makes it easier to achieve the proper ratio between the materials in the various layers.

Melt Pump

Another device that helps to provide a highly accurate output is the melt pump. In a melt pump the stability of the output is almost completely independent of input variations.

A melt pump is a gear pump, which by design is a positive displacement device, meaning that for every revolution of the pump, a fixed volume of melt is delivered, regardless of the material being extruded. The melt pump is mounted between the extruder and the die. The melt flows from the extruder into the pump and out into the die. The infeed pressure into the melt pump is normally very low compared to the pressure that the extruder would have to develop in order to extrude the melt directly through the die. At the same output, an extruder feeding a melt pump operates against a lower pressure and therefore provides higher output, at a lower melt temperature, than an extruder without a melt pump. However, lower back pressure on the extruder also reduces the level of mixing of the melt.

The pressure required to extrude the material through the die is provided by the melt pump. The pump therefore isolates the output from any variation that may occur in the extruder output.

Melt pumps have gained popularity in the last several years in extrusion applications. However, they add a significant amount to the cost of a coextrusion machine.

It is not within the scope of this chapter to fully explore the design of extruders and screws and the operation of extruders, although the aspects of extrusion in blow molding becomes more critical than ever in coextrusion. Several texts can be found on this subject [14, 15], and much more is available in various technical papers.

Melt Pipes

Because of their size and configuration, extruders cannot all be connected directly to the die. Instead, melt pipes, also called delivery tubes, connect the extruders to the die. The design of the tubes is very simple, but some precautions must be taken. First, the melt passages need to be properly sized to prevent undue pressure drops and the resulting increase in temperature of the melt. The flow channel must be well streamlined to prevent any stagnation or hang-up areas where material could degrade. The thermal mass of the tube is another important aspect of the design. The temperature of the tube must be kept uniform at all times. Very low thermal mass will result in fluctuations with every cycle of the temperature controller. Excessive mass will result in long heat-up time when the extruder is first heated, or a very slow response when the temperature setting is purposely changed.

Coextrusion Dies

The heart of the coextrusion process is the coextrusion die, in which the various materials are combined in the proper order and the desired thicknesses to form a parison. The ability of several layers made of different materials to flow together is based on the high viscosity of plastic melts. Because of the high viscosity, little if any mixing will occur between two materials that flow together when their boundary is on a stream line. Only flow disturbances, intentional or otherwise, will enhance mixing.

The basic parameters for good coextrusion blow molding die design are the same as for any monolithic extrusion blow molding die. However, in a coextrusion die the design parameters for all materials in the system must be met simultaneously.

First, each material must flow uniformly at the desired rate and the flow passage must restrict the flow sufficiently to provide the desirable back pressure to the extruder. Additionally, the pressure drop in the die for each material has to be such that when two materials are combined in layers their pressures match. Also, the combining has to be done with both materials at the same velocity.

Three major requirements need to be addressed in the die: first, the physical arrangement to accommodate multiple feeds of materials into the die; second, the formation and location of weld lines where the melts recombine to form the annular shapes of each layer; and third, how best to combine the layers into the parison, once inside the die.

Most coextrusion dies are of a center post design rather than a spider head design. That is, a material being fed into the die must flow around a center post to form an annual configuration in each layer before it starts to flow along the die to the die exit. Although a spiral die is different from a center post die, the mandrel assembly is supported through the center member as in a center post die.

Dies can be classified by various attributes. The most important ones are:

1. Body construction/layers combination
2. Manifold design

Die Body Design

Die construction can be divided into two types, shell design and modular design. Each has unique advantages and disadvantages.

Shell Design

The Bekum die design (Fig. 5-15) is a good representation of shell design. In this design, the manifolds feeding all the layers are assembled in parallel within the die shell to form the layers that are combined in the extruded structure. The major advantage of this design is its relatively short height. The performance of the die is based on the proper design of the assembled elements. Typically there is no control of an individual manifold to adjust flow or melt behavior. The temperature of the die is controlled by means of heaters mounted on the outside of the shell. As the manifolds are arranged one within another, there is virtually no way to provide different temperatures to different materials within the die. This in turn forces a careful choice of materials so that they can be extruded together properly. A shell design requires a fixed number of elements, determined by the number of layers the die is designed to combine. When not all layers are needed for a certain product, some of the elements can be fed with the same material to form a single layer, or spacer elements can be used to take up the space in the die. However, the use of spacer elements requires disassembly and reassembly of the entire die for each job change. The design is shown in Fig. 5-15 is for five layers. For a larger number of layers a two-tier design is used (Fig. 5-16). In this design the middle layer, the functional barrier layer, flows a relatively long distance before joining the structure. This length is similar to the flow length of the inside structural (body) layer, which flows against the center of the die. This situation demands that a single temperature, the die temperature, accommodate both materials at the same time.

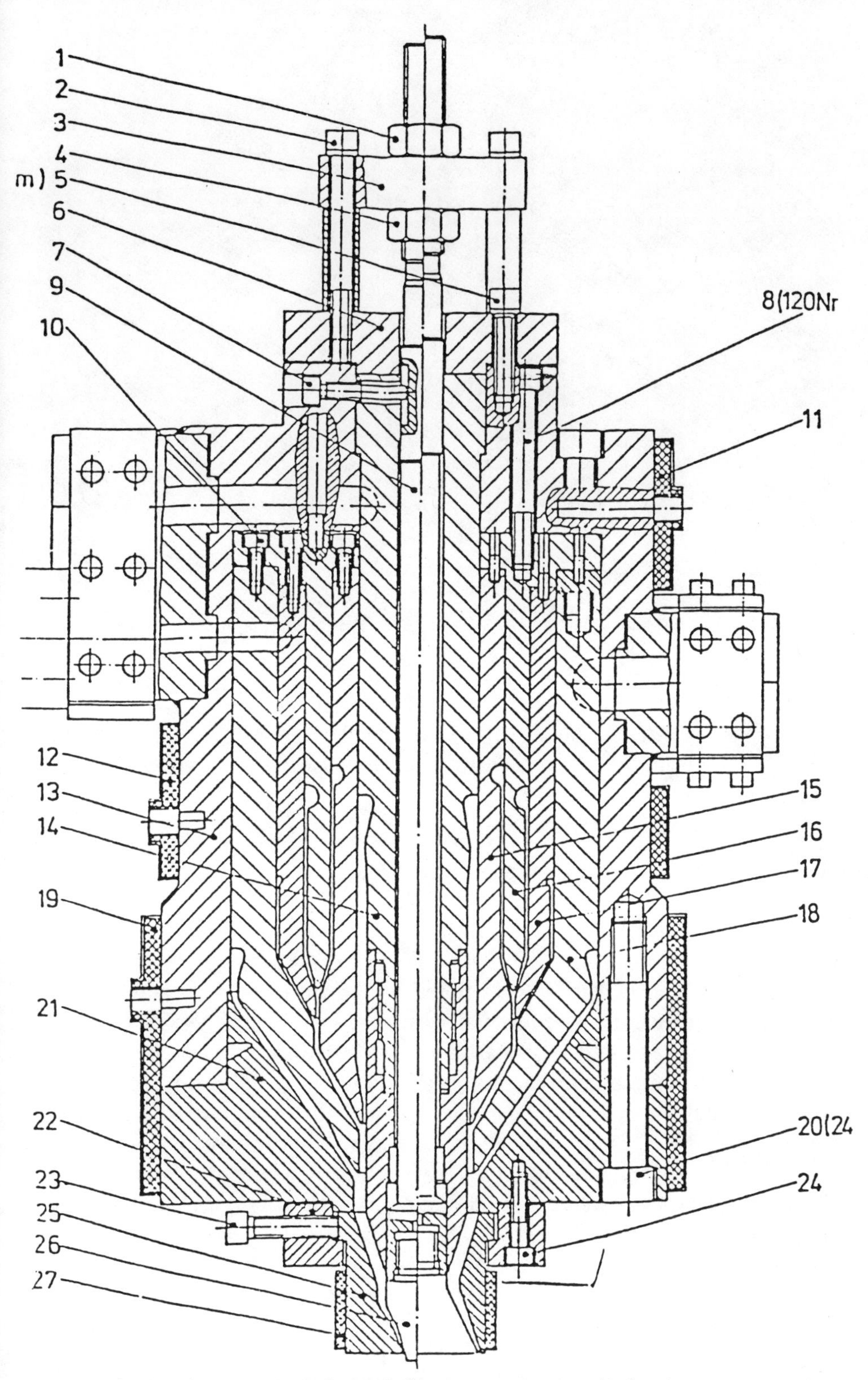

Figure 5-15 Bekum shell type coextrusion die head.

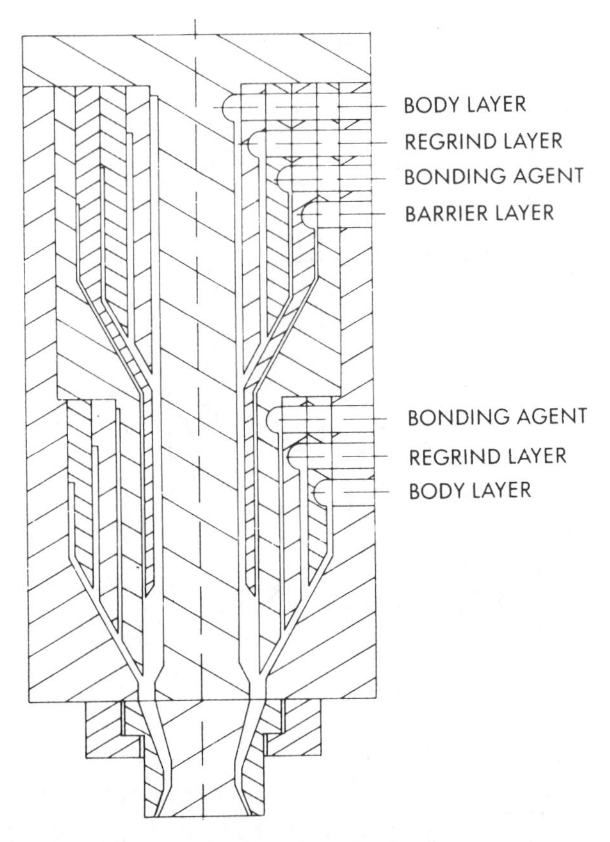

Figure 5-16 Bekum seven-layer coextrusion die head arranged in two tiers. Courtesy: *Plastics Packaging* (Jan/Feb 1988).

Modular Die Design

In a modular design die additional layers are created by adding new manifold elements in a sequential configuration. Figure 5-17 shows the propagation of layers in a typical modular coextrusion die [16]. Extrusion starts with the innermost layer, at the highest manifold of the die. Additional layers are added to the outside of the parison being formed as it progresses through the die. This structure is utilized in various designs. Figure 5-18 is one of the earliest die designs, by Toppan Printing Company of Japan [17]. Variations on this design can be found in the Ando die (Fig. 5-19) [18] and the MAC die (Fig. 5-20).

The sequential build-up of layers makes it possible to construct the die by stacking up modular pieces, one for each layer of the structure. The modules all have basically the same outside design, but each one contains a manifold specifically designed for the material being extruded from the element and the thickness needed. The number of modules used is the same

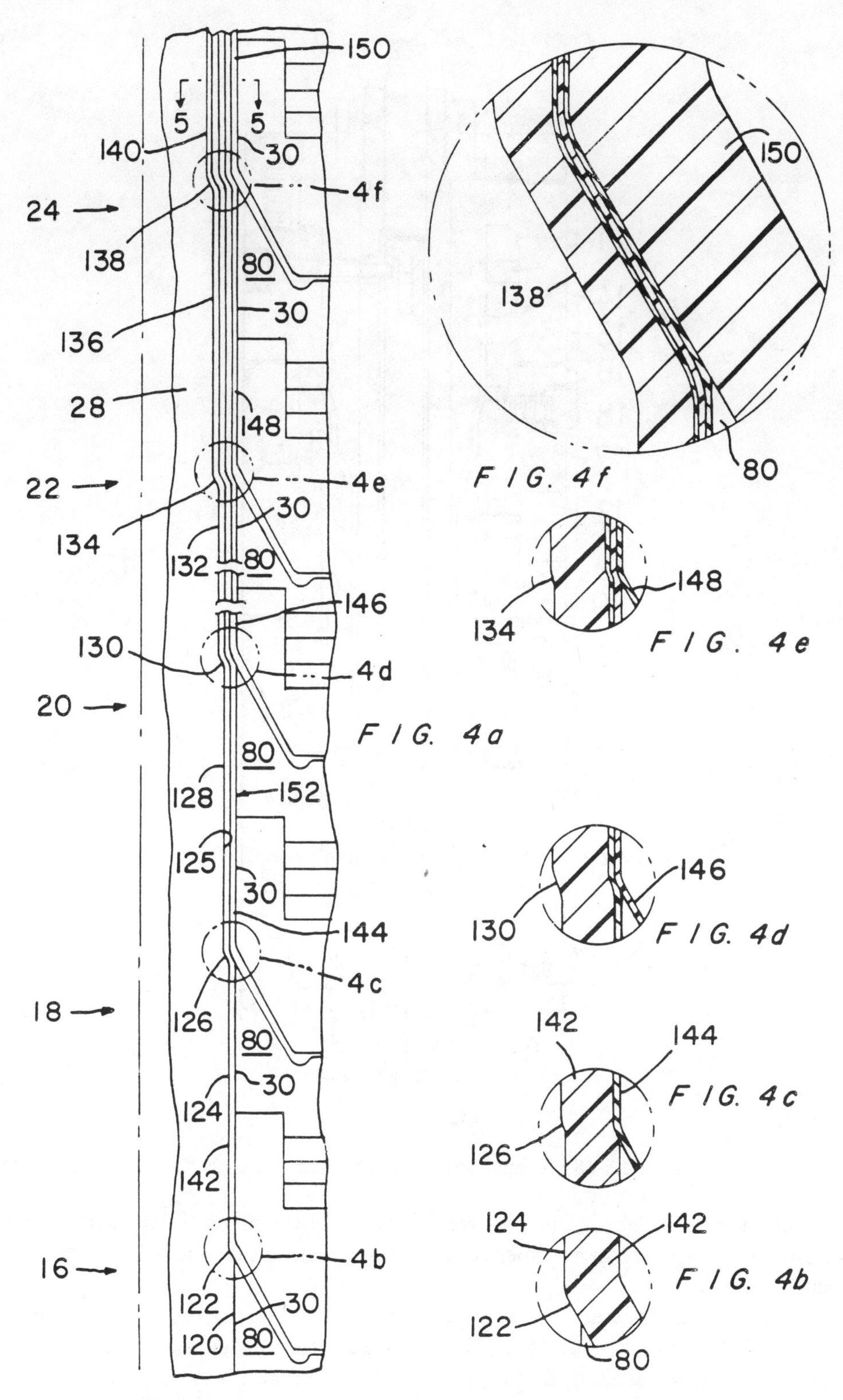

Figure 5-17 Sequential build-up of layers in a coextrusion die.

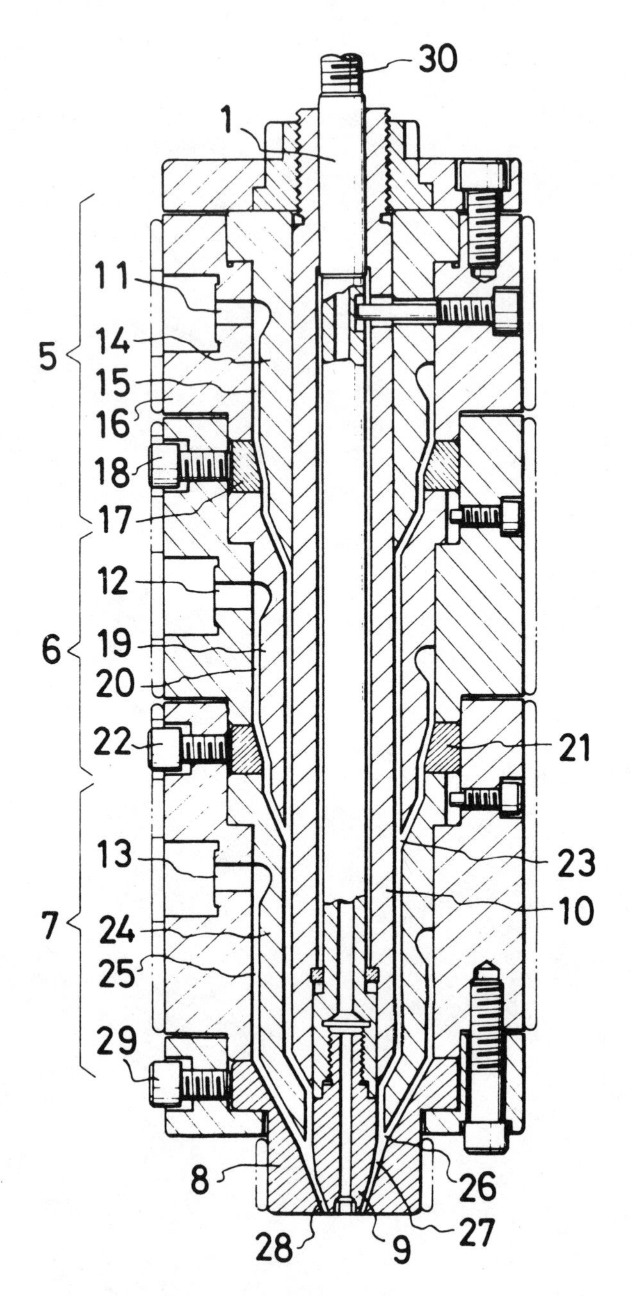

Figure 5-18 Toppan Printing coextrusion die head.

as the number of layers, and the overall height of the die will vary with this number. When a large number of layers is used the die can become quite tall.

Modular die heaters are arranged on the outside of the die body and can be divided in such a way that heat is directed to each element. Therefore the temperature of each module can be controlled at a level appropriate to the particular material being extruded. This feature is quite important in

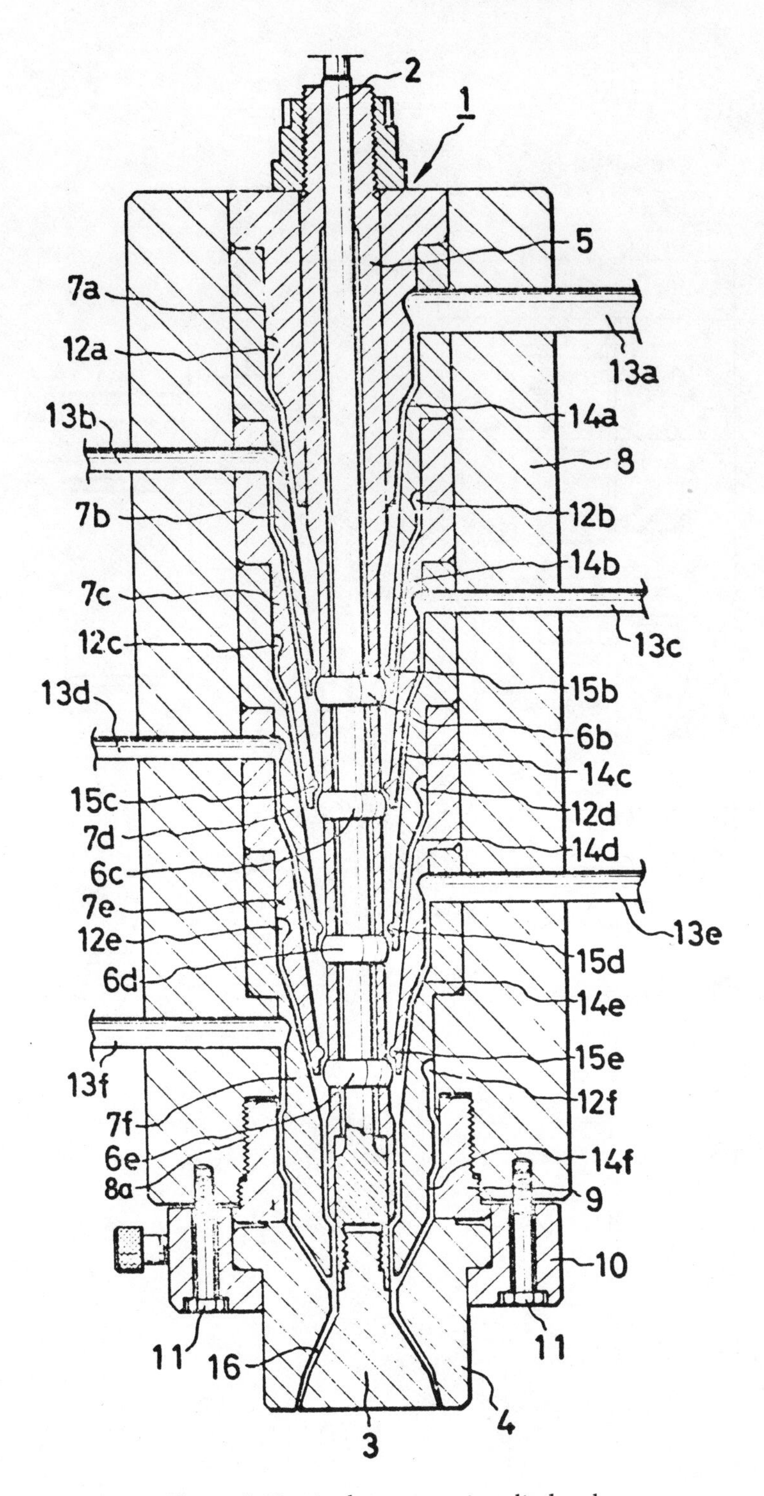

Figure 5-19 Ando coextrusion die head.

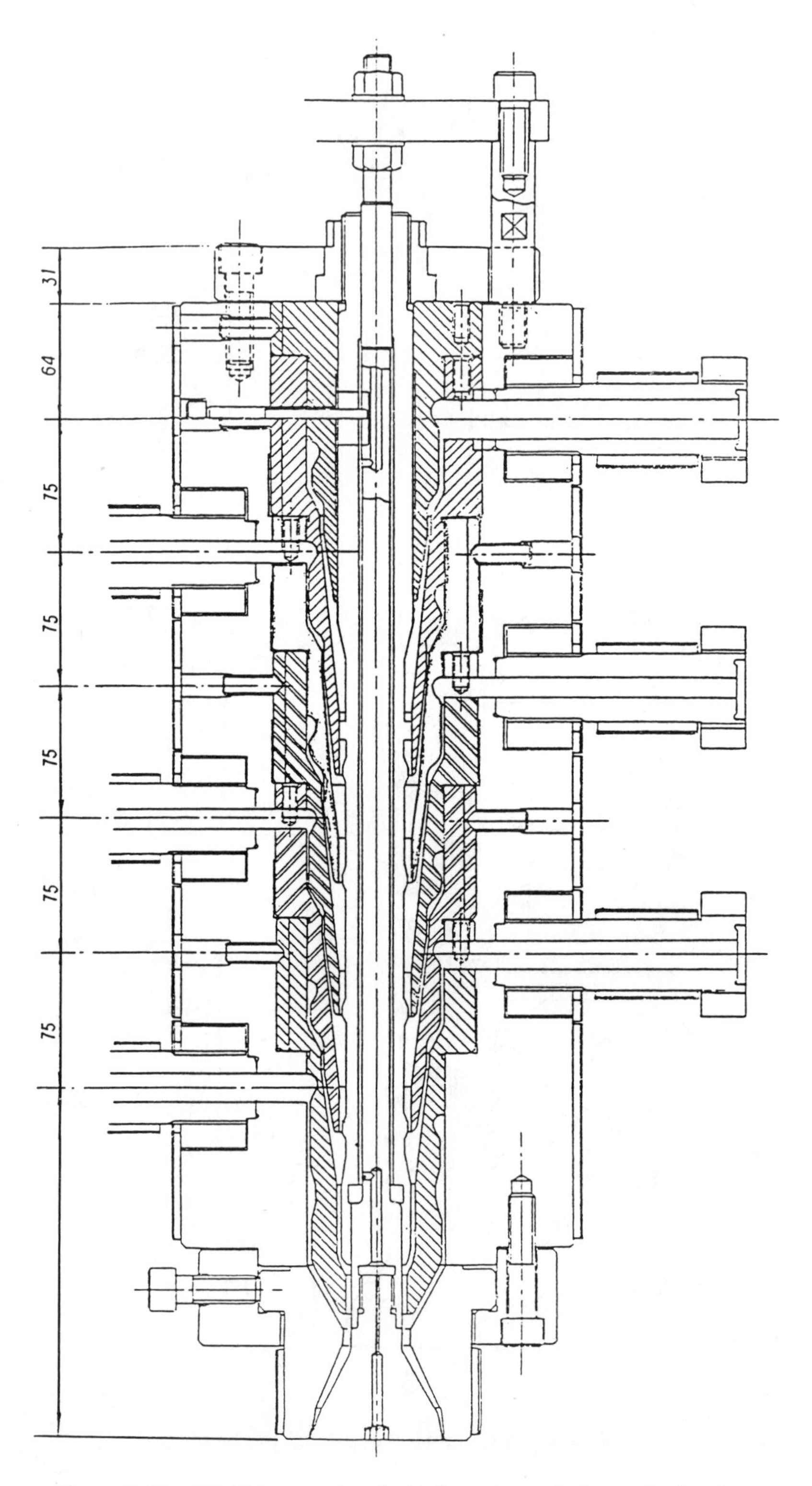

Figure 5-20 MAC International single parison six-layer die head.

the modular design because the outside of each new layer flows in contact with a die surface for some distance. The inside of a new layer flows in contact with the oncoming layers already created.

Another feature of die designs that must be considered is the location of the weld lines. Each layer has a weld line and several factors are involved in its location. In a shell design die proper spacing of the feed ports for all materials makes it hard to arrange all weld lines in the same locations. The appearance of the finished product, especially when it is clear or translucent, makes it preferable to line up the weld lines, because they may be noticeable. Another preference may be to line them up with the parting line. In an opaque wall the visual effect is less likely to be a factor, and staggering the weld lines is preferable from the standpoint of strength. This is because the mechanical strength of the weld line is sometimes inferior to the rest of the wall.

Modular die design makes it easier to arrange the weld lines in any way required because melt tube interference is less severe. Splitting the melt from one extruder to form two layers will restrict the ability to select the location of the feed port. However, internal passages to rearrange the location of the manifold feed port can be incorporated into most designs.

Manifold Design

A second aspect in which manifolds can vary is the design of the flow passage. The manifold distributes the melt from the feed port of the die to form the annular shape needed for the parison. This is one of the most important features of the die, as it affects the ability to control the circumferential uniformity of each layer as well as the entire parison and the quality of the weld line. Because of the number of layers involved and because some of them are very thin, each layer must be uniform in order to produce a quality article. Any decrease in thickness, especially in thin layers, may result in the need to significantly increase the average thickness of those layers in order to achieve the minimum thickness required everywhere in the blown product. This will have an adverse effect on the cost of manufacturing.

A survey of the various die designs available from blow molding equipment manufacturers is presented in Table 5-7 [19].

Toroidal Manifolds

In a toroidal manifold, like the one shown in Fig. 5-21, designed by Solvey [20], the main chamber to circumferentially distribute the melt is in the form of a torus. The torus is relatively low in resistance because of its large cross-sectional area, which allows the melt to flow quickly around the

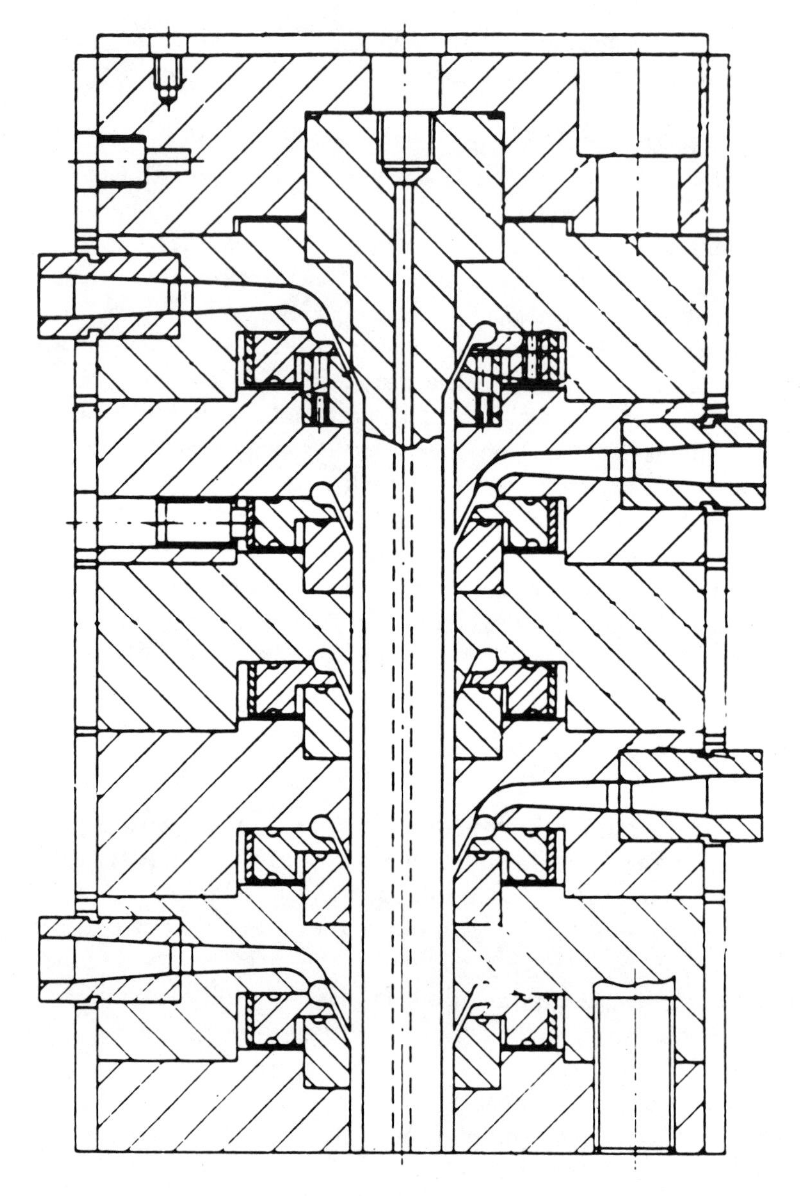

Figure 5-21 Solvey coextrusion die head with toroidal manifolds.

center post of the die and weld again to form the annular shape. The circular distribution is done in one plane. Correction to compensate for the smaller amount of material needed at the far side from the melt entry position is achieved by reduction of the cross-sectional flow area from the inlet point to the farthest point around the circumference. Reduction of the flow channel area is illustrated in Fig. 5-22. However, a pressure variation is developing from the inlet point, where it is the highest, to the far point, where it is the lowest. This pressure difference results in different flow rates barring pressure compensation down-stream in the die. The advantage of this design is its compactness, especially in the height of the die. A typical manifold element for such a die design is shown in Fig. 5-23 [20]. However, this approach is for a predetermined flow rate and operating conditions, with little freedom to deviate from them.

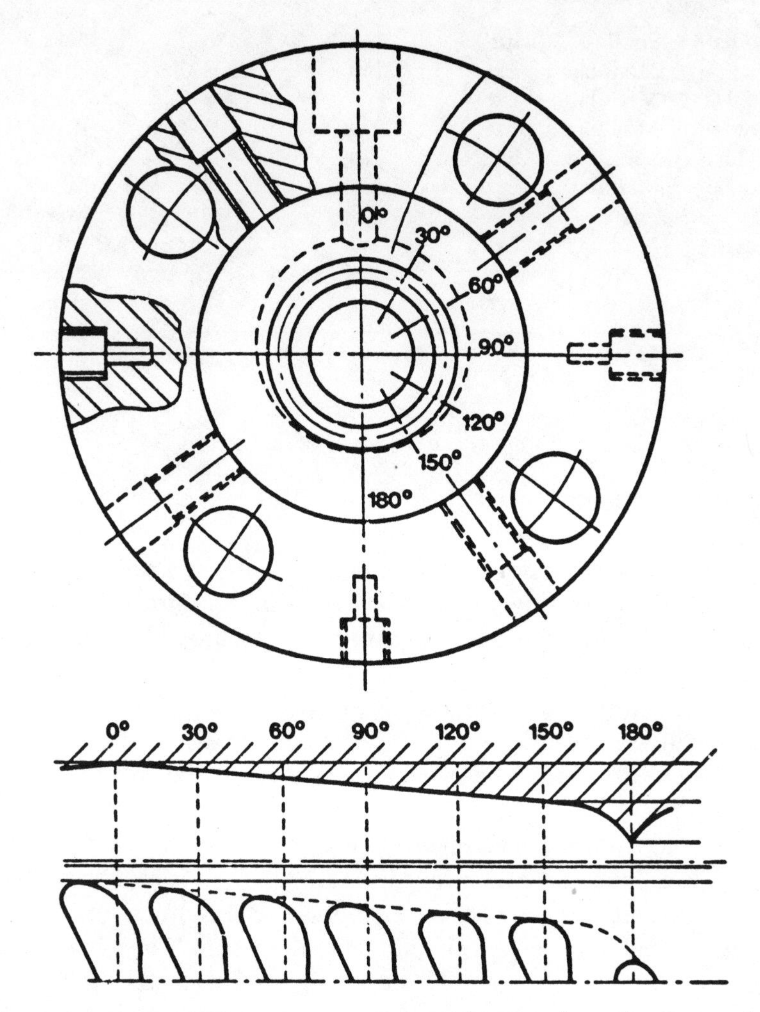

Figure 5-22 Manifold cross section: area reduction along the flow path.

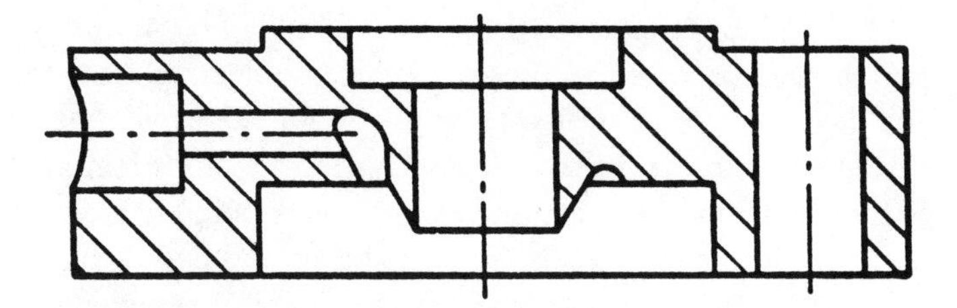

Figure 5-23 Manifold element from the Solvey die head.

Table 5-7 Coextrusion Blow Molding Die Heads

Channel Type	*Body Type*	*Layer Division*	*Adjustable*	*No. of Layers*	*Manufacturer*
Spiral	Modular	Ext	No	2–7	Battenfeld-Fischer
Cardioidal	Shell	Int	No	3–7	Bekum
*	Modular	Ext	No	3–7	Graham
Toroidal	Modular	Ext		3–7	Johnson/Uniloy
Cardioidal	Modular	Ext	Yes	3–6	Kautex
Cardioidal	Shell	Ext	No	5	Kureha
Cardioidal	Modular	Ext	No	2–7	MAC
*	Shell	Ext	*	6	Metal Box
Toroidal	Modular	Ext	Yes	Up to 6	Solvey
Cardioidal	Shell	Ext	Yes	Up to 6	Toyo Seikan (Toppan)

*Unknown

Cardioidal Manifolds

The cardioidal manifold, shown in Fig. 5-24 in a design by Krupp Kautex, consists of two main flow areas. The first, which is similar to the toroidal manifold, distributes the melt circumferentially around the center post to form the annular flow. But while flowing, the melt is also directed toward the die exit. This sets up the condition for pressure drop equilibration in the second flow area, the land area. The land is actually a narrow gap through which the melt flows. The gap opening is fixed in size but the flow length is longest near the feed port, where the pressure is the highest, and shortest at the far end from the feed opening, where the pressure is the lowest. As a result, after flowing through the land area the melt is at the same distance from the die exit, at equal pressure around the circumference, and flowing at a uniform velocity toward the die exit, all around. The advantage of this design compared to the toroidal design is its better ability to achieve a uniform flow. The compensation for flow and the associated pressure drop from the feed side to the far side is achieved by changing the flow length at the land. The land length affects the flow rate in a linear manner, whereas the flow gap opening affects the flow in a cubic power relationship. Additionally, this design leads to more streamlining of the flow channels and a smaller volume of melt within the die. High melt velocity is maintained at all points; along with the streamlined design, that helps to minimize dead spots in the flow, which could lead to stagnation and degradation—factors that can be of great importance for all materials and especially for those with limited heat stability. The cardioidal design also is important in normal operations during a change of materials or colors: better cleaning is possible

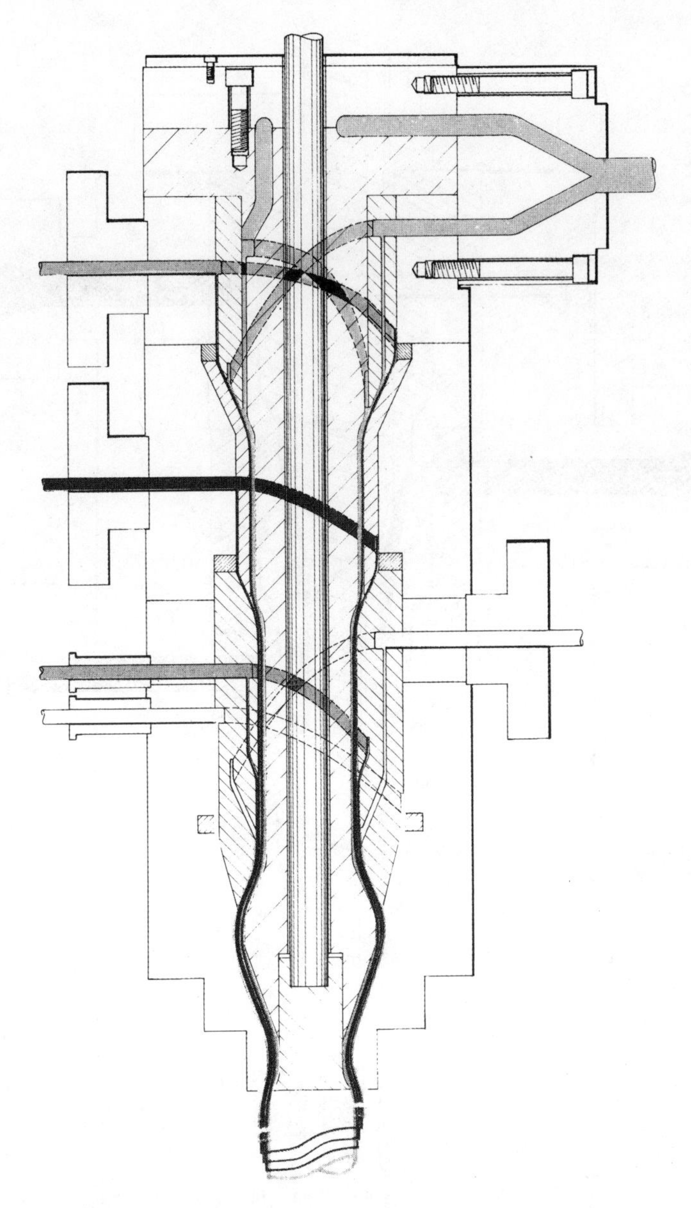

Figure 5-24 Krupp Kautex six-layer coextrusion die with cadoidal manifolds. Courtesy: Werksfoto Krupp Kautex. Krupp Plastics & Rubber Machinery, Inc., 330 Talmade Rd., Edison, NJ 08818.

by pushing the old material out of the die with the new, which reduces the amount of purge required and the time involved.

The disadvantage of the cardioidal design is the increased length of each manifold, and the more complicated design and fabrication. The additional length required can be reduced by nesting one manifold within another, as was done by Ando in the design shown in Fig. 5-25. However, this limits the ability to control the temperature individually fo. each manifold and material.

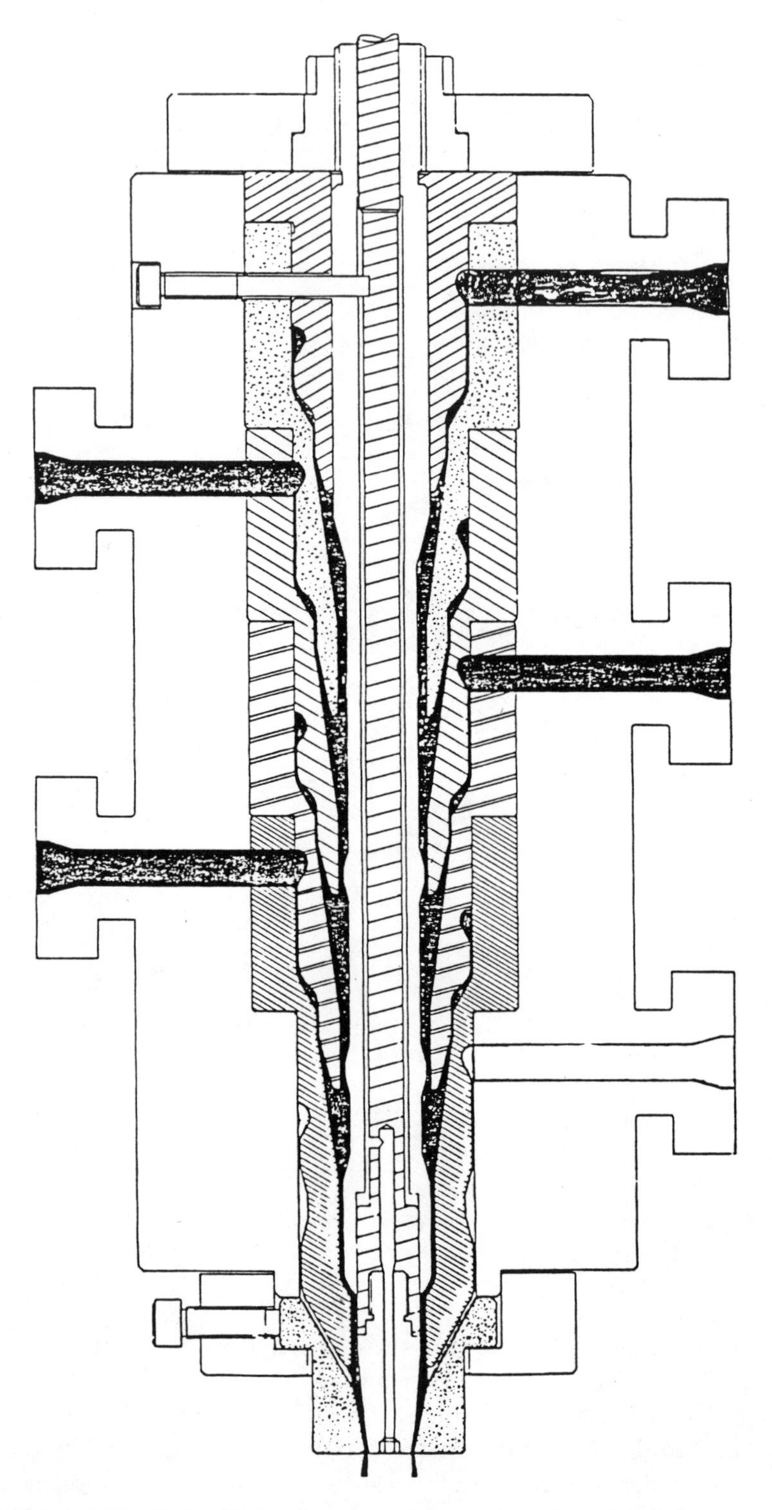

Figure 5-25 Ando die head design with nesting of manifold elements.

Spiral Manifolds

The spiral manifold, introduced by Battenfeld (Fig. 5-26) [21], was developed from a similar design used extensively in blown film dies.

In the spiral manifold the melt is distributed circumferentially by a spiral flow channel. The spiral path makes several turns as the melt is let out into a narrow passage that forms a land area and which is followed by a relaxation zone. One way to visualize the formation of layers in the spiral

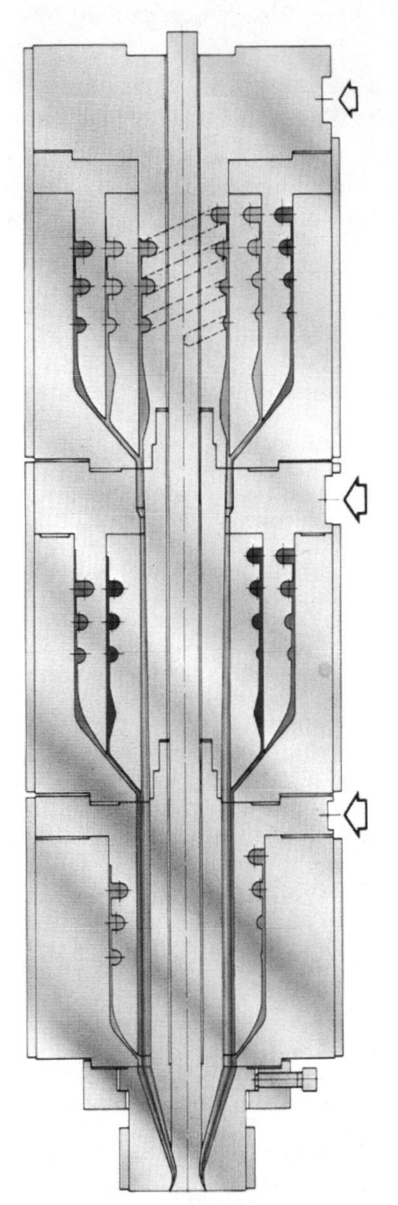

Figure 5-26 Battenfeld spiral coextrusion die head for six layers, utilizing three modular elements. Courtesy: *Kunstoffe German Plastics* 77 (July 1987).

manifold is to consider each layer as a combination of several sublayers wrapped spirally around one another. In such a combination there is no real weld line in the conventional sense—the beginning and end streams just blend into the total layer.

The output of the spiral manifold is linear with respect to the pressure (Fig. 5-27). The total pressure drop of each spiral manifold can be controlled by the length of the spiral and its cross-sectional area. The circumferential thickness distribution from each manifold improves with the length of the spiral (Fig. 5-28) and can be reduced to uniform thicknesses regardless of the viscosity of the material when sufficient length is provided. The required length and cross section of the spirals for all materials are determined by a computer simulation of the flow in the die.

The die is a combination of shell and modular design. Each modular section in Fig. 5-26 can create up to three layers. The manifold core elements (Fig. 5-29) are internally arranged in a nesting manner, each designed specifically for the material being extruded and the thickness needed. Layers are combined sequentially, and the flow path of a material against the die

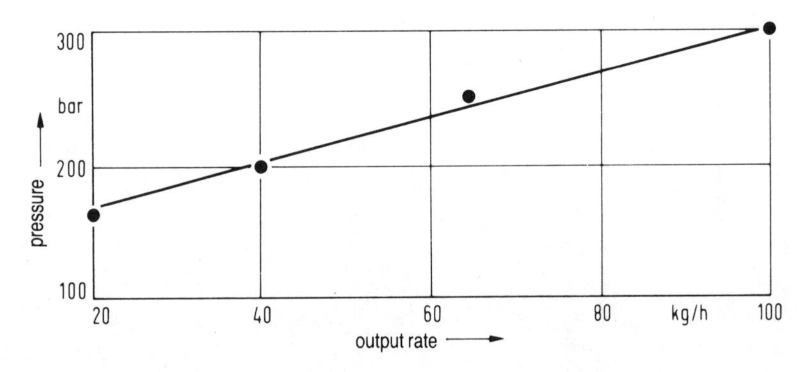

Figure 5-27 Relationship of pressure and output in a spiral mandrel distribution system. Courtesy: *Kunstoffe German Plastics* 77 (July 1987).

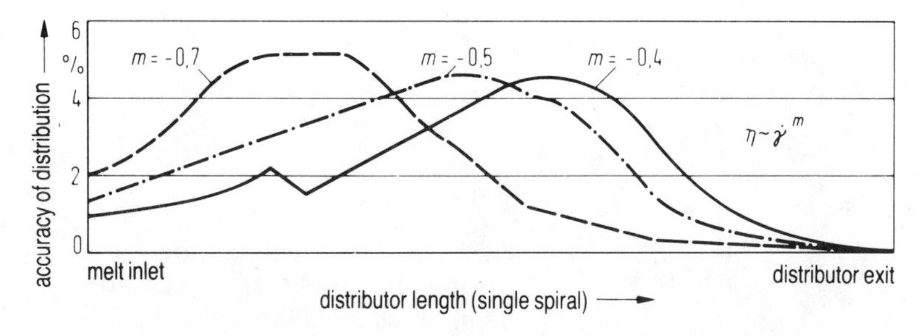

Figure 5-28 Distribution behavior in a spiral mandrel distributor for various viscosity exponents. Courtesy: *Kunstoffe German Plastics* 77 (July 1987).

Figure 5-29 Manifold core elements of the Battenfeld spiral die head. Courtesy: *Kunstoffe German Plastics* 77 (July 1987).

surfaces is kept short except where the transition is made from one section to another.

An extrusion head for six or seven layers is constructed of three sections. The arrangement shown in Fig. 5-26 is for six layers. The highest section contains the manifold for three materials, the second section adds another two materials, and the last section can add either one or two layers. This arrangement provides flexibility in setting up the temperature of each section, similar to a modular die. The mechanical arrangement of the manifolds needs only a very small diameter center post in the die, therefore it is possible to extrude small-diameter parisons, which may be difficult to achieve in some other designs.

Die Controls

Circumferential Thickness Control

One of the most important factors in achieving a quality product is material distribution in the blown product, which in turn depends on the control of the material in the parison prior to blowing. The desired shape of the parison cross section—and the thickness at each location—is related to the final shape of the product, but is most frequently circular.

In coextrusion two levels of control are needed. The first is individual layer control and the second is total parison control. The total parison is controlled in the conventional manner: a ring is moved relative to the mandrel to correct for any imbalance. If the overall shape is not circular the orientation of the ring to the mandrel needs to be maintained. Individual layer control is more difficult to achieve because of mechanical considerations.

Modular die design is the one design that provides the access needed for individual circumferential layer thickness control. To facilitate this, a restriction ring is incorporated in the design at the exit of each manifold so it can be moved in any direction by turning adjusting bolts from outside the body of the die. By changing the flow gap inside the flow channel the amount of material flowing by is adjusted. This feature is found in the very early design by Toppan Printing (Fig. 5-18), and later in the Solvey design (Fig. 5-21).

If the die construction does not allow the necessary access for adjustment, the uniformity of each layer depends on the accuracy of the die construction and the uniformity of flow through it, which is highly dependent on the melt temperature, its uniformity, and the die temperature uniformity. This may restrict the ability to use the same elements for more than one material and a single flow rate. Such a restriction may be costly in terms of extrusion tooling.

Layer Thickness Control

The Ando design [18] incorporates an additional flow ring at the exit from each manifold (Fig. 5-19). This ring, unlike the adjustment ring discussed above, cannot be moved from the outside. However, it does provide a simple element that can be machined easily and put into the die to help control the flow out of each manifold and therefore widen the range of materials that can be extruded from it. The actual arrangement within a single manifold section is shown in Fig. 5-30.

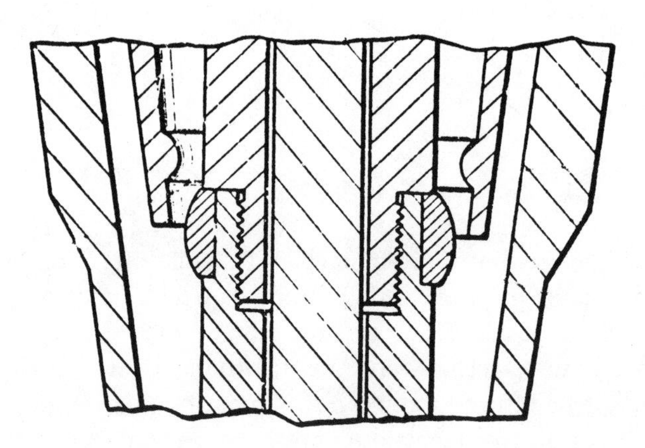

Figure 5-30 Ring flow adjustment element in the Ando die head.

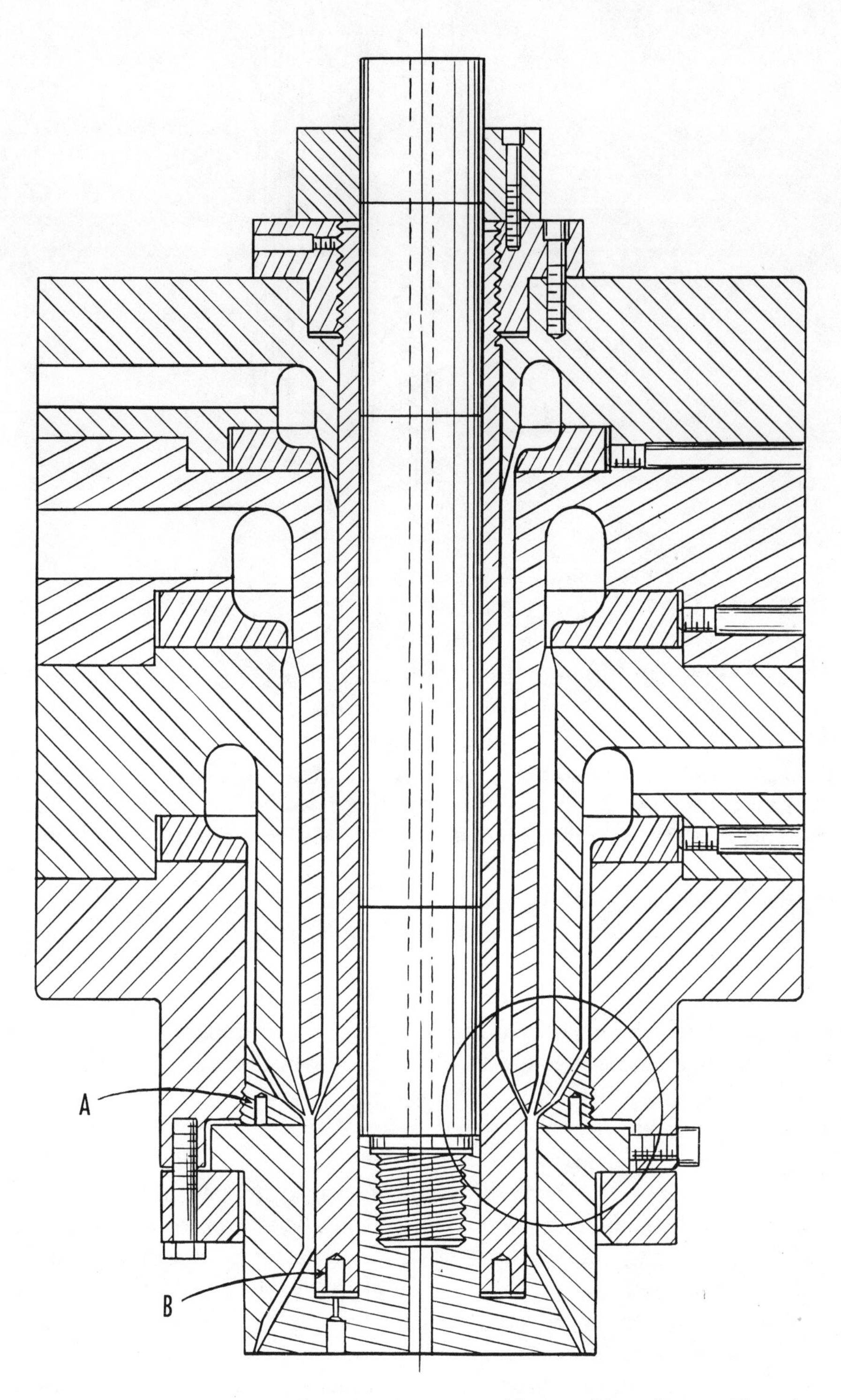

Figure 5-31 Johnson Controls, Inc., coextrusion blow molding die provides adjustment of flow channel gaps by notching part A for the outside layer or part B for the inside layer.

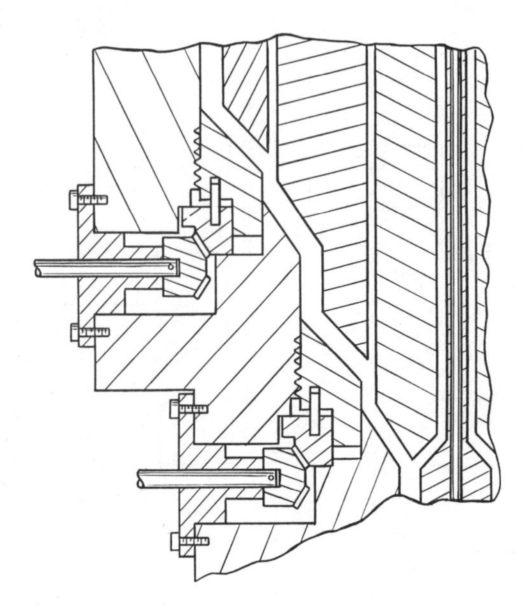

Figure 5-32 Gear train provides adjustment outside the die for flow passages in the Johnson Controls, Inc., coextrusion die.

A new design introduced by Johnson Controls (Fig. 5-31) provides for the control of the exit passages of all the manifolds of the die but one. Through axial movement of the die elements (marked A and B) the gaps of the inner and outer manifold layers are varied. Changing the gaps adjusts the flow from these two layers relative to the flow from the fixed-gap middle-layer passage. This ability to adjust the flow broadens the range of materials and viscosities that can be extruded from the die without replacement of elements.

In the manual system the adjustments are reached by removing the ring and tip, and rotating the elements that are threaded. A second design provides for the rotation of the adjusting elements from the outside of the die via a gear arrangement shown in Fig. 5-32. A larger number of layers can be handled by adding flow junctions farther down the die.

Temperature Control

The importance of controlling melt temperature was discussed previously. In coextrusion, temperature control should be applied to each of the melts. However the desired temperature may differ from material to material, and the compact mechanical structure of the die makes it difficult to accommodate this need.

In the discussion of modular dies it was pointed out that this design provides greater freedom to control the temperature of each section by providing individual heaters and temperature controllers, whereas the shell design encompasses all manifolds. However, depending on the construction of adjoining sections, the manner in which the flow channel is created, and the flow requirement of the melt against the metal surfaces of the channel, the temperature separation may not be sufficient to accommodate a large temperature range.

Several dies incorporate various means to separate the manifolds and thermally insulate them from one another. The GE-Graham Engineering design (Fig. 5-33) [16] incorporates a means of improving the ability to get

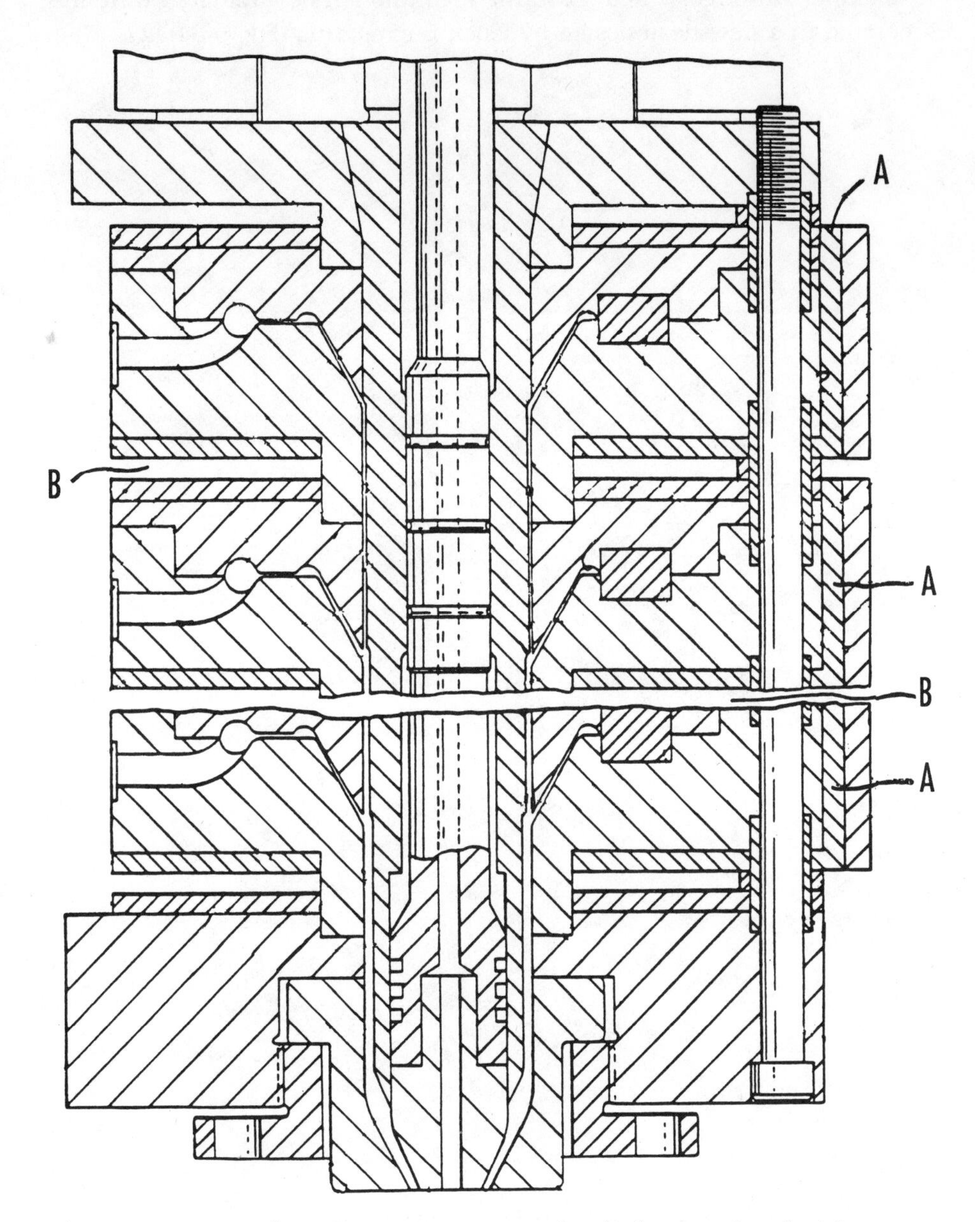

Figure 5-33 GE-Graham Engineering coextrusion die head equipped with copper heat conductors (A) and air gaps (B) for thermal isolation.

several temperatures at various parts of the die head. Air gaps (B) are created between adjoining manifold sections, and special copper conductors (A) placed under the heaters extend around the outside of the modular section. Those conductors enhance the heat transfer from the heater to the inside of the section where the flow manifold is located. Cooling is achieved through convective air flow in the gaps.

To provide a capability to lower the temperature of some sections of a die, channels through which a cooling medium can be circulated were incorporated in a new head design by Coex Engineering (Fig. 5-34) [22]. The

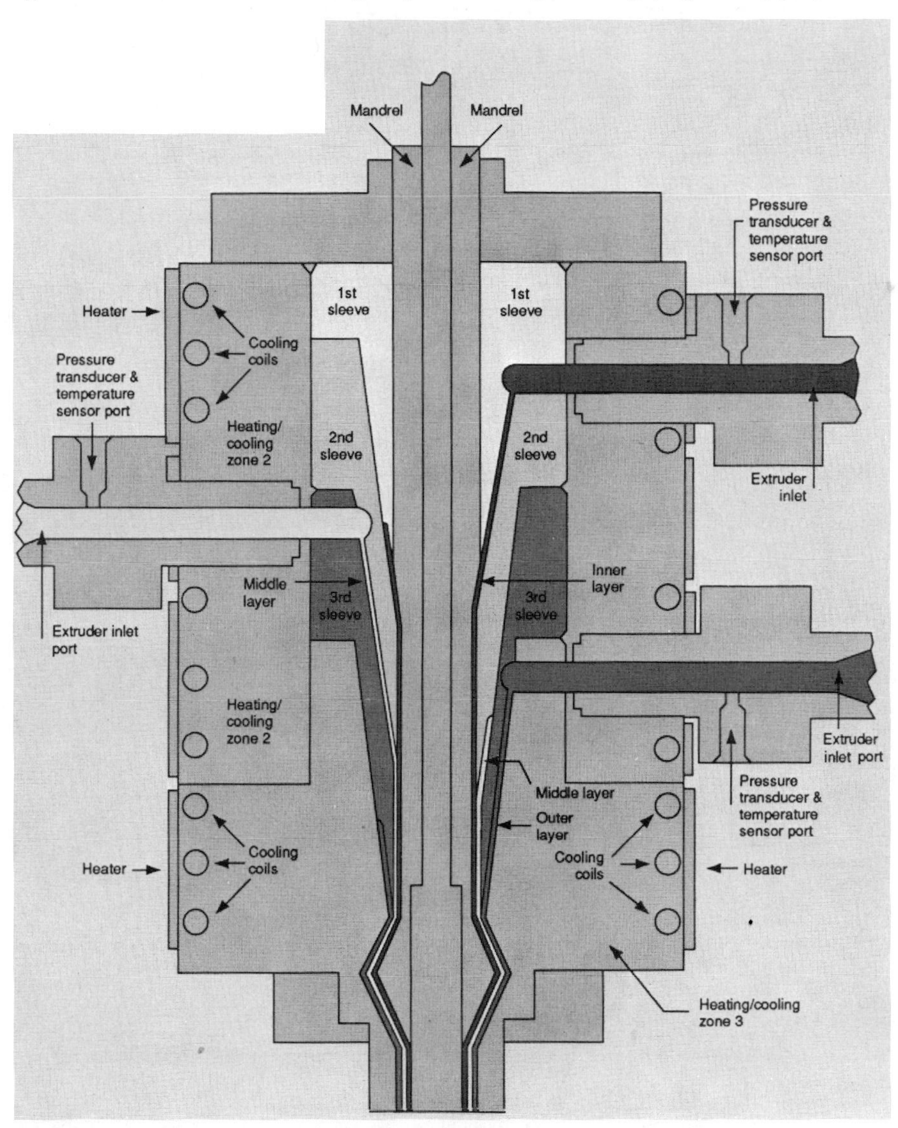

Figure 5-34 Coex Engineering die head utilizing cooling circuits to provide widely different processing temperatures between various manifolds. Courtesy: Coex Engineering, Inc., Libertyville, Ill. *Plastic World,* Cahners Publishing Co., Newton, Mass.

combination of heaters on the outside of the modular elements and cooling channels within provides a larger temperature range over which individual manifold temperatures can be maintained.

Other Dies

Interrupted Multilayer Die

The JSW (Japan Steel Works) accentuated blow molding machine is equipped with a head in which the flow of the inner layers can be interrupted without disturbing the total extrusion [23]. In this design (Fig. 5-35), the inner layers, usually the barrier layer plus two adhesive layers, are intro-

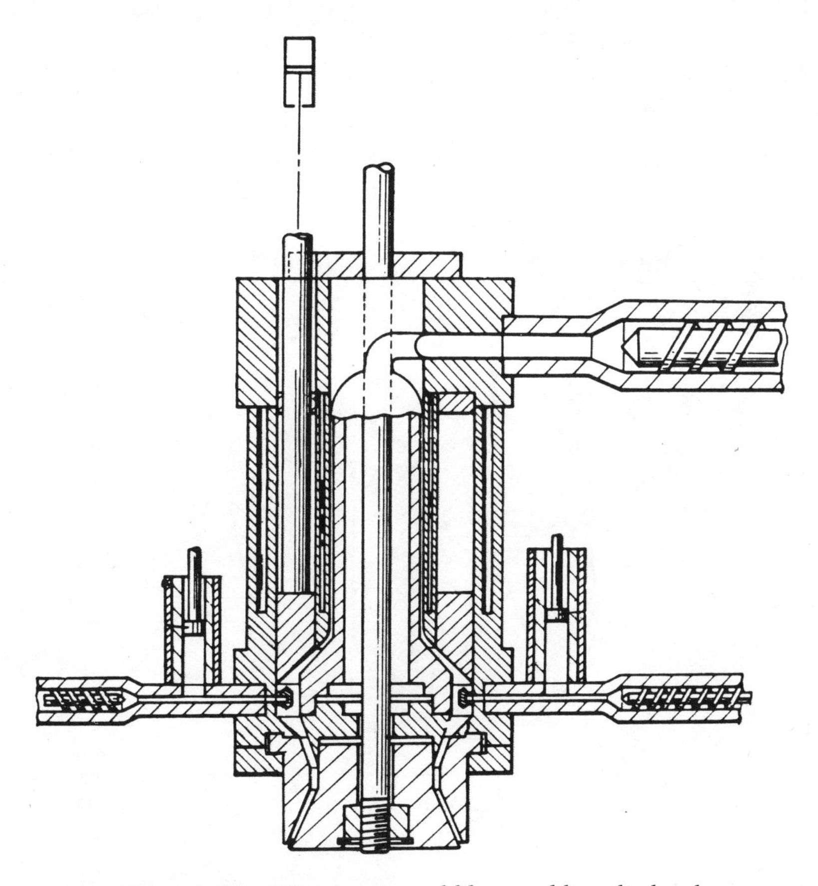

Figure 5-35 JSW accentuated blow molding die head.

duced through a spider. Within the core of the spider three manifolds create three layers, all meeting at a common discharge point (Fig. 5-36). The materials are fed to the center of the spider through the spider legs. The inner layers are extruded from the spider into the middle of the structural material, which was separated into two concentric layers when passing the spider section. The extrusion of the inner layers is done by the use of an accumulator rather than directly from the extruder screw. This makes it possible to create discontinuous flow of the inner layers, by programming the accumulator action. The total extrusion can thus be programmed to eliminate the barrier material and the adhesive from the pinch area or any other area as desired. This can result in a better pinch, as it consists of joining only the same material from opposite sides of the parison. Additionally, pinched-off areas that are to be later trimmed off can in some cases be made of structural material only. This makes it easier to reclaim the material as regrind and return it to use blended with virgin material, without the need to form a special layer to "bury" the regrind.

The extrusion of the inner layers directly between the structural layers means that even materials with vastly different viscosities can be accommodated, as the flow is controlled only by the inside and outside materials, which in many cases are the same and require the same die temperature.

The proper functional performance, for example, barrier, in the pinch area is solely dependent on the thickness of the pinch, which by design will be several times thicker than the normal thickness of the wall.

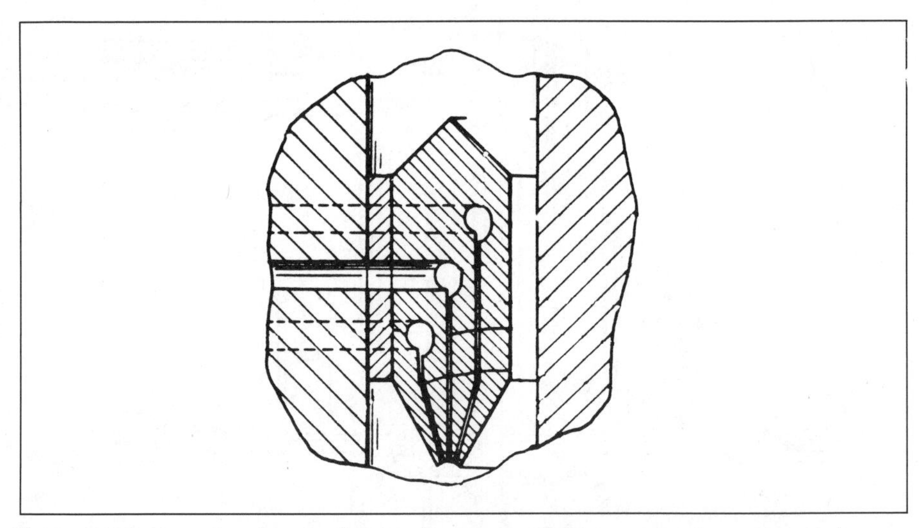

Figure 5-36 Functional and adhesive layers manifolds within the spider of the JSW die head.

Accumulator Dies

The use of an accumulator in blow molding is well known and its place is very important, especially in industrial applications where the blow molded parts are rather large. The accumulator makes it possible to extrude a large parison in a very short time, which otherwise would have a significant parison sag problem, making axial material distribution control very difficult if not impossible. A patent assigned to Ishikawajima-Harima and Kabushiki Kaisha [29] describes the construction and operation of an accumulator type head for multilayered extrusion blow molding, sometimes referred to as the IHI coextrusion head. The accumulators are concentric rings and pistons nesting one within the other; they are all activated simultaneously through a common hydraulic control system to form the multilayer structure (Fig. 5-37). Through the programming of the extrusion of the various melts from their accumulators, the ratios between the layers can be changed and therefore so can the thickness of each layer. The simultaneous extrusion of all layers may create some circumferential distribution problems due to dynamic interference of one stream of melt with another.

Another approach to the accumulator head design can be found in the design by Hartig (Fig. 5-38) [25]. The materials flow from the extruders through individual manifolds to the single accumulator chamber already in the proper order. Because of the single action of the ram, the proportion of flow between the various materials is predetermined by the output ratios of the extruders.

Multiparison Dies

Enhanced productivity of extrusion blow molding machines is usually achieved first by adding molds that are supplied parisons from the same die head, and then by increasing the number of die heads. Today, some companies offer twin and triple parison dies. Figure 5-39 is a triple parison die for two-layer coextrusion manufactured by Battenfeld Fischer Blowmolding Machines. Figure 5-40 shows the twin parison die offered by Mac International Associates. It is a single-body design; the splitting into two parisons is internal in the die at each feed manifold. This arrangement provides a compact design as to the centerline distance between the two parisons. Another way of splitting the flow is shown in Fig. 5-41 [13]: the extruder feed is split into two passages when entering the die from the melt pipe.

The splitting of the melt flow either from the extruder to two layers or to a dual parison die requires the addition of a flow valve in at least one of the streams to help balance the flow between the two directions. A valve in such a splitting manifold is shown in Fig. 5-42 [13] and in Fig. 5-41. The valve action increases the resistance to flow in the flow passage. Care must

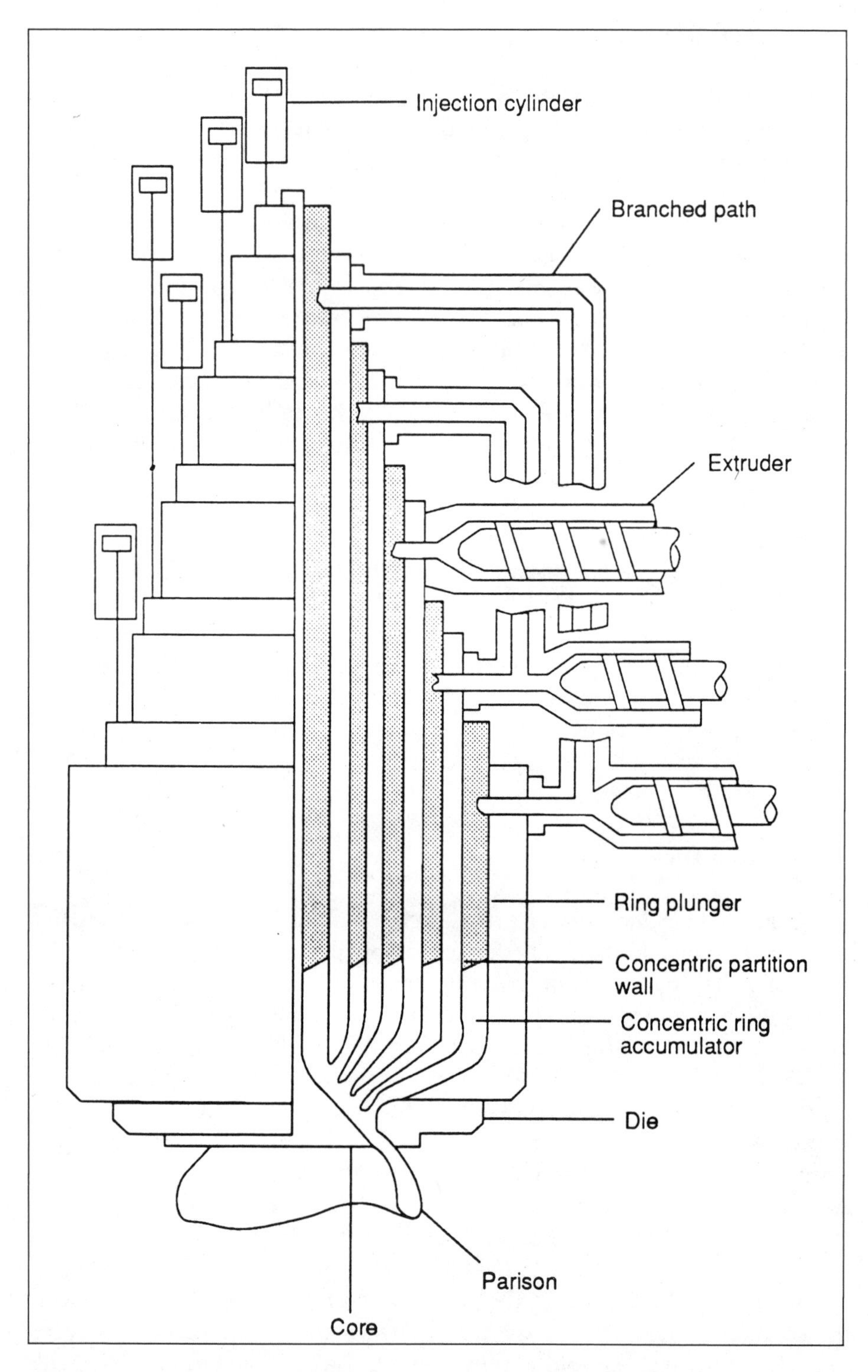

Figure 5-37 IHI accumulator die head. Courtesy: The Packaging Group Inc. (USA).

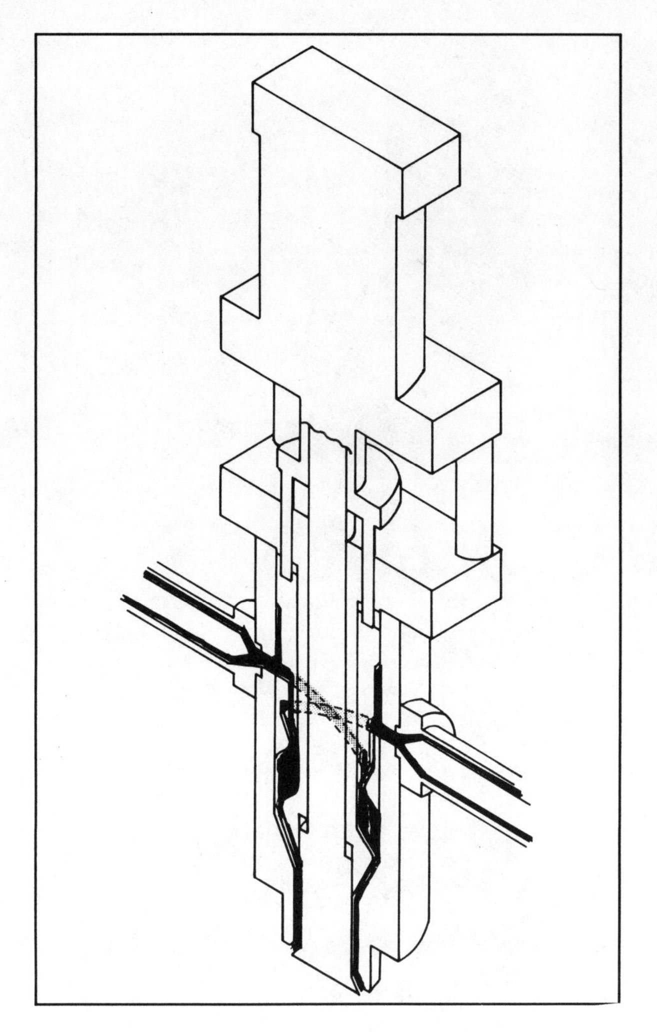

Figure 5-38 Hartig two-layer single-ram accumulator die head. Courtesy: *Plastics Machinery & Equipment* (July 1989).

be exercised in the design of a restrictor valve to reduce the chance of hang-up points and stagnation of the melt, which will result in material degradation.

Trimming

Standard trimmers used in blow molding production can also be used for trimming coextruded blow molded articles. Where the additional functional and adhesive layers are similar in their chemical composition to the structural materials (e.g., all olefinic based), the trimming operation is very similar to that used when the product is made from the structural material only. However, some part of the structure could be of a brittle material. In such a case the conditions at the trimmer may need to be modified so that

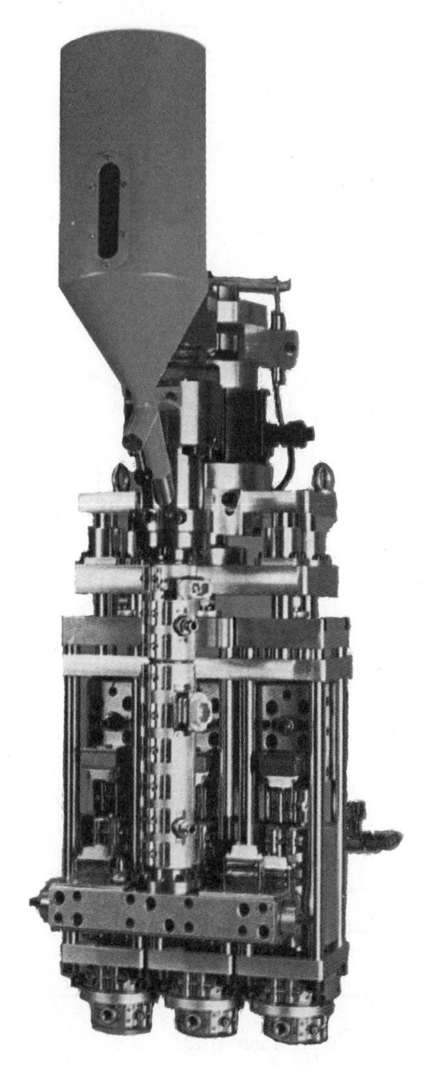

Figure 5-39 Battenfeld Fischer triple head for two-layer coextrusion blowmolding machines. Courtesy: *Plastics Machinery & Equipment* (July 1989).

the brittle material does not crack during trimming. This is normally done by operating at a higher temperature at the trimmed area. The need to modify the conditions will depend on the characteristics of the material and especially its thickness. Microcracks in the trimmed area in a brittle component of the wall can lead to premature failure of the product if, during use, some stresses are applied to it.

Where reaming is used to produce a final close-tolerance opening in a product, special care must be taken in the design of this area so that the functional layer is not removed. Either sufficient thickness is provided by the inside structural layer or the wall design in this area needs to be inverted so that after trimming there is a functional layer all the way to the cap. This applies especially to containers in which the functional layer is on the inside.

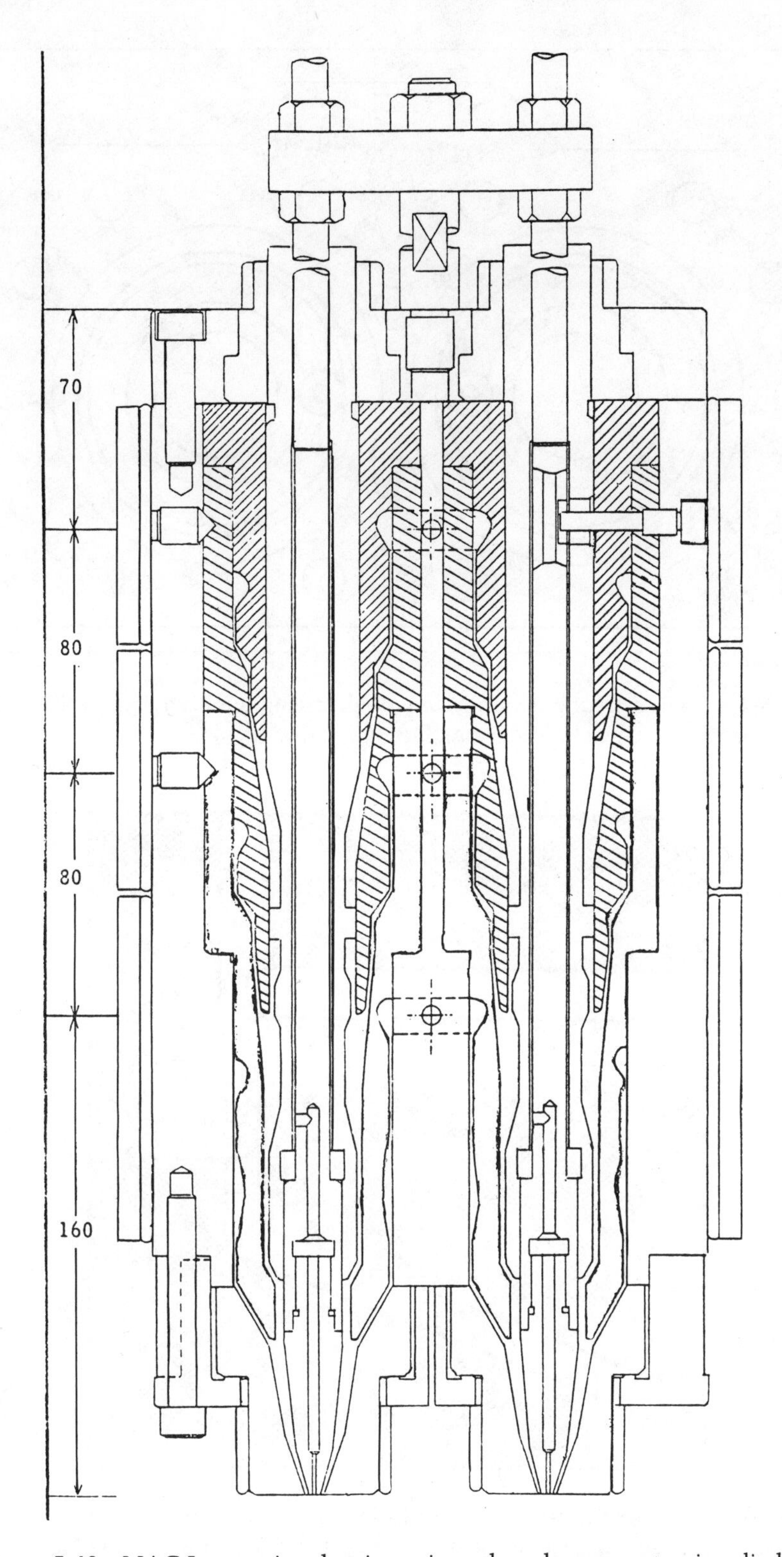

Figure 5-40 MAC International twin parison three-layer coextrusion die head.

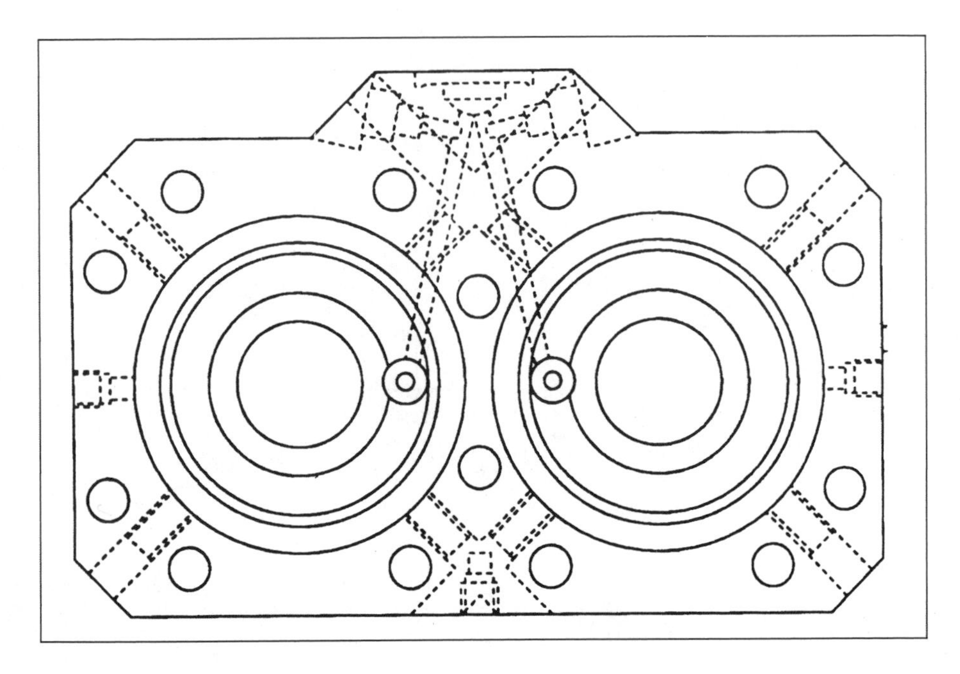

Figure 5-41 Melt splitting manifold in a dual parison die head equipped with two flow valves. Courtesy: The Packaging Group Inc. (USA).

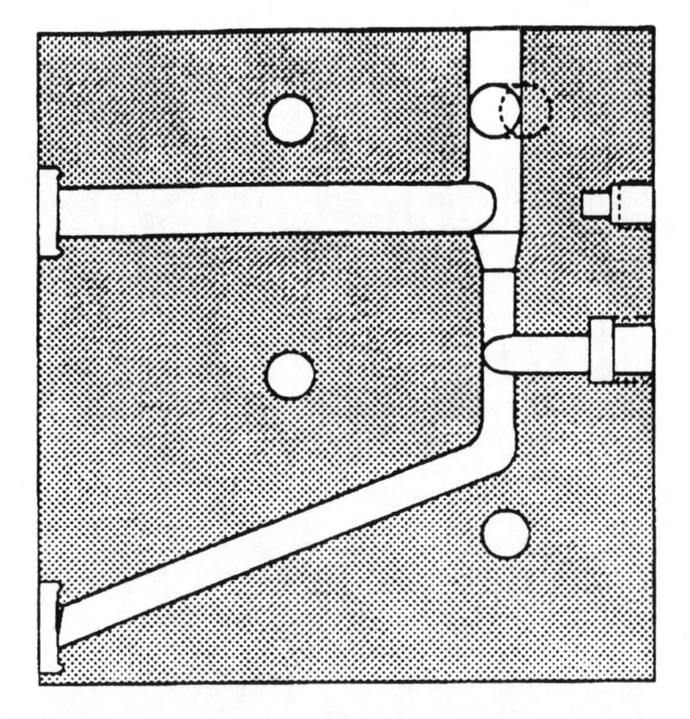

Figure 5-42 Manifold to split the melt from one extruder into two layers, equipped with a single flow-adjusting valve. Courtesy: The Packaging Group Inc. (USA).

PROCESS

The coextrusion process does not differ much from the standard extrusion blow molding process. The same practices that result in good production customarily need to be practiced and maintained to produce acceptable coextruded products.

The simultaneous extrusion of several materials mandates modifications in the operation of the extrusion end of blow molding. These changes are especially required during start-up and shutdown. During those periods the differences between the various materials are more noticeable, especially their flow characteristics as a function of the melt temperature and the die temperature.

During production the main concern is to maintain the proper feed rate of all materials in the coextruded structure. A stoppage of the functional layer or an adhesive layer may not be readily detected in the final product because of the relatively small amounts of those materials in the structure. The loss of an adhesive layer will lead to delamination that will normally appear when the wall is subjected to flexing, but will not necessarily affect the functional performance of the product, especially for a product with a short service life, as in packaging. However, the loss of the functional layer will result in a product that cannot perform as designed. Depending on the material being stored inside the blown product, a safety or health problem could also be created.

Another factor that requires close attention during normal production is the temperature of the various extruders and the die. Because of the need to operate at an overall temperature suitable for all the materials, the actual temperature may be extreme for one or more of them. This means that the tolerance for temperature drift or increased residence time is reduced drastically if no material is to degrade.

Start-up and Shutdown

Difficulties during start-up and shutdown depend on the materials being combined in the coextrusion. The basic die temperature is set to the temperature required by the structural materials, which make up the bulk of the extrusion. Where a combination of materials requiring various tempertures is used, first the extrusion temperature of the structural layers is established, then the die temperature can be changed slightly as needed for the functional and adhesive materials. In designs where the inner layers are immediately encapsulated within the structural material, a high melt temperature for the inner layers can be sometime tolerated.

When extrusion is started, only the structural layers are extruded first; after extrusion stability is reached for those layers, the functional layer and the adhesive layers are added. At shutdown, most functional layers—either because of thermal stability or because they require changes in die temperature—are normally purged out of the die and the extruder, especially if the shutdown is to be longer than two or thee hours. When purging is done first, the temperature of the die can be set, as would be done for shutdown with monolithic material.

QUALITY CONTROL

The manufacture of a product always involves the need to establish quality control procedures, which when implemented properly will help to maintain the product within specifications. The basic quality control process for coextruded products is the same as for standard extrusion blown production, except that one additional measurement is needed: measurement of the thickness of each layer within the total thickness of the wall. It is also possible that the products commonly being packaged or stored in the coextruded articles may generate the need for special testing. As such special testing is likely to be unique to a particular packaged product, it is outside the scope of this discussion. Only the measurement of layer thickness is described in this section. Chapter 14 is devoted to quality control testing.

In some cases, where barrier layers are concerned, only the presence of the barrier in at least a minimum thickness can insure proper performance in protecting the packaged product. Additionally, lack of adhesive could result in delamination of the layers.

Each coextrusion blow molded product probably has one or more critical locations. These are most often the result of design factors, and usually they are thinner than the rest of the product wall. Verification of the present of the functional layer and adhesive at a sufficient thickness at these locations, through a sound quality control procedure, will be indicative of sufficient thickness elsewhere in the product. Monitoring of the consumption of the barrier resin and adhesive resin by their extruders will also provide indications that sufficient material is being provided for the needed performance.

Manual Layer Measurements

The manual procedure for measuring individual layer thickness is to examine a cross section of the wall of the blown product with a magnifying device. In order to be able to see the various layers a very thin sample is used. A microtome—a device that provides for the clamping of a small wall sample and cutting a thin shaving from it—is best for obtaining the sample. The sample is cut at a plane perpendicular to the wall. Due to small differences in light transmission and refraction, the layers will appear in slightly different colors, (Fig. 5-43). By use of a calibrated microscope or a high-power eyepiece the thickness of all layers can be determined.

Figure 5-43 Photomicrograph of a multilayer structure. Courtesy: EVAL Company of America.

Electronic Measurements

One type of electronic system measures the thickness of each layer as well as the total wall thickness simultaneously. Such a system can be used on line or off. The measurement is based on a multiwave length infrared (IR) technique. Through the use of multiple filters, several IR frequencies are generated, each geared to a material in a certain layer. A detector receives the IR after it has passed through the wall and a microprocessor-based decipherer converts the detector output to thickness readings [26].

Another device, made by Top Wave of Finland, is based on the use of visible and near infrared wavelengths that are reflected from the interface of any two layers when there is a difference in the refractive index of the two materials. A detector, an interferometer, picks up the light reflected by the sample and compares it with a reference beam of light that passes along an adjustable length path. When the two beams hit the detector, interference results because the wavelengths from the sample are phase-shifted by the characteristics of the layer materials. The length of the reference beam path is adjusted until the interference is canceled. Since the amount of adjustment required is directly proportional to the thickness being measured, it can be electronically converted to a thickness measurement [27, 28].

Several other measuring devices are available to measure film or tube thickness. Various techniques based on beta and gamma radiation, ultrasound, capacitance, and electric fields are used. These devices and techniques often can be adapted for measurements of coextrusion of blow molded articles.

REFERENCES

1. Naureckas, E. N. "Coextrusion Blow Molding Developments." *Preprints, 2nd Blow Molding Conf.* (Itaca, IL) Soc. Plastics Eng. (1984). pp. 37–45.
2. Graebner, L. S. "Food Packaging: An Opportunity for Barrier Coextrusion." *Proceedings, Coex '84.* (Princeton, NJ) Schotland Business Research, Inc. (1984). pp. 25–74.
3. Estes, A. E. "Market Outlook for Coextrusion Blow Molding Barrier Bottles." *Proceedings, Coex '84.* (Princeton, NJ) Schotland Business Research, Inc. (1984) p. 347.
4. Schreiber, P. "Delivering the Goods in Plastics Packaging." *Plastics Engineering* (May 1986): 29–33.
5. Drennan, W. C. "Tie-Layer Resins: More than 'Ties that Bind'." *Plastic Packaging* (September/October 1988): 32–35.
6. Wachtel, J. A., B. C. Tsai, and C. J. Farrell. "Retorted EVOH Multilayer Cans with Excellent Oxygen Barrier Properties." *Proceedings, Future-Pak '84.* (Atlanta, GA) Ryder Association, Inc. (1984). pp. 5–33.
7. Hind, V., and M. Weiss. "Coextrusion Widens the Market for Blown Bottles." *British Plastics and Rubber* (January 1985): 23.
8. Naureckas, E. "Nonsymmetrical Layer Structure in Coextrusion Blow Molding Containers." *Proceedings, Coex '84.* (Princeton, NJ) Schotland Business Research, Inc. (1984). pp. 361–372.
9. Topolski, A. S. "Multiple Layers for Multiple Markets." *Preprints Coextrusion IV.* (Arlington Heights, IL) Soc. Plastics Eng. (1987). pp. 9–30.
10. Graumann, R. E. "Barrier Materials Used in Multilayer Structures." *Preprints, Coextrusion III.* (St. Charles, IL) Soc. Plastics Eng. (1985).
11. Ikari, K., and T. Moritani. "Recycling of EVOH-based Composite Materials." *Proceedings, Coex '84.* (Princeton, NJ) Schotland Business Research, Inc. (1984). pp. 289–319.
12. "Processing EVAL Resins." Tech. Bulletin No. 120. EVAL Company of America.
13. Mignin, M. "Flexibility Consideration in Coextrusion Blow Molding Equipment." *Proceedings, Barrier Pack '88.* (Chicago, IL) The Packaging Group, Inc. (1988).
14. Tadmor, Z., and I. Klein. *Engineering Principles of Plasticating Extruders.* New York: Van Nostrand Reinhold (1970).
15. Rauwendaal, C. *Polymer Extrusion.* New York: Oxford Univ. Press (Munich: Hanser Verlag) (1986).
16. Briggs, M., and E. Teutsch. Modular Extrusion Head. U.S. Patent No. 4798526, assigned to General Electric Company and Graham Engineering Corporation. January 17, 1989.
17. Kudo, K., and T. Mizutani. Multilayer Parison Extrusion Molding Machine for Blow Molding. U.S. Patent No. 4047868, assigned to Toppan Printing Co. September 13, 1977.
18. Ando, K., and Y. Matsuo. Die Assembly for Extruding Multilayer Parison. U.S. Patent No. 4578025. March 25, 1986.
19. Strum, W. L. "Fundamentals of Coextrusion Blow Molding." *Proceedings, 4th Annual Meeting.* (Orlando, FL) The Polymer Processing Soc. International (1988).
20. Dehennau, C., M. Kerger, and L. Dubuisson. U.S. Patent No. 4657497, assigned to Solvey and Cie. April 14, 1987.
21. Onasch, J., and P. B. Junk. "Coextrusion Blow Molding—Processing, Application, Materials." *Kunststoffe German Plastics* 77 (July 1987): 663–668.
22. "New Coex Head Gives Tighter Bottle Control." *Plastic World* (August 1988): 27–30.

23. Motonaga, T., K. Fukuhara, T. Tamura, and T. Yoshida. Extrusion Molding Equipment for Multilayer Parison. U.S. Patent No. 4717326, assigned to Mazda Motor Company and Japan Steel Works. January 5, 1988.
24. Iwawaki, A., K. Kojima, H. Fukase, Y. Shitara, A. Nomura, T. Sato, and H. Shibata. U.S. Patent No. 4149339, assigned to Isikawajima-Harima and Kabushiki Kaisha. April 17, 1979.
25. Galli, E. "Two- and Three-Layer Blowmolding Developments." *Plastics Machinery and Equipment* (July 1989): 26–27.
26. Sneller, J. A. "Selective Gauging: New Tool to Keep Cost of Coextrusion Way Down." *Modern Plastics* (August 1984): 42–44.
27. Törmälä, S. "Cost Savings with Off-line Thickness Gauge for Multilayered Barrier Films and Coatings." *Preprints, Coextrusion IV.* (Arlington Heights, IL) Soc. Plastics Eng. (1987). pp. 149–163.
28. Mäkelä, J. "Layer Thickness Profile Measurement in Multilayer Bottles or Films." *Proceedings, Coex '84.* (Princeton, NJ) Schotland Business Research, Inc. (1984). pp. 373–388.

CHAPTER

6

Other Blow Molding Processes

CHRISTOPHER IRWIN

Several special blow molding processes have evolved through the years to solve unique process problems or product applications. The multilayer and biaxial orientation processes, described in previous chapters, are two very important examples. The neck ring, drape, and dip or displacement techniques are three other such processes; they are described here.

NECK RING PROCESS

This process, originally developed by Owens-Illinois in the mid-1950s, is an important part of the history of blow molding and is still in use today. It is a clever process in which the finish of the bottle is injection molded to provide quality and the body of the bottle is extrusion blow molded to provide process flexibility. This is the only practical process that is capable of producing hollow handleware bottles with a precision injection molded neck. Figure 6-1 describes the basic sequence of operation.

Many modifications and improvements have been made by Owens-Illinois, such as process control and in-mold labeling. The equipment is ideal for short to moderate production run requirements.

The primary weakness of the neck ring process is a slightly longer cycle time for some bottle applications. This is caused by a sequence of operation

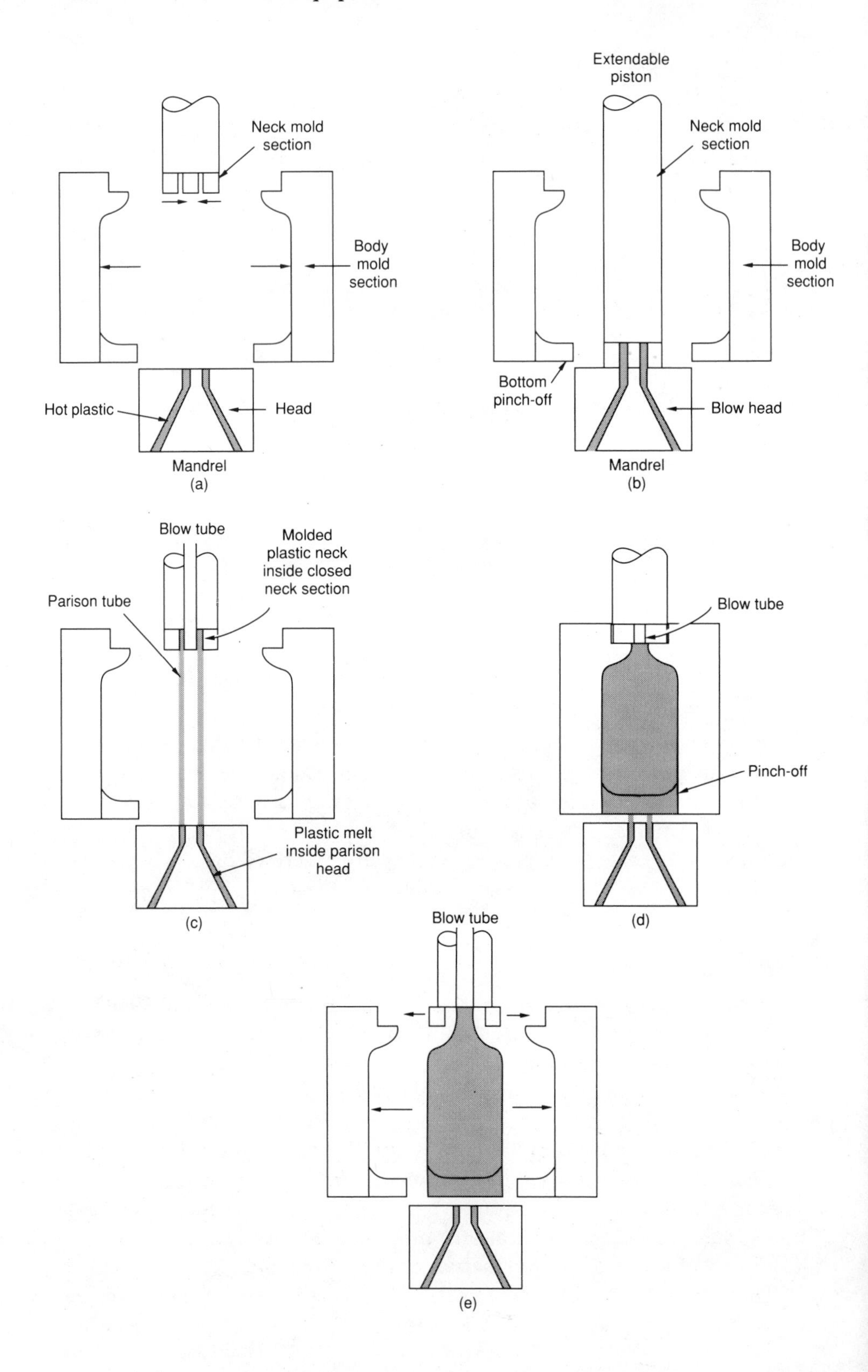
Neck mold
section
Body
mold
section
Hot plastic
Head
Mandrel
(a)
Extendable
piston
Neck mold
section
Body
mold
section
Bottom
pinch-off
Blow head
Mandrel
(b)
Blow tube
Molded
plastic neck
inside closed
neck section
Parison tube
Plastic melt
inside parison
head
(c)
Blow tube
Pinch-off
(d)
Blow tube
(e)

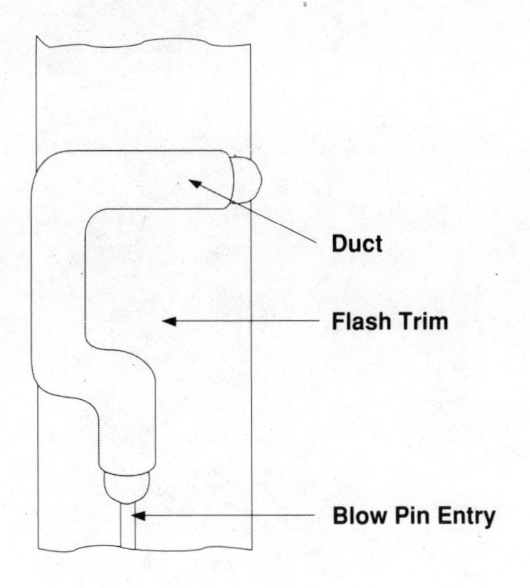

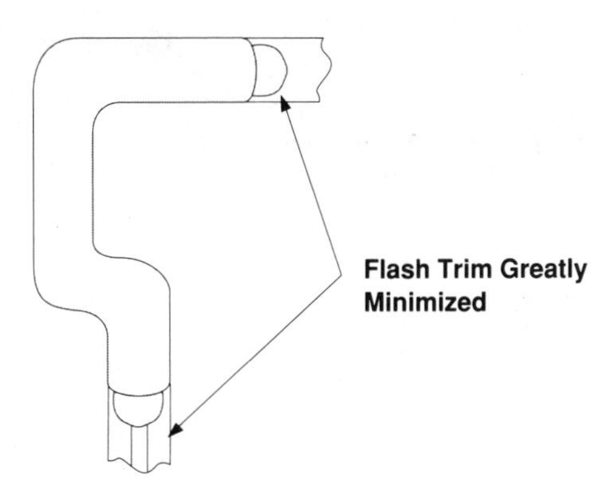

Figure 6-2 Drape vs. conventional blow molding. (a) Duct produced on conventional extrusion blow molding equipment. (b) Duct produced by drape blow molding process.

that requires the finish of the bottle to be molded before the extruded parison is formed for blow molding the bottle. Bottle removal at the end of the cycle is also time consuming.

The equipment used by Owens-Illinois is not commercially available; however, a small German firm, Ossberger-Turbinenfabrik, has developed a small machine version of the process suitable for vials and squeeze tubes.

←

Figure 6-1 Neck ring process. (a) Body section open, neck section closed, neck section retracted. (b) Neck section extended to mate with parison nozzle (plastic fills neck section). (c) Neck section retracted with parison tube attached. (d) Body section closed, making pinch-off (parison blown to body sidewalls). (e) Body molds open, neck molds open, bottle about to be ejected. Courtesy of John Wiley and Sons.

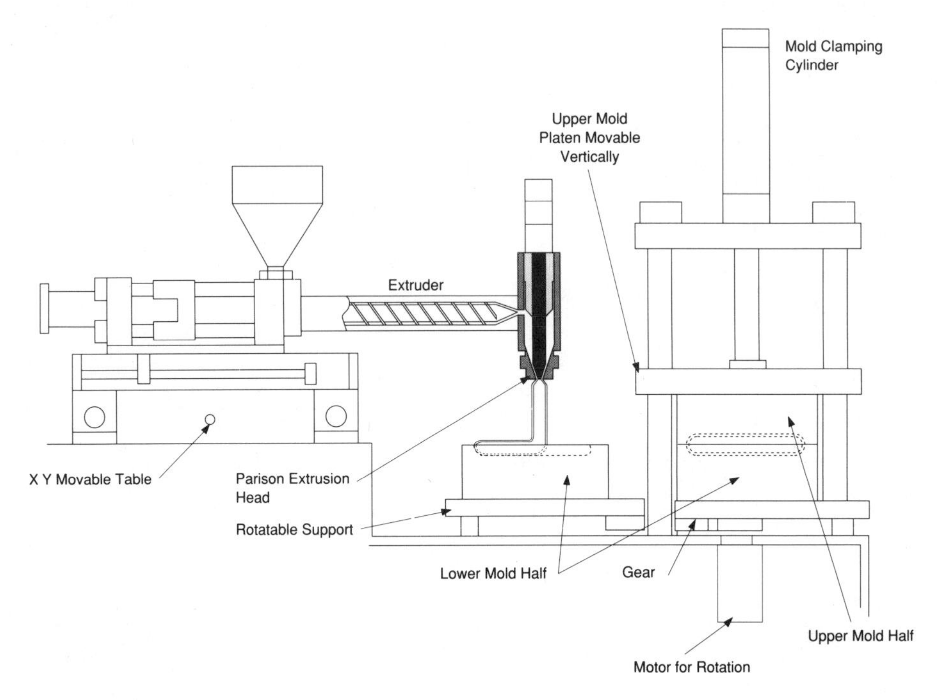

Figure 6-3 Multidimensional blow molding. Courtesy of Excell Manufacturing.

DRAPE PROCESS

Drape blow molding is a general term referring to several multidimensional blow molding techniques. Developed mostly in Japan, this approach seeks to minimize the amount of flash or trim material that is often a part of complex shapes molded on conventional extrusion equipment (Fig. 6-2).

The first approach, developed by Excell Corporation, Tokyo, uses an extruder/accumulator head combination mounted on a numerically controlled carriage. The carriage moves horizontally in both X and Y directions (Fig. 6-3). A sealed parison is extruded into an open lower mold half while the moving die head follows a curved path dictated by the part/mold cavity shape. A small amount of support air inside the parison is provided to prevent it from collapsing.

As soon as the parison is in position, the open lower mold half is indexed 180 degrees into a vertical press where the upper mold half closes onto the parison. The parison is now expanded into the part shape using conventional needle blow techniques.

An alternative approach developed by Sumitomo Mold Manufacturing Company can be adapted to conventional extrusion blow molding equipment. In this approach the mold tooling is divided into three sections. The center section closes independently of the upper and lower sections.

The sequence essentially begins with the center section closed. A small-diameter parison is extruded through the exposed cavity opening (Fig. 6-4).

1 Mold Cavity Open

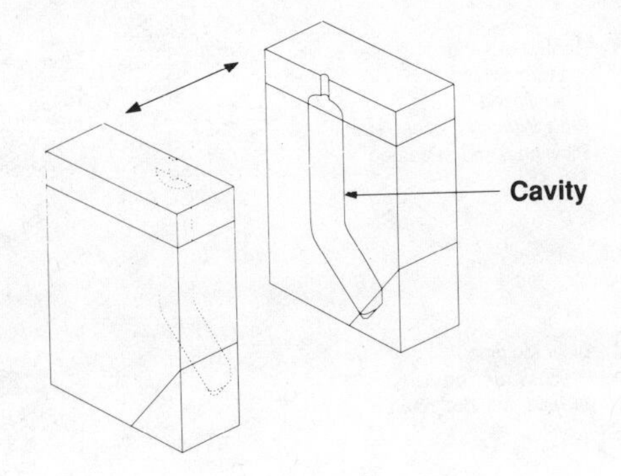

2 Mold center portion closed, top and bottom portions are left open, parison is extruded through opening at top while vacuum helps pull it along from the bottom.

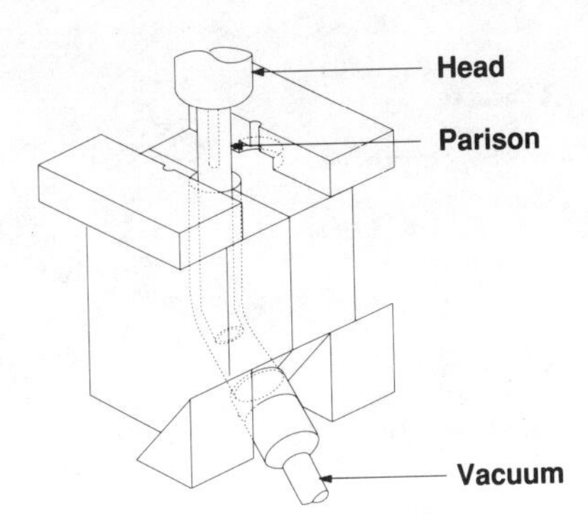

3 Vacuum source is removed, mold top and bottom portions are closed pinching each end of parison and sealing blow pin entry.

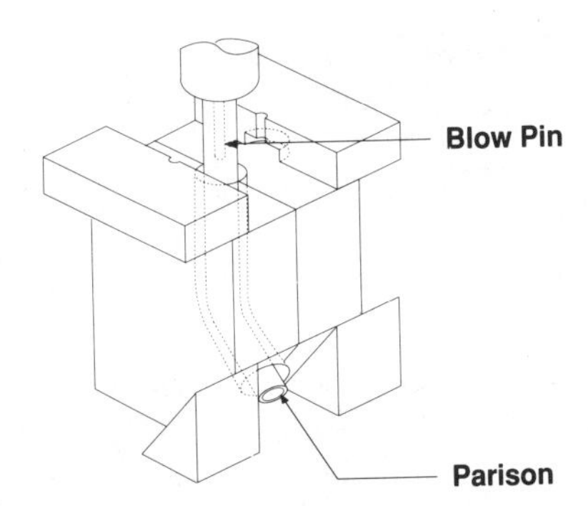

4 Three dimensional blown part is removed with small amount of flash at each end.

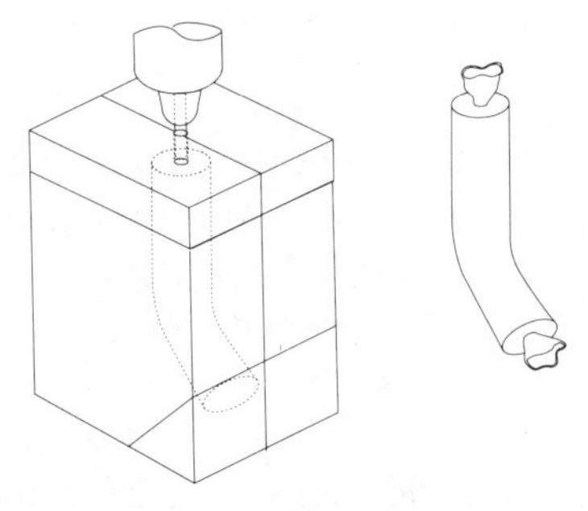

Figure 6-4 Multidimensional blow molding, alternative approach.

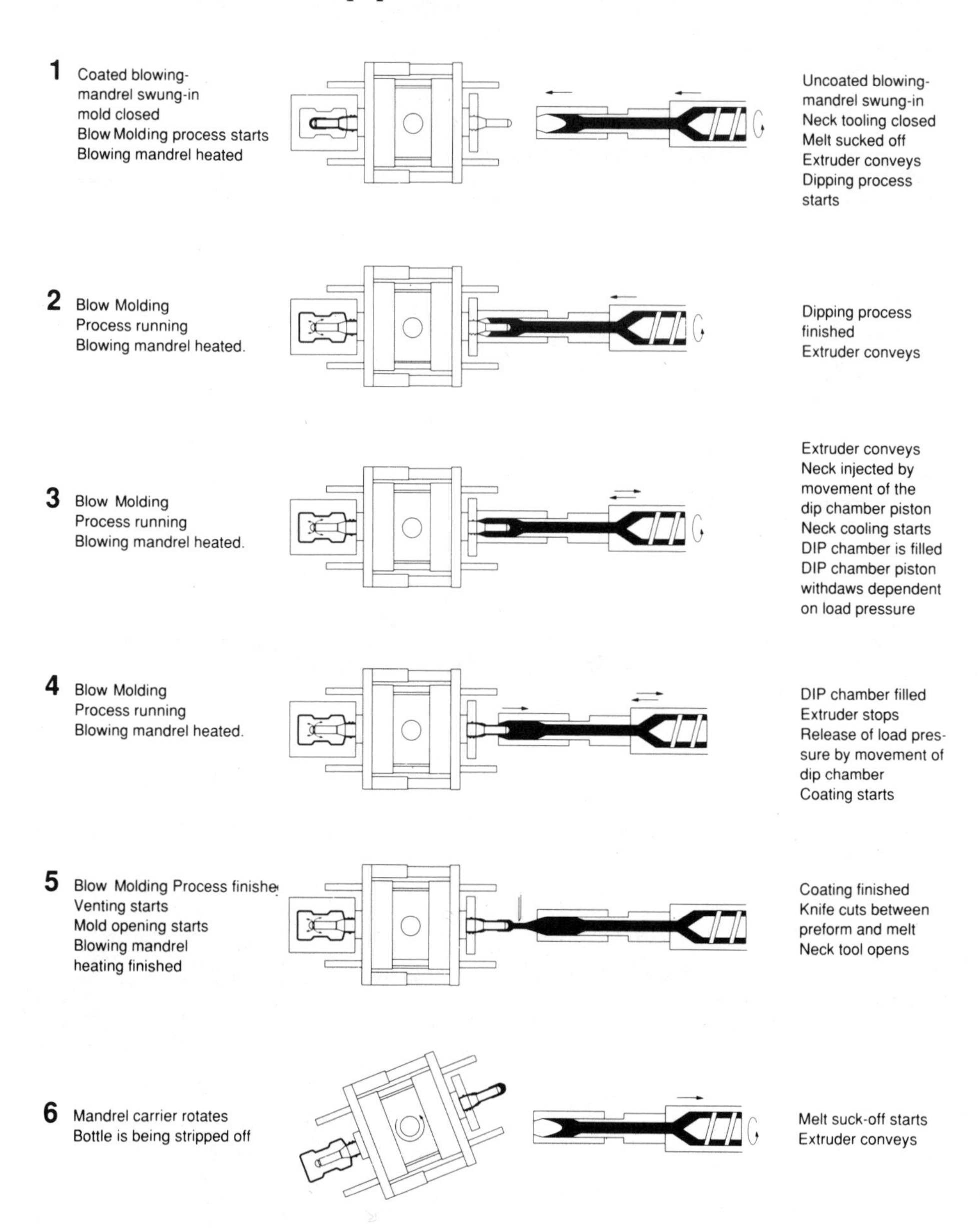

Figure 6-5 Dip blow molding. Courtesy of Staehle Maschinenbau.

A vacuum at the opposite end helps pull the parison through the cavity as it is extruded. At critical corners or other areas in the part shape, additional air is injected into the cavity to provide a "pillow" of support as the parison is pulled/pushed along.

After the parison is in position, exposed at both ends of the cavity, the upper and lower mold sections close to seal the parison. The parison is now expanded into the part shape using conventional techniques.

1 Step 1 - Vertical extruder deposits shot of material in open cavity.

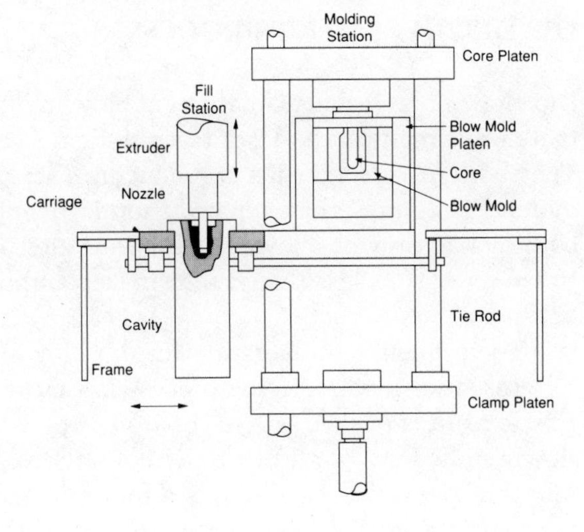

2 Step 2 - Cavity containing melt indexes between platens; ram forces up against core, which displaces the melt, forming a preform shape. Packing piston advances upward, displacing enough melt to pack the preform and neck area.

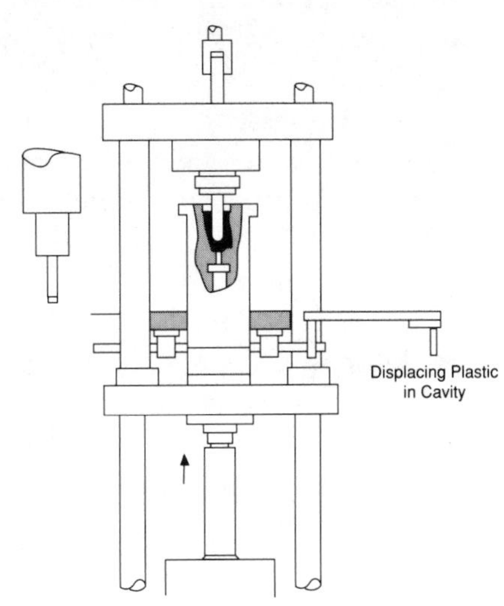

3 Step 3 - Cavity decends, leaving preform on core; blow mold closes around preform and bottle is blown.

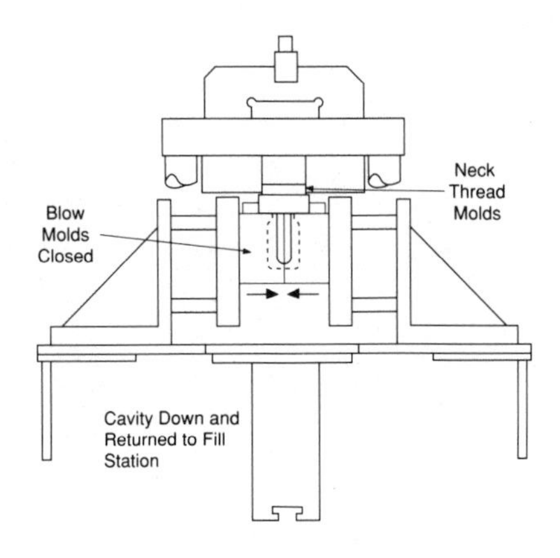

Figure 6-6 Displacement blow molding.

DIP/DISPLACEMENT PROCESSES

Dip blow molding and displacement blow molding are two similar processes that mimic many of the advantages of the injection blow molding technique. They both feature a means of producing a scrap-free preform with far lower molding pressures than typically found in injection blow molding. They both seek to provide a lower cost tooling alternative. The processes are best suited for bottles less than 8 fl oz in size, although larger bottles have been made.

The sequence of operation for dip blowmolding is shown in Fig. 6-5. The essential feature of the process is that a chamber filled with melted plastic resin is pushed onto a core rod or blowing mandrel. The chamber is slowly pulled away, allowing the mandrel to be coated with plastic resin. The thickness of the coating is a function of the mandrel size relative to the chamber opening as well as the speed at which the chamber withdraws. Some adjustment of part weight is possible.

The sequence of operation for displacement blow molding is shown in Fig. 6-6. The essential feature of the process is that a premeasured amount of melted plastic is deposited in an open preform cavity that is then lifted and pushed onto a core which displaces the melt pool. Eventually, the core and cavity form a closed unit that completely defines the preform shape. Because the melt is deposited in the cavity beforehand, a preform is created without a melt entry gate.

Although the dip and displacement blow molding processes have intriguing advantages, neither has achieved widespread commercial success. There are many reasons for this lack of success. Most are related either to nontechnical issues or to a lack of further applications development.

SECTION II

AUXILIARY PROCESSES AND EQUIPMENT

CHAPTER

7

Auxiliary and Materials Handling Equipment

ROBERT JACKSON

Materials handling and the use of ancillary equipment is thought by many blow molders to be of very low importance compared with the skills required to actually blow mold a quality part. This may well be true in terms of the skill levels required. However, from an expense/cost standpoint the correct implementation of auxiliary systems can make the difference between bankruptcy and profit.

The formula for profitablity in molding is quite simple:

Costs	–	Selling price of product	=	Profit before taxes

Costs include: raw materials (plastic, resin); equipment (principal, interest); production labor/time, and administrative overhead (including rent and utilities).

The machine, its applicablity and speed, and the skills of those running it are some of the primary ingredients in this formula. So is efficiency. For example, if 3 to 10 percent of the raw material goes out in the trash, there is a real problem. If parts are being processed at less than the fastest possible cycle time because there is not enough mold chilling capacity, it will be impossible to make a profit no matter how skilled the molder. If high heat exchanger temperatures are cooking the O rings in the machine's hydraulic system, there will be leakage and eventually a problem with machine downtime.

Another formula that is useful for quick calculations is:

$$\text{Resin cost} \times \text{Markup} = \text{Selling price}$$

In blow molding a common mark up is between two and four times resin cost. Thus the true cost of resin loss or waste is not just the cost of the resin itself but a part of the potential profits.

$$\text{Resin waste} \times \text{Markup} = \text{True loss}$$

If a plant is losing 5 to 10 percent of the total plastic purchased it is a drain on the expected profits by as much as 10 to 40 percent. This is a paper number but it tells us how important closed material handling systems can be.

In addition, cycle times and labor and machine untilization can be negatively affected by as much as 10 to 20 percent if the proper ancillary equipment is lacking. A well-run blow molding shop should have an efficiency rate of at least 90 to 95 percent of available uptime. This can only be accomplished by having the right secondary as well as primary equipment.

Ancillary Equipment Needed

1. Material handling systems and hopper loaders (including coloring and drying)
2. Water cooling systems (including both part cooling and the machinery's own heat exchanger's cooling)
3. Grinder and regrind processing system
4. Post molding part-handling system

The cost of these items can be 30 to 50 percent of the total equipment budget. This chapter discusses the functions, capabilities, selection criteria, and costs of ancillary equipment. In addition, a few rules of thumb for sizing each of the items are offered.

This kind of equipment tends to be acquired first as stand-alone items from which systems evolve with later additions. That evolution is one of the more expensive aspects of secondary equipment. But if a systems approach to acquiring ancillary equipment is taken from the onset, some of that expense can be curtailed.

MATERIALS HANDLING

Hopper Loaders

The simplest labor saving device is a hopper loader. It lifts the resin out of the gaylord into the hopper. More importantly, it does not spill the resin on the floor. It usually comes with a built-in self-cleaning filter (for fines removal). A hopper loader can later become part of the system. Some loaders can also mix in regrind and add color concentrate or other additives. (See Figs. 7-1, 7-2, 7-3.) Depending on the supplier and the type of unit, the average cost of a hopper loader is from $1000 to $10,000. (All costs in this chapter are in 1989 dollars.)

Dryers

Some plastic materials are hygroscopic, others take on surface moisture. With such materials drying is usually required. Drying equipment can be as simple as an oven in which the plastic sits in flat trays for a period of time to drive off the moisture. This is a very labor-intensive approach, and most molders use a dehumidifing drying system instead. Such a system

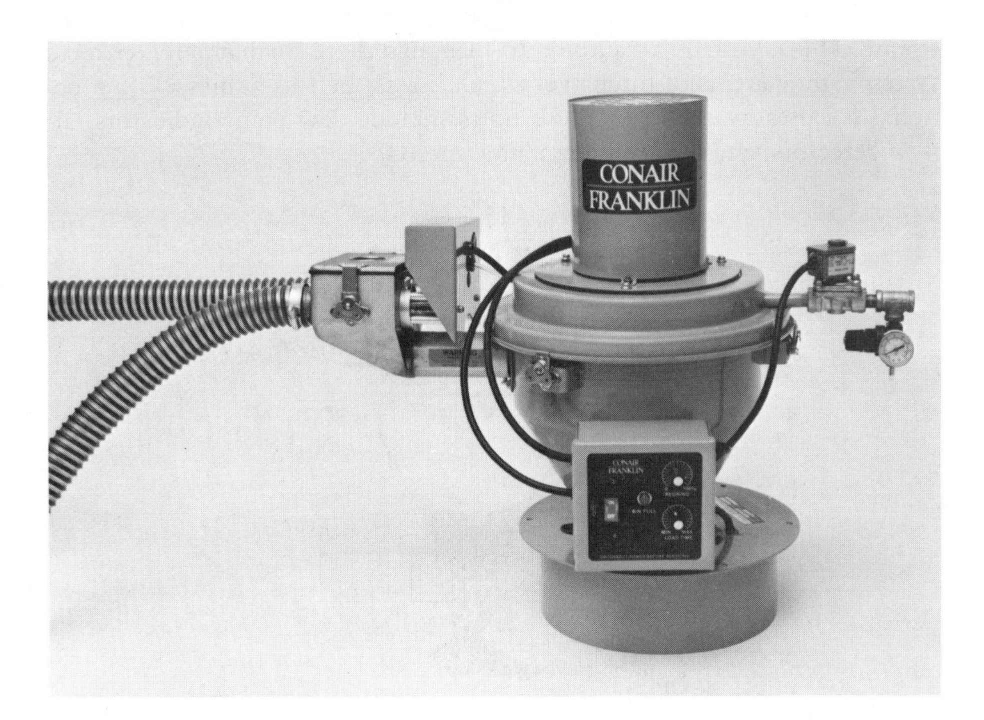

Figure 7-1 Ratio hopper loader. (Courtesy of Conair Franklin, Franklin, PA.)

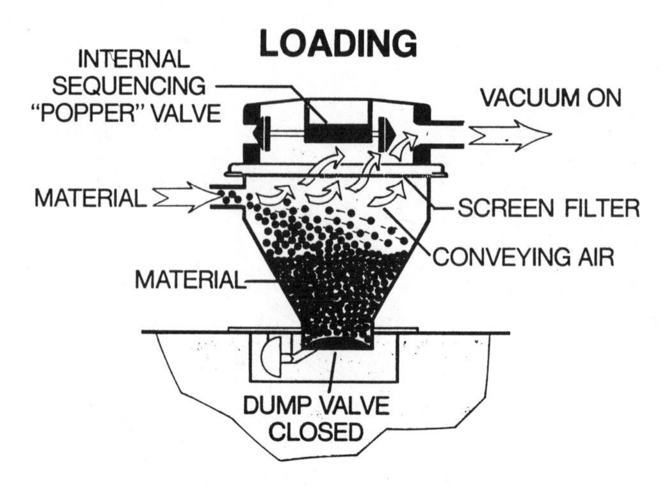

Figure 7-2 Cutaway diagram of a typical hopper loader during its loading sequence. (Courtesy of Conair Franklin, Franklin, PA.)

consists of a large hopper with a diffuser in the bottom and a second unit that contains two or more beds of a molecular sieve material. One bed is on line with the hopper and a self-contained blower circulates dry air between the molecular sieve and the material in the hopper. While that is occurring, a second bed is off line, being dried out with its own internal heaters. There is then a cool-down period before the dry bed comes on stream. This is usually all automatic; although there are manually operated systems, they are labor intensive, which limits their usefulness. Other optional features available on these units include dew point indicators, air flow detectors, and electronic machine interfaces.

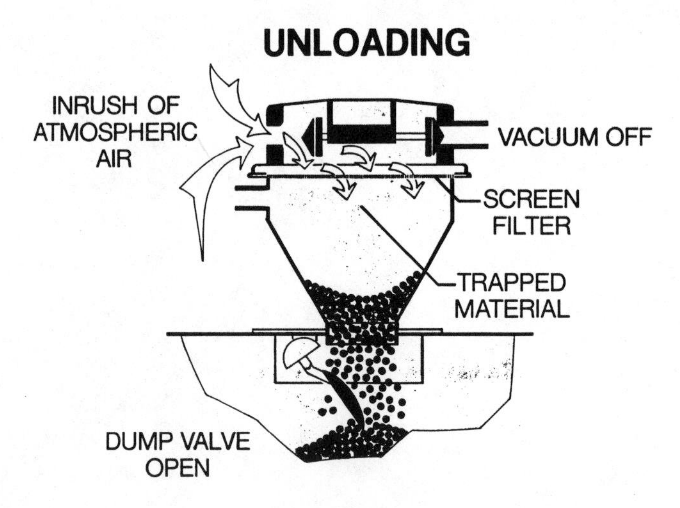

Figure 7-3 Cutaway diagram of a typical hopper loader during its unloading sequence. (Courtesy of Conair Franklin, Franklin, PA.)

Dryers are sized according to pounds per hour of throughput of a particular resin. As an example, to dry polycarbonate at 200 pounds per hour would require approximately a 600-pound hopper and three hours initial drying time. Thereafter, it would be on line and operate continuously. Dryers range in price from $6000 to $20,000, depending on capacity.

A hopper loader can be mounted on top of a dryer to futher automate the process. If drying is planned in the future, then when the hopper loader is originally purchased it should be a bottom-sealed type to work with a dryer.

Silos

Plastic resin purchasing has several cost break points. The first is bags or gaylords. Bags contain fifty pounds of resin, gaylords 1000 pounds. The next break point is bulk truckloads of 40,000 pounds, and then railcar quantities of 100,000-plus pounds. The difference in the three break points is up to five or six cents a pound. Unless at least one million pounds per year of a single type of resin is purchased, railcar loads are not practical. Truckloads are very realistic if for nothing else than space savings.

Silos to receive railcar and truckload deliveries of resin are approximately $15,000 for 100,000-pound, outside stand-alone sizes. A truck-fill silo has a tube up the side; the truck has its own blower to deliver the resin. A railcar silo must have its own vacuum receiver and vacuum pump, which makes it more expensive. Silo foundations are fairly extensive, but are dependent on soil compaction tests at the plant site. Silos can also be the bolt-together farm type; these are less expensive and can be put inside the building. This of course wastes manufacturing floor space, but is a practical start-up solution.

Systems

After the plant has expanded to ten or fifteen machines, a complete materials handling system is in order. This is expensive but in the long run will provide a good return on investment. If used properly, systems are the most efficient and will allow the least material to be lost. The material price can then be negotiated in annual contracts at the most attractive level.

The system required depends entirely on the layout of the plant, and the cost cannot be estimated without a thorough on-site investigation. The silos and the machines will be interconnected by conveying lines with basically automatic operation.

WATER SYSTEMS

A water system can have even more impact on the profitablity of a plant than the materials handling system if it is improperly implemented. The simplest form of water system simply connects the machine's heat exchangers and the molds directly to city water. This could be very expensive, depending on the location of the plant. Other possible drawbacks include:

- The normal city water temperature may be too high (anything above 55°F [13°C]
- It may have too high a mineral content, especially of lime, that can continually foul the heat exchangers
- There may be an expensive sewer tax for returning the clean warm water

In many areas of the continental United States city water is just not suitable for use because of temperature, hardness, cost, and/or availablity.

Water systems are most easily understood if one thinks of closed loops of circulating water running at a particular temperature. The machines must run at 110–120°F (43–49°C). The molds will run anywhere from 30° to 90°F (1–32°C), depending on the material and part configuration, but generally between 40 and 50°F (4–10°C). The ability to maintain a particular set point of temperature in each loop is important. Since a plant will run various materials with various temperature requirements, adjustablity becomes equally important.

Portable chillers are self-contained refrigeration units with recirculating pumps. They can provide maximum flexiblity of use in maintaining the required temperature settings. Assigning one portable chiller to each machine makes it possible to run any job in any machine, but using a chiller to cool the heat exchangers of the machines does not make the most efficient use of the chilled water. Generally, chillers are used for the mold loop (cooler temperature range) and cooling towers used for the machinery loop (higher temperature range).

The cooling tower is a fairly inexpensive unit that uses the change in temperature of evaporation to cool the water, typically from a 95°F (35°C) entrance temperature to an 85°F (29.5°C) exit temperature. (Fig. 7-4) This water is then pumped to the various heat exchangers that need cooling. The tower works most efficiently in areas where the average wet bulb ambient temperature is no greater than 76°F (24.5°C).

Chillers are available as portable or fixed, air-cooled or water-cooled units (Fig. 7-5). An air-cooled unit should be connected to a duct to the outside of the building during the summer months. During the winter it should be connected to an inside duct to help to heat the plant. Water-cooled units require tower water or city water to cool their own heat exchangers. If exit temperatures below 40°F (4.5°C) are required, ethylene glycol will

Applied Tower Cooling Systems

Capacities 20 thru 265 tons

Dependable, Compact
Broad Range of Capacities

Figure 7-4 50-ton cooling tower and associated equipment. (Courtesy of Applied Process Equipment, Wood Dale, IL.)

have to be added to the water lines. Portable chillers come in various sizes from 3- to 30-ton capacity.

In large plants central water chillers can be used. These are usually 20- to 50-ton chillers located in one or two fixed positions to provide the cooling water for all of the machinery and molds. A tower is usually used to cool

Figure 7-5 A 10-ton advantage portable chiller, air-cooled with solid state load matching controls. (Photo courtesy of Advantage Engineering, Greenwood, IN.)

the heat exchanger of a central chiller. If a central chiller is used to cool the machines, a water temperature control valve must be employed to maintain the oil in the machinery at the required 110–120°F (43–49°C).

The cost of portable chillers is approximately $7,000 for a 5-ton unit to $14,000 for a 20-ton unit. Towers are about $75–100 per ton of cooling. Thus a typical tower of fifty tons would cost $5,000. Central chilling units are in the same $1,000 per ton price range as portables, but with cost savings because of their size.

The choice of a system will depend on the nature and cost of the local water supply. The system should be sized for the hottest day of the year. Two rules of thumb for sizing the system are reasonably useful:

1. For every 30–50 pounds of material cooled, one ton of chilling is necessary

2. For every horsepower of the machines, 0.11 ton of chilling is necessary

The important factors in evaluating water systems are all related to the temperature difference or change, Δt, between measurements at two relevant points in a medium. Significant quantities of heat—measured in BTUs—can be transferred using water; how fast and how efficiently is one of the main determinants of profitability. A Δt of 3°F (1.6°C) across a mold is excellent; a Δt of 10°F (5.5°C) is terrible. The rate and character of the chilling water flow affects the Δt: turbulent flow picks up heat better than laminar flow. Since turbulence generally increases with flow rate, the flow of chilling water in gallons per minute (gpm) is as important as its temperature. Chillers and towers all have their own pumps, which are almost always too small to produce sufficient turbulence for optimum cooling. Larger pumps mean more turbulent flow and a better Δt. The calculation of required pump size is very involved, and involves knowing the mold water channel diameter, the number of turns in the channel, and the desired Δt.

Basic cooling need calculations are more straightforward. As an example, suppose that blow molding a part uses 80 pounds of HDPE per hour and the water is to be at 50°F (10°C). Using the rule of thumb that 30 pounds of material requires one ton of chilling, 2.6 tons of chilling are required to cool the mold. If the mold temperature is to be maintained with a Δt no greater than 2.5°F (1.4°C), the calculation is as follows [formula supplied by Pat Ozo, president, Cooling Technology, Inc.]:

$$\text{One ton of cooling} = 12{,}000 \text{ BTU/hr [basic equivalence]}$$

$$12{,}000 \text{ BTU/hr} = \text{Flow, gpm} \times \Delta t \times 500 \text{ [constant]}$$

$$\frac{12{,}000 \text{ BTU/hr}}{\Delta t \times 500} = \text{Flow for one ton of cooling}$$

$$\frac{12{,}000}{2.5 \times 500} = 9.6 \text{ gpm/ton of cooling}$$

$$9.6 \times 2.6 \text{ [cooling tons required} = 24.96 = 25 \text{ gpm for } \Delta t \text{ of 2.5°F for mold throughput]}$$

The only remaining question is at what pressure (psi) the water would have to be. That will depend on the line diameter and number of curves in the line. In the example the psi would have to be sufficient to achieve a flow of 25 gpm through the mold cooling channels. The normal range in this kind of system is 60–120 psi.

If turbulent flow is more efficient than laminar flow, would not very cold water also be better? Not always. If the molds are run very cold on a humid day the hold halves will sweat, and sweating will mark the surface of the finished parts. So, maintaining the required Δt by controlling the flow rate of cool or moderate-temperature water is the real answer.

Another important water system consideration is treatment of the water. The pH may need to be balanced initially or on a continuous basis, and in many areas of the country a softening treatment, especially of the make-up water for the power system, will be required. This is not only a matter of maintaining operating efficiency, but also one of protecting an investment by trying to preserve the life expectancy of the equipment.

GRINDER AND REGRIND USE

A grinder (or granulator) reduces scrap from molded parts to a usable form. This scrap is then fed back into the molding machine on a percentage basis. The hopper loader can be programmed in a ratio configuration to add the regrind at a set rate.

Grinders come in many types, sizes, and cutting blade variants. The two types most commonly used in blow molding are the upright feed and the auger feed. An auger feed grinder sits under the machine and has a side feed. An upright feed grinder is free-standing and must be fed either with a conveyor or manually. Both types are produced by several manufacturers, with a great variety of blade arrangements. Each has a sizing screen and a method of changing the screen to obtain a particular granulate size; 5/8 in. (0.625 in., 0.159 mm) is the most common.

The basic grinder selection criteria are type, size, and horsepower. The choice of blade configuration is largely a matter of preference. The feed throat of the grinder should be slightly larger than the product being ground. The horsepower will determine the throughput in pounds per hour. The purpose of a production facility is to turn out product, not scrap, so the scrap available for throughput will consist primarily of the tails and moils that are a normal part of blow molding. Grinding of product overrun and rejects will occur, but infrequently enough (one hopes) to not be a factor in grinder selection. A 7 HP grinder with a 10 in. × 12 in. feed throat is a good overall choice.

Regrind material is money that should find its way back into the product, not into the trash. An important consideration is keeping the grinder blades sharp and properly aligned. This will save energy, and will insure maximum usability of the regrinds. A second concern is that the regrind not affect production cycles or product quality. This is best accomplished by making sure that it is reintroduced directly into the machine it came out of as quickly as possible. If a ratio of more than a 10–30 percent regrind to virgin (unprocessed) material is used, it may affect the balance of the molding machine. Regrind tends to run stiffer than virgin material and if different ratios are added at different times the machine will run inconsistent cycles, resulting in more scrap.

Noise is an unwanted attribute of a grinder. All grinders should be bought with as much soundproofing as possible. The cost of a 10 in. × 12 in. 7 HP grinder with soundproofing is approximately $6,000.

POST-MOLDING PART HANDLING

This category includes conveyors, pack-off tables, leak detectors, and flame treaters or electric discharge tunnels. Conveyors deliver finished products to pack-off tables, where they can efficiently be packed for shipment without interfering with the production operation. Leak detectors are a great labor saving method of insuring part quality. They may have their own conveyors or can be a part of the machine. Flame treating is necessary if silk screening or decorating is to be done in-plant. The flame treater is a natural gas jet directed at the container or part for a second or two to burn off the ozone that collects on the part surface. Many producers are becoming increasingly aware of the cost of the gas as well as the potential danger of an open flame and are switching to electric discharge tunnels. These are initially more expensive, but over time are less expensive to operate and safer.

Conveyors are not very expensive, about $1,000 for twenty feet. Leak detectors are $8,000 for a unit that will check a container every two seconds. Flame treaters are $500, and electric discharge tunnels $20,000. A tunnel will treat many bottles in a random group and is very cost competitive with a flamer over a one-year run.

CHAPTER

8

Reclaiming Blow Molding Plastic Scrap

VIRGINIA A. SHARKEY

Blow-molded reclaim consists of:

- Start-up scrap
- Flash trim (necks and tails), which makes up a high proportion of the overall input
- Parisons (hollow purgings or bladders) that continue to be extruded during mold changes or molding cycle problems, and which can become globs of appreciable size
- Reject parts, which range from small cosmetic type containers and PET bottles to 55-gallon drums and other very large blow-molded pieces, such as furniture sections

Granulators to handle this diversity of scrap are classified into four types:

1. Beside the press
2. Auger
3. Central
4. Shredder/granulators

It is essential that they be of a size to handle the bulkiness of the reject products, which constitute up to 10–15 percent of the production rate in blow molding, as well as the necks, moils, and tails that are generated on a continuous basis and granulated immediately for a closed-loop system.

GRANULATION SYSTEMS

Beside the Press

Granulator systems that are placed beside the press are most widely used for blow molding scrap granulation. They range in size from 8 in. ×

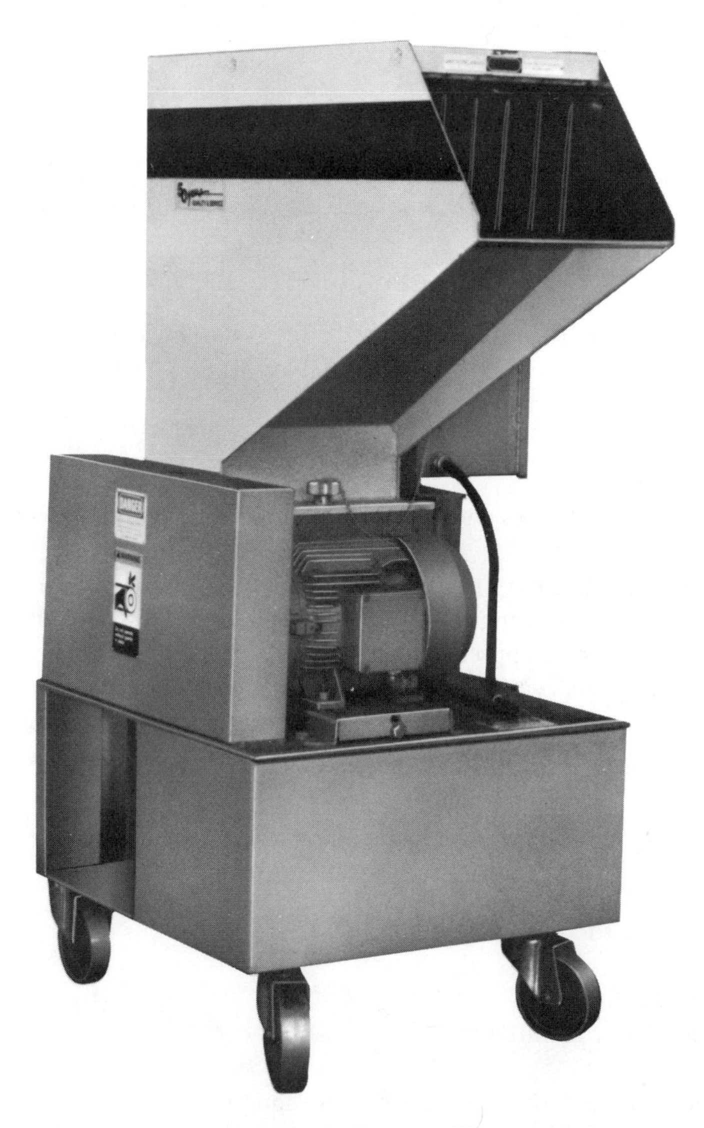

Figure 8-1 8 in. × 12 in. infeed beside-the-press blow-molded parts granulator.

Figure 8-2 18 in. × 36 in. infeed beside-the-press blow-molded parts granulator.

12 in. infeeds with 5 HP, up to 18 in. × 36 in. with 60 HP (Figs. 8-1, 8-2). Certain granulator essentials provide greater efficiency:

1. Either a scoop rotor or tangential feed will permit the granulator to ingest large parts without having them bounce or bridge on top of the rotor (Fig. 8-3).
2. A large cutting circle is needed to allow more open area between the blades to ingest rejects, and open rotors to allow air-flow cooling of the cutting chamber. This prevents plasticizing during heavy granulation of hot/warm materials (Fig. 8-4).
3. A large screen area that follows the circumference of a generous cutting circle will not only deliver higher throughput, it will also reduce the fines and the smearing that is generated when processing polyolefin materials.
4. Slow rotor speeds will reduce the bouncing effect of blow-molded parts for faster penetration into the cutting circle. A slow rotor speed also reduces heat and noise generation.
5. The granulator must have sufficient horsepower to handle parisons, energy-absorbent materials, collapsed parisons, purging, and fused parts.

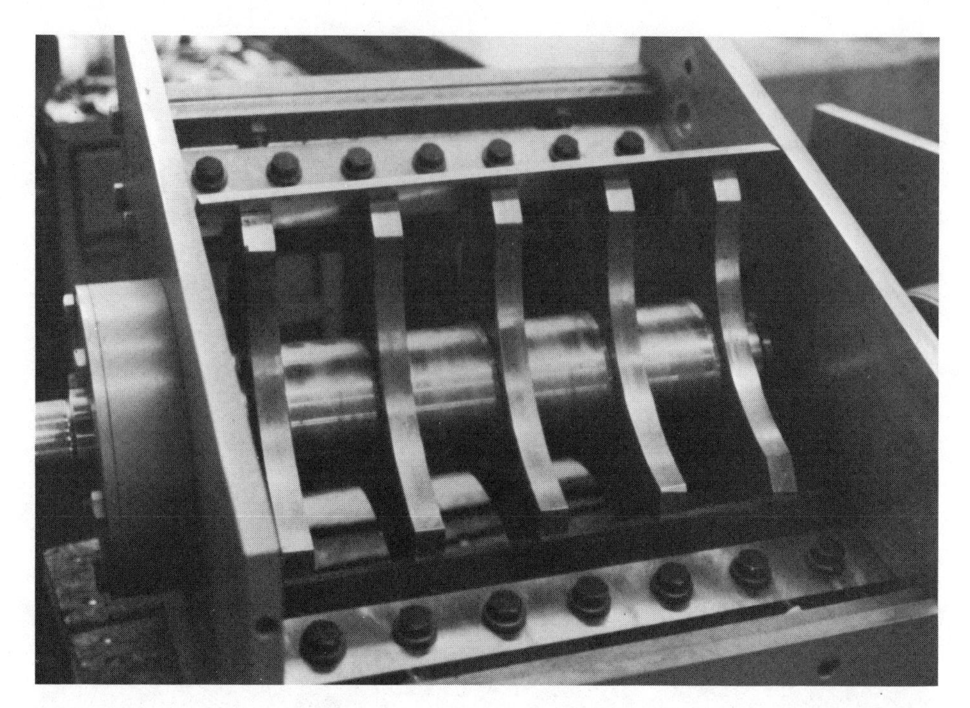

Figure 8-3 Scoop rotor design accommodates a variety of blow-molded shapes for ease of ingestion.

6. A two-knife rotor for small parts, a three-knife rotor for larger sizes, and a scissor cutting action will provide a clean cut with less fines.
7. Easy-access features are desirable to allow for quick color and polymer changes and services (Fig. 8-5).

Beside-the-press granulators are especially effective in a closed-loop system where the scrap is fed by hand or automatically fed directly into the granulator, the granulate is air-conveyed to a mixing station where it is proportioned and mixed with virgin material, and then is fed back into the blow molding hopper for reprocessing. Where applications vary and changes are made regularly, the closed-loop system is preferred because it eliminates product contamination and material handling problems. It is essential for maintaining high-quality regrind when processing barrier layer products and materials with many different colors and melt indexes.

Beside-the-press granulators should be rugged enough to support the horsepower needed to handle bulky parts. And, if the application is high-impact materials with heavy cross sections, dual flywheels and heavy-duty knife carriers should be included. Dual flywheels add inertia and deliver the equivalent of additional horsepower to cut through heavy cross sections. Heavy-duty knife carriers will support the entire knife length so that the knives can withstand high-impact stress that causes knife damage and downtime.

Figure 8-4 Open rotor design allows air flow through chamber for a higher quality granule.

Hopper configuration is important to allow material to pass rapidly through the hopper to reduce flyback and noise. Adding a muffled air flow vent assists the material flow. In a closed-loop system, an oversized blower on the discharge lends itself ideally to handling warm/hot parisons directly from the blow molder. For example, polystyrene, 180°F PVC, and similar materials are not a problem as long as there is sufficient flow of air through

Figure 8-5 Segmented screens and cradles allow easy removal and completely clear access for service.

the granulator system, the knives are sharp, and recommended rotor bed knife clearances are held. A top-vented hopper allows ambient air into the granulation chamber adding additional cooling. Most beside-the-press granulators have integral sound reduction for noise restraint when they are placed directly beside the molding machine. Accessibility for on-line servicing and cleanout are important considerations.

Auger Granulator Systems

Systems of this type are in use for blow molding applications, but they are limited to smaller-size, light-wall parts (three-pint containers and less). The necks and tails can drop directly into the auger trough and are then conveyed by the auger to the cutting chamber. A secondary hopper can be provided for manual feed of rejected bottles.

Central Granulator Systems

These are used when the part size exceeds the 18 in. × 36 in. beside-the-press maximum infeed capability. Infeed sizes range up to 32 in. × 50 in., with up to 150 HP. They are more widely used by captive molding houses for applications such as 55-gallon drums, fuel tanks, and other very large molded pieces such as automobile sections. Central granulators also reclaim parisons that have been allowed to accumulate into very large lumps. Because of the size of the parts that are handled in central granulator systems, storage and contamination can be a problem.

Shredder/Granulators

A combined shredder and granulator is an effective means to handle very large, bulky parts, for example, 60-gallon drums, auto and truck body parts, enormous parisons, baled postconsumer bottles, and the like. Ram feeding assists ingestion of the large parts into a slow-speed, high-torque shredder (Fig. 8-6a) for preprocessing into 1–2-in. wide pieces of various lengths. The pieces are then automatically discharged directly into a granulator below for secondary, high-speed size reduction (Fig. 8-6b). The shredder/granulator system reduces the size of the granulator needed. It allows the processor to use less total horsepower and it increases equipment uptime because knife servicing is greatly reduced. An integral shredder/granulator package saves space and prevents material contamination. But, conveyor systems equipped with metal detectors are also in wide use for feeding material from the shredder to the granulator. The capital equipment investment is not as great for a shredder/granulator as it would be for less efficient systems (e.g., precutting or dual granulators).

GRANULATOR INFEED AND EVACUATION SYSTEMS

The infeed system is one of the most critical parts of the granulation system. It must be designed to correctly feed material with controlled flyback and noise suppression. Automated feed can be accomplished by pneumatic con-

Figure 8-6a Slow-speed rotary shredder blades preprocess large bulky parts into narrow strips.

Figure 8-6b Strips from the shredder are discharged into a granulator below for high-speed size reduction.

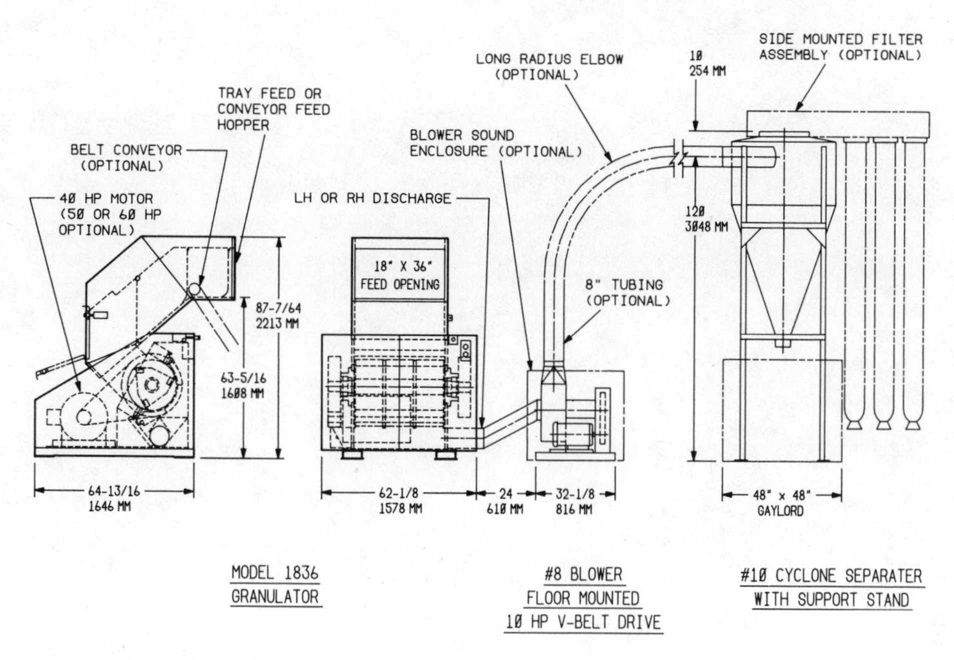

Figure 8-7 Blow-molded products granulator with floor-mounted blower, cyclone separator and stand, and filter assembly.

veying or belt conveying. Both types of systems can feed from multiple stations. Pneumatic systems are effective for transporting bottles and/or trim scrap within a limited size range. Belt conveying offers a greater part size range and also the ability to be fitted with a metal (ferrous and nonferrous) detection system to protect the granulator from costly damage due to tramp metal.

Material evacuation systems (Fig. 8-7) consist of vacuum conveying, blower-cyclone systems for continuous material evacuation. Sound enclosures, mufflers, and filter assemblies provide quiet, efficient evacuation. Auger conveyors can also be utilized for continuous evacuation. Flow-through metal detection/separation systems can offer further protection after granulation is completed.

CHAPTER

9

Mold Temperature Control

ROSS H. DEAN

Greater blow molding efficiencies may be attained with an improved understanding of heat transfer mechanics. When effective heat transfer engineering practices are applied, improved molding economics can result. Effective part cooling can be achieved by observing heat transfer basics as they relate to:

1. Type of plastic being molded
2. Wall thickness of the molded part
3. Type of material used to construct the mold
4. Mold wall thickness between the cooling channels and plastic being molded
5. Temperature of the mold cooling water
6. Volume of mold cooling water (lb/hr)
7. Design of the mold cooling water zone (length and shape for turbulent flow)
8. Blow air pressure
9. Type of intramold cooling used
10. Nonmodular mold construction methods (modular assembly creates thermal interfaces between the mold cooling water and the plastic being molded)
11. Molds having thermal covering on the outside

12. Ideal surface roughness of the mold cavity with adequate air vent reliefs to relieve trapped air
13. Accurate mold temperature control capable of regulating the mold cooling temperature to attain best possible molding conditions

The production rate for almost all thermoplastic parts produced by injection molding, extrusion, blow molding, vacuum or thermoforming, or other processes depends to a great extent upon the efficiency with which the plastic material is heated and cooled. This applies especially to blow molding because the time required to cool the molded part primarily determines the rate of production and controls to a great extent the quality of the part being produced. The shape, thickness uniformity, and design configuration are all vital factors in the efficient production of components in economical cycles. Nonuniformity results in uneven cooling, which produces warpage and size variations. Cooling is the most significant factor in this cycle.

Blow molding of polyolefins is the major means of manufacturing blown products. Recently, the development of engineering resins has caused changes in some of the techniques in order to utilize the resins much more satisfactorily and to produce a fine surface finish that can be painted or maintained in the blown condition for subsequent operations. Currently, high-density polyolefins represent approximately 70 percent of the blow molding resins, with PET the second largest at 20 percent.

High blow air pressure (greater than a minimum of 60 psi) is necessary to achieve good surface smoothness and mold reproduction. Low blow air pressure increases plastic shrinkage and increases cycle times because cooling is less effective. With some resins, a pressure of 100 psi has been found to be very satisfactory.

MOLD CONSTRUCTION

The most common material used for blow molds is aluminum. Aluminum has good conductivity, is lightweight, and has a low mold cost. In considering the thermalconductivity, as measured in calories per square centimeter per centimeter per degree Centigrade per second ($cal/cm^2 \cdot cm \cdot C \cdot s$), aluminum has a thermoconductivity of 0.37, beryllium copper 0.21 to 0.61, Kirksite 0.25, and steel 0.12 to 0.14. Aluminum is soft, however, and to protect against wear to specific points in the molds, steel or beryllium copper inserts are used in production applications. The introduction of dissimilar materials, no matter how closely they are machined and mated, affects the heat transfer characteristic of the molds. Kirksite is less frequently used for inserts because of its weight and mass.

The porosity of cast aluminum and the possibility of leakage makes it desirable to machine molds and channels in the mold for carrying heat transfer fluids where practical. Fortunately, the quality of large aluminum castings has greatly improved. Cast-in tubing is not recommended for production tooling because of the reduction of heat transfer between the mold and the tubing. However, because of its accessibility to the back side of the mold, cast-in tubing is an excellent method of cooling in prototype molds. Changes can be made easily without having to remove gasket-sealed back plates. For small or experimental projects, the back of the cast mold can be cooled by spraying, providing recognition is given to the porosity factor. Bubblers, or fountains, in the machined channels can also be used in areas not accessible to complete channels. Fillers to reduce porosity can affect the heat transfer rates of the mold.

For all molds, but especially large molds, channels large in diameter (e.g., 1 in.) and as close to the surface of the molded part as possible are highly desirable to insure proper circulation through the work area to heat and to cool the mold as rapidly as possible. To assist in the movement of entrapped air to the mold vents and to improve the heat transfer rates, a roughened cavity surface is necessary when blow molding polyethylene and polypropylene. The degree of roughness depends upon the material and the size of the part.

Water, with its excellent heat transfer characteristics, is used primarily as the heat transfer medium for polyolefin blow molding, but for the newer engineering resins a synthetic heat transfer fluid that operates up to temperatures of 250° to 300°F (121–149°C) or more at low pressure is highly desirable because of the safety factor. Because the heat transfer characteristics of the synthetic fluid are not as effective as water, care must be exercised to use proper fluids in a safe manner to obtain as efficient production cycles as possible. Compromises sometimes are necessary, depending upon the required temperatures.

COOLING CALCULATIONS

The heat extraction load or the amount of heat to be removed from the product must be determined. This is important, as the amount of heat taken out by the blow mold must be known if the process is to be economically predictable. The amount of heat to be removed, Q, is determined by the material's temperature and the amount of plastic being delivered to the mold. It is calculated as follows:

$$Q = C\, m\, \Delta t\ (0.003968)$$

where: Q = total change desired during molding, BTU

C = specific heat of the plastic material being processed, cal/g °C

m = amount of plastic per hour to be cooled expressed, g

$\Delta t = (T_i - T_f)$ = initial plastic (parison) temperature into the mold minus final (demolding) temperature of the plastic, °C

0.003968 BTU = 1 calorie (constant)

As an example for determining the heat load for a typical mold, the following data are used:

- C, specific heat for polyethylene = 0.55 cal/g °C
- m, amount of PE to be cooled = 32 shots/hr × 18.75 lb/shot × 0.80 shot reduction length = 480 lb/hr
- 1 lb = 453.59 [conversion factor]
- T_i, parison temperature = 420°F (215.6°C)
- T_f, demolding temperature = 100°F (37.8°C)
- Δt, material temperature change = $T_i - T_f$ = 215.6 − 37.8 = 177.8°C
- 1 calorie = 0.003968 BTU [constant]

Calculate heat extraction load:

$$Q = C\,m\,\Delta t\,(0.003968)$$
$$= (0.55)\,(480 \times 453.59)\,(177.8)\,(0.003968)$$
$$= 84{,}483 \text{ BTU/hr per mold, avg.}$$

Assume 75 percent efficiency for heat transfer between chilled water and polyethylene. Then:

$$\text{Cooling required} = \frac{84{,}483}{0.75} = 112{,}644 \text{ BTU/hr}$$

With polyolefins it frequently is desirable to run the molds as cold as possible, 40° to 60°F (4.5–15.5°C) or lower. Condensation or moisture on the mold can cause outside surface defects when mold cooling temperatures are below dew points. To reduce or eliminate these, either the mold heat transfer fluid temperature can be increased, or it is possible with recently developed techniques to dehumidify the immediate blowing area to eliminate the condensation and maintain good surface appearance at a fast cycle. Effective dehumidification systems can be installed on existing equipment very satisfactorily, permitting ready access to the blow area while providing the dehumidification necessary to prevent condensation. Savings of 20 to 30 percent have been reported through the use of this system.

Utilization of existing water temperatures in plants can be supplemented to improve cooling conditions through the use of turbulent flow of fluid through the mold channels. Depending upon the channel sizes, the greater the volume and the higher the pressure of the fluid put through the channels, the greater the heat transfer. A turbulent flow wipes the side walls of the channels, permitting better heat transfer than that obtained with laminar flow of the fluid through the channels. This laminar flow characteristic tends to have reduced heat transfer at the channel circumference, whereas a turbulent flow enables more heat to be removed quicker with higher fluid temperatures, which also reduces the possibility of condensation.

A large-capacity temperature control unit (e.g., with 2 to 7½ HP pumping capacity, depending on the mold and channel sizes) can not only provide the desired turbulent flow, but can also assist in maintaining uniformity of control throughout the mold on an automatic basis. Production reports indicate that temperature variations of no more than 1° to 2°F (0.6–1.1°C) are readily obtained with this approach, even with intricate molds.

To determine the proper flow for each mold, the Reynolds number should be determined. The Reynolds number is a nondimensional parameter used to determine the nature of flow along surfaces. Numbers below 2100 represent laminar flow, numbers from 2100 to 3000 are transitional flow, and numbers above 3000 represent turbulent flow.

To determine the Reynolds number, N_{Re}, the following calculations illustrate values derived for flows of 45 and 140 gallons per minute (gpm) and are based upon a large-capacity temperature control unit controlling five zones per mold half of a blow mold:

$$N_{Re} = (DV)\frac{P}{M}$$

where: D = pipe dia. = 1 in. = 0.08333 ft

V = fluid velocity = ft/sec

P = fluid density = 72.3 lb/cu ft

M = fluid viscosity = 0.01002 poise × 0.0672 lb/ft-sec/poise

= 0.000673 lb/ft sec

$$V = \frac{\text{Flow rate, gpm}}{\text{No. of zones}} \times \text{Volume} \times \text{Time, min} \times \begin{matrix}\text{Cross-sectional}\\ \text{area of pipe}\end{matrix}$$

Volume of water = 0.1337 cu ft/gal

Time: 1 sec = 1/60 min

Cross-sectional area = $\pi D^2/4$

$$V \text{ for 45 gpm unit} = \frac{45}{5}(0.1337)\frac{1}{60}\frac{\pi(0.08333)^2}{4}$$
$$= 3.68 \text{ ft/sec}$$

$$V \text{ for 140 gpm unit} = \frac{140}{5}(0.1337)\frac{1}{60}\frac{\pi(0.08333)^2}{4}$$
$$= 11.44 \text{ ft/sec}$$

Then, the Reynolds numbers are:

$$N_{Re} = DV\frac{P}{M}$$
$$= 0.08333\ (3.68)\frac{72.3}{0.00673} = 32{,}944 \text{ for 45 gpm unit}$$

and

$$= 0.08333\ (11.44)\frac{72.3}{0.00673} = 104{,}416 \text{ for 140 gpm unit}$$

TEMPERATURE EFFECTS

While it is difficult to generalize on many of these factors because of the wide variations in part designs, mold designs, and production requirements, the importance of stock temperature and mold temperature on cooling time, and the effect of these temperatures on impact and environmental stress crack resistance (ESCR) properties are extremely important.

To illustrate the typical effect of temperature, a 10 oz high-density polyethylene bottle was used to develop the data presented in the accompanying figures. Extruder output rates and design can greatly affect stock temperatures.

The importance of wall thickness in cycle time is illustrated in Fig. 9-1. The effect of stock and mold temperatures on cooling time is shown in Figs. 9-2 and 9-3, respectively. The effect of mold temperature on the bottle ESCR is shown in Fig. 9-4.

Each 20°F (11°C) change in stock temperature represents approximately one second of cooling time in bottles; therefore, the fastest cycle is highly desirable to operate at a stock temperature as low as possible consistent with molding a good quality part. The cooling time of parts is related to the wall thickness. Tests have shown that wall thickness increases of 50 percent can increase the required cooling time as much as 200 percent.

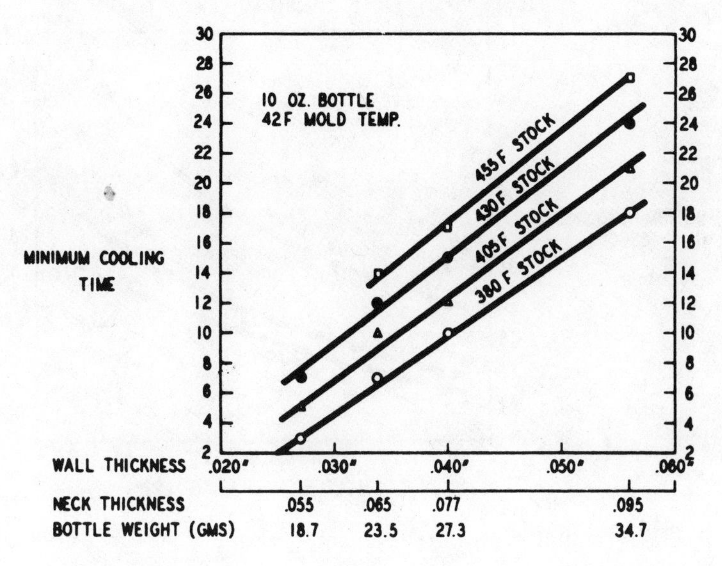

Figure 9-1 Effect of thickness on cooling time. Courtesy of Phillips 66 Company.

CONTROL METHODS

One of the common means of reducing cooling cycles in blown products is the use of internal cooling, within the blown product itself, using water, liquid CO_2, nitrogen, circulated air, supercooled air, or some other similar product or mix that can be integrated with the blowing cycle to provide cooling on the inside as well as on the outside. Water spray is of course

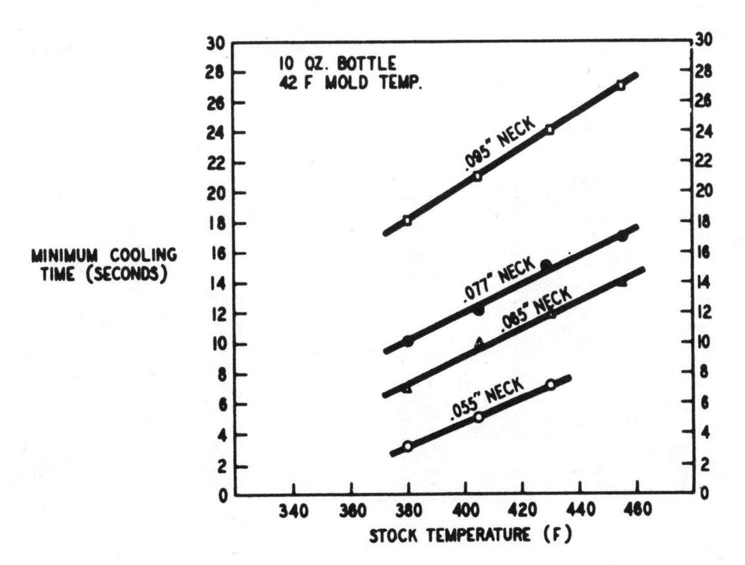

Figure 9-2 Effect of stock temperature on cooling time. Courtesy of Phillips 66 Company.

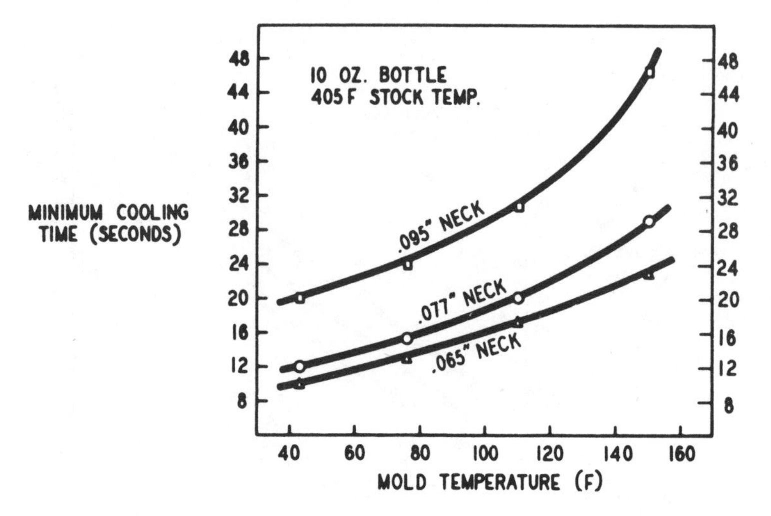

Figure 9-3 Effect of mold temperature on cooling time. Courtesy of Phillips 66 Company.

least costly, but CO_2 has received great acceptance because it is clean, chemically inert, and can effect more economical production of many types of blown parts. Care must be exercised designing the equipment to direct the blown cooling medium against the heavier sections of the product, since these will be the slowest to cool. Cooling fluid should be injected along the long axis of the part if at all possible, and in all cases, regardless of the fluid being used, it should be circulated as uniformly as possible through the

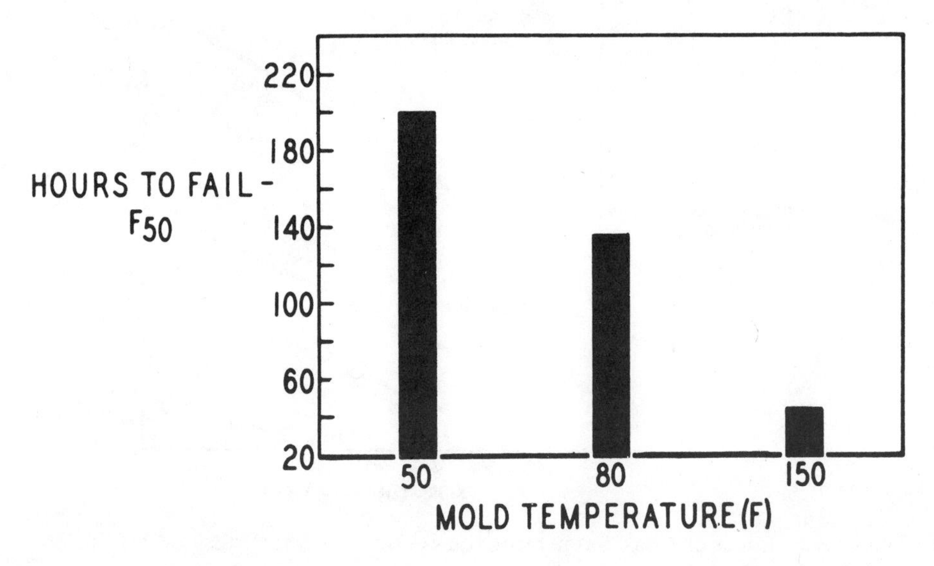

Figure 9-4 Effect of mold temperature on Bottle ESC 23.5 ± 0.5 grams bottle weight. Courtesy of Phillips 66 Company.

product. Double-wall parts are examples of part geometry that is difficult to cool internally because uniform distribution of the coolant is almost impossible.

Refrigerated air blown into the product is also advantageous under some conditions, depending on the shape and configuration of the part.

Internal cooling with water is based upon introducing an air/water mixture into the blown molded part. The water cools the inside part wall and is vaporized into steam. Incoming high-pressure air and water droplets then push the steam and hot air mixture out of the blown item through a preset pressure relief valve. This continuous process results in the circulation of fine water droplets in the blown product, maintaining a given pressure, and allows the water heat of vaporization to remove heat very rapidly. Metering of the amount of water and the timing of the introduction of the water/air mixture must be established for each part design. To prevent surface blemishes the inside skin of the blown part must be solidified prior to introduction of the water particles. The final cycle action is to blow air through the product to finish the cooling and remove the remaining moisture.

Systems to provide an air/water cooling sequence that is automatically timed with the planned production are available and can be used to reduce substantially the blowing and cooling cycle. Depending upon the design of the blown product, cooling cycle reductions of two-and-a-half to three times have been obtained. There are some instances when the failure to remove a drop or two of the water in the blown product is objectionable, but unless this is a special product, air/water cooling definitely can be advantageous.

Another technique frequently used to reduce cycle time is to remove the blown product from the mold and place it in a fixture for cooling with cold air or an air mix blown into the part. This does not restrict the machine time and still provides a properly cooled part without warpage. The additional handling cost frequently is more than offset by a reduction in scrap, increased production, and improved part quality.

Experimentation with cellular foam as a thermal covering on the exterior of the mold has not provided an improvement in heat transfer conditions or in production cycles.

In some mold designs the same channels are used for heating and cooling fluids, with a timed sequence developed for the flow of the various temperature fluids through the channels in keeping with the blow cycle. There are applications with engineering resins in which separate volumes of heated and cool fluids are maintained constantly and regulated valving insures a flow of the proper temperature fluid through the channels in synchronization with the molding cycle. The heated fluid initially provides the mold temperature required to achieve a good surface finish. The cool fluid then is introduced into the same channels to cool the mold and part as quickly as possible. Reports indicate that approximately three to five seconds are required for the change from hot to cool fluid, with the same time to reheat the mold.

With engineering resins it is essential that the mold temperature be maintained sufficiently high for the material to remain fluid, so as to produce the detail necessary for the product and to provide a fine finish for either subsequent painting or for use in as-blown condition. In order to produce parts with a class A finish, the cooling cycle is slightly more than one-half of the full cycle, while the mold heating is approximately one-fourth of the cycle.

For quality purposes primarily, in order to insure the desired finish on the final blown product, cycles for engineered resin blown parts are normally longer than those for polyolefins.

It is obvious that to maintain good temperature control of the work area, a system with good pumping capacity and precise sensitivity is essential. Maintenance of the desired temperature for heating, cooling, and holding of high temperatures for engineered resins can be accomplished automatically through the use of good temperature control equipment.

RESIN MOLD TEMPERATURES

Examples of thermoplastic resins now being successfully blown in a variety of products include high-density polyethylene such as Phillips Marlex HHM 6007, HHM 5502, HXM 50100, and others. These require a mold temperature of 30 to 60°F (1–15.5°C) for proper production.

Certain of the General Electric thermoplastic materials are also being blown successfully; the Noryl GTX family of resins are blown with mold temperatures ranging from 100 to 190°F (38–88°C), with a few running as high as 250°F (121°C). General Electric recommends Xenoy mold temperatures between 125 and 175°F (51.5–79.5°C), Lexan mold temperatures between 75 and 85°F (24–29.5°C), and GE Geloy mold temperatures between 80 and 140°F (26.6–60°C).

DuPont has developed class A exterior finish resins, such as Arlyon, Zytel, Hytrel, and Bexloy especially for automotive exteriors. These resins require a mold temperature ranging from 200 to 400°F (93–204°C), depending on the part design. The mold temperature for these materials must be high enough to permit flow at the surface of the mold, and yet low enough for part removal without deformation or stress. This is also essential for the reproduction of the part surface from the mold.

In summary, proper mold temperature control produces:

- Parts with greater dimensional stability and improved quality
- Most efficient cycle times with increased production
- Reduced scrap
- Increased predictability of blow molding process
- Reduced energy consumption
- Greater control over variables in the process

ACKNOWLEDGMENTS

Donald L. Peters
Larry Moser
Plastic Technical Center, Plastics Division
Phillips 66 Company
Bartlesville, Oklahoma

Andrew J. Stoll
Manufacturing Engineering
Plastics Products Division
Ford Motor Company
Milan, Michigan

Katrina J. Branting
Development Representative
Engineering Polymers
E. I. DuPont de Nemours & Co., Inc.
Wilmington, Delaware

CHAPTER

10

Postmolding Equipment

PAUL E. GEDDES

Several blow molding producers have discovered that it is as difficult to design and install postmolding equipment and get it into regular operation as to get a blow molding machine itself into production. Unfortunately, many blow molding machines have been purchased with little thought given to the postmolding operations. A systems approach should be taken to the problem of selecting a blow molding machine. Such an approach regards the entire plastics manufacturing process as an integrated line in which the major factors are the blow molding machine's capabilities, the blow mold design, and the postmolding operations. Consideration of postmolding operations should extend beyond the molded parts handling, trimming, or secondary operations to such phases as leak testing, product filling, labeling, packaging, palletizing, warehousing, and marketing.

In many instances the cost of postmolding equipment, as well as the required factory floor space, is far greater than the cost of the blow molder and its associated equipment. However, if a complete systems approach is used at the time of design, the best economic results will be obtained. The key to getting the best results is thorough planning that considers all aspects of the component operations and equipment and the possible alternatives. Careful attention is given to selecting a blow molding machine for its abilities to mold product, to correct sizing of support equipment, such as a water chiller, and to mold design. The same kind of attention should be given to product extraction or removal from the blow mold, its postmold cooling, handling, orientation, and subsequent procedures.

Good systems approach planning selects features not only for themselves, but for how they can contribute to a more efficient and more economical operation overall. For example:

- With containers having closures, if the molding machines can prefinish, no secondary trimming may be required, only deflashing. Some machines automatically deflash, either in the blow mold or at an intragel postcooling-deflash position within the machine.
- In industrial containers of 5–55 gallon capacity, neck threads of both the internal and external types can be formed by the machine and mold, thus eliminating the need for secondary operations for this purpose.
- Several machines that produce small containers (up to five-gallon capacity) commonly utilize a tail flash puller device, which eliminates the need to deflash this portion in a secondary step.

Similarly, work has been done with in-mold labeling, in-mold drilling and threading operation, in-mold internal treatment to enhance permability, and several other techniques, all in an effort to reduce postmolding operations. These various techniques are only a sampling of the many and varied ideas employed.

As machines are improved and in-machine operations increase, so does the complexity of the process. Therefore, not all schemes are successful from an economic viewpoint, which indicates the importance of collecting information from equipment and material users as well as the suppliers. Proper evaluation starts with the product: its design performance requirements, the quantity to be manufactured, and each manufacturing step from the raw material to shipping and marketing. Then equipment can be selected to meet all those requirements in the most economical system configuration.

This chapter discusses a variety of postmolding operations equipment and techniques. It should be noted that in some of the accompanying photographs safety guards or other protective features have been removed from the equipment to provide a clear view of the item or feature under discussion. This was done only for purposes of illustration. Equipment should only be installed and operated with all recommended safety features in place.

PARTS HANDLING

Parts handling in postmolding operation starts at the point when the blown part has been formed and is ready to be taken from the machine. The oldest method of part removal and handling is by hand. In many factories manual product removal is employed effectively and economically. However, as product quantity increases this technique usually loses any economic appeal, especially when ever-increasing labor costs, labor shortages, and safety regulations are taken into consideration. Nevertheless, manual part removal is a simple, effective method that is in wide use and should at least be considered for low-volume operation.

A stripper is a simple device for parts handling at the blow molder. It is typically activated by either a mechanical linkage or an air cylinder interconnected with the molding machine. Its operation is commonly a vertical downstroke that pushes the molded part from the die to strip it from the blow pin and separate the part with its flash from the molten parison at the extrusion die face. This approach is currently in wide use on such machines as the Uniloy 350, the Impco B-13 and B-30, and others (Fig. 10-1). Safety devices with a stripper will avoid mold damage as well as possible operator/maintenance personnel injury. Most stripper-equipped machines have safety interlocks and a fully guarded and enclosed mold clamp area. The stripper runs automatically, discharging the parts onto a conveyor or into a chute for further handling. This method of parts handling is inexpensive and has the advantage of allowing a fully automatic machine molding cycle. It is quite suitable for container type products, which release from the blow mold easily. Some more complex industrial shapes may require in-mold ejectors to assist part removal, or may even require manual part removal. However, in most instances automatic part removal from the blow mold and subsequent part ejection may be designed and utilized economically. (See Figs. 10-2, 10-3, 10-4.)

Figure 10-1 A horizontal stripper plate above blow molds moves vertically downward to eject parts. Courtesy of Walron Industries.

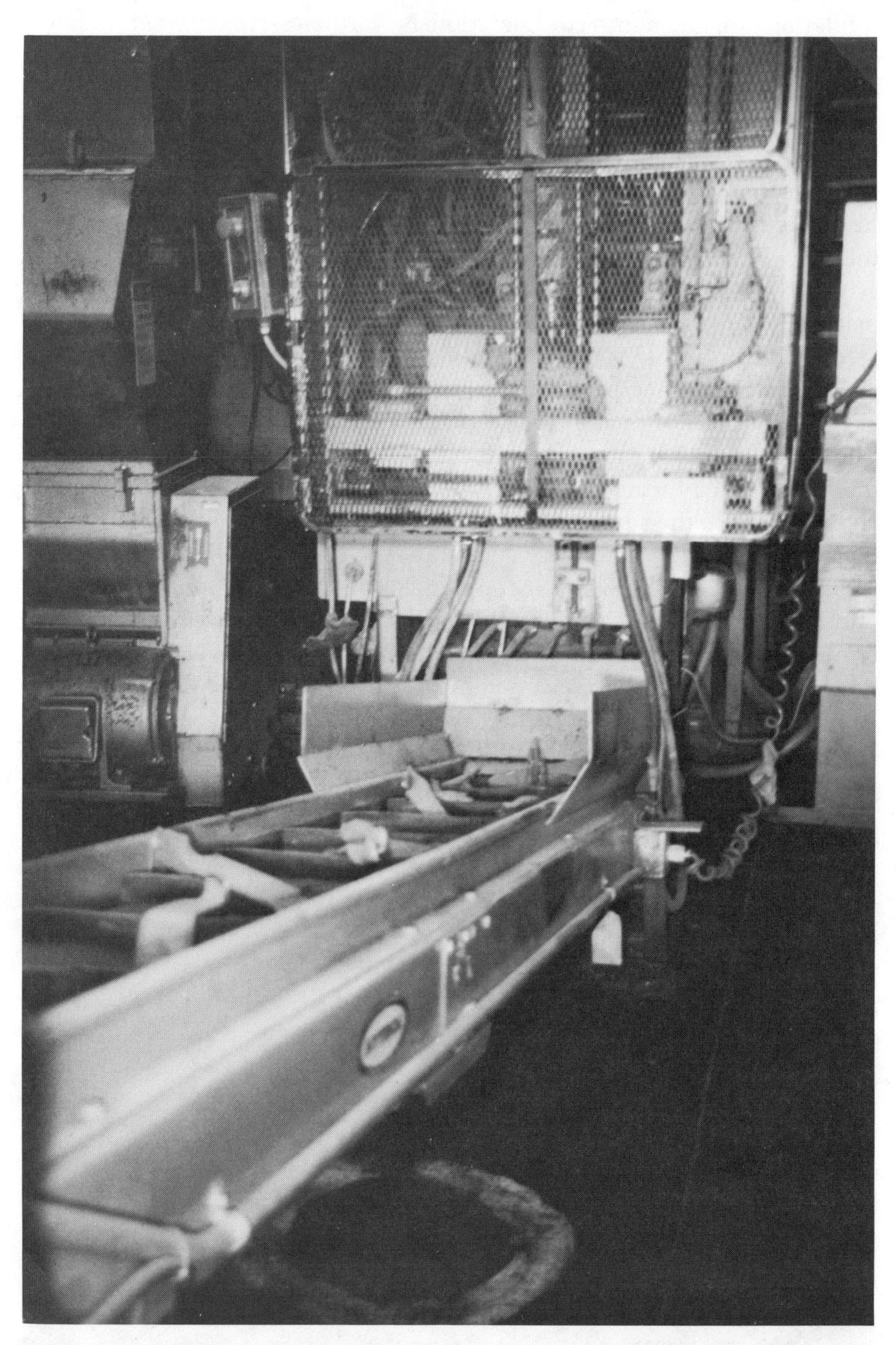

Figure 10-2 A cleated inclined belt conveyor receives parts ejected from the blow molds above and carries them to the next trim operation. The conveyor may run continuously or stop and start with each blow mold machine cycle for additional cooling time. Courtesy of Walron Industries.

Figure 10-3 Parts handling may include some packaging. This specialty film wrap machine provides a way to handle twelve highway safety barrel bases. Courtesy of Crocker Ltd.

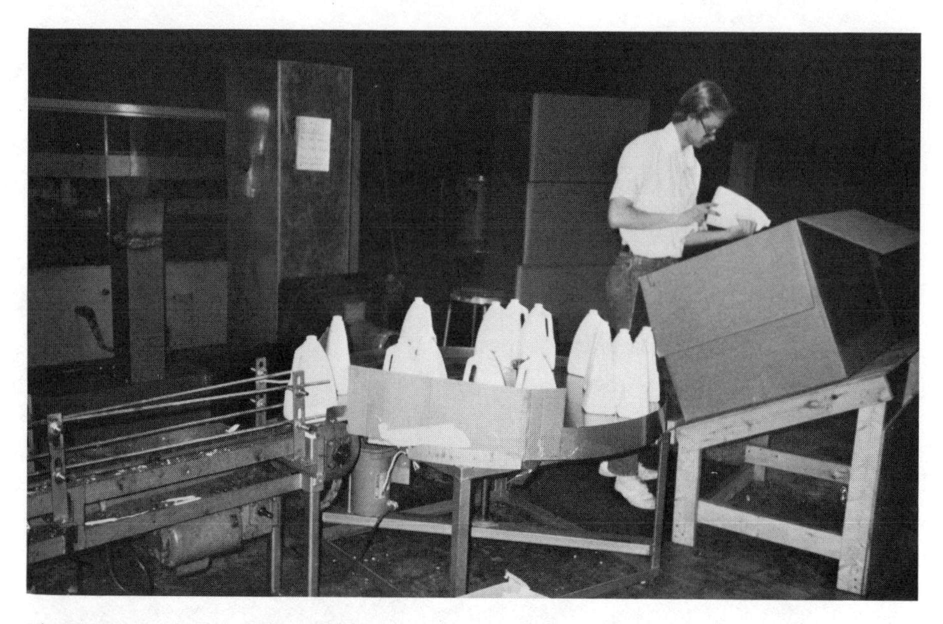

Figure 10-4 A conveyor supplies product from the blow molder to a manual packing station. The turntable revolves to bring product within easy reach to be put into a carton on a packing buck. Courtesy of Ameritainer, Inc.

Other part extraction devices and techniques include:

- Pinchers that seize the flash area to remove the part from the mold
- Moving the forming mandrels used for industrial part filler necks outside of the machine's platen area
- Mold ejectors that push the part from the mold onto a conveyor or into a chute
- Motorized chains that clamp the flash on two sides of the parison above the mold and bring the part out of the machine platen molding area

Of course, large industrial parts require considerably heavier and stronger mechanisms than do lightweight containers.

Most molded products are removed from the blow mold and machine platen area while either warm or semimolten inside; cooling is completed outside the mold. This decreases the machine molding cycle time and increases product yield. Any parts handling method must take into account the warm state of the molded product and allow sufficient cooling time prior to any dimensionally critical secondary operation. This may be accomplished simply by extending the cycle time, by use of a conveyor, cooling fans or tables, or other fixtures, by use of cold gases such as CO_2 or nitrogen, or by several other heat transfer techniques. (See Figs. 10-5, 10-6, 10-7.)

Figure 10-5 Cooling and parts handling combination fixture. Courtesy of American Seating Co.

Figure 10-6 Water-cooled nests of aluminum maintain product dimensions during cooling. Courtesy of Crocker Ltd.

Figure 10-7 Water-cooled fixture for holding 120 mm thread flat and round while cooling. Courtesy of Walron Industries.

Parts handling also may include automatic case packaging. This is typically employed for high-volume container operations. An automatic corrugated container machine erects a carton, glues one end, and feeds the carton into an automatic case packing machine where the product is loaded and the carton is sealed. Most machines are semicustom manufactured for a particular product application. Next, the packed cases are automatically palletized, sometimes shrink wrapped on the pallet, and automatically placed in warehouses. Subsequent removal from the warehouse for shipment may also be accomplished automatically. Mechanical material handling in these later stages is complex and requires considerable engineering; it is mentioned here only to suggest how far systems approach planning may need to extend.

FINISH MACHINING

Finish machining encompasses an extensive and many-faceted area. Here, finish machining of the most common blow molded products is considered. Containers constitute the largest percentage of such products, but the proportion of industrial items is growing rapidly.

Neck Sizing and Fly Cutting

Most containers blow molded on early machinery required neck finishing of their openings by a secondary process carried on outside the mold or molding machinery. Many current products still are finished in this way, but for many others the required operations are performed in the mold.

Neck finishing is typically the facing or machining of the sealing surface at the top of the thread opening. This surface must be correct in dimension height and squarely cut so as to not have a cocked top edge. Sometimes the inside diameter must also be sized or finish machined. Containers with openings larger than 1 in. may be fly cut to obtain an opening cut accurately to inside dimensional size. Many industrial parts are fly cut when round openings or holes are required to complete the product. These finishing operations are accomplished by several types of machines, among which some are custom designed and built while others are stock machines available for purchase. (See Figs. 10-8, 10-9, 10-10.)

Generally, neck inside diameter sizing and top spot facing is accomplished in one step with a dual purpose cutter. Cutter designs range from a single flute with no spiral to multiple spiral flutes, made of hardened or unhardened tool steels or carbide compounds. Most plastic materials are rather abrasive and a high-quality hardened steel or carbide cutter is recommended. Slightly dulling the sharp edges of a new cutter with a fine stone will reduce its tendency to bite or grab the workpiece, which typically results in chatter marks. A spiral cutter will pull the chips away from the cut and

Figure 10-8 Blown containers entering a secondary deflashing machine for removal of top, handle, and bottom flash. Courtesy of Ameritainer Inc.

Figure 10-9 Containers exiting the deflash trimmer shown in Fig. 10-8. This machine can be set up to deflash containers that have been prefinished in the blow molding machine, as shown, or to perform top sealing surface machining along with the deflash operation. Courtesy of Ameritainer Inc.

out of the container, a desirable feature because most containers may not have chips inside them when supplied to the filling line. Another technique to reduce chips inside the container is to perform the operation with the product inverted (neck down) and/or provide an air passage through the spindle and cutter assembly. A small amount of air is supplied to slightly pressurize the container, and the air escaping around the cutter as it operates blows chips out of the container. Finish machining of an inside diameter of a container or opening may also be accomplished by reaming.

Figure 10-10 Fly cutting of 110 mm container neck is performed here after parts are cooled as shown in Fig. 10-7. Courtesy of Walron Industries.

Drilling and Reaming

Drilling of diameters is very common on many industrial parts, and on some container products. The operation is quite varied in nature and may be accomplished by such techniques as freehand drilling, hand drilling assisted by a drill bushing fixture, automatic sequence operations in special fixtures or machines, as well as by mold punches and/or drills. Generally, drilling is considered for holes 1.0 in. or less in diameter. For larger holes fly cutting or machining with a router or sawing is more practical. Fly cutters have one center point about which one or two cutters are rotated to cut larger diameters. For small-diameter holes, $^1/_4$ in. and less, a standard drill bit is adequate, although drill bits especially designed for plastics, with minimal spirals, are available. For $^1/_4$ to $^1/_2$ in. diameter holes a spad-point drill bit is recommended. Taken from the woodworking industry, the spad-point bit has a sharp center point that contacts the work piece first to hold the bit in position when the edges at the outer diameter cut into the material—a sort of dual fly cutter in effect. For larger diameters, $^1/_2$ to 1 in. multiple-tooth cutters and common hole saws are used. The best results are obtained with multiple-tooth and hole saw cutters that have minimal or no set to the teeth.

Reaming is a somewhat difficult operation unless the workpiece is held firmly in place. However, it can produce an extremely high quality finish with excellent accuracy. For example, suppose a fuel tank filler neck of 2.0 in. inside diameter is to be machined 1.70 in. in depth with a combination inside chamber of $^{1}/_{16}$ in. $\times$ 45° and a top facing operation with specifications of a 16 rms microfinish that has a ± 0.005 diameter variation. To achieve this, the molded part must be held in storage for a minimum of seven days after molding to reach maximum firmness for cutting. The cutter is a four-fluted, straight flute, carbide unit with 0.010 in. cylindrical land design. Spindle speeds of 200–400 rpm are typical and this operation is performed in a Bridgeport milling machine with a heavily constructed clamping fixture (rigidity is a must).

Tapping, or cutting internal threads, may be accomplished either within the blow mold or as a postmolding operation; the choice depends on economics and product performance requirement. In-mold internal threading is typically accomplished by blowing thread forms—as on containers—forming threads around a mandrel, or blowing threads over and around an unscrewing core. For applications requiring large-diameter thread forms of a tight tolerance, postmolding internal tapping may be required. For example, garden sprayer tanks have an internal thread used to hold, seat, and seal a hand pump. Typically 2.0 in. or larger in diameter, these thread forms must be accurate in form as well as major and minor diameters. To tap or cut in and then reverse cutter rotation to bring it out drags chips back through the cut threads, sometimes damaging them. To avoid this, a collapsible tap is used. It cuts threads as it goes into the workpiece to a prescribed depth, then mechanically cams (collapses) its three or four thread cutting blades into its shaft so it can be withdrawn without touching the cut threads. This technique is also used for such products as plastic utility cans, safety containers, and disposal containers. (See Figs. 10-11 through 10-16.)

Figure 10-11 Drill jig for blow-molded stadium seat with drill bushings for location. Hand drill is used (lower right). Courtesy of American Seating Co.

Figure 10-12 Holding clamp (closed) and combination single straight flute reaming tool with spiral single-flute top-facing tool. Courtesy of Crocker Ltd.

Figure 10-13 Multiple-spindle drill head and drill press with holding fixture on table. Courtesy of Crocker Ltd.

Figure 10-14 Multiple air drills, with product positioning jig, quickly drill several holes. Courtesy of ACM Plastic Products.

Figure 10-15 Four air-operated cutters quickly place drain holes in this blow-molded flowerpot. Courtesy of Zarn, Inc.

Figure 10-16 Drilling, reaming, and spot facing operations are accomplished by this one machine.

Cutting

Cutting can be accomplished by hand, using standard knives or specially contoured or ground knives. In spin trimming, common utility knife blades mounted in jigs or fixtures are advanced into rotating parts or the parts are rotated into blades. So-called hot knife cutting actually uses wires that are heated by electrical resistance; the wires are typically nichrome. While not very fast, hot knife cuts are smooth and burr-free. Generally, no chips are generated but there may be some air pollution, depending on the type of material being hot cut.

Routing is another widely used method of cutting. A router produces a cut of reasonably good quality, and is adaptable to irregular shapes and large parts. The router may be used freehand or to follow a template, or may be fully automated. Bits with standard $^1/_4$ in. diameter shafts are typically used with industrial electric and air-driven router motors. (See Figs. 10-17 through 10-21.)

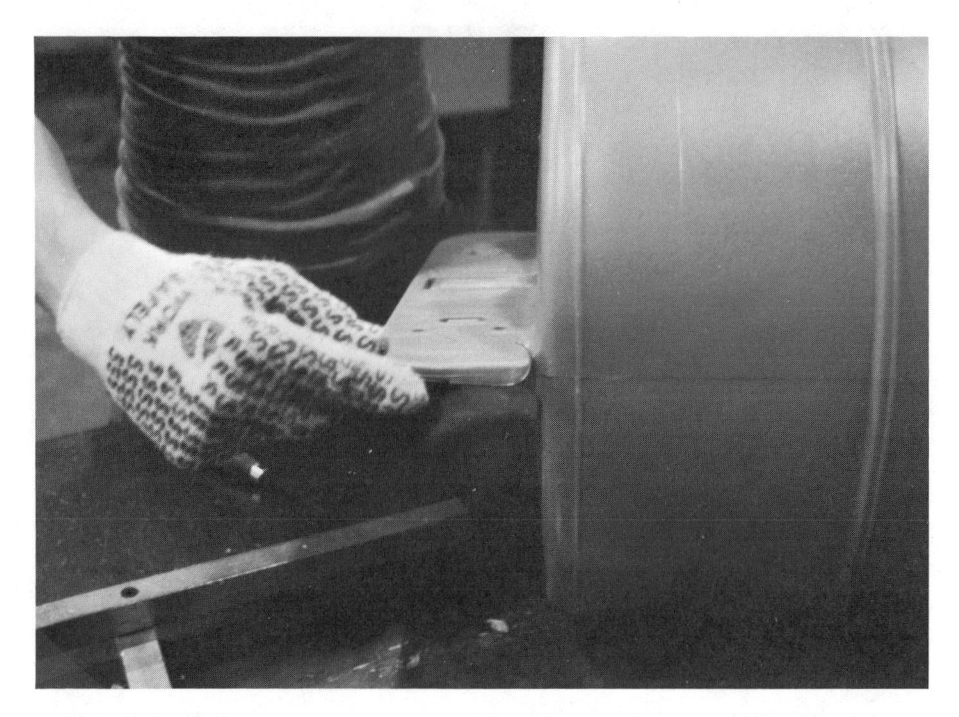

Figure 10-17 Hand trimming (cutting) of parting line flash. Note the thin-bladed, specialty shaped radius knife. Courtesy of Crocker Ltd.

Figure 10-18 A vertical thin-blade cutting machine is sometimes referred to as a guillotine. Courtesy of Walron Industries.

Figure 10-19 Cutting under the sprinkling can handle is accomplished in this fixture-cutting machine. Courtesy of Zarn, Inc.

Figure 10-20 For router cutting, large parts such as this highway safety barrel must be held securely in position. Then a hand-held router is moved along a fixed surface around the part's circumference. Courtesy of Crocker Ltd.

Figure 10-21 This hand-operated router fixture is first loaded with a part that is held in position by vacuum; then the top template is dropped down and its contours are followed with a hand-held router. Courtesy of Crocker Ltd.

Sawing

Sawing of blow molded parts and flash is very common; tools range from a small handsaw to a large chainsaw. The most common device is a table saw equipped with a nest to hold the workpiece; the part may be moved through the saw manually or by a variety of automated arrangements. Some automated machines utilize more than one saw and may also drill, rout, assemble, leak test, and perform other operations. Band saws are also commonly used, some equipped with sliding tables, automatic part feed, brakes to stop the blades, and other operator protection features. In general, sawing quality is best with blades having no tooth set. Circular saws of 8 to 12 in. diameter, having 100–150 teeth and running at 3600 rpm generate good quality cuts. Band saw blades of 14–16 teeth per inch with no (or minimal) set perform well. Frictional heat generation or buildup may be reduced by such techniques as coating blades with Teflon, but most techniques shorten blade life. Carbide blades will last longer but are not required. (See Figs. 10-22 through 10-25.)

Figure 10-22 Bandsawing operating with expanded table and throat to accommodate large-size parts. Courtesy of Crocker Ltd.

Figure 10-23 Saw fixture for large industrial trash can top. Courtesy of Zarn, Inc.

Figure 10-24 Sawing machine with multiple motors and multiple blades mounted in angular position. Semiautomatic table moves in and out with part positioned in nest held by toggle clamp. Courtesy of Walron Industries.

Figure 10-25 Manual-loaded, manual-feed saw fixture. Courtesy of Walron Industries.

TRIMMING AND DEFLASHING

Good-quality trimming and deflashing begins with adequate machine clamp tonnage and proper mold design (pinch-off and lay-flat area). Secondary equipment or techniques cannot make up for deficiencies in these factors. Manual operators will have difficulty in trimming flash from industrial products or containers that are not first molded correctly, and job-related medical complaints will increase significantly, especially if hand knives are used.

Trimming in most high-volume applications, such as containers under one gallon, is typically done automatically. Some blow molding machinery trims the product in the mold or in a postmolding station; other installations do so in a downstream in-line secondary machine, a trimmer (see earlier, Figs. 10-8, 10-9). Industrial products are very often trimmed manually with hand-held knives. This is often practical with products having long cycle times, or with shorter cycle products if more operators are employed. Air and hydraulic presses also may assist with deflashing operations. These units typically have a die shoe into which an operator hand loads each part; the unit is activated and stopped with pushbutton switches, and the trimmed product is removed manually. An indexing type conveyor may also be used to feed parts into and out of a press for automated deflashing. Containers may stand upright or lie horizontal, as necessary.

The deflashing operation is quite rigorous, for the forces required are often greater than anticipated. Deflashing problems may arise from insufficient machine mold clamp tonnage and/or deficient blow mold design. Typical compensations to make up for these deficiencies include using an oversized deflashing clamp or increasing the speed of deflashing machine operation for greater shock effect. (See Figs. 10-26 through 10-29.)

An advanced part trimming technique is laser cutting. While the laser may be somewhat expensive, and the generation of fumes during use must be considered, the quality of the trim is superior. Fumes will depend on the types of material being trimmed or cut and may be handled by air filtration units. Laser trimming is used for the tops of PET containers for tennis balls. The containers are blown with a top that is laser trimmed to provide a flange onto which a metal closure is installed using can seaming equipment. An alternate method is to shear the top and flange by mechanical dies. Since the trimmed edge is covered by the metal closure top during the seaming operation, no rough trimmed edge is exposed.

Water jet trimming is somewhat similar to laser trimming. A high-pressure stream of water is used to cut or trim various shapes. Typically, the table on which the workpiece is situated is moved by a computer numerical control (CNC) program to duplicate an operation many times; the water jet head could be controlled instead in a similar manner. An alternative to CNC control is a template that is traced by a mechanical feeler pin to guide the movement of the table.

Figure 10-26 Flash is sheared from the product at the mold-produced pinch-off lines in this air press. Courtesy of ACM Plastic Products.

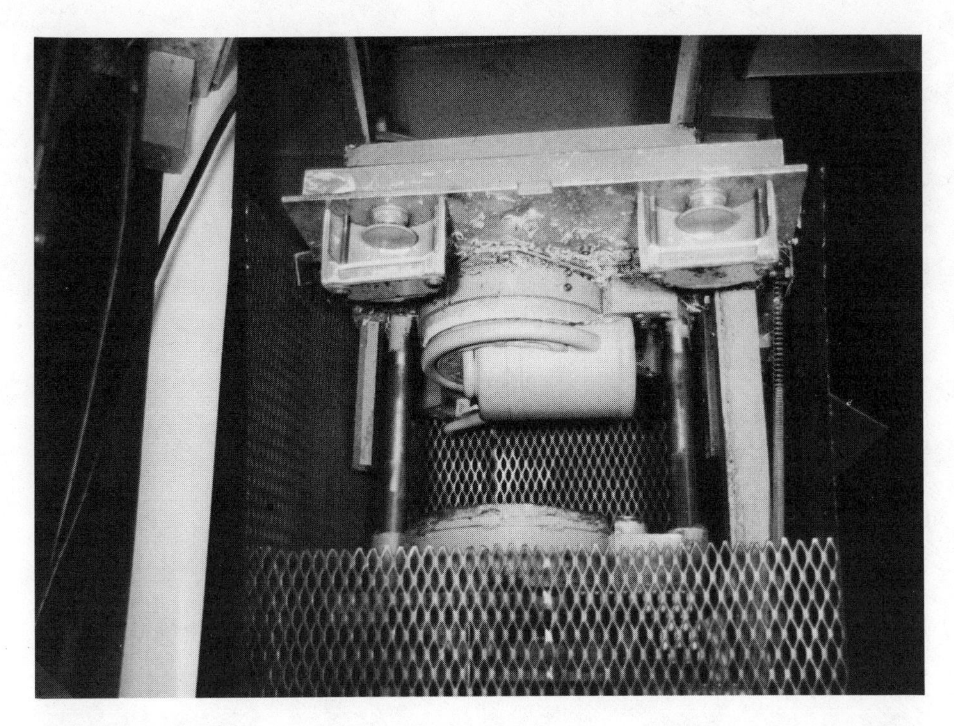

Figure 10-27 Deflashing of a watering can's handle area. Courtesy of ACM Plastic Products.

Figure 10-28 This complex machine drills, punches, shears, and dimension-checks a right and left version of a part simultaneously. Courtesy of Walron Industries.

Figure 10-29 Air press machine with interchangeable punch-shear die shoe fixture. Large dual air cylinders used to operate the press are supplied from an air surge tank. Courtesy of Walron Industries.

ROBOTICS

Part Handling

The use of robots for part removal from the machine and subsequent placement is increasing for several reasons. Robot devices have become less expensive; there are more off-the-shelf, easy to program units available; there are economic incentives with respect to health and safety issues and labor costs; and robot part handling produces improved quality.

Routing

Generally, robots are most desirable for repetitive, difficult work. Thus, in addition to part removal and handling such procedures as router operation may be ideal for robot control. Routing sometimes produces unwanted environmental noise. In a typical application the robot router is enclosed in a small sound-restricting or soundproofed compartment within the factory. An operator loads and unloads products, but is not present in the compartment during the routing operation.

Trimming

The operation of a trim machine is also a repetitive task that could well be converted to robot control. The robotics need not be complex: a few items such as air cylinders, mechanical arms, and the like interconnected by an electrical controller is in essence a functional robot. From this sort of simple configuration one can progress to designing more complex robot machines programmed to perform many kinds of specific tasks.

There are many electric controllers for robot machines on the market today, offering simple to complex controls. (See Figs. 10-30, 10-31.)

Appendix: Definitions of Terms

CAD Computer-aided design.
CAM Computer-aided manufacturing.
CNC Computer numerical control.
closure The top portion of a container, often threaded.
controller Device that manages machinery functions-sequences.
deflash To trim or otherwise remove flash from a part.
downstream Usually the equipment and operations following actual molding.
ejector Device that removes a part from a mold or die.
flash Excess, unwanted material in a product before finishing.

Figure 10-30 This control panel includes the high-voltage motor controls, computer, and computer interface workstation required to operate a highly complex automotive gasoline fuel tank finishing machine. Such a control station may cost as much as many blow molding machines.

Figure 10-31 Typical controller racks with plug-in cards allow easy access, repair, and changes.

fly cut A cutting operation where cutter(s) are rotated around one center point. Normally used to cut diameters in excess of 1.0 in.
forming mandrel A shaped support post typically used to form and size container necks and threads.
IML In-mold labeling.
lay-flat The flash area of a blow mold.
line A complete system to manufacture a product.
parison The molten plastic tube extruded from a die head that is placed in the blow mold to be formed into a product.
permeability The degree to which a material allows seepage of a given liquid or gas.
pinch-off The edges of a blow mold that seal the container's parting line.
post molding After or following the process of actual molding.
prefinish Finishing of a container's threads and top sealing surface by the molding machine in the mold. Container may be a bottle, plastic drum, fuel tank, etc.
sealing surface Normally the top edge of a container's threaded neck
spin trimming Trimming method in which a container is revolved into and against a fixed knife.
spot facing To machine finish a surface such as a small area or top of a container.
stripper Device used to extract product form the mold and/or molding machine.
tail flash The lower portion of flash on a blow molded part.
trim To remove flash from a part or machine in some manner.
trim nest A jig or form used to hold a part to be trimmed.

CHAPTER

11

Decoration Equipment and Processes

DONATAS SATAS

Blow molded plastic products, such as bottles, are most frequently finished by applying a coating or print over the surface. Decoration is required for both aesthetic and functional purposes: the decoration often carries a message, such as printed description of the container contents, in addition to imparting a distinctive change of color or gloss required to meet the designer's idea of the finished product.

Decoration may be carried out by coating, printing, or labeling. These are usually separate operations performed after molding. Sometimes, however, decorating is combined with molding in such operations as in-mold labeling or in-mold coating. The decoration of plastic surfaces has been covered in detail by several books devoted specifically to the subject [1,2,3] and less extensively in chapters in books covering other aspects of plastics technology [4,5].

SURFACE WETTING, ADHESION, AND COATING LEVELING

A coating or print that is applied over a plastics surface must adhere to the surface in order to provide a useful and functional product. The coating must wet the surface adequately for satisfactory adhesion to occur. A liquid drop placed on a solid surface may spread over that surface or may remain a drop.

If the drop spreads, then it wets the surface; if it remains as a distinct drop, the surface is not wetted. With wetting, adequate adhesion of a coating to a plastic surface can be achieved; without wetting the adhesion will be unsatisfactory. In addition to being wettable, a surface must be relatively clean: it should not be covered with a film of contaminant which, even if it does not interfere with wetting, will act as a so-called weak boundary layer. Such a layer constitutes a barrier that prevents contact with the surface and therefore prevents adhesion.

For a coating or ink to wet a plastic surface adequately, its surface tension must be lower than the critical surface tension of the plastic. Table 11-1 lists the critical surface tension of various plastics. Such data have been developed by many researchers and they are summarized in a single article [6]. Examination of the table shows that the plastics we associate with having bonding problems—fluorocarbon polymers and polyolefins—have low critical surface tension values.

A simple way to get some information about the critical surface tension of a plastic surface is to apply a liquid of known surface tension and to observe whether it spreads or retracts (forms beads or drops). When the liquid barely wets the surface, we assume that the liquid surface tension is equal to that of the critical surface tension of the solid. Bottles or felt pen kits containing liquids of various surface tensions are available and are useful for such an evaluation. Table 11-2 shows surface tensions of formamide/Cellosolve blends covering the surface tension range of main interest for plastics. The test details are described in standard ASTM test procedures [7] and in other published literature [8].

The surface tension of a liquid coating or ink may be decreased by the addition of surface active agents (surfactants). Water-soluble surfactants are used for aqueous coatings, and solvent-soluble surfactants are used for solventborne coatings and inks. Surfactants may affect other coating properties and therefore it is advisable to evaluate their effect carefully. Figure 11-1 shows the effect of wettability. A well wetted surface (*a*) exhibits a continuous coating with a good adhesion, while a poorly wetted coating (*b*) may exhibit poor coverage, because the color of the background may show through. The coating will also have poor adhesion.

Coating adhesion may be tested by several methods, although it is difficult to obtain reliable quantitative data. The most frequently used qualitative test involves applying pressure-sensitive adhesive tape over the coat-

Figure 11-1. Coating-surface interaction. (a) Coating wets the surface. (b) Coating has receded from the surface.

Table 11-1 Critical surface tension of various plastic surfaces

Material	*Critical Surface Tension (dynes/cm at 20°C)*
Poly(l,l-dihydrofluorooctyl methacrylate)	10.6
Polyhexafluoropropylene	16.2–17.1
Poly(tetrafluoroethylene-*co*-hexafluoropropylene)	17.8–19
Polytetrafluoroethylene	18.5
Poly(tetrafluoroethylene-*co*-chlorotrifluoroethylene)	20–24
Polytrifluoroethylene	22
Silicone rubber	22
Polydimethyl siloxane	24
Polyvinylidene fluoride	25
Poly(tetrafluoroethylene-*co*-ethylene)	26–27
Butyl rubber	27
Polyvinyl fluoride	28
Polyvinyl butyral	28
Polybutyl acrylate	28
Polyvinyl methyl ether	29
Polyurethane	29
Polyisoprene	30–31
Polyethylene	31
Polychlorotrifluoroethylene	31
Polypropylene	31
Polystyrene	33–35
Polyethyl methacrylate	33
Polyvinyl chloride	33–38
Polymethyl methacrylate	33–34
Polyacrylamide	35–40
Cellulose	35.5–41.5
Rubber hydrochloride	36
Polyvinyl ethyl ether	36
Polyvinyl alcohol	37
Polyvinyl acetate	37
Polyisoprene, chlorinated	37
Chlorosulfonates polyethylene	37
Nylon 6	38–42
Chloroprene	38
Polyvinyl chloride, rigid	39
Cellulose acetate	39
Polyvinylidene chloride	40
Polyphenol acrylate	40
Polyoxyphenylene	41
Polysulfone	41
Nylon 6,6	42.5–46
Polyethylene terephthalate	43
Casein	43
Polyacrylonitrile	44
Cellulose, regenerated	44–45
Phenol-resoreinol adhesive	52
Polyamide-epichlorohydrin resin	52
Urea-formaldehyde resin	61

ing, which is sometimes cross-scored through to the surface below, and peeling the tape off at a sharp angle to see whether any coating comes away. This test allows the rejection of poorly adhered coatings, but does not yield any quantitative information about the bond strength. Other methods involve adhering a probe to the coating surface and then measuring the force required to pull the coating away, or employing a probe to scratch the surface of the coating and to measure the resistance to scratching [9].

The rheological properties of the coating and ink have an effect on the ease of application and the appearance of the coating. Figure 11-2 shows the viscosity of various coating types as a function of shear rate. Viscosity is independent of shear rate for a Newtonian coating (curve *a*). Fortunately, most solventborne coatings exhibit Newtonian behavior.

Many aqueous coatings exhibit either pseudoplastic or plastic behavior, characterized by the decrease of viscosity with increasing shear rate and by a high viscosity at low shear rates. High viscosity and the existence of a yield value at low shear rates cause poor leveling of the coating. The coating surface is usually not smooth immediately after its application, because of the mechanical disturbances introduced by the coating technique: roll, spray, brush, or some other method. These surface irregularities are expected to disappear in a coating that levels well. Good leveling properties require low viscosity at the low shear rates at which the leveling process takes place. The leveling process has been extensively discussed in connection with paints [10,11].

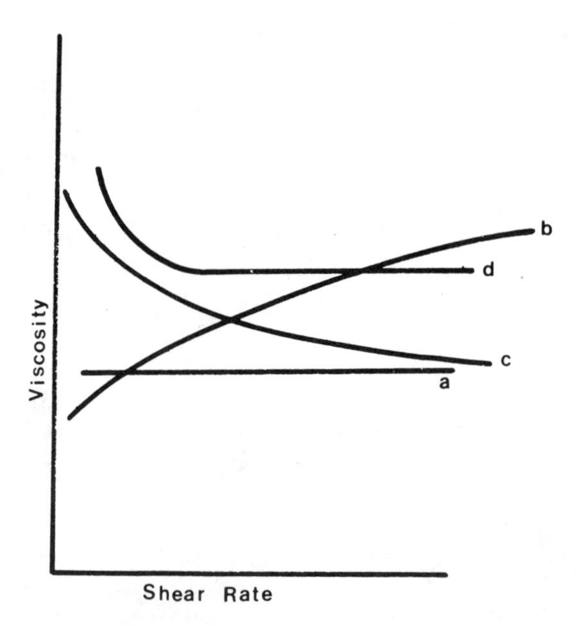

Figure 11-2 Viscosity vs. shear rate for various coatings. (a) Newtonian. (b) Dilatant. (c) Pseudoplastic. (d) Plastic.

Table 11-2 Surface tension of various formamide-Cellosolve mixtures

Formamide (volume percent)	*Cellosolve (volume percent)*	*Surface Tension (dynes/cm)*
99.0	1.0	56
96.5	3.5	54
93.7	6.3	52
90.7	9.3	50
87.0	13.0	48
83.0	17.0	46
80.3	19.7	45
78.0	22.0	44
74.7	25.3	43
71.5	28.5	42
67.5	32.5	41
63.5	36.5	40
59.0	41.0	39
48.5	51.5	37
42.5	57.5	36
35.0	65.0	35
26.5	73.5	34
19.0	81.0	33
10.5	89.5	32
2.5	97.5	31
0	100.0	30

SURFACE TREATMENT

In order to obtain acceptable surface wetting and adhesion, the liquid surface tension must be lower than the critical surface tension of the plastic surface. We can decrease the liquid surface tension by the addition of surface active agents, or increase the critical surface tension of the solid surface by surface treatment. In addition, surface treatment may involve surface cleaning—removal of foreign material that can contribute to poor adhesion because it interferes with coating contact with the surface.

The following processes are used for the surface cleaning and treatment of plastics materials:

- Water washing
- Solvent cleaning and etching
- Mechanical abrasion
- Chemical etching
- Additives

- Flame treatment
- Corona discharge
- Plasma treatment
- Ultraviolet and other irradiation

The most often used methods are flame treatment for sheets and irregular molded items and corona discharge for films. Surface treatment is especially required for plastics of low critical surface tension, of which polyolefins constitute the largest volume and fluorocarbon polymers constitute the most difficult surfaces to treat. Surfaces of other plastics may also be treated or at least cleaned.

Washing and Solvent Cleaning

Water washing removes various surface impurities, such as release agents, electrostatically or otherwise attracted dust particles, and additives that have migrated to the surface. A typical cleaning cycle may consist of several steps: treatment with a cleaner followed by several rinses and finished by a rinse with deionized water.

Solvent cleaning may consist of wiping, immersion in solvent, spraying, or vapor degreasing. Wiping is the least effective process and may result in distributing the contaminant over the surface rather than removing it. Immersion, especially if accompanied by mechanical or ultrasonic scrubbing, is a better process. It is even more effective if followed by either immersion or a spray rinse. Vapor degreasing is the most efficient process, because the surface does not come in contact with the contaminated solvent bath. Vapor degreasing is carried out in a tank with a solvent reservoir on the bottom. The solvent is heated and vapor condenses on the cool plastic surface. The condensate dissolves surface impurities and carries them away. Cleaning is usually completed in one minute.

In addition to cleaning, solvent may swell the plastic surface, causing changes that may result in improved ink or coating adhesion. Toluene and chlorinated solvents are the most effective for the treatment of polyolefin plastics.

Abrasion

Cleaning by abrasion removes surface contamination and increases surface roughness. Removal of surface contamination eliminates the weak boundary layer and increased roughness increases the bonding area; both have a positive effect on adhesion.

Mechanical abrasion is carried out by several processes: dry blasting with nonmetallic grit (flint, silica, aluminum oxide), wet abrasive blast (slurry of aluminum oxide), hand or machine sanding, and scouring with tap water and scouring powder.

Chemical Etching

Chemical etching is the exposure of the plastics surface to a solution of reactive chemical compounds. This treatment causes a chemical surface change, such as oxidation, thereby improving surface wettability (increasing its critical surface tension). It may also remove some material, introducing a micoroughness to the surface. Chemical etching requires immersion of the part into a bath for some time period, then rinsing and drying. The process is more expensive than most other surface treatment processes and therefore it is used only when other methods are not sufficiently effective. Fluorocarbon polymers are often etched chemically, because they do not respond to other treatments; ABS parts may be chemically etched for metallic plating.

Etching solutions are oxidizing chemicals, such as sulfuric and chromic acids, or metallic sodium in naphthalene and tetrahydrofuran solution. Such solutions are highly corrosive and difficult to handle and to dispose of.

Additives and Primers

The surface characteristics of a plastic material may be changed by the addition of incompatible additives that migrate to the surface. Additives such as fluorocarbon polymers, low molecular weight fluoroacrylates, polyamides, waxes, silica, and others are used for this purpose.

Surface characteristics may also be changed by applying a primer to the surface. Many polymers are useful as primers; one example is polyethylene imide, which is a widely used primer for polyolefins.

Flame Treatment

Flame treatment is widely used to improve ink adhesion to various polyolefin and other plastic containers. Often the print is also flame treated to dry the ink and to further improve its adhesion. Flame treatment is especially useful in treating irregularly shaped objects. Figures 11-3 and 11-4 illustrate the adaptability of flame treatment for such products.

Flame treatment oxidizes the surface and makes it more easily wettable. The plastic surface is contacted for a period of less than one second with the oxidizing portion of the flame. The gas is burned using 10–15 percent excess air over the stoichiometric ratio in order to obtain an oxidizing flame with a temperature of 1100–2800°C (2012–5072°F).

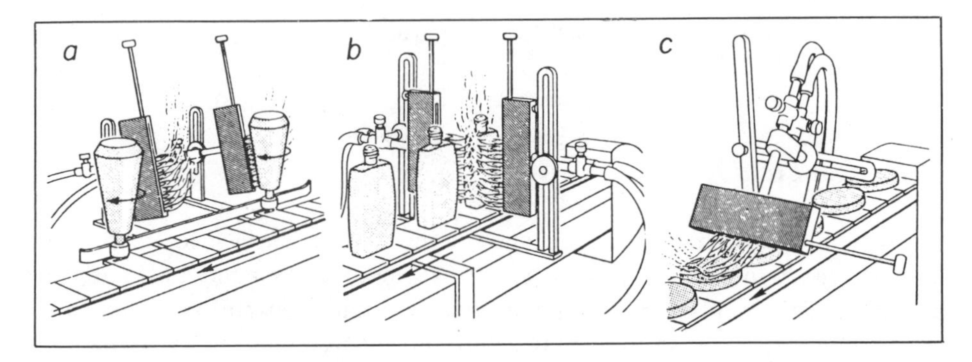

Figure 11-3 Burner arrangement for irregularly shaped bottles. (a) Conical bottles. (b) Flat bottles. (c) Lids. Courtesy of U.S.I. Chemicals.

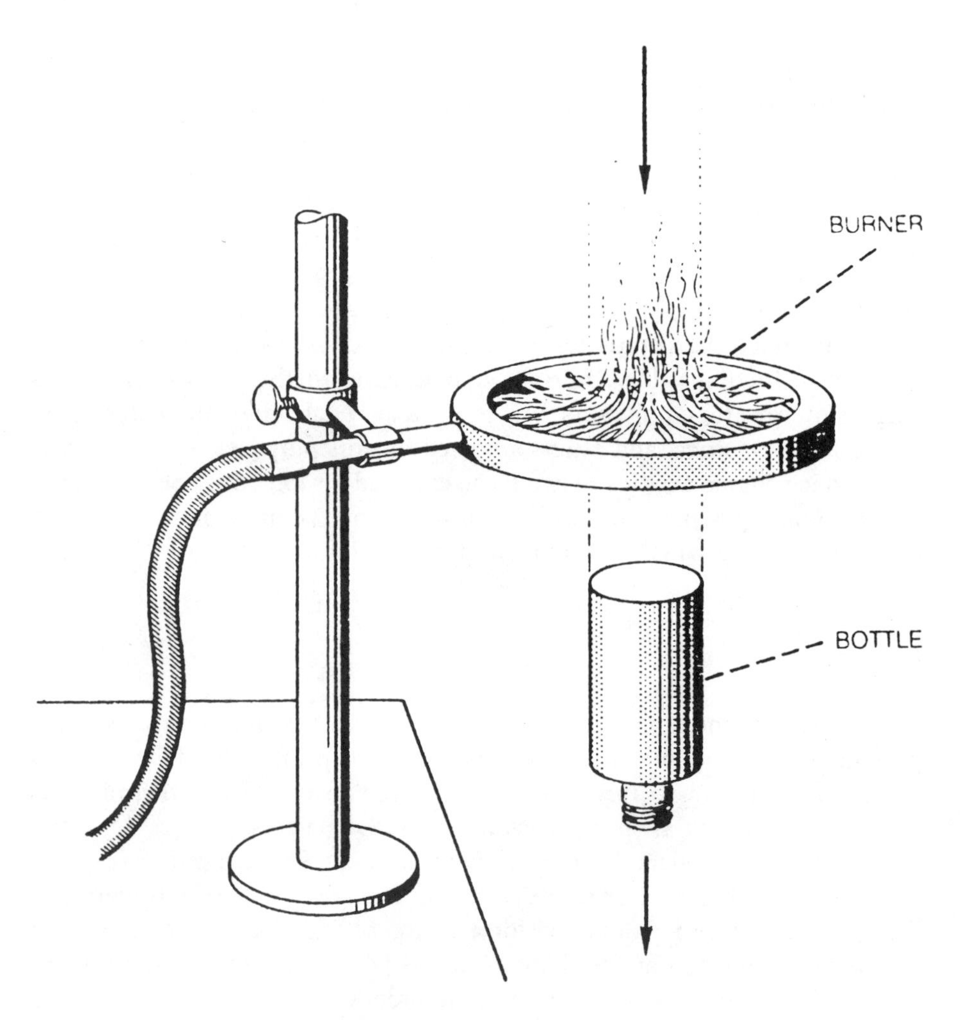

Figure 11-4 Ring burner for round bottle treatment.

Corona Discharge

Corona discharge treatment consists of passing the polymer surface through a plasma produced by corona discharge taking place between electrodes. The method is mainly used for flat plastic materials, such as films, but it can also be used for irregularly shaped objects, although such applications are still infrequent.

Plasma Treatment

Cold gas plasma treatment is effective but expensive and rarely used for mass produced packaging containers. It is a batch process: the parts to be treated are placed in a chamber that is evacuated and plasma is produced inside the reactor.

SPRAY PAINTING

The application of clear or colored coatings (paint) over plastic articles by spraying is a widely used method, mainly for its suitability in handling irregular molded articles. Spraying is rarely used for films or other flat-surface materials; there are more suitable methods available for such products. Spray painting equipment and processes are discussed in more detail in several publications [12,13].

Spraying consists of atomizing a liquid stream by a spray gun. Small paint particles produced by the gun are then delivered to the plastic part, contact its surface, and remain adhered after the liquid vehicle is removed by drying.

Several atomization techniques are used:

- Air atomization
- Airless atomization
- Air-assisted airless atomization
- Atomization by centrifugal forces
- Electrostatically assisted atomization

Spraying is adaptable to automation, and robots have been successfully introduced to handle repetitious, and usually dirty, spraying operations.

Air Atomization

Air atomization is carried out by an air gun where a high-velocity air stream breaks up the liquid coating. Air atomization yields a smaller particle size than other spraying methods: the lower particle size limit is about 5 μm [12]. It is the only method to produce a fine spray in which the diameters of all the droplets are less than 15 μm. Fine particle size allows good control of the coating thickness. The main disadvantage is considerable overspray (poor transfer efficiency) unless electrostatic assist is used. Pneumatic spraying is also inefficient from the point of view of energy utilization. Air guns not only atomize the paint, they also shape the atomized stream. Usually a fan pattern is used because it allows better control of coverage.

Airless Spraying

Airless atomization involves forcing the paint through a small nozzle at a sufficiently high velocity or by impinging the paint stream against an obstacle, causing a sudden change in flow direction. In either case, the fluid stream disintegrates into small particles. The degree of atomization is rather poor, the particle size is large, and the fluid must be of sufficiently low viscosity. This spraying method, however, is more efficient in terms of energy consumption than air atomization, and the transfer efficiency is much better.

Paint viscosity may be lowered for airless spraying by heating it. A heater utilizing hot circulating water may be used to decrease the paint viscosity.

An airless operation can be improved by providing a low-pressure air stream to assist atomization and to direct the paint to the target. While the transfer efficiency is slightly decreased, the degree of atomization is substantially improved.

Centrifugal Atomization

Centrifugal atomizers consist of a rotating disk or bell. The paint flows to the periphery of the rotor and disintegrates as it leaves the rotor's edge. The atomization is poor, the paint direction is difficult to maintain, but the process is very efficient mechanically. In order to improve the atomization, the rotor speed must be increased to very high levels, which is mechanically difficult, or electrostatic assist must be used. If not for the latter technique, centrifugal coating would not be used as widely as it is today.

Electrostatic Assist

While electrostatic forces may cause the disintegration of a fluid stream, the electrostatic method is not used alone. It is extremely useful, however, as an assist with other spray methods, especially with centrifugal spraying.

Table 11-3 Transfer efficiency for various spray equipment

Air atomization	35%
Air atomization with electrostatic charge	55%
Airless atomization	50%
Airless atomization with electrostatic charge	70%
Disk and bell atomization with electrostatic charge	90%

An electrostatic force improves the disintegration of the electrically charged paint stream but, most importantly, it directs the charged paint particles to the oppositely charged target (usually at ground potential), and increases transfer efficiency. The surface of the target must be conductive. Although plastic surfaces are not conductive, this is easily corrected by a light spray of conductive salt solution sufficient to maintain the required charge on the object to be painted.

Transfer Efficiency

Transfer efficiency is a measure of the percentage of paint that reaches the target. Table 11-3 lists transfer efficiencies for various spraying methods. It is obvious from this table that the transfer efficiency of air atomization is poor and that electrostatic assist is very helpful in improving the transfer efficiency. Transfer efficiency is very sensitive to the geometry of the object to be painted: a much higher transfer efficiency is expected for large flat objects than for small irregularly shaped parts. Therefore, the values listed in the table are only approximate averages obtained at standard spraying conditions.

Masking

Spray paint often must be deposited only in selected areas and the areas not to be coated must be protected by masking. The simplest masking method involves employing tape and paper to cover areas of the plastic surface. The choice of a masking tape, usually a creped paper pressure-sensitive tape, is important. The tape must be resistant to the solvents used in the spray paint, it must be able to withstand the drying conditions, and it should separate cleanly upon removal. The sharpness along the edge of tape largely depends on the tape quality.

Hard masks are reusable metal or plastic parts placed over the object to be painted to prevent paint deposition on selected parts of the surface. Cast polyurethanes and silicone rubber are often used for plastic masks. Metal masks may be fabricated by regular mechanical methods or may be electroformed. Such masks are used repeatedly, followed by cleaning, usually in a solvent wash tank. Duplicate or triplicate masks may be needed for continuous operation and may include automated mask washing for the most mechanized versions. A typical spray mask is shown in Figure 11-5.

Paints

Numerous paint compositions are used for spray painting plastics surfaces. Both solventborne and aqueous paints are used. Paints are usually classified on the basis of binder used. The most often used paints are as follows:

Acrylic: solventborne enamels, aqueous acrylic emulsions, melamine and other modified acrylic enamels
Polyurethanes: one- and two-component solventborne urethanes, aqueous polyurethane emulsions
Alkyds and modified alkyds
Epoxies, including various modifications
Polyesters
Vinyl and modified vinyl, either solventborne or latex systems
Nitrocellulose: solventborne paint
Polyamide: solventborne paint especially suitable for polyolefins

Solventborne paints may be ambient air dried or dried with the application of some forced air; aqueous latexes require higher drying temperatures; some catalyzed paints (epoxy, polyurethane, some alkyds) dry at room temperature by chemical crosslinking, not by vehicle evaporation. Catalyzed paint has a limited pot life, of course, and should be used up within a limited time period. The current emphasis on new developments of aqueous and high solids coatings resulted from environmental restrictions. Developmental work still continues and improved coatings are used to replace solventborne systems.

Some plastics may be sensitive to certain solvents and care must be taken that such a solvent is not used in the formulation in an amount sufficiently large to cause damage to the plastic surface. Such solvents should be removed from the surface quickly. Sometimes employing an active solvent may help the adhesion. Table 11-4 lists solvent suitability for various plastics. Polyolefins and nylon exhibit good solvent resistance and solvent selection is not critical for these plastics.

Table 11-5 shows the suitability of paints for various plastics. Paints that are suitable for spraying may also be applied by other methods.

Table 11-4 Solvents for plastics for reducing paints*†

Solvent Group	*Aliphatics*	*Aromatics*	*Alcohols*	*Ketones*	*Esters*
Thermoplastic Resins*					
ABS[1]	R	NR	R	NR	NR
Acrylics	R	NR	R	NR	NR
PVC[2]	R	NR	R	NR	NR
Styrene	R	R	R	NR	R
PPO[3]/PPE[4]	R	R	R	R	R
Polycarbonate	R	NR	R	NR	R
Nylon‡	—	—	—	—	—
Polypropylene‡	—	—	—	—	—
Polyethylene‡	—	—	—	—	—
Thermoset Resins					
RIM[5]	R	R	R	R	R
Polyester	R	R	R	R	R

*Reducers selected for lowering paint viscosity should be determined in conjunction with the paint manufacturer.

†R = recommended, NR = not recommended.

‡The superior solvent resistance of nylon, polypropylene, and polyethylene makes it difficult to generalize solvent groups for paints. Urethane and polyester paints have been effective.

[1]ABS: acrylonitrile-butadience-styrene polymer.

[2]PVC: polyvinyl chloride.

[3]PPO: polyphenylene oxide.

[4]PPE: polyphenylene ether.

[5]RIM: reaction injection molded urethane.

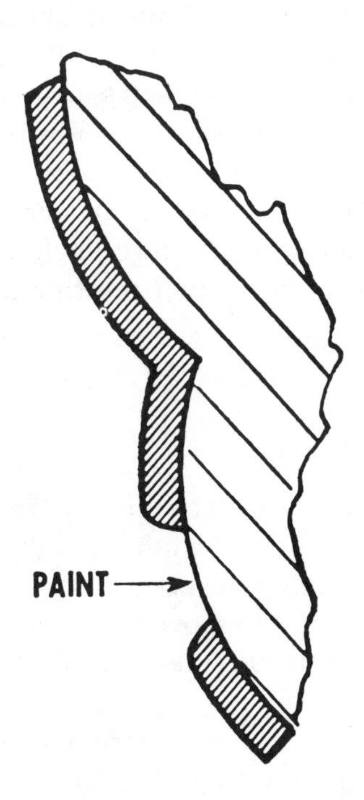

Figure 11-5 Plane surface mask.

Table 11-5 Recommended paints for plastics

Plastic	*Urethane*	*Epoxy*	*Polyester*	*Acrylic Lacquer*	*Acrylic Enamel*	*Acrylic Waterborne*
ABS	R	R	NR	R	R	R
Acrylic	NR	NR	NR	R	R	R
PVC	NR	NR	NR	R	R	NR
Styrene	R	R	NR	R	R	R
PPO/PPE	R	R	R	R	R	R
Polycarbonate	R	R	R	R	R	R
Nylon	R	R	R	NR	NR	NR
Polypropylene	R	R	R	NR	NR	NR
Polyethylene	R	R	R	NR	NR	NR
Polyester	R	R	R	NR	NR	NR
RIM	R	NR	NR	NR	R	R

R, recommended; NR, not recommended.

In order to obtain adequate paint adhesion, primers may be required. Polyimide-based primers are often used for polyethylene; polyurethane primer is useful for polypropylene.

SCREEN PRINTING

Screen printing is a stencil process in which ink is forced through the open areas of a screen. The open areas determine the print pattern. The principle of screen printing is illustrated in Figure 11-6 [14]. The ink is pushed through the screen openings by a squeegee and then spreads out, giving a continuous ink coating within the desired areas. Screen printing allows the deposition of relatively heavy ink coatings, which is not possible with other printing methods. Although the process is versatile and simple—and the equipment may be inexpensive—it is also slow and therefore not suitable for large-volume operations. In such cases, other printing methods will usually be favored.

Basic Equipment

Screen printing equipment is rather simple; it consists of a screen mounted in a frame, and a squeegee to force the ink through the screen. The printing process can be carried out in a simple manual way. It can also be automated by providing mechanical means to move the screen and the squeegee, thus complicating the equipment requirements although the basic principle remains the same.

The screen determines the print pattern and the ink thickness. In its function it is similar to the printing plate or printing cylinders in other printing processes. Screens were made from silk fabric originally, thus the

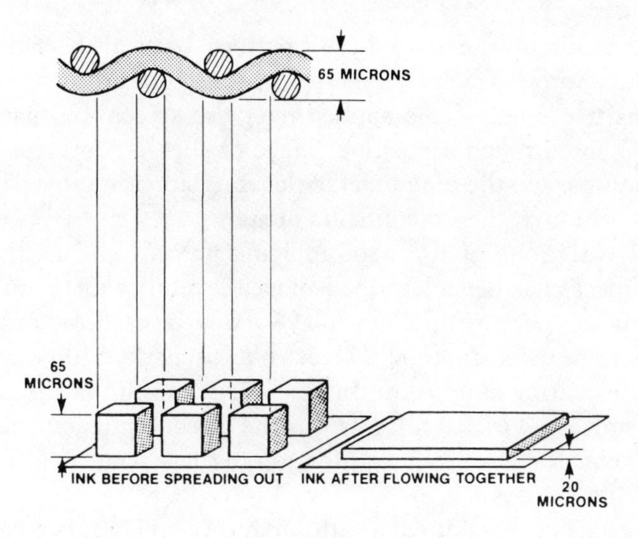

Figure 11-6 Formation of film from ink dots. Courtesy of Tedko, Inc.

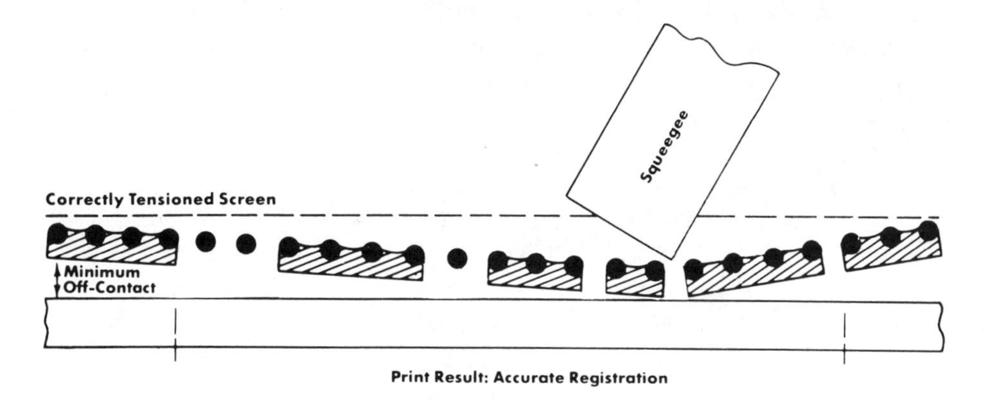

Figure 11-7 Mechanics of screen printing process. Courtesy of Tedko, Inc.

name "silk screening" for the process. Synthetic polymer monofilament fabric or metal mesh has replaced silk. They are available in various fiber diameters and counts (number of filaments per inch). Filament diameter and the number of openings per unit area determine the amount of ink deposited.

Polyester monofilament fabrics are most often used for screen printing. Various mesh sizes are available, but 240–305 mesh fabrics are the most popular. The fabric may be calendered to flatten out the filaments, thus improving the contact of the screen with the surface to be printed. Stainless steel wire mesh yields an accurate and long-lasting screen. A nylon monofilament screen has a long service life but poorer dimensional stability.

The screen is stretched over a frame, now usually made from metal tubing rather than wood. The screen must be properly tensioned over the frame. Adjustments for tension may be provided. Figure 11-7 shows the importance of screen tension in the printing operation.

In order to obtain the desired print pattern, parts of the screen must be blocked. The photoimage process is the most important for stencil making. A photosensitive emulsion is applied over the screen and parts of it are exposed to light through a positive image of the art work. Exposed areas become insoluble and the remainder of the emulsion is removed by washing. Ink can pass through the no-emulsion areas.

Several variations of the photoimaging process are used. The direct emulsion process has been described above. Several coats of emulsion may be applied and the coating thickness may vary in the case of stencils prepared by the direct emulsion method. Precast photosensitive films are available and provide a coating of uniform thickness. Such a film is laminated to the screen and processed in a similar way as the direct emulsion coatings. Film may be also exposed prior to its lamination to the screen. This is an indirect stencil making process.

A squeegee is used to force ink through the screen. It is a flexible blade made from an elastomer, usually polyurethane because of its durability and good elastomeric properties. The hardness of the rubber used in the squeegee has an effect on the printing: hard rubber (above 50 Shore A durometer) squeegees produce a cleaner, more crisp definition; softer squeegees deposit more ink and are more suitable for rough surfaces, but print definition is not as good. Textured surfaces will generally be printed by using a squeegee of durometer below 50.

The elastomer used should be resistant to solvents in the ink, or in the case of UV-curable inks, to the monomers used. The squeegee's blade usually has a rectangular profile and the edge is rubbed against the screen. The edge should be kept sharp. For heavy ink deposits, a rounded squeegee profile is used.

Screen Printers

Screen printers vary from simple hand-operated units to fully automated machines. There are three basic printer designs: flat bed, rotary, and cylinder. Schematic diagrams of these machines are shown in Figure 11-8. The flat bed machine is the simplest design. The printing stock and screen are stationary and the squeegee moves along the screen to deposit the ink in the required pattern. Flat bed screen printers are used most often and they may be equipped with automatic stock feeding and printing. Such printers can decorate up to 100 parts per minute. Figure 11-9 shows a flat bed screen printer.

A rotary screen printer consists of a drum constructed from a metal screen. The squeegee is located inside the drum and the ink is pumped inside. In the process both the screen and printing stock are moving, while the squeegee is stationary. Rotary screen printing was developed in the 1950s and is used for printing textiles, but it is also finding applications for plastic goods [15].

Cylinder screen printers are the least common. They consist of a stationary squeegee and moving printing stock and screen. The process is similar in principle to that of the rotary machine, except that the screen is flat and therefore the stock is fed intermittently. Cylinder machines are reported to produce sharp image and can be run at a rate of 3000–4000 cycles per hour. Some have printed 8000 parts per hour.

Both rotary and cylinder screen printers are suitable for flat printing stock: films or sheets. A flat bed screen printer can handle a variety of geometrical shapes, and many plastic bottles and other containers are screen printed.

Drying of the screen-printed pattern is an integral part of the printing process and is often the slowest step. Convection, infrared, or a combination

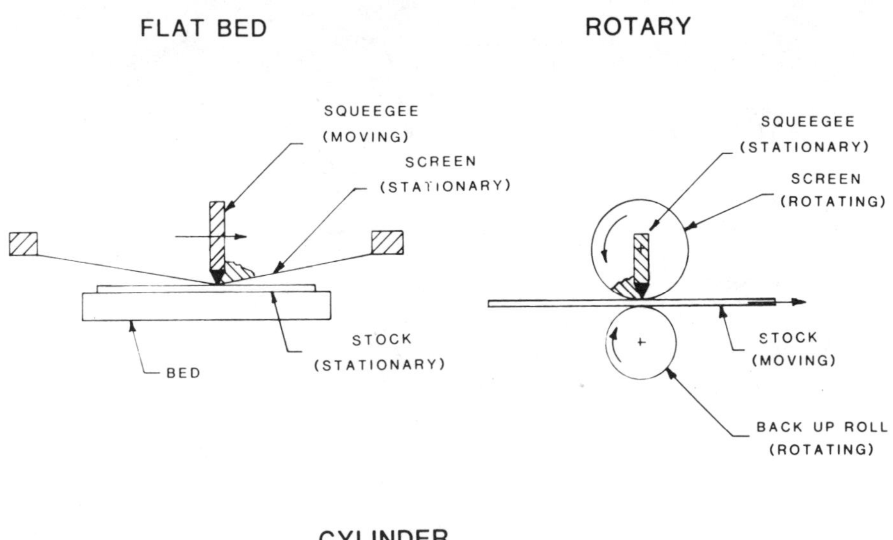

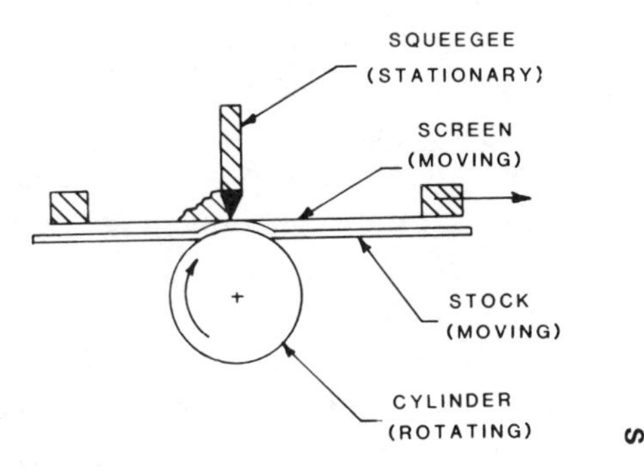

Figure 11-8 Basic types of screen printers. Courtesy of Sheldahl, Inc.

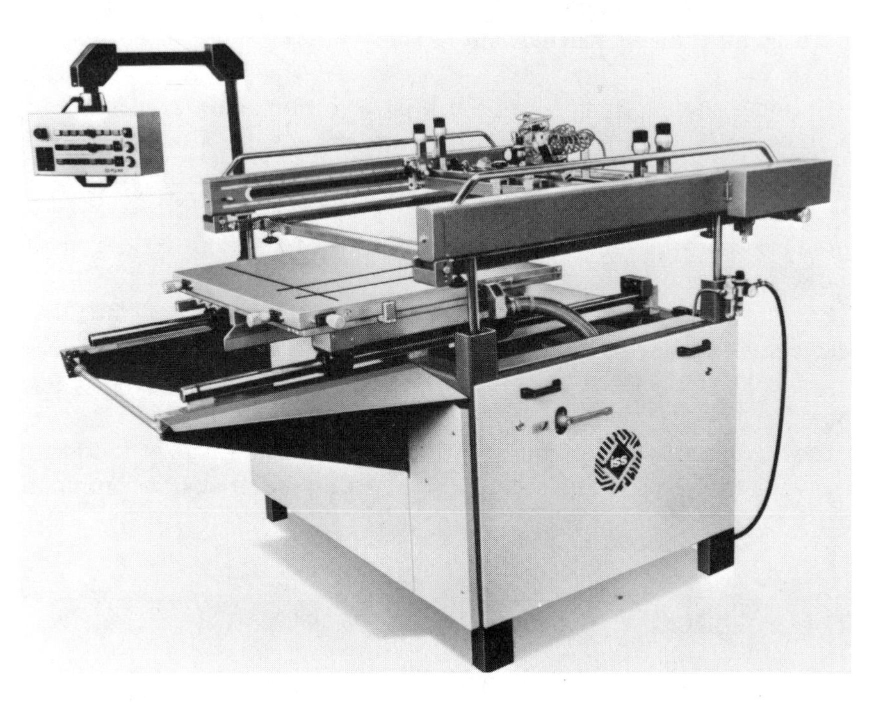

Figure 11-9 Automatic flat bed screen printer. Courtesy of Fineline, Inc.

of the two may be used to dry and cure printed parts. Static eliminators may also be included as optional equipment.

Ultraviolet-curable inks are gaining acceptance in screen printing, because they cure quickly upon exposure to UV radiation [16]. The introduction of UV curing helped the expansion of screen printing for plastics: heat-sensitive materials are easily cured and no bulky drying ovens are required.

Inks

Screen printing requires inks with special properties. The ink should have pseudoplastic flow properties. When placed on the screen it should possess a sufficiently high yield point to not flow because of gravity. When a shear force is applied by a squeegee, the yield point is overcome and the ink should flow easily through the screen openings. The flow should continue after the ink is deposited onto the substrate to assure that a continuous and smooth ink film is formed. Yet the flow should not be excessive, otherwise sharp images will not be obtained. The ink should also have a good adhesion to the plastic surface. Compounding a good screen ink is not an easy task, since some of the requirements are mutually incompatible and the best compromise must be sought.

Several ink types are used for screen printing. Solventborne inks are used most often. While they will dry at room temperature, accelerated drying is most often required in order to be able to run the operation at an acceptable speed. Ink may dry on the screen, requiring frequent cleaning.

Reactive inks are either cured by oxygen in the air (some alkyds), or are two-part systems that cure by chemical reaction between the ink components. Epoxy- and urethane-based inks are often used as two-component systems. While such inks may have good properties, they are more difficult to use and the drying of ink on the screen is much more troublesome than the drying of soluble solvent-borne inks.

Aqueous emulsion based inks are not used often, but they are becoming more important. This trend is expected to continue because of environmental requirements. Foaming and emulsion drying on the screen are difficult problems with aqueous inks.

Ultraviolet-curable inks consist of oligomer and monomer blends that contain both the required colorants and photo initiators, which start the curing after irradiation. UV inks do not dry on the screen, cure fast, and show low shrinkage. However, they also have some disadvantages: Some of the monomers may be irritants, and it is difficult to obtain a high degree of opacity, because UV light has only limited penetration and highly opaque thick coatings may not cure with UV irradiation.

In addition to these types of inks, various specialty inks may be used. Inks to obtain special surface effects, such as texturing, including imitation wood grain, are available. The gloss of the ink surface may be regulated by the addition of dulling agents, usually silica or sometimes metallic soaps. Highly abrasion-resistant inks and coatings may be applied by screen printing, especially by employing UV curable inks. Dead-front inks are used to hide printed displays until illuminated from behind. They are made by the addition of a small amount of black pigment to a clear vehicle. Such a coating provides sufficient hiding power to mask the print when viewed by reflected light, but not when viewed by transmitted light.

Printing Process

The most important part of screen printing is a well-imaged and -developed stencil supported by a correctly tensioned fabric. Screen mesh size is selected according to the job requirements (see Table 11-6). For a flat bed process, the image should be half as long and half as wide as the screen. The screen mesh should be oriented at 22.5 degrees to the frame to obtain good print definition. The screen frame is locked into the holder, positioning the screen. Stock is aligned with respect to the screen by several methods. The registration of the first color to be printed is not too important, but subsequent colors must be accurately aligned to the first image.

The distance between the screen and the print stock is most critical.

Table 11-6 Screen printing mesh size applications

*Mesh**	*Use*
195–240	Background color
204–240	Heavy, wide characters
240–280	Average to slightly coarse characters
305	Fine detail, high quality
305–355	Four-color process
305–420	Fine designs, halftone
395–420, calendered	Recommended for UV color graphics

*Medium polyester screen fabric.

The starting point of about 6 mm (0.25 in.) should be decreased until the screen will just break away from the stock as the squeegee moves across the screen. A larger distance will contribute to image distortion.

Squeegee pressure and angle affect the image. The minimum pressure that yields full intensity printing should be used. The angle between the squeegee and screen should be less than 90 degrees. Rectangular squeegee blades are inclined sufficiently so that printing is done by one edge. The squeegee speed should be such that the screen peels away from the stock just behind the squeegee. Too high a speed will produce smudging.

Screen printing is usually carried out by depositing a solid color, rather than employing the dot halftone process used in other printing methods. Therefore, suitable images rely on well-defined areas of single colors, like a poster, rather than areas of varying color and tone, as in a photograph. As many as seven or eight different colors may be deposited in register by screen printing. A four-color process employing halftone dots is difficult in screen printing.

FLEXOGRAPHIC PRINTING

Flexographic printing, or flexography, developed mainly as a method suitable for printing on plastics [17,18]. It is basically a letterpress process, since it prints from the raised image, except that instead of employing a hard metal plate, a soft, rubbery plate is used. For a comparison of different printing methods see Figure 11-10. Printing ink is transferred from the inked relief surface to the product by applying some pressure. While flexography was considered to be a crude printing process some time ago, it has been refined and developed to the point where high-quality printing is possible. In some areas it has become the dominant printing method. Flexographic printing is widely used for identification printing on plastics. A diagram of the process is shown in Figure 11-11.

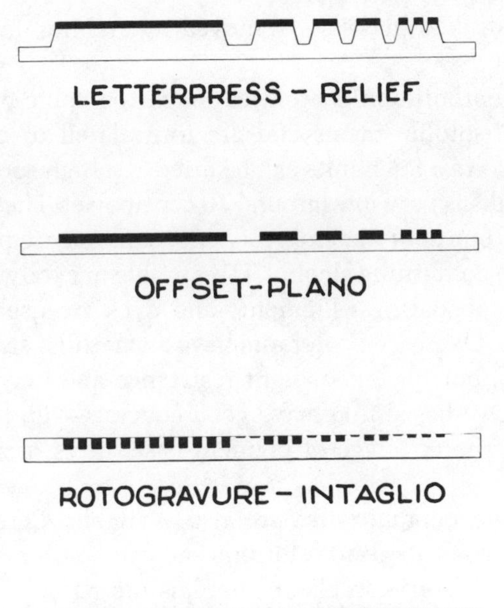

Figure 11-10 Profiles of various printing plate surfaces.

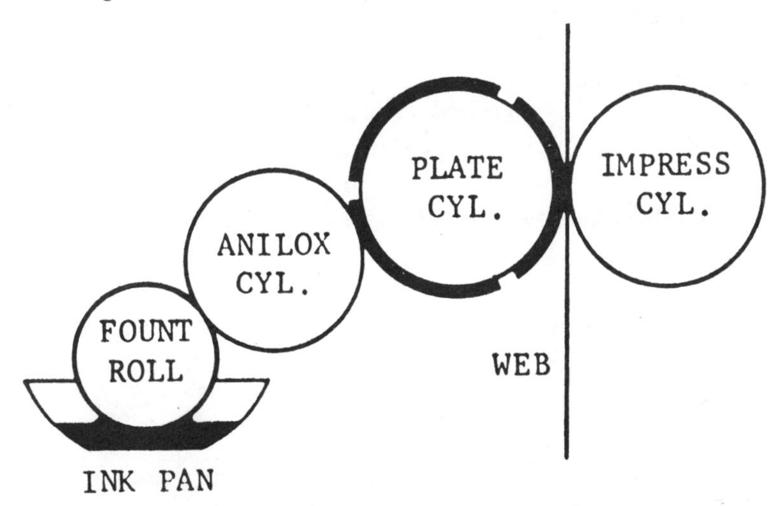

Figure 11-11 Flexographic printing.

Rubber and photopolymer printing plates are used for flexography. Rubber plates are made by duplicating the original photoengraving in metal by means of plastic matrices. When properly made, rubber plates are quite accurate in thickness. Rubber plates release various ink types well. Photopolymer plates represent a more recent improvement available since 1974. Light-sensitive plastics are used for these plates. The image is transferred through photographic negatives and the areas affected by light are cross-linked. Noncrosslinked plastic is removed by etching, leaving a photoengraved plate.

Various solventborne inks are used for flexographic printing. They are based on alcohol-soluble resins and are formulated to contain about 30 percent or more solids. Such inks can be dried at a high speed. The viscosity of flexographic inks is very low (around 70 centipoise). These inks have good water resistance, but poor alcohol resistance and may require topcoating if used on a package containing alcohol. UV-curable or catalytic cured coatings may be used as topcoatings. Pigments and dyes are used as colorants in flexographic inks. Dye-based colorants have a very high strength, give clean and bright colors, but have poor light resistance and may fade in sunlight in a few hours. Dye-based inks have good coverage—up to 400,000 sq in./lb—compared to the coverage of pigment-based inks of about 150,000 sq in./lb.

Waterborne flexographic inks are also available. Certain acidic resins are soluble in aqueous alkaline solutions and can be used as a base for such inks. Aqueous emulsions can also be used as ink base.

Pressure-sensitive adhesive labels are printed largely by flexography. Figure 11-12 shows a multistation press for label printing.

DRY OFFSET PRINTING

This printing method is a combination of two processes. A raised-surface printing roller, such as in letterpress, is used to transfer the ink to a smooth offset roll from which the ink is transferred to the product surface. The distinguishing feature of this printing method is the offset roll. Dry, pasty inks are used for offset printing and this method is widely used for plastic products: 90 percent of all plastic tubes and many jars, cups, bottles, and other containers are printed by offset. Figure 11-13 is a diagram of a dry offset printing unit, including a feeding station for plastic containers.

PAD PRINTING

Pad printing has grown into an important method for printing plastic products. It is basically an offset gravure process. The ink is applied in excess over an engraved steel plate called a cliché. The excess is removed by a doctor blade, leaving the ink only in the engraved recesses. A soft silicone rubber pad picks up the ink from the cliché and transfers it to the part to be printed. These steps are shown in Figure 11-14.

The process has been used for a long time, in a limited way, for the printing of watch dials, for which a fragile gelatine pad was used. The method gained wide acceptance only after the introduction of the silicone rubber pad in the late 1960s. Pad printing is a reciprocating process, although a rotary modification was introduced recently. The reciprocating process is

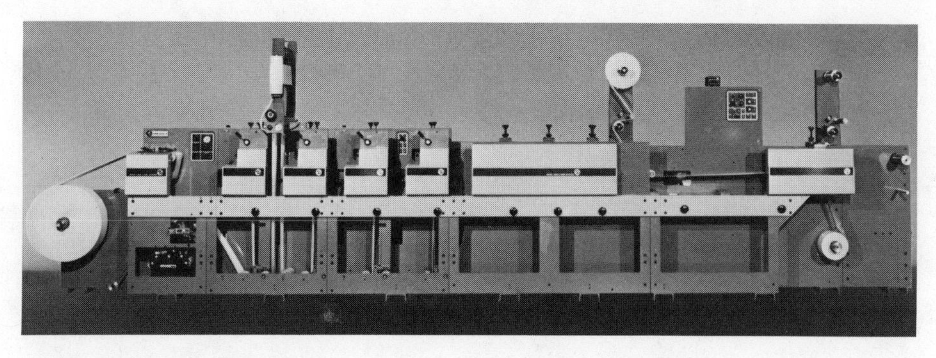

Figure 11-12 Multistation pressure-sensitive label printing line. Courtesy of Mark Andy, Inc.

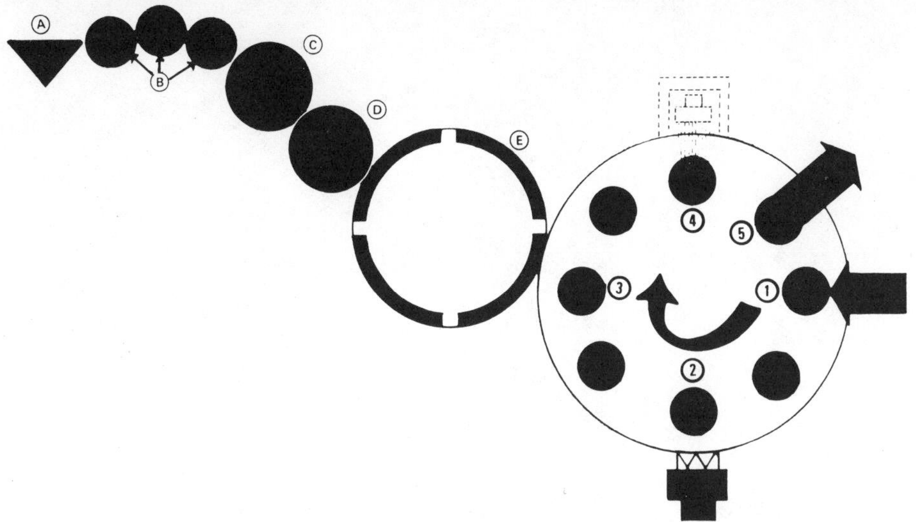

Figure 11-13 Dry offset printing station for round containers. Inking station: (A) ink fountain, (B) steel ductor and dab rolls, (C) roller system, (D) form roll, (E) plate cylinder, (F) blanket platen. Container handling station: (1) feed station, (2) surface treatment, (3) print station, (4) UV curing, (5) transfer to next operation.

inherently slow; it operates at a rate of 30–40 cycles per minute, but microcomputer-controlled units capable of 200 impressions per minute with an accuracy in registration of ±0.0001 in. have been built. The rotary process reportedly can print 300–400 bottles per minute.

The main advantage of pad printing is its suitability for printing irregularly shaped and rough surfaces. This capability is only limited by the conformability of the soft silicone rubber pad. Pad printing can reproduce fine line engraving with a high level of fidelity. Pad printing is generally used only for applying the ink to small areas. Areas up to 40 sq cm (6 sq in.) are easily covered. It was demonstrated more recently that much larger areas can be covered as well. Wet ink can be printed over wet ink. The main limitation is that only thin ink layers can be applied, which affects the ink opacity.

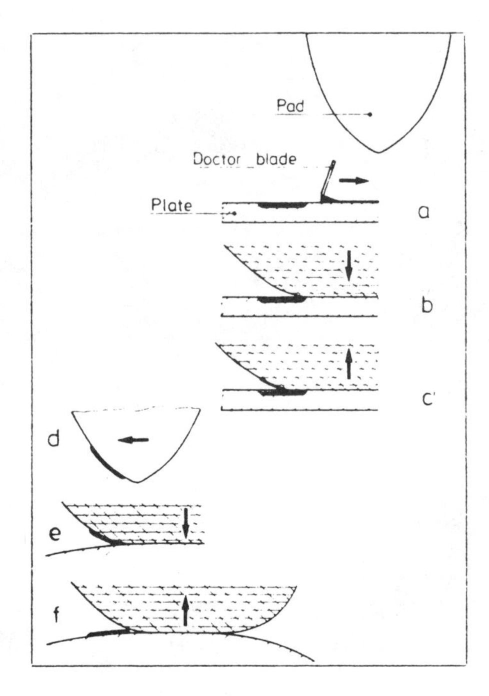

Figure 11-14 Stages of pad printing. (a) Doctoring to remove excess ink. (b) Initial contact with cliché (plate). (c) Ink removal from cliché. (d) Pad moving to printing surface. (e) Initial contact of printing surface. (f) Removal of the pad.

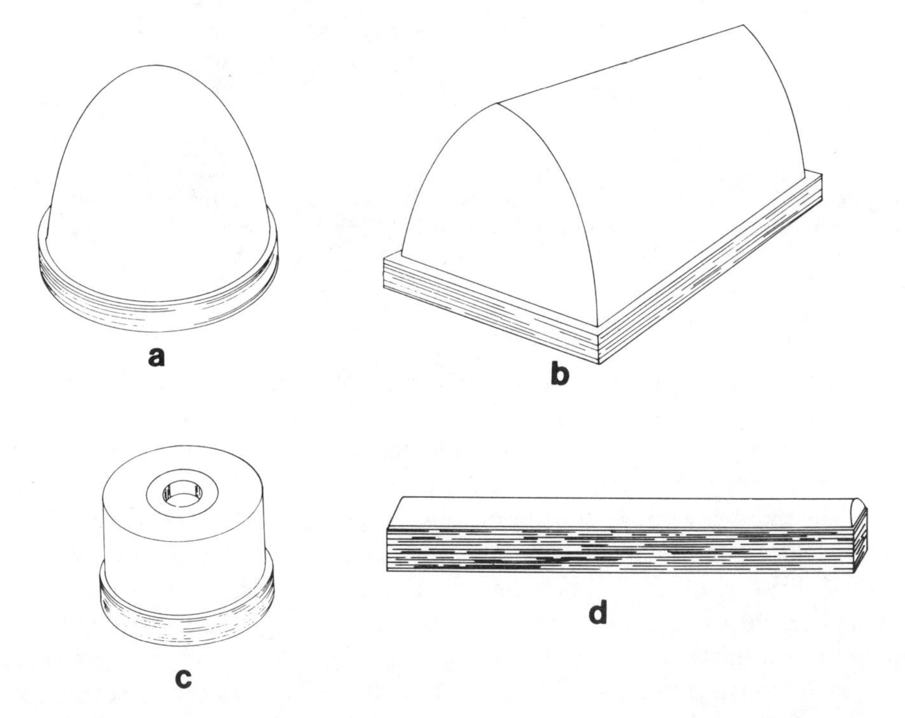

Figure 11-15 Basic pad shapes. (a) Standard. (b) Wedge. (c) Doughnut. (d) Ribbon. Courtesy of Midwest Technical Service.

Equipment

Pad printers are more expensive than hot stamping or screen printing equipment.

Clichés may be made from various metals, depending on the expected usage and the durability required. Steel clichés are the most durable but are also the most expensive to engrave. They are made from 0.4–1 cm (0.16–0.39 in.) thick hardened or chrome plated steel and are engraved mechanically or by photoetching to a depth of 0.02–0.04 mm (0.0008–0.00016 in.).

Copper, zinc, and magnesium plates are also used. They are easier to engrave, but do not last as long as steel clichés.

Plastic clichés are prepared from nylon or polyethylene by employing a photoresist emulsion, irradiating with UV, and then etching. Plastic plates may last for up to 20,000 imprints. Sometimes plastic clichés may be used to develop the proper art work for irregular surfaces where distortion must be incorporated into the design to compensate for the distortion of the silicone pad when printing.

A brush or spatula is used to apply the ink to the cliché. A steel or phosphorus bronze blade is used for the doctor blade that removes excess ink. The finer the art work, the thinner must be the blade. Ink application and doctoring systems can be automated if necessary.

The pad is manufactured from a soft silicone rubber by plasticizing the elastomer with silicone oils. Pad shape is selected according to the geometry of the part to be printed. Figure 11-15 shows several standard pad shapes. Pad *a* is the most often used shape. The pad is compressed during printing until the side walls are almost vertical. Many other shapes are available: one pad manufacturer offers 400 different combinations of pad shape, size, and hardness. Pads are available in four hardness levels in the Shore A scale range of 30 to 60. The hardest pad that can be used should be selected; harder pads are more durable and have better ink release properties.

Work holders or jigs must be designed to allow fast loading and unloading. They must retain items to be printed firmly and in register. For applying several colors in register, an indexed rotary table is most often used. Figure 11-16 shows a pad printer with a single printing head.

Inks

Inks for pad printing must have high opacity, since the ink layer deposited by the printer is thin. Various ink types may be used: modified screen printing inks, two-part reactive inks, acrylic- or polyurethane-based inks, or inks that diffuse into the plastic.

Ink viscosity is very important. A highly viscous ink may not be picked up by the pad from the cliché. If the viscosity is too low, the cohesion may also be too low and the ink may not transfer completely from the pad. In

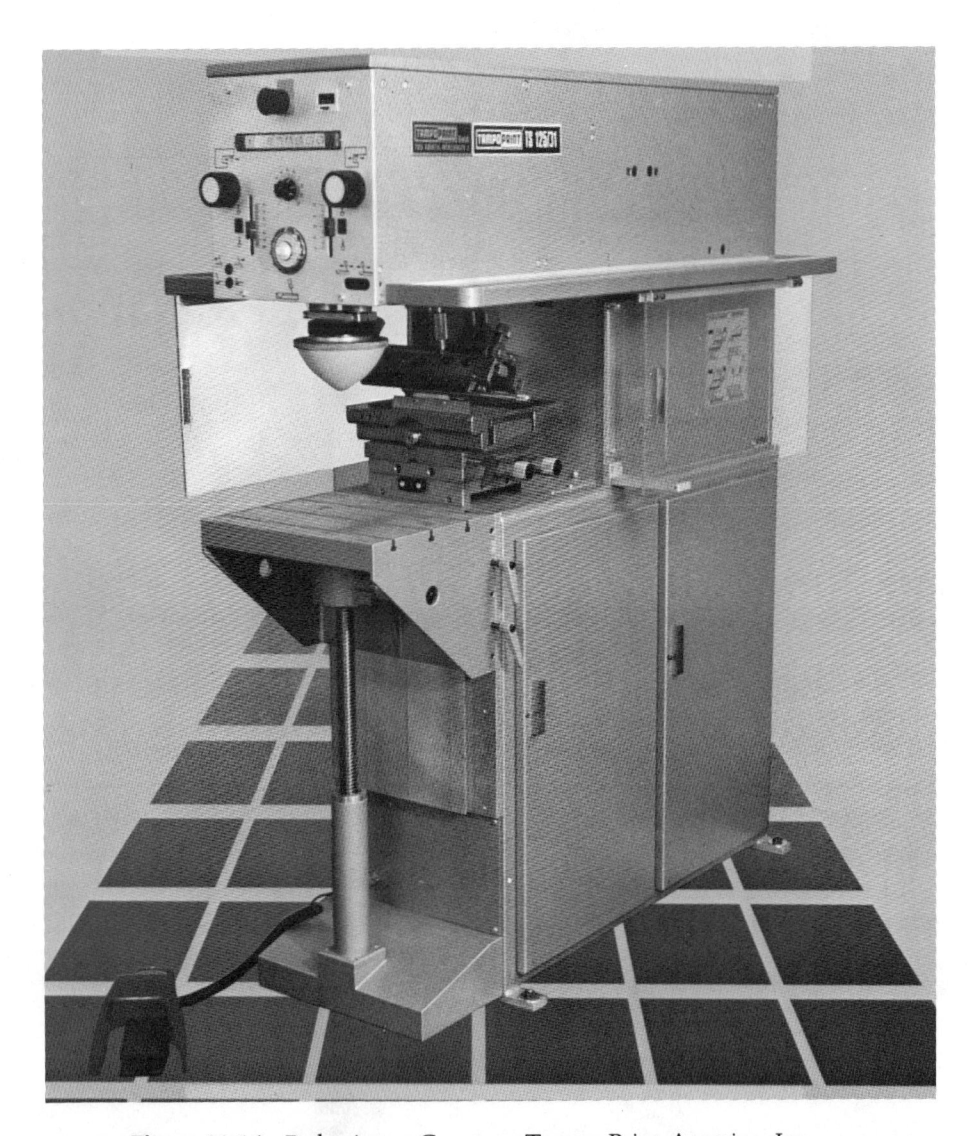

Figure 11-16 Pad printer. Courtesy Tampo Print America, Inc.

pad printing there is no ink splitting: ink is transferred completely from the cliché to the printed surface.

HOT TRANSFER PROCESSES

There are several methods that fit into the category of hot transfer processes. The three methods used in plastics decoration are:

- Hot stamping
- Decal transfer
- Sublimable dye transfer

Hot Stamping

Hot stamping involves the transfer of a pigmented coating from a carrier to the part to be printed by applying heat and pressure. The method originated with gold leaf application to leather book bindings and even now silver or gold metallic colors dominate hot stamping. Hot stamping is carried out by bringing a heated die in contact with the product to be decorated while interposing a stamping foil between the two. The die may be engraved in order to transfer the mirror image of the engraved pattern. A flat die may be used to transfer the foil only to the high points of a molded plastic article. The color print may be deposited below the surface (inlaid effect) and the metal printing die may also be used as an embossing tool. The print may also be deposited on the surface usually by employing a silicone rubber die.

Hot stamping foils (also called roll leaf) are composed of several layers:

Film carrier
Release coating
Decorative coating
Adhesive coating

Polyester film (0.1–1 mil thick) is used as a carrier most often, although cellulose acetate and cellophane film or paper may be occasionally used. Film thickness ranges from 10 to 40 μm, depending on the application. Polyester film has sufficient strength and temperature resistance to withstand the transfer operation.

A release coating is the first coating that is applied to the film. It consists of a wax or resin that has a low melting point and liquifies when contacted by a hot die. This allows the separation of the film from the decorative coating. The release coating should have a sharp melting point so as to form a sharp line at the edge of the die. A feathered edge may be produced otherwise.

A decorative coating is the pigmented coating that forms the required image. It may consist of several coats. The coating often is a thin aluminum layer deposited as vapor under vacuum. Such metallic coating needs protection, and a clear protective topcoating may be used so the silver color is visible. For metallic coatings that look gold or other colors, a coating containing a transparent dye of the desired color is first applied over the release coating and the metallic layer is subsequently vacuum deposited. Most of the foils used in the hot transfer process are single colors, but patterned foils are also used, especially for covering large surfaces. Wood grain foil is available in imitations of various wood surfaces. Such foils may consist of several coatings required to reproduce the wood surface. Typically, wood grain foils consist of release, protective topcoating, grain, base, and adhesive coatings. Foils with other patterns are also used. Such specialty foils should actually be considered as decals, but usually they are grouped with regular hot stamping foils.

An adhesive coating is required to secure the image to the surface of the decorated article. The adhesive must be activated in the temperature range used in the stamping operation. The temperature of the stamping die and the contact time are adjusted to properly melt the release coating so as to obtain a clean transfer, and to activate the adhesive coating to obtain good adhesion. Specific adhesive coatings are used according to the surface to be decorated.

Decal Transfer

Decal transfer is an extension of hot transfer printing. A decal is prepared in a way similar to a transfer foil, except that a complete multicolor design is printed on the foil and transferred as a unit.

Hot transfer decals may be screen printed and are subject to the general limitations of screen printing—limited tonal effects and primarily fine line printing—but also have the advantages of screen printing, such as good opacity and low first cost in preparing screens and setup. Transfers can also be gravure printed, which can reproduce photographic effects as well as line drawings. Transfers of this type have a much higher print quality, but the first cost is higher and relatively large runs are required to make it economically feasible.

Decal transfers are widely used to decorate plastic bottles, compacts, and other molded products. Standard hot stamping equipment is used, and a special registration feed system may be required.

Sublimable Dye Transfer

Dye transfer resembles hot stamping, but instead of a foil, a paper liner printed with sublimable dyes is used. These dyes sublime upon the application of heat and are transferred to the plastic. The pattern is not deposited on the plastic surface, but diffuses into the plastic, thus providing a considerably more abrasion-resistant image. The diffusing dye, however, gives a slightly blurred image. The usual operating temperature range is 300–450°F (150–230°C); the dwell time is between 15 and 30 sec. Thus, the plastic used should have a sufficiently high temperature resistance. This process is limited to only a few select plastic materials.

Sublimable dye transfer is an adaptation of a similar process developed for textile printing, especially for the printing of polyester fabrics. Its use for plastics has never developed to a large-volume business because the process is rather slow as compared to hot stamping or decal transfer and the processing temperature is high.

Hot Stamping Equipment

A hot stamping machine brings the heated die, plate, or roll into contact for a time period of less than one second with the surface of the product, with the foil or decal placed between the two. Reciprocating or rotary machines are used. Reciprocating machines are used for flat or moderately curved surfaces and for relatively small stamping areas. Rotary machines can decorate cylindrical parts where the area of coverage is greater than 90 degrees and also large surfaces. Figure 11-17 shows the differences between reciprocating, rotary press-cylindrical die, and rotary press-flat die machines.

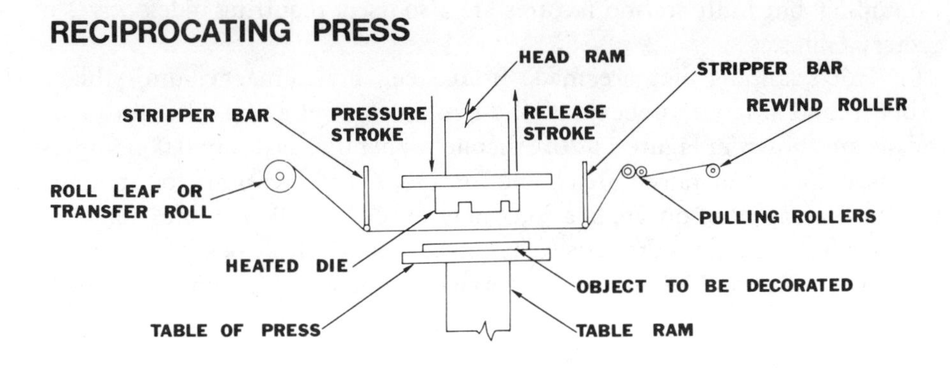

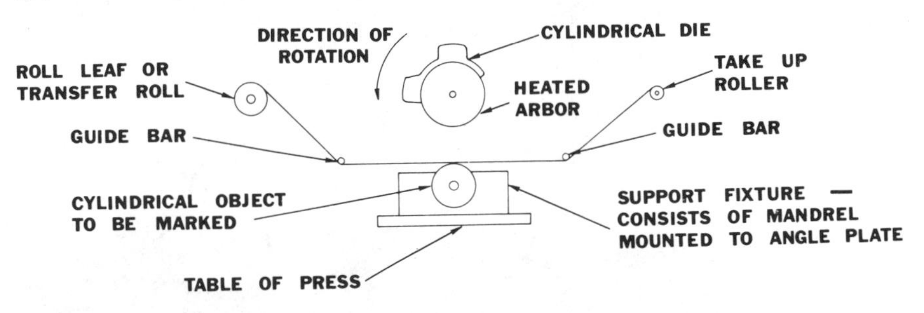

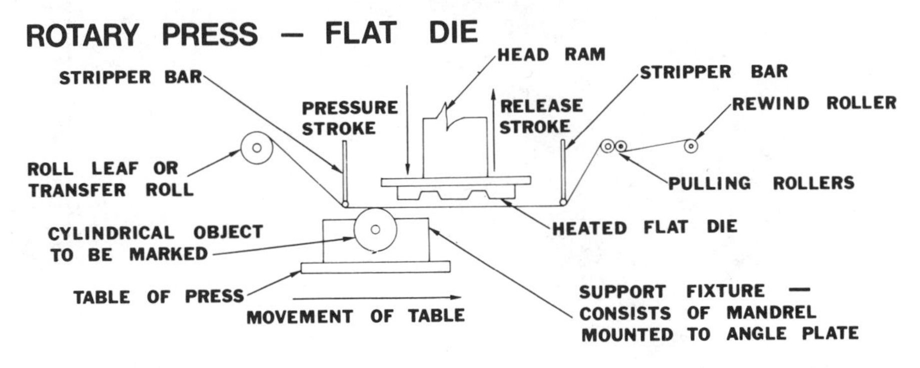

Figure 11-17 Hot stamping processes. Courtesy Kensol-Olsenmark.

Figure 11-18 shows a schematic drawing of a two-stage hot stamping machine intended for stamping cosmetic bottle caps and Figure 11-19 is a photograph of a hot stamp label printer.

The pressure is achieved mechanically by employing pneumatic or hydraulic cylinders to apply pressure on the die. Direct-acting air presses are most often used. If the actuating controls require manual operation, both hands should be used for safety reasons.

A special mechanism to advance the foil is required. Fixtures (or tooling) are provided to hold and support the part under the printing area. Fixtures are most often produced from aluminum, although plastics such as urethanes and epoxies are also used. Most fixtures are for single-station hot stamping, but multistation fixtures are also used, requiring index drives or rotary tables.

Hot stamping dies are made from steel, brass, magnesium, silicone rubber, and silicone rubber/metallic laminates. Metal and silicone die designs are shown in Figure 11-20. Silicone rubber dies last for 50,000 impressions if used over raised areas, and for 150,000 impressions for stamping smooth areas. Roll-on applications may exceed 200,000 impressions. Excessive heat usually is the cause of premature bond failure of silicone rubber/metal laminates. The maximum service temperature is considered to be 600°F (315°C).

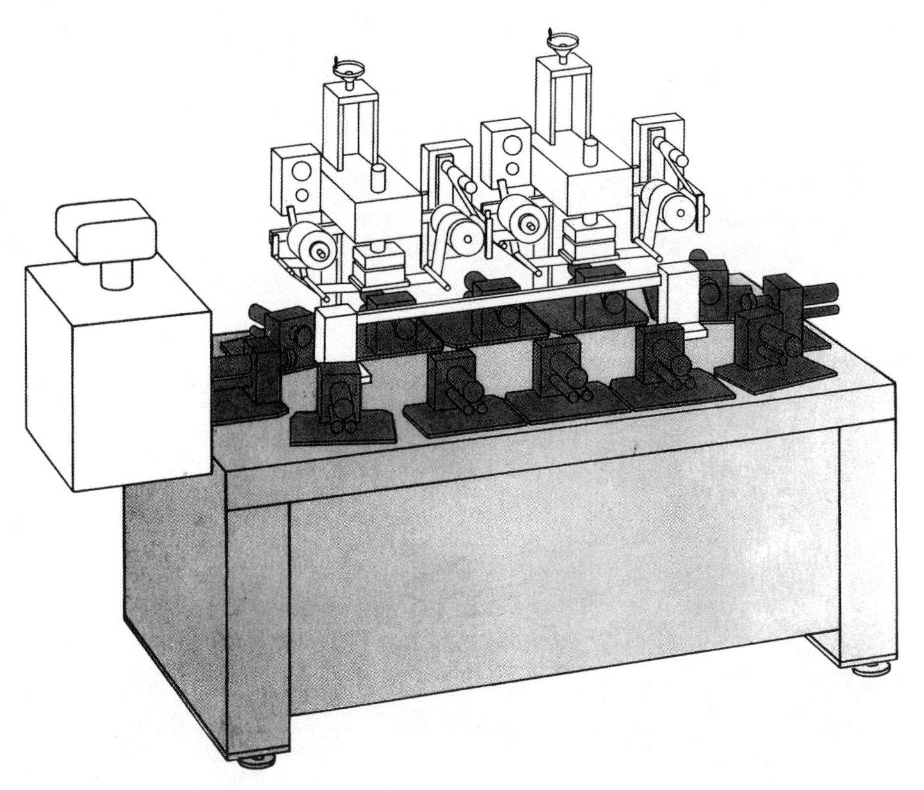

Figure 11-18 Two-stage hot stamping machine for cosmetic bottle caps. Courtesy of Kurz Hastings.

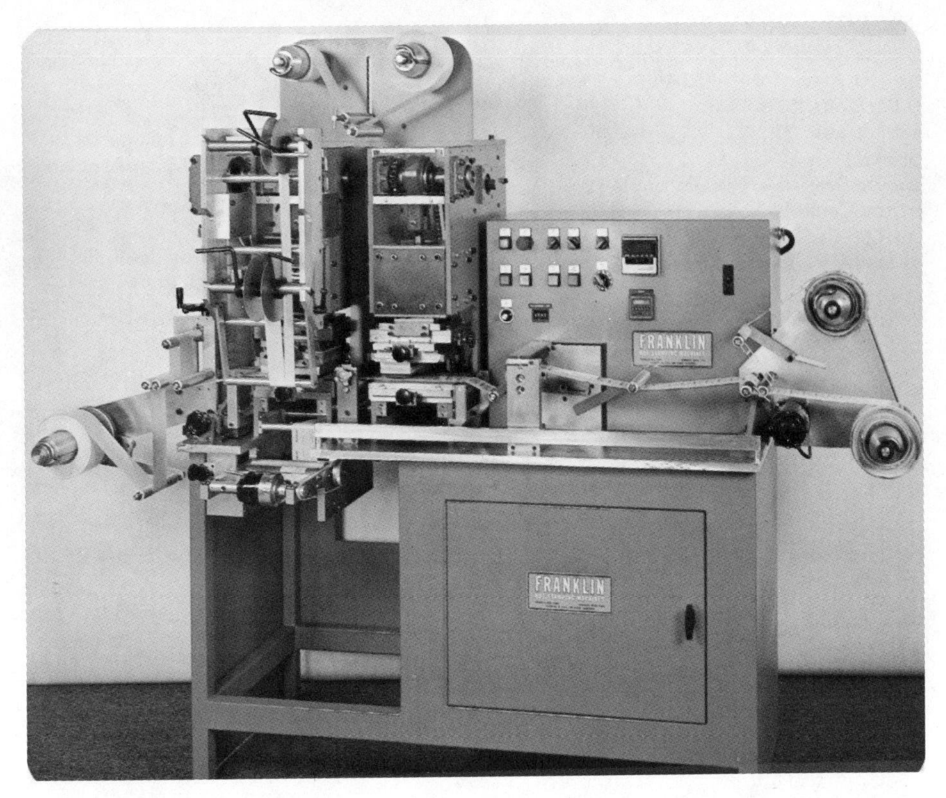

Figure 11-19 Hot stamp label stock printer. Courtesy of Franklin Manufacturing Co.

The primary methods of plastics decoration—direct screen printing, pad printing, hot stamping, and heat transfer—are compared in Table 11-7 [19].

LABELS AND DECALS

Labeling is the process of affixing a precut, printed, flexible material to the surface of the product. Thus labeling eliminates the need to print on the plastic directly, which may be difficult in the case of irregular parts. Labels are printed in large quantities on flat label stock: paper or film. If the function of a label is purely decorative, it is called a decal, an abbreviation of decalcomania, which is the process of transferring designs from specially prepared paper.

Labeling is the least expensive and therefore very widely used method for affixing informative messages and decorations to various packages, including plastic bottles and other containers.

Label Stock

Label stock is the material used for label printing. The most often used material for label stock is 50–60 lb bleached kraft paper coated on one side with a clay coating for improved printability. For high-quality labels the

Table 11-7 Comparison of plastics decoration methods

Feature	*Direction Screen Printing*	*Pad Printing*	*Hot Stamping*	*Heat Transfer*
Learning curve time frame for system start-up	Days to weeks	Days to weeks	Hours to days	Hours to days
New operator learning curve time frame	Hours to days	Hours to days	Minutes to hours	Minutes to hours
Set-up skill level	Skilled	Skilled	Semiskilled	Semiskilled
Operator skill level	Semiskilled	Semiskilled	Unskilled	Unskilled
Part chargeover: new part, new design	Minutes to hours to change tooling (nest), change screens, change inks, reregister (longer for multicolor)	Minute to hours to change tooling (nest), change clichés, change pads, change inks. Reregister (longer for multicolor)	Seconds to minutes, rarely hours to change tooling (nest), change application head, change foil. Reregister (longer for multicolor). Recheck pressure, dwell and temperature	Seconds to minutes, rarely hours to change tooling (nest), change application head, change roll of heat transfers. Reregister (time same as for single color). Recheck pressure, dwell and temperature
Process variance causing defective printing	Ink viscosity somewhat critical	Ink viscosity extremely critical	Process is quite stable. Silicone rubber die forgives mild surface blemishes	Process is quite stable since less pressure is required, softer rubber (50–60 durometer), forgives many surface blemishes
Part variance causing defective printing	Poor surface finish and blemishes, e.g., sink marks (some defects)	Surface blemishes usually unaffected unless blemish is extreme (fewer defects)	Direct ram overcomes thickness changes. Toggle machines very sensitive to thickness changes (some defects)	Direct ram overcomes thickness changes. Toggle machines very sensitive to thickness changes (some defects)
Cost of equipment	Small to large area (100 sq in): not very cost sensitive; multicolor: very cost sensitive	Small to large area: cost sensitive; multicolor: cost sensitive	Small to large area: cost sensitive (multicolor more than one pass or more than one machine)	Cost sensitive to size, not sensitive to multicolor tooling (nest and heads). Cost same for single or multicolor

Cost of tooling	Single color: low to moderate; multicolor: moderate to high depending on tolerances required	Single color: low to moderate; multicolor: moderate to high depending on tolerances required	Single color: low to moderate; multicolor: moderate to high depending on tolerances required	Low to moderate cost as for single color even though graphics may be multicolor
Part input–output equipment cost	Approximately same for all processes; may be more costly for mulitcolor	Approximately same for all processes; may be more costly for mulitcolor	Approximately same for all processes; may be more costly for multicolor	Approximately same for all processes; NOT more costly for multicolor
Cost of inks, foil, transfers	Inks not very cost sensitive to size	Inks not very cost sensitive to size	Foils cost sensitive to size; costs for multicolor increase linearly per color and area	Transfers cost sensitive to size, not as sensitive to additional colors
Recognition factors	Thicker and more opaque, no clear/adhesive at edges	Thinner and less opaque fine copy; large color areas may look weak	Colors usually debossed can be bright gold or silver	May have tooling halo around design, usually multicolor
Image size and limitations	Screens can be made any size	7″ × 14″ usual limit, special machines can print 10″ × 20″	Limited by pressure of machine and tendency to trap air, usual range 300–500 psi. Roll-on solves air entrapment and can apply 12″ × 24″.	Limited by pressure of machine and tendency to trap air, usual range 100–300 psi (soft goods as low as 30–50 psi. Roll-on solves air entrapment and can apply 12″ × 24″.
Resolution of detail	Medium	Fine to medium	Medium	Fine, including 133 line 4-color process
Large areas of solid color	OK with good equipment and operator	Not good without multiple prints of color	Possible, but trapped air can be a problem; roll-on machine helps	Possible but trapped air can be a problem; roll-on machine helps
Opacity	Good	Poor; with multiple prints only fair.	Good	Good if screen printed; fair if gravure printed
Color match	In-house responsibility or ordered from outside supplier	In-house responsibility or ordered from outside supplier	Use closest color of foil available; get custom formulated for long runs	Inks must be custom formulated when transfer is printed

(continued)

Table 11-7 *(continued)*

Registration of colors if multicolor	Fair to good depending on equipment, tooling, operator and size stability of plastic from first to last print	Fair to very good depending on equipment and whether part stays in same nest through all color prints and quality of the tooling	Fair to good depending on equipment, tooling, and size stability of plastic	Very good; 4-color process, 133-line screen demands tight registration; so does multicolor and fine detail
Part shape and limitations	Flat or single curve (cylinder)	Can be irregular and/or compound curve but art distortion requires trial and error to correct	Flat or simple curve (cylinder)	Flat, single curve or slight compound curve; carrier paper wrinkling limits shape and size on compound cuves
Arc limits on cylinder or a cylinder with draft or taper 1° or less	Almost 360°, avoiding ink to screen contact on the wrap if it will be a problem	Approximately 100° arc for reciprocal machine or 360° for special wrap machines	Approximately 90° arc for reciprocal machine, 360° (with slight overlap preferred) for wrap machines	Approximately 90° arc for reciprocal machine; 360° with slight overlap preferred for wrap machine except if wax release, then 360° minus $\frac{1}{8}$
Process wet ink or dry	Wet-drying or curing required between and/or after final color	Wet-drying or curing required between and/or after final color. Some ink systems can be printed wet on wet.	Dry; proceed to next process	Dry; proceed to next process
Inventory required	Screens of various meshes, inks to make custom colors for partiuclar substrate chemistry, solvents, cleaners, retarders, squeegees, or order match color inks as needed	Pads of various sizes, shapes and durometers, inks to make custom colors for particular substrate chemistries, solvents, cleaners, retarders, or order match colors as needed.	Various hot stamping foils in various colors of chemistries compatible with surfaces to be marked, or order as needed	Heat transfers for specific job
EPA and fire safety consideration	Flammable materials must be stored and insured accordingly	Flammable materials must be stored and insured accordingly	None	None

Courtesy of the Meyercord Co. [19].

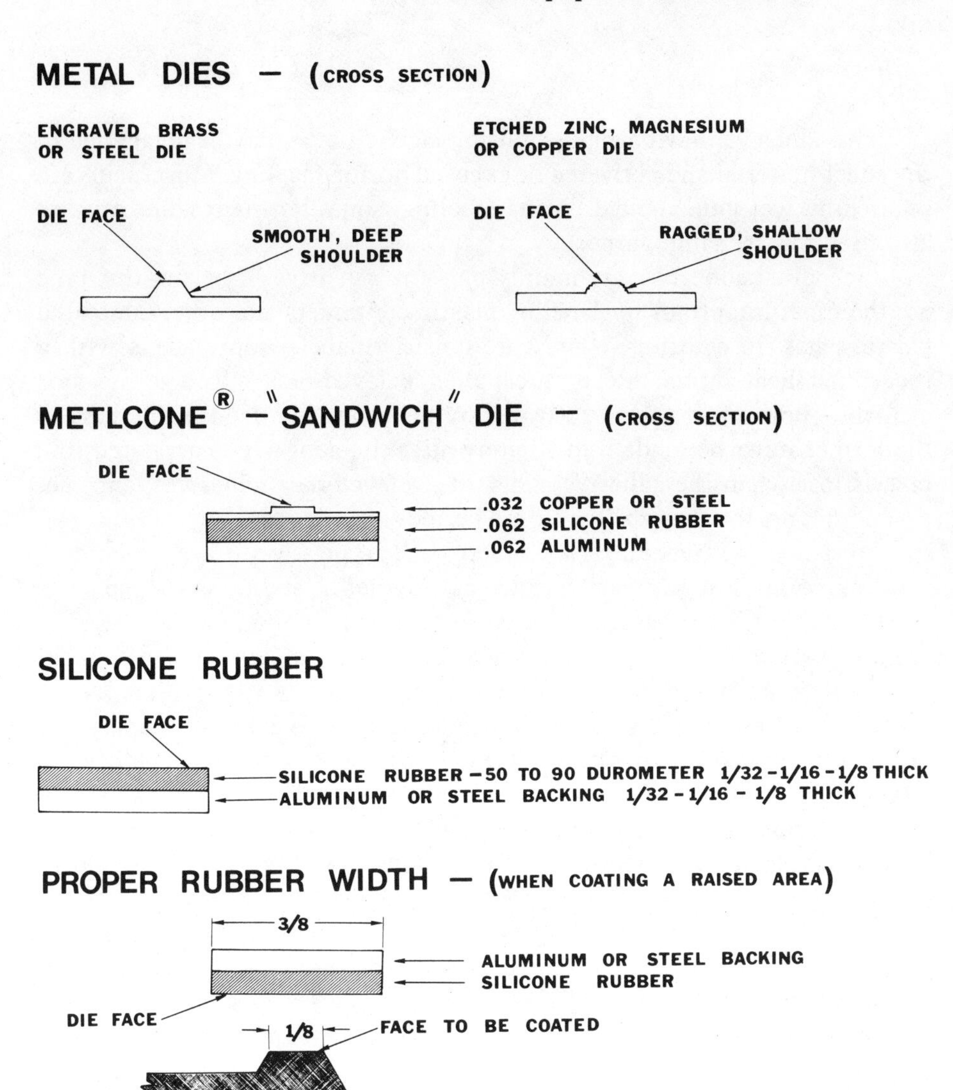

Figure 11-20 Metal and rubber dies. Courtesy of Kensol-Olsenmark.

coating may be a high gloss cast coating. Metallized papers or paper/aluminum foil laminates are frequently used to enhance label appearance.

Plastic films are also used as label stock. Many plastic containers are flexible and film stock, such as plasticized vinyl, is capable of flexing along with the container, avoiding the formation of wrinkles that might occur in a paper label. Transparent label stock may be used to imitate screen printing: only the label printing is visible against the background of the product surface.

Adhesives

The adhesive may be applied during labeling or it may be preapplied to the sheet material and activated or exposed during labeling. Most labels are secured by wet glue applied during labeling; some hot melt adhesives are also used for the same purpose.

Preapplied adhesives are mainly pressure sensitive. Labels of this type are the most important for labeling plastic containers and other fabricated plastic parts. In addition to pressure-sensitive labels, some labels with a preapplied heat-activatable or preapplied delayed-tack adhesive are also available. Both of these are activated by heat, but the bond in the case of the former must be made immediately after the adhesive is activated; the tack is lost when the adhesive cools off. Delayed-tack adhesives, once activated, retain the tack for long time periods.

Pressure-sensitive adhesives adhere well to plastic surfaces, including some adhesion-difficult plastics, such as polyolefins and fluorocarbon polymers. Furthermore, pressure-sensitive adhesives do not require drying after the application, unlike wet glue labels. Therefore, impervious label stock, such as films and metallized or laminated papers, may be used with pressure-sensitive adhesives. Pressure-sensitive adhesive labels are easy to apply and only a light pressure is required to secure the bond. Before use, adhesives are protected with a release sheet from which the labels are removed prior to their application. Labels of irregular geometry can be die cut easily on release paper and they are much easier to handle when mounted on release paper.

The label industry divides adhesives into several categories: permanent, removable, and low temperature. A permanent adhesive is expected to secure the label firmly in place so that a paper label will be destroyed upon removal. A removable adhesive allows clean label removal without its destruction. In some cases only temporary removability is required, such as when the label is not properly applied and requires a correction. Such adhesives are described as repositionable.

Low-temperature adhesives should retain tack at refrigerator or at freezer temperatures. Some of these adhesives may be applied directly to food products and they must be FDA approved for either direct or indirect food contact.

Pressure-sensitive adhesives may consist of several polymer types and many of their properties depend on the adhesive's chemical composition. The most often used adhesives are acrylic emulsions (and sometime solutions). These are single-component adhesives; that is, no additives such as resins are required to make them tacky. Most acrylic adhesives are polar polymers and their adhesion to polyolefins and other low-energy surfaces is not always the best. This deficiency is easily corrected by formulation. Acrylic adhesives have excellent aging properties; they are clear and do not yellow on aging or exposure to sunlight. The next category, about equal in volume to acrylic emulsions, is hot melt adhesives compounded from styrene/isoprene/styrene block copolymers and tackifying resins. Block copolymer based adhesives have good adhesion to polyolefins. The next, less important adhesive type, is SBR latex based adhesives, which are compounded with tackifying resin emulsions. Other adhesive types may also be used for labels, but less frequently.

Release Sheet

Pressure-sensitive adhesive labels and decals are supplied on a silicone-coated release carrier. Calendered, machine finished, polyethylene coated, and clay coated kraft papers of 50–60 lb weight are usually used. Polyethylene-coated paper exhibits a lower degree of curl when exposed to humidity changes. This is especially important for labels that are printed and handled in sheet form.

Release-coated films are also used. Films give a much smoother adhesive surface; they are stronger and cause less tearing during high-speed label dispensing.

Label Manufacture

Labels are produced in three major steps: manufacture of label stock, label printing, and label die cutting.

Label stock is manufactured by the transfer coating technique: the adhesive is applied to the release liner, dried, and then laminated to the face stock, transferring the adhesive from the liner to the face stock. This method, rather than direct coating, is preferred because the face stock is less resistant to oven handling. Figure 11-21 is a diagram of transfer coating.

Label printing is usually a narrow-web operation. Therefore, the large label stock rolls are slit to the required width. Printing is done by flexography or by letterpress on multicolor narrow-web machines. Some large-volume label production may be carried out in wide width on rotogravure printers.

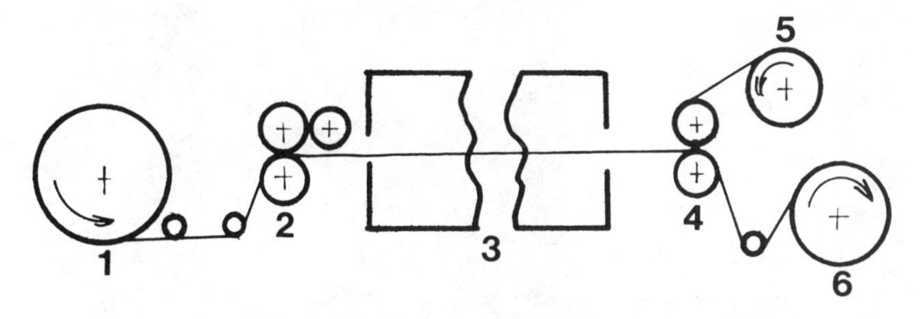

Figure 11-21 Transfer coating of pressure-sensitive label stock. (1) Release paper unwind. (2) Coater. (3) Drier. (4) Laminating rolls. (5) Label face supply roll. (6) Rewind.

On the narrow-width machines, die cutting is done in line with printing. Such a label producing machine is shown in Figure 11-22. Reciprocal or rotary dies may be used for die cutting. The die-cut matrix is removed and the labels remain attached to the release liner.

Label Application

Large decals are usually applied by hand, but automated equipment is used for applying small labels. Labels may be secured by wet glue or hot melt adhesive application at the time of labeling, or most frequently in the case of plastic packaging, by pressure-sensitive adhesive preapplied to the label stock. While pressure-sensitive labels are more expensive than wet glue labels, the application equipment and operating costs of pressure-sensitive labeling are lower and therefore it is preferred for low-volume runs. Table 11-8 shows the comparison of labeling speeds and equipment costs for most commonly used labeling and marking processes.

If the release paper with attached labels is run over a sharp edge, the label disbonds and continues to move on the straight pass. This is illustrated in Figure 11-23. Pressure-sensitive label application is based on this behavior. After removal from the release sheet the labels are deposited onto the package by one of the several methods illustrated in Figure 11-24.

Figure 11-22 Label and stock printer. Courtesy of Mark Andy, Inc.

Table 11-8 Labeling and marking speeds for various methods

Operation	*Speed (pieces/min)*	*Equipment Costs*
Glue labeling	350–400	High
Pressure sensitive	300	Low
Heat seal labeling	200–300	High
Screen printing	100–125	Moderate
Heat transfer	60–80	Moderate

IN-MOLD LABELING

Various in-mold decorating and labeling processes have been developed in order to eliminate an operation that otherwise must be carried out separately. In-mold labeling challenges pressure-sensitive and heat transfer labeling.

In-mold inserts have been used for a long time in a limited volume. These inserts consist of a plastic film preprinted with the required decoration and usually made from the same material as the molded article. Such an insert is placed inside the mold and becomes an integral part of the molded article.

The tremendous growth of plastic bottles created the opportunity to increase the use of in-mold labeling, which is only a slight modification of the earlier insert process. In-mold labeling is suitable for rectangular and oval containers, and usually employs paper labels produced from a medium base weight stock coated on one side for printability and on the other side with a heat seal coating. Die-cut labels are placed in magazines from which a robotic arm picks them up and inserts them into the mold. In-mold labels have several advantages: they are more attractive than postapplied labels, they have no wrinkles, and they are inlaid into the plastic bottle surface. It is also claimed that a reduction of 10–15 percent in container weight is possible, because the label becomes an integral part of the bottle and reinforces its walls [20].

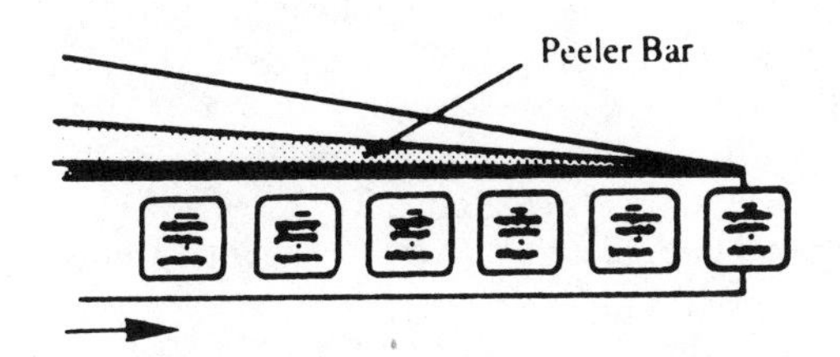

Figure 11-23 Pressure-sensitive label removal.

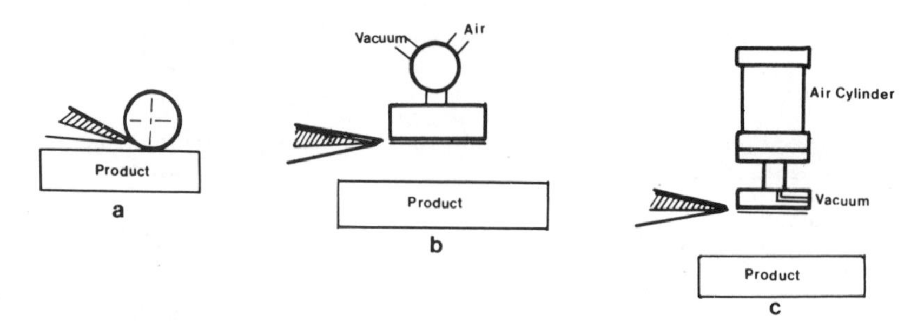

Figure 11-24 Pressure-sensitive label application methods. (a) Roll-on. (b) Blow-on. (c) Impresser.

The recent introduction of plastic labels eliminates the label bulge problem experienced with paper labels, which act as insulators and decrease the cooling rate of the plastic in the area covered by the label, causing a distortion in that area. Another advantage of plastic labels is that they are directly recyclable to the blow molding process. Plastic labels are more expensive (polyethylene labels cost three times as much as paper ones), but the savings in materials because of recycling offset this cost difference [21]. Polystyrene label stock is also offered for this application [22]. In-mold labels are printed by any one of the common printing methods: offset lithography, rotogravure, flexography, or letterpress.

REFERENCES

1. Satas, D., ed. *Plastics Finishing and Decoration.* New York: Van Nostrand Reinhold, 1986.
2. Margolis, J. M., ed. *Decorating Plastics.* Munich, Vienna, New York: Hanser Publishers, 1986.
3. Stoeckhert, K., ed. *Veredeln von Kunststoff-Oberflaechen.* Muenchen Wien: Carl Hanser Verlag, 1974.
4. *Modern Plastics Encyclopedia.* New York: McGraw-Hill, 1988.
5. Satas, D. Decorating Plastics. *SPI Plastics Engineering Handbook,* M. K. Berins, ed. New York: Van Nostrand Reinhold, 1990.
6. Shafrin, E. G. *Handbook of Adhesives,* I. Skeist, ed. New York: Van Nostrand Reinhold, 1977. pp. 67–71.
7. ASTM Standard D2578-67.
8. Markgraf, D. A. "Practical Aspects of Determining Level of Corona Treatment." *Polymers, Lamination and Coatings Conference,* TAPPI, Boston, September 24–26, 1984. pp. 507–511.
9. Mittal, K. L., ed. *Adhesion Measurements of Thin Films, Thick Films, and Bulk Coatings.* ASTM STP 640, Philadelphia, 1978.
10. Beeferman, H. L., and D. A. Bergren. *Journal of Paint Technology* 38(492) (1966):9–17.
11. Matsuda, T., and W. H. Brendley, Jr. *Journal of Coating Technology* 5(658) (1979):46–60.

12. Von Hor, R. C. "Spray Painting Plastics." *Plastics Finishing and Decoration,* D. Satas, ed. New York: Van Nostrand Reinhold, 1986. pp. 130–145.
13. Satas, D. "Spray Coating." *Web Processing and Converting Technology and Equipment,* D. Satas, ed. New York: Van Nostrand Reinhold, 1984. pp. 97–104.
14. Gilleo, K. B. "Screen Printing." *Plastics Finishing and Decoration,* D. Satas, ed. New York: Van Nostrand Reinhold, 1986. pp. 156–173.
15. Bell, J. E. "Rotary Screen Coating." *Web Processing and Converting Technology and Equipment,* D. Satas, ed. New York: Van Nostrand Reinhold, 1984. pp. 81–96.
16. Satas, D. "Ultraviolet Irradiation." *Web Processing and Converting Technology and Equipment,* D. Satas, ed. New York: Van Nostrand Reinhold, 1984. pp. 319–330.
17. Anthony, G. H. "Flexographic Printing." *Plastics Finishing and Decoration,* D. Satas, ed. New York: Van Nostrand Reinhold, 1986. pp. 174–192.
18. *Flexography Principles and Practices.* 3d ed. New York: Flexographic Technical Assoc., Inc., 1980.
19. Janco, R. A. *Technical Papers,* vol. XXXIII. 45th ANTEC, Soc. Plastics Eng., 1987. pp. 1136–1138.
20. Brockschmidt, A. *Plastics Technology* 30(13) (December 1984):63–66.
21. Lodge, C. *Plastics World* (May 1989):39–42.
22. Dudley, C. A., and R. S. Harfmann. "Opticite Film: The Plastic In-Mold Label Stock." *In-Mold Labeling: Advances and Innovations.* North Branch, NJ: IML Technology, Inc., Sept. 23, 1986. pp. 22–33.

SECTION 3

PLANT OPERATION

CHAPTER

12

Variables Affecting the Blow Molding Process

THOMAS E. DOUGLAS

The blow molding processes currently in use have been described in the preceding chapters. With the growth of statistical process control and statistical quality control (SPC, SQC), and all other computer-integrated manufacturing (CIM) programs it is imperative to define those variables that affect machine cycle time and part quality, and to establish the interrelationships of these variables in order to properly monitor and control the process and product.

MACHINE SETUP

The initial equipment setup is the most critical step in efficient controlled production. Setup sheets with all critical information relating to mold position, head tooling, cooling line routing, and peripheral equipment setup must be followed exactly once the optimum setup has been established.

If records are not kept, a mold will not be set up the same way twice in a row and all previously established CIM data will be useless. Many new machines have memory cards or tapes or disks that store the actual instrument setting from a given machine run and reset the machine exactly each time a new setup is completed.

The initial job setup is critical because it affects all other variables of the process. The following items should be included in the setup:

- Grinding and blending equipment
- Photos if used
- Trim equipment
- Testing equipment
- Packaging equipment
- Labeling and identification equipment
- Vacuum assists
- Prepinch systems
- Tooling design and program profile
- Extruder temperature profiles and operating speed
- Linear positioning controls
- Head, accumulator, and tooling temperatures
- Machine sequence timer settings
- Melt temperature and pressures
- Speed controls
- Hydraulic pressure

RESIN

Probably the most uncontrollable variable in the blow molding process is the base resin. Variations from supplier to supplier of like resins, and even variations from lot to lot and box to box within a lot from one supplier, can cause havoc in the blow molding process. While the molder may not be able to control the base resin entering the process, he can do much to reduce the variations within the process.

Elimination of contamination of resin by dirt, trash, and/or other resins can be eliminated by commonsense control and cleanliness in the process. Proper, consistent proportioning and blending of new resin, color concentrate, additives, and regrind will reduce variations to a minimum.

Resin temperature affects viscosity and part quality by causing variations in the parison extrusion time, length, thickness, and cooling time. This temperature is a critical parameter that must be monitored carefully and the source of any variations must be accurately identified. Extruder factors that can cause resin temperature variations include:

- Extruder heat profile
- Screw speed
- Screw and barrel condition
- Barrel cooling conditions and setup
- Temperature settings for controls in manifolds
- Heads and tooling

- External factors (e.g., air currents caused by fans, open window, air leaks, etc.)

MOLDS

Variations in mold conditions have a direct effect on part quality. Important factors are:

- Mold temperature, chiller performance
- Mold venting, design
- Mold closing speed, setup
- Mold surface condition, wear or damage
- Mold accessories (hot knives, neck inserts prepinch systems, core slides, punches, etc.)

The production of defective parts can be greatly reduced by recognizing the interrelationships of process and material variations and properly analyzing problems to provide timely, accurate solutions. Chapter 13 deals with the latest methods used to provide real-time information and control of the blow molding process.

The blow molding cycle is depicted in Figure 12-1. As can be seen, some of the cycle atributes must take place sequentially while others overlap, with some functions being performed during the same time period.

Care should be taken when adjusting cycle timing controls because some equipment, especially older models, is set up in such a way that one timer can overlap another and cause sequence errors. Opening the molds before the blow air is exhausted can be a safety hazard as well as a cause of defective parts production.

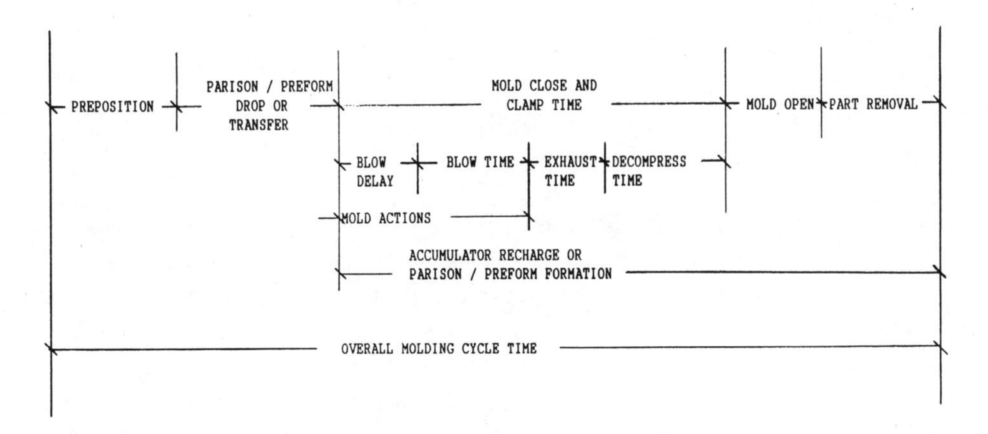

Figure 12-1 Blow molding cycle.

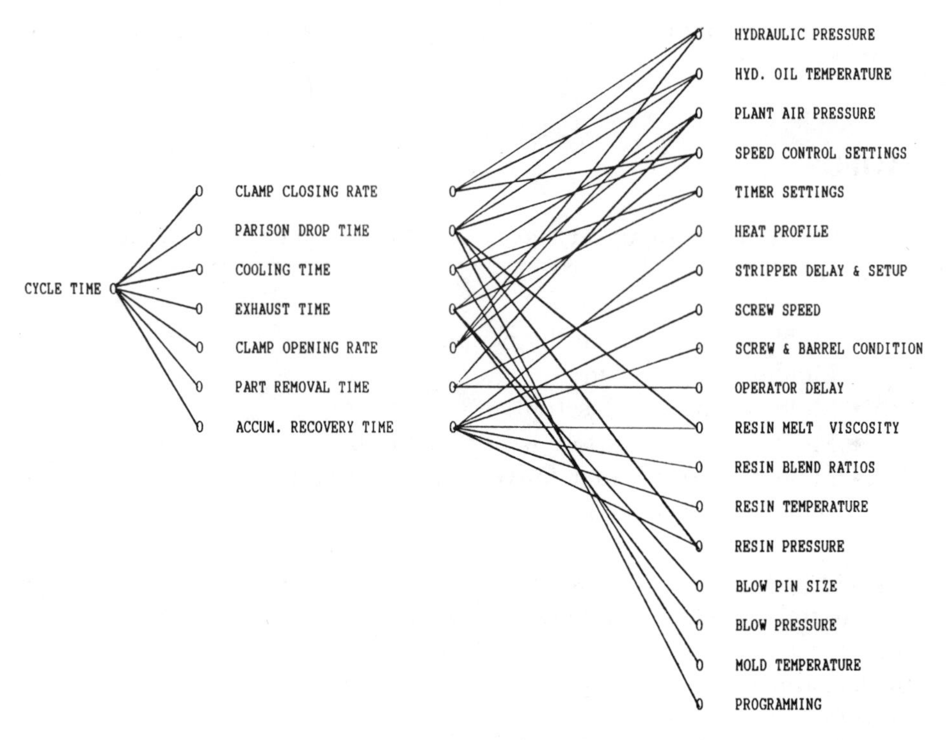

Figure 12-2 Interrelationships of cycle time variables.

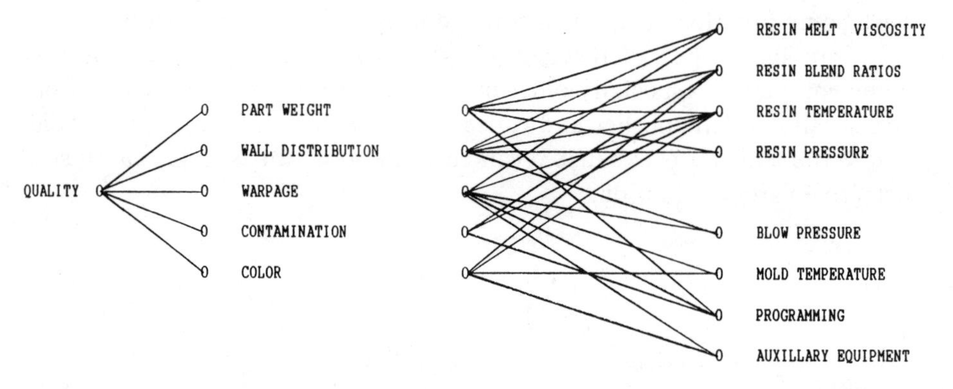

Figure 12-3 Interrelationships of part quality variables.

Figures 12-2 and 12-3 list the variables in the blow molding process and show their interrelationships. As can be seen, the evaluation process can become very complicated when just one parameter varies. The effect of one variable on other variables must be understood in order to properly analyze and correct problems in a timely manner.

Chapter 13 discusses methods of isolating, evaluating, and controlling these variables in order to assure quality part production at the most economical cycles.

CHAPTER

13

Computer Integrated Manufacturing in Extrusion Blow Molding

DENES HUNKAR

The following CIM plan is presented as a concept, but it can be fully implemented with currently available hardware and software products. The CIM plan described here allows the blow molder to use molding and auxiliary equipment of his choice; it does not tie him to any one manufacturer. It employs technologies that can be retrofitted to existing blow molding and auxiliary equipment. While the blow molding process is taken as the model for this conceptual presentation, the plan is also being implemented in injection molding and transfer molding for semiconductor encapsulation. Its principles should be valid for other processes, as well.

ELEMENTS OF THE CIM PROGRAM

CIM begins with the integration (connection) of all equipment into a computer network (see Fig. 13-1). All equipment and processes that have an effect on productivity or quality will be monitored and controlled by a central computer. This CIM plan addresses five functional aspects of plant operation that have a major impact on productivity and quality:

- Process control
- Quality control

- Response
- Equipment maintenance
- Production control

Statistical Process Control

Statistical process control (SPC) forces all production machines to operate and produce parts according to the standards established by a previously approved production sample. The SPC program operates as an actual process control by calling for corrections of the machine or process in the event that the process deviates from the accepted standard.

The benefit of this program is that part quality will correspond to the quality of the standard sample, as will the machine cycle time and other cost factors. The net result is consistent and predictable productivity and an improvement in yield.

Automated Statistical Quality Control

Statistical quality control (SQC) is a derivative practice based on the results of SPC. Conceptually, SQC can reject parts that do not conform to

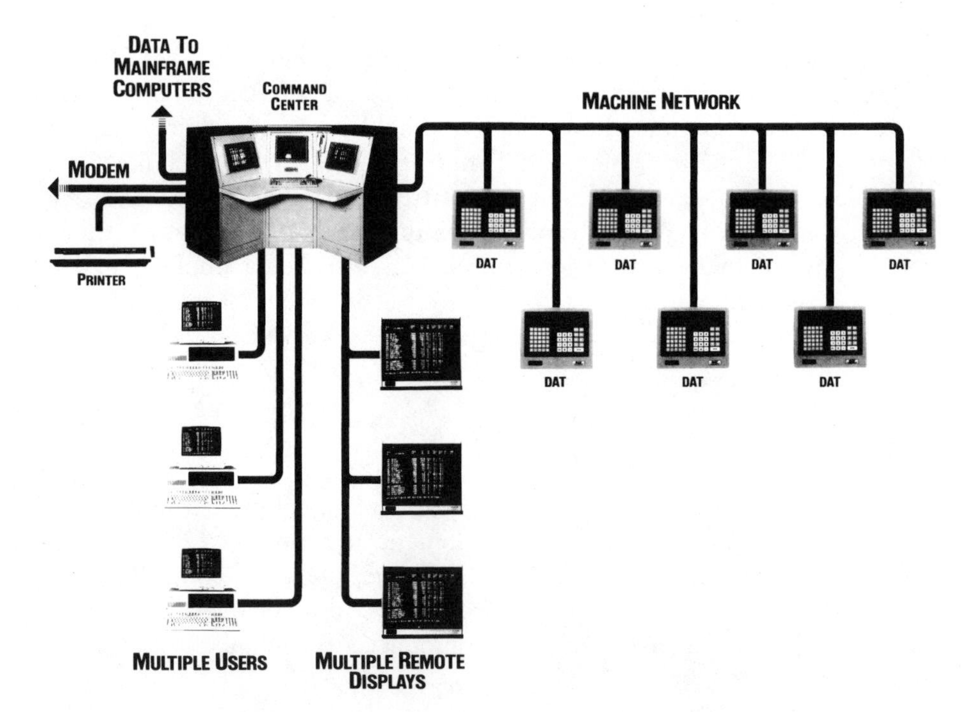

Figure 13-1 The CIM communication network provides for almost instant reporting from all equipment on all processes, to all departments.

the approved standard sample. In practice, parts are physically rejected and diverted into reclamation systems or review bins for potential reclamation, and alarms are provided at the machine and at the central computer to inform management personnel.

The benefit of this program is in the prevention of defective parts. The savings can be considerable to molders who might otherwise have to bear the expense of secondary finishing or reprocessing of defective parts. Additional cost savings are promoted through rapid implementation of remedies that will prevent further manufacturing of defective parts. Also, customer relationships will improve, since zero-defect shipping can be implemented on the basis of 100 percent quality control of all parts or containers.

Computer-aided Response Program

The central computer is used as a means of prompting corrective action in the event that equipment fails or a process deviates. This is a powerful tool, since all blow molding machines and auxiliary equipment are connected to the network. A single, central supervisory person can monitor the behavior of the entire plant.

The benefits of monitoring are clear, since warning occurs immediately when the process deviates. Industry statistics indicate that the average blow molder discovers that something is wrong fifteen to twenty minutes after it has happened. With the real-time alarms of the CIM system alerting management to a problem as soon as it is happening, considerable time can be saved through reduced downtime and elimination of confusion, and through reduced or eliminated running time while producing unacceptable parts. This program has been a key requirement for those manufacturers who are attempting fully automatic operation.

Computer-aided Equipment Maintenance Program

The current tools available to a maintenance man in a plastics shop will not allow him to diagnose the behavior of the molding machine in depth. Accordingly, shop maintenance generally occurs when something breaks down, resulting in downtime that could have been prevented by planned preventive maintenance. The monitoring capabilities of the CIM program provide maintenance personnel with a new and sophisticated tool that allows them to look into the behavior of machine hydraulics and of the process to a depth not previously possible. With this new insight, maintenance managers can review daily the operation of all machines and auxiliary equipment in a matter of minutes, generally on the first shift. Downtimes will be considerably reduced and surprises almost totally eliminated.

Experience indicates that new, as well as old, machines benefit greatly from the computer-aided maintenance program. Productivity increases without added capital equipment. This program has proved to be another essential element in implementing "lights out" operation—in which the plant runs unattended, under automated control, for some period—since all maintenance programs can be carried out during one shift of each day.

Production Control with MIS-Qualified Data

In addition to quality and production control, the CIM program provides a management information system (MIS). Many processors realize that manually collected production information is no longer adequate for manufacturers to survive in today's competitive conditions. Manually collected data are often inaccurate, and manual data gathering is always too slow for real-time plant management. With the CIM program, management data are available immediately and at any time. This allows management to make critical decisions quickly and to see the results of their decisions on a real-time basis.

With the SPC program qualifying parts as good or bad while they are being produced, production data are similarly qualified by the management information system. This prevents under- and overruns, permits more accurate assessment of real-time productivity, and significantly improves the forecasting of job completion and the scheduling of new jobs.

HARDWARE/SOFTWARE OVERVIEW

The following is a summary of the main physical components required to implement the functional elements listed above.

Central Computer

The command center shown in Fig. 13-1 is essentially a data collection device facilitating two-way communication with all monitoring terminals. This central computer supports the machine network, to which all data collection systems are connected, and the remote data display network, to which production monitors are connected, facilitating dissemination of information throughout the plant.

The central computer must be a high-speed unit, equipped with optically isolated network controllers and heavy-duty disk drives, since on the order of 2.5 megabytes of SPC data can be collected in twenty-four hours from just a twenty-machine shop. Configuring this computer for multiuser operation will be required for plants that intend to let a number of persons draw on the data collected by the central computer. Subsidiary terminals might be located in the maintenance manager's office, in production management, process engineering, and in quality control. Many users have also found it advantageous to connect a model to the central computer to enable them to transmit data to another plant or to an equipment supplier's service facility for on-line troubleshooting assistance.

Machine Network

Each machine network (Fig. 13-1) can service a substantial number of machines (more than 30) through their monitoring systems—either suitably designed process controllers or dedicated monitoring terminals. A high-speed (9600 or 19,200 baud) data communication protocol ensures reliable communication between the computer and the machines, which can be up to 4000 feet away from the central computer. Transmitted data include SPC data, MIS data, alarm messages, and operator help requests or messages. Highest priority for data transmission is assigned to SPC and alarm messages; MIS data are collected as a second priority. Accordingly, the central computer updates all MIS files as time allows. However, no MIS data can be lost, since the machine-monitoring terminals hold all information until there is a chance to update the records of the central computer. If data communication is interrupted, automatic alarms warn personnel at the central computer so that communication can be reestablished as quickly as possible.

Local Area Network

A local area network (LAN) (Fig. 13-2) is often described as a "cell" network. Each molding machine monitoring terminal can support data from several auxiliary devices, such as chillers, heaters, dryers, and robots. Data from these devices are funneled through the monitoring terminal to the machine network and thence to the central computer. Newer auxiliary equipment is often equipped with a serial communications port that enables direct data transmission. This permits monitoring the equipment's behavior directly through its own sensors, and transmission of equipment setups from the central computer directly to the auxiliary equipment in a total recipe format. Note, however, that the communication protocols designed into the various pieces of auxiliary equipment must be compatible with the overall CIM system in order to facilitate direct digital control and data communication.

Older auxiliary equipment can also be monitored by the machine monitoring terminal, through addition of the necessary direct sensors that can measure temperature, humidity, pressure, velocity, and other factors. With-

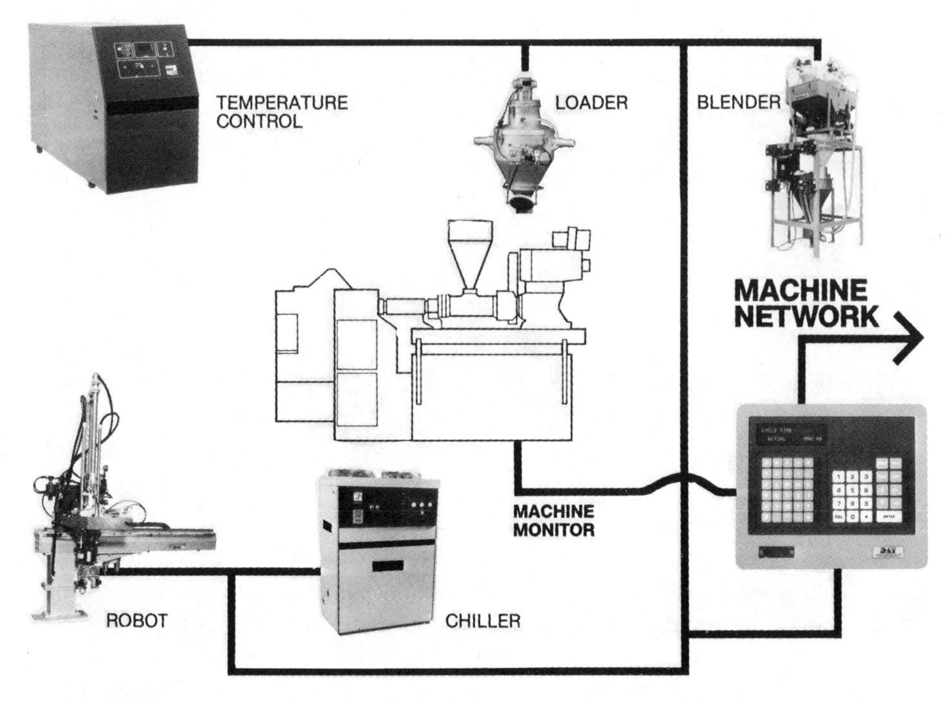

Figure 13-2 The local area network monitors all processes around the machine that have an effect on productivity and quality, and holds this information until it is picked up by the central computer.

out these, the local area network monitor would not be able to perceive and diagnose failures in the auxiliary equipment, nor issue alarms to warn personnel. Auxiliary equipment without built-in communications capability cannot be set up remotely or automatically in a recipe manner.

Remote Data Display Network

The central computer disseminates information throughout the plant. The remote data display network (Fig. 13-1) can support numerous color CRT displays at strategic locations thousands of feet away from the central computer. In addition, the remote display driver that converts digital data into a color picture is able to communicate through serial ports and recognize messages sent to it individually. Thus, customized applications are possible where different displays are to be sent to various work areas or departments, or reports are printed locally.

Data Collection Terminals

Two types of terminals may be connected to the machine network, providing CIM-related information in an identical fashion. One of these is a dedicated device for monitoring the molding machine and auxiliary equipment in the local area network. Conceptually, this is a generalized device that can be used without modification on different types of blow molding machines, utilities, or secondary processes, such as finishing lines. It incorporates digital and analog monitoring capabilities, as well as process monitoring software capable of comparing process data obtained from the molding machine with the upper and lower control limits downloaded from the central computer's SPC file for that job. The data terminal can thereby make good/bad part decisions in real time, before the mold opens, and provide an output to part sorters or grabbers for good/bad part sorting on the basis of decisions made through the SQC program.

This terminal has three data communication ports. Typically, one port is used to communicate via the machine network with the central computer. The second is used to communicate with molding machines or auxiliary equipment in a digital manner, assuming they have these facilities. The third is used as a data inlet port, for bar code readers, weight scales, or gauging equipment that might be used for quality control. These data are automatically added to the SPC database and transmitted to the central computer. This permits comparison of physical measurement data with molding parameter data, which can be extremely useful in optimizing the molding process.

The data terminal is capable of supporting a complete cell local area network and acts as the gateway for messages between the central computer and the molding machine or auxiliary equipment. Although it is not necessary for it to have more than a forty-character display, the keypad can be used to send either coded or English language messages to the central computer.

The second type of data collection terminal is a microcomputer machine controller that provides all the data collection functions in addition to the closed-loop process, temperature, and sequence control. Many older machines do not have such controllers, nor do many newer machines need them to operate satisfactorily. Thus, the dedicated data terminal provides a more economical alternative in many cases. But only with the intelligent machine controller can complete setups be downloaded from the central computer to the molding machine (and auxiliaries) via the machine network. Similarly, the intelligent machine controller provides the convenience of storing completed setups in memory for future production runs.

CIM System Databases

All data provided by the central computer are available in the data interchange format convention used by most spreadsheet programs and statistical programs on the market today. Under this concept, the data are represented by type on a given row and samples are recorded as columns of data. With this method, the intercept point of a column and row can locate a record corresponding to a sample and the specific data collected. This allows personnel to use conventional spreadsheet programs to analyze the raw data or to create their own reports from the information in the database. A number of generic, preformatted reports—the minimum needed to manage a plant effectively—are also available as a starter kit that includes:

Production reports by day and by shift
Downtime reports by day and shift
Reject reports by day and shift
Daily job setup records
Consolidated material requirement reports
Consolidated mold and machine history and status reports
Universal job scheduler
Year-to-date job reports

For those who already have reporting capability through a mainframe, production information can now be fed to any mainframe system real-time, but a bridge must be designed between the collection hardware and the mainframe.

PROCESS CONTROL WITH SPC

To many molders, statistical process control implies an empirical concept by which parts can be accepted or rejected. SPC provides a means by which changes can be identified so that the program or the control can be reset to its original status. SPC is practiced by first establishing a standard production sample. This sample is approved by quality-control and production-control personnel and becomes the standard against which all subsequent production must be compared.

The data obtained during the sample standard run are applicable to all machines that subsequently run that same product in production. Accordingly, SPC forces all machines running the job to the standards of the sample. The benefits that are a meaningful, reliable standard is established, based on the capabilities of the mold, machine, and material. Quotations can be given confidently to customers based on this information, since plant management can rely on the same performance being repeated on other machines.

Function of the SPC Database

The practice of useful SPC demands monitoring of critical parameters of the molding machine that can affect part quality and machine productivity. In practical terms, process parameters related to part quality are those that influence part weight and dimensions. For extrusion blow molding, a number of factors must be identified as the minimum necessary for monitoring. Software exists today that can monitor all these factors and maintain a real-time SPC database during production. However, such software cannot be expected, with today's technology, to identify problems associated with material contamination, deviations in color, or changes in additives of too small a magnitude to affect melt density. Some of the latter factors can be controlled by monitoring upstream auxiliary equipment (e.g., feeders, proportioners), and others can be eliminated by proper manufacturing practices (e.g., to prevent contamination).

The Standard Sample

The standard sample can be obtained either at mold runoff or when the job has already been run, during a convenient production period. Sampling begins with the machine optimized and warm, as it is expected to be during production. When sampling begins, quality control personnel confirm that good parts are being produced. Sampling is started by the central computer as the machine paces through its normal cycles. A statistically significant number of samples (at least sixty-four) must be taken.

Quality control personnel determine that the parts produced during the sampling period are acceptable. Production control personnel determine that machine cycle times and efficiencies have been acceptable. The sampling data are then analyzed in order to determine the upper and lower control limits for each molding parameter, against which all future production cycles must compare.

The upper and lower control limits will constitute the boundaries of acceptable production conditions. Establishing the limits for each critical variable is a vitally important step in the SPC program. The limits may be set tighter or looser for particular variables, as a way of weighting them according to their degree of influence on quality or productivity parameters. This can be done manually by personnel with considerable molding experience; however, there exists today so-called expert software that does this much faster and more scientifically. Using current principles of artificial intelligence, such rule-based software draws on its own body of empirical "experience," derived from statistical analysis of actual production data, analogous to intelligent decision making by a human expert.

Once set, the upper and lower control limits remain on file in the central computer. When the job is scheduled to a machine, the control-limit file is downloaded to the machine monitoring equipment (whether the machine process controller or a separate data-gathering device), which, from that point on, will qualify every machine cycle against those control limits.

Benefits of SPC

Since all production is forced to conform to the approved standard sample, it is no longer possible for production machinery to operate with efficiency levels below that of the standard. This generally will mean an overall increase in productivity.

New insights will be gained toward purchasing new machinery, focusing on its capabilities of handling specific jobs according to the standard sample. With the use of monitoring equipment and a portable computer equipped with the same expert software, it is possible to test out a new machine's capability to run acceptable parts on a production mold at the machine builder's lab. Or, when it is impractical to ship the mold to the machine builder, some molders have instituted a policy of not accepting delivery of a machine until it has been test run and produced qualified parts in their own plants. This sort of procedure will force capital investment to be placed on those types of machines that can produce for the user in a reliable and predictable manner.

In many plants, the behavior of personnel (e.g., unauthorized or unjustified "dial twiddling") can be a source of problems affecting productivity. The SPC program will immediately sense that changes have been made to the process and warn the supervisory personnel who monitor the central computer. Thus the SPC program will identify responsibility for the actions of floor personnel.

AUTOMATED QUALITY CONTROL WITH SQC

It is important to recognize that SPC is, in fact, a highly effective form of machine (process) control. The control limits obtained from the SPC sample are loaded into the monitoring terminals attached to each molding machine. This establishes an upper and lower boundary for each critical process parameter, within which the machine must operate. Statistical quality control (SQC) starts when machine data are compared to the process control limits during each molding cycle. SQC can identify good vs. bad cycles while the part is molded so that this information is available when the mold opens. The result is that either the cycle passes as good, or alarms are generated, indicating that the cycle is bad, and the shot is rejected (possibly with the aid of part diverters or grabbers). In this manner, 100 percent quality control is provided automatically, and zero-defect shipping comes closer to practical reality.

In this CIM concept, an alarm vocabulary is an integral part of the machine monitoring system. It is essential that alarms be received both at the machine, so that operating personnel can recognize when their actions or adjustments cause a deviation beyond acceptable limits, and at the central computer, where management personnel will be alerted that a change is taking place.

Practitioners of CIM in blow molding have learned to rely on this automated SQC information to the extent that defective parts are automatically reprocessed and returned to the machine hopper to prevent contamination of the material. In many operations, this has proven to be a more cost-effective alternative than using human quality control to requalify the part sometime after it has been molded and perhaps had further value added to it through secondary operations.

Benefits of SQC

SPC closes the loop with SQC. Practicing SQC will mean happier customers and, more importantly, will reduce or eliminate surprises. The process-alarm program for SQC becomes, in turn, the driving factor in the computer-aided response program aimed at the reduction of waste.

COMPUTER-AIDED RESPONSE PROGRAM

One of the major advantages of CIM lies in continuous surveillance of all equipment that can affect part quality or productivity. This provides real-time information at the central computer when the process deviates or equipment fails.

The costliest segment of waste is operating equipment to produce unacceptable parts. Machine time is wasted, as are utilities, manpower, and material. The computer-aided response program is aimed at stopping waste in this sector. It does so by providing real-time alarms, complete with diagnostic messages prompting plant personnel to action.

Each machine monitoring terminal on the machine network can have its own unique, customizable vocabulary of alarms. Besides molding machines and closely related auxiliaries, even labeling, decorating and filling machines have been outfitted with monitoring terminals. Also, it is not widely recognized how much plant utilities—such as the main voltage, plant cooling tower or chiller, compressed air, and the like—affect the productivity of individual machines. These utilities have been found responsible for differences in production between day and night shifts, for example, as environmental conditions change. Accordingly, inclusion of plant utilities is essential in the monitoring program of the CIM plant.

Alarm Displays and Logging

Each machine monitoring system displays a process alarm when it occurs. This prompts action by floor personnel when they are at the machine. These alarms also serve as a feedback to setup personnel, who can correct the process, based on the instruction provided by the monitor, so as to keep the process between the SPC limits.

The likelihood is high, however, that floor personnel will not see an individual alarm as it occurs. Nor can the floor alarms be seen by anyone in a "lights out" plant. Accordingly, process and machine alarms are sent to the central computer, where the console printer provides a record of all alarms that occur. A color display provides a visual log, color coded to prompt the observer to action, depending on the importance of a particular alarm. The SPC expert software can prioritize alarms, as some deviations beyond SPC limits may have less significant consequences for product quality than others. Alarms are also logged in the central computer's memory in an alarm file that can be analyzed by plant management or maintenance personnel on the next day to determine inefficient equipment or personnel behavior.

Benefits of Faster Response

A computer-aided response program can be highly effective in reducing reaction time to a failure. The conventional plant relies on roving floor personnel to be at a machine when a failure occurs, or on the quality control personnel to determine after the fact that bad parts have been produced. Since all information is now concentrated in the hands of the central computer operator, he or she can dispatch manpower immediately to a machine that needs help. Practitioners of this program are using FM radios for this purpose, which has enabled them to cut response time significantly.

Experience indicates that far more efficient use of manpower will result from such a program. In addition, the alarm messages contain diagnostic information that cuts down further the troubleshooting time required to solve a problem.

One important result is much less machine time spent producing bad parts, and significantly reduced downtime due to breakdowns. Another major benefit is the identification of the real reasons for downtime. Some management systems depend on the operator to identify these reasons, which results in inaccurate data or no data at all. The computer-aided response program envisioned here requires no manual input, and provides a complete historical log of all downtime and the reasons for it, enabling management to plan and to diagnose production bottlenecks in a plant.

COMPUTER-AIDED EQUIPMENT MAINTENANCE

The objective of this program is to ensure that all machines operate in an optimal manner, producing the fastest cycle possible consistent with good quality part production. Even new machines exhibit inconsistent behavior, primarily due to hydraulics or worn components that will significantly affect productivity. This program will reveal inconsistencies, regardless of machine age, and will allow maintenance personnel to correct them quickly.

A summary of machine performance data for the previous twenty-four hours, produced by the central computer, can be analyzed by the maintenance manager in the morning to determine action to be taken. Graphs can be produced of any SPC variable, which can enable him to diagnose faulty hydraulic components, machine wear, and other factors that cause the machine to produce bad parts periodically. This analysis can be undertaken in a few minutes for the entire plant.

In a "lights out" plant, this program is critical, since it will insure that the machines operate correctly for the next twenty-four hour period. The maintenance analysis will also provide data for long-term actions to be undertaken during a scheduled maintenance shutdown of a machine. Daily analysis and maintenance will significantly increase the productivity of all machines in the plant. The chief benefit of this program is that it does not require capital investment to increase plant production in a meaningful manner.

SPC Data for Diagnostics

Today's software can generate, with just a few keystrokes, a graph of the behavior of any SPC variable over twenty-four hours of production, with upper and lower limits shown for comparison. This offers every maintenance manager information available at his desk, almost instantaneously, that heretofore would have required a technically sophisticated molding expert to stand by the machine on the shop floor, laboriously recording cycles with an oscilloscope. Furthermore, use of plastics-specific expert software can provide more useful information than can standard X-bar and R charts generated by off-the-shelf, general-purpose SPC software packages. The latter often tend to identify more cycles as bad than is truly justified, because they cannot take into account how machine variables are damped in their ultimate effect on part-quality variables. Software that has learned how molding processes function can set control limits that correctly distinguish between significant and insignificant variations.

Observation of process variables over twenty-four hours lets a maintenance manager see how machine performance correlates with time of day, and how it is affected by ambient temperature, availability of utilities, and the functioning of auxiliary equipment around it. This will provide a new understanding of what effects the machine's environment has on production.

Unwarranted process adjustments by plant personnel are logged in the SPC files, together with time of day, allowing the person making the adjustments to be identified. This can result in better training programs for floor personnel.

Historical SPC data stored in computer memory can be used in several ways. When a job is to be run on a new machine, the productivity can be compared with past performance. Likewise, when machines become old, they may not be able to meet the standards of the SPC program. Therefore, the maintenance manager is in an excellent position to determine when a major overhaul is required or when a machine should be replaced. Accordingly, overhaul or buy decisions can be made on the basis of solid historical data rather than empirical knowledge, assuring management that the investment will be justified.

PRODUCTION CONTROL WITH MIS-QUALIFIED DATA

Machine-mounted SPC data collection terminals also serve as collection devices for management information system (MIS) data. The system can be designed so that only minimal human data entry is required to obtain meaningful management information. Management will no longer have to depend on manual entry to identify reasons for problems such as downtime or bad-parts production, since the SPC database contains this information in the first place. About the only information the operator needs to enter is the number of cavities run, since this cannot be monitored automatically.

Automatic collection of production data in a qualified manner is the heart of the program to ensure plant management with solid information. With real-time production totals of good vs. bad parts clearly identified, the possibility of over- and underruns is greatly reduced, and far more accurate job completion forecasts are possible. Gaps in management information data can be eliminated by the distributed intelligence approach to data collection. In this concept, MIS data are collected by each terminal and are periodically sent to the central computer to update its files. Accordingly, even if network communication breaks down, or if the central computer fails, data continue to be collected at each machine. When communication is reestablished, the central computer's files are simply updated with the latest information available at the machine.

Management information data can be retrieved from the central computer files on a regularly scheduled pass, such as each day or shift. An up-to-date production summary is also available immediately from the central computer at any time it is desired.

Management Displays; Reports

In concept, production management requires two types of information:

1. Real-time information on what is happening now
2. Historical information on what has happened over a period of time

Production reports generally provide historical information; production management screen displays provide real-time information.

In the CIM program, real-time color displays are provided at the central computer or, optionally, throughout the plant. These displays typically show where the job is, how much longer it has to run, production efficiencies, and whether or not the machine produces good or bad parts. Color highlights call attention to problem areas, such as machines that are down or producing below standard levels. Other displays include the performance of individual jobs in a job-to-date, shift-to-date format. And further displays at the central computer provide production scheduling, machine status indication, and specific files relating to job, mold, and material management.

Preformatted reports are automatically printed at the end of each shift. The MIS files are also automatically recorded on floppy discs for historical storage. These reports can be formatted by the user to suit his own needs. Although several preformatted reports are available as a starter kit, off-the-shelf formats may not take best advantage of the user's experience in managing his own plant. Reports should reflect one's individual management style; therefore, preformatted reports offer no competitive edge. Custom-formatted programs, on the other hand, accommodate established methods and procedures that have been used and accepted in the past. Also, everyone can have a report to his liking.

DOES CIM PAY

Fourteen molding plants that use CIM have reported the following yield increases resulting from CIM. Yield increases are here defined as increased revenue without additional capital investment. The results show the minimum and maximum improvements reported by these users:

- Through closed-loop implementation of SPC: 5 to 12 percent improved yield
- Through computer-aided maintenance: 7 to 14 percent improvement
- Through fast response prompted by alarms: 4 to 11 percent improvement
- Through improved production control information: 2 to 6 percent.

CHAPTER

14

Blow Molding Quality Control

W. MICHAEL JAYCOX

In establishing a comprehensive quality control inspection and testing program for blow molded plastic containers, one must recognize several important facts in order that the program will be embraced by both the manufacturing and quality control staffs. These facts are:

- Because most blow mold operations incorporate fast-paced production equipment, it is imperative that labor and inspection costs be kept to a minimum.
- The quality control department carries out inspection and tests of blow molded containers as an auditing function, thus making the tests and inspections performed by the actual manufacturing personnel of equal, and more often greater, importance.
- Whether a blow mold operation is designed to provide containers for use in house or to outside packaging operations, the finished blow molded container will be looked upon as a packaging supply component, not an actual finished commodity. For that reason the container will be closely scrutinized on the packaging line and may disrupt the production schedules of the packaging departments if defects are not discovered early in the blow mold process.
- The ultimate acceptance criteria for blow molded containers is whether or not they will perform as expected and will not meet with adverse

consumer reaction. This is not to say that container specif
should not be legitimately and strictly observed, but rather th
is an element of leniency that may be built into specificatio
guidelines which can work to the advantage of both the manufac er
and inspector.

- Finally, if the manufacturing personnel directly responsible for
ducing the blow molded containers do not assume responsibility
the quality of those containers during the manufacturing process, n
after the fact, a quality control program will be at best minimall
effective.

With those points in mind, there are two aspects of establishing a workable quality control program that must be examined. One is the tests and inspections acquired by the manufacturing personnel in the blow molding department; the other is the test and inspection of the quality control personnel who perform audits on the finished containers. Each of these groups must perform certain types of tests and inspections, following specified methods and schedules.

EQUIPMENT CAPABILITIES

Prior to designing a quality control program for any given blow molding operation, it is highly desirable to perform process capability studies of each blow molding extruder to determine the averages, ranges, and standard deviations of test areas deemed critical to producing a quality container. Those areas will usually be:

1. Bottle weight
2. Wall thickness
3. Overflow capacity
4. Neck/thread dimensions:
   Inside diameter (ID)
   Outside diameter (OD)
   H dimension (overall height of neck area)
   T dimension (thread area measurement)
   S dimension (sealing area of bottles that require capping)
   E Dimension (same as OD)
5. Drop-impact resistance
6. Characteristics perceptible to visual inspections for contamination: stress lines and die lines, unmelted plastic particles, and others
7. Adequate colorant (if applicable)

For specific information on performing process capability studies (PCS), consult the *Quality Control Handbook* [1], chapters 22–24. This handbook is an excellent source of basic information about PCS and about methods of acquiring statistical data relative to a blow mold operation.

This chapter considers the relative importance of the various product physical properties and dimensions. Their importance depends on the final use of the container: what substance it will be used to hold, whether it will be capped with a threaded or friction-fit cap, its size, whether or not pigment will be added to the resin for color, and other factors.

Quality control testing can be approached in two ways. One is to test every possible area, and soon learn that some tests will not be used for acceptance/rejection criteria and generate nothing more that a great deal of technical documentation. The other approach is to test selectively, choosing those tests from which it is possible to derive an accurate indication of the container's overall quality level. This chapter is devoted to logical tests, those that will produce the greatest amount of information in the least amount of time.

Because blow molded containers are most often the product of extremely fast paced manufacturing equipment, any appropriate quality control efforts are always a race against time: the containers to be tested have usually already been scheduled for use by the time tests can begin. It is quite appropriate, then, to first discuss the quality control program applicable to the manufacturing process. It is here that the first line of defense against container defects is established. The ultimate responsibility for setting up control tests and measures falls upon the quality control expert. That person must possess some basic knowledge of blow molding and a working knowledge of what tools the typical blow mold operator must have in order to make adjustments to the equipment.

The following tools are those most often needed by the blow mold operator:

- A mandrel bar, usually a 5/8 in. drift punch
- A die adjustment wrench, usually a 3/8 in. × 7/16 in. 12 point box-end wrench, although some extruders use differing sizes of die retainer bolts
- Brass probes, used for probing the die/mandrel. Probes must be soft enough to prevent damage to the fragile die/mandrel assembly; steel or other hard materials should never be used on dies or between molds. A probe and/or brass knife can be fashioned from brass stock or from a brass core welding rod with the outer slag surfaced buffed off.
- A complete set of Allen wrenches
- A pair of channel-locking pliers

An assortment of other tools is certainly desirable, including screwdrivers of several sizes (standard and Phillips), standard knife, pliers, an assortment of fixed and adjustable wrenches (crescent wrench, etc.), and a machinist's hammer.

OPERATOR TESTS

There are various tests that the blowmold operator should perform at certain intervals on a regular basis, as listed below.

TEST	FREQUENCY
Organoleptic Inspection	Each 30 min period; more often as needed

Many container defects can be identified by direct inspection by human physical senses. The two senses most useful for such organoleptic inspection are vision and touch. Visual inspection can be used to identify:

- Proper colorant (if applicable) in container
- Holes or stress lines caused by contaminants in the plastic resin
- Container warpage caused by excessive heating/cooling during the blow mold process
- Container blemishes such as water marks, mold deformities, and others
- Die lines caused by burrs on the die tips (appear as stress lines)
- Neck/thread deformities caused by defective, malformed, misaligned, warped, incorrectly sized or damaged blow pins or poor equipment adjustment
- Other general problems of a visual nature

Simply feeling a container from neck to bottom can give information about plastic distribution in the walls and along the lowermost part of the container where most drop/impact failures occur. Weak spots in the container become apparent. Hot spots that will result in warpage or distortion of the container can be felt if the container is checked very soon after being ejected from the molds.

TESTS FOR PHYSICAL ATTRIBUTES

Drop-Impact Test	Once every hour
Leaker-Seeper Test	Once every two hours
Container Weight Check	Once every two hours
Weld Strength Test	Once every two hours

To supplement the operator testing, the quality control department should perform each of the preceding tests every four hours, or upon special request when a specific problem is suspected. A list of common defects that can be detected by such testing follows.

1. Ragged, pitted, saddled or uneven sealing surface.
2. Excess flash or plastic shavings on container neck area or on pinch-off areas on the body of the container.
3. High or low *H* dimension.

4. Underblown areas, most recognizable in the neck area or in contour areas.
5. Holes or contaminant "windows" in the container.
6. Excessive flash, dents, or punctures on or near the label panel area, or near the heel of the container.
7. Weak areas appearing transparent in the handle slug area.
8. Significantly overweight or underweight containers.
9. Containers with far too much or not enough colorant. Too little dye will allow the product fill level to be easily seen. Too much dye will contribute to excessive brittleness, increasing the potential for stress cracking or drop test failures.
10. Warpage, caused by bottles being too hot when ejected from molds or from poor container wall thickness distribution.
11. Dirty or greasy containers.
12. Water marks: orange peel or herringbone appearance.
13. Webs or hairline cracks across handle; webs may also be blockages of the handle, which can cause filling difficulties.
14. Weak particles in container walls caused by unmelted or partially melted plastic particles.
15. Poor mold alignment; appears as two halves not matching up properly, either vertically or horizontally.
16. Defective container seams; may appear as excessively wide seams.
17. Poor plastic distribution in container; may appear as a container having one thin side and one thick side.
18. Deformed thread area; may be caused by improper air pressure on extruder or by damaged mold inserts.
19. Nicks or burrs on container parting lines; appears as blemishes along the seams and is a major cause of pinhole leakers; usually caused by mold damage.
20. Cocked or crooked necks.
21. Delamination of plastic, usually on colored containers.
22. Parison, or flash, folds; usually detected as lumps of plastic on bottom seam, but may also show up along side seams; may cause nozzle blockage.
23. Rocker bottoms: rounded bottom that prevents container from standing upright; usually caused by improperly adjusted air pressure on extruder or by poor mold venting.
24. Stress lines; may be caused by die tip burrs or other die deformities, or by contaminants being trapped somewhere in the die/mandrel areas.
25. Bottles appearing to have proper weight and wall distribution, but still show signs of weakness; may have too much regrind being introduced to the extruder.

BLOW MOLDED CONTAINER TEST RECORD

Plant__________ Date__________

Machine________ Container Type/Size__________ Color__________

Shift	Test		Mold & Cavity Identification					
1	Finish Dimension	GO						
		NO-GO						
		x						
	CYP @ 73 F	Filled						
	Drop/Impact							
	Weld Strength							
	Overflow Capacity							
	Bottle Weight							
	Visual Inspection							
	Other:							
	Comments							

Shift	Test							
2	Finish Dimension	GO						
		NO-GO						
		x						
	CYP @ 73 F	Filled						
	Drop/Impact							
	Weld Strength							
	Overflow Capacity							
	Bottle Weight							
	Visual Inspection							
	Other:							
	Comments							

Shift	Test							
3	Finish Dimension	GO						
		NO-GO						
		x						
	CYP @ 73 F	Filled						
	Drop/Impact							
	Weld Strength							
	Overflow Capacity							
	Bottle Weight							
	Visual Inspection							
	Other:							
	Comments							

Visual Inspection

(record defect by #

1. Contamination
2. Thin Walls
3. Webbed Handles
4. Stress Lines
5. Excess Flash
6. Tear/Dent/Crease
7. Color Lines
8. Rocker Bottoms
9. Warpage
10. Bad Seams
11. Water Marks
12. Weld Misalignment
13. Underblown Area
14. Nozzle Blockage
15. Uneven/Ragged Tri
16. Other__________

Drop/Impact Test:

drop each bottle three times-record as indicated:

P - Passed

F-1 Failed 1st drop

F-2 Failed 2nd drop

F-3 Failed 3rd drop

Figure 14-1 Quality testing report form.

This list certainly does not encompass all the defects that may be encountered, but it can serve as a starting point. Figure 14-1 is a quality report form that is to be used when performing quality control tests and inspections. The other tests described in the remainder of this chapter will be carried out by the quality control staff in a lab environment and will be performed as indicated in the test procedure, or as the particular plant operation dictates.

BLOW MOLDED CONTAINER TEST METHODS

The test methods described in the following pages are not intended to be all-inclusive; rather, they are the test procedures most often used in the blow molding trade, and are most applicable to containers for consumer applications. The procedures outlined may be modified to be more suitable for your specific container or test needs.

It is not suggested that each of these tests need be performed. The tests that are appropriate for a given operation are only those that provide needed, useful information and that can be performed in an expedient manner, thus providing a measure of control in the operation.

Many of the following tests can and should be performed by the manufacturing staff at the time of container production to ensure quality at the point of manufacture. Those tests that cannot be performed on the production floor must therefore be conducted in a lab environment. (The lab should conduct *all* tests, but manufacturing should do as many as possible during the production process.) Keeping in mind that tests performed by the quality control department will serve as an auditing function, it is desirable to select random container samples from warehouse stock so that there has been ample curing time: blow molded containers reach approximately 95 percent of their attributes after twenty-four hours of curing.

Container Weight Test

Container weight must be tightly controlled in order to control operating costs, and so that the container will perform as intended. Both overweight and underweight containers pose defect problems specific to the weight variation.

Apparatus

Any good scale capable of weighing in grams, and allowing weights to be indicated to two decimal places.

Procedure

1. Weigh container samples representative of each cavity from each extruder and record all weights.

2. Note extreme weight variations from containers produced on a common extruder. Note weight variations over and/or under the specified weight tolerances.

Container Overflow Capacity Test

The overflow capacity test determines whether the container is capable of holding the desired quantity of product for which it is intended. The overflow capacity must be great enough to allow sufficient head room when the container is filled to the proper product quantity or volume. A standard filling temperature of 23°C (73°F) is specified. This is necessary since the larger containers tend to expand and contract noticeably with temperature changes.

Apparatus

Balance, 5 kilogram capacity, 0.5 gram sensitivity.

Procedure

1. Weigh the container to the nearest gram.
2. Fill the container nearly to the top with tap water at 23 ± 2°C. Hold the container at an angle to avoid forming air bubbles, which tend to cling to the inside surface and distort the test results.
3. Dry the outside of the container and place it on the balance. Add additional water using a dropper until the water is level with the container sealing surface, but not raised above it.
4. Lift the container by its neck momentarily and then return it to the balance. If the water level has diminished, add more water until the water is again level with the sealing surface.
5. Repeat the gentle lifting and refilling three or four times if necessary so that the water level at the sealing surface is constant.
6. Record the gross weight.
7. Calculate the weight of the water (net grams) by subtracting the container weight from the gross weight recorded in step 6.
8. Calculate the capacity to the nearest 0.1 fl oz as follows [2]:

$$\text{U.S. fl oz} = \frac{\text{Net grams water} \times 128 \text{ fl oz/gal}}{3772.16 \text{ g/gal water/23°C}} = \frac{\text{Net grams water}}{29.470}$$

$$\text{Brit. fl oz} = \frac{\text{Net grams water} \times 128 \text{ oz/U.S. gal} \times 1.0409 \text{ Brit. fl oz/U.S. fl oz}}{3772.16 \text{ g/gal water/23°C}}$$

$$= \frac{\text{Net grams water}}{28.312}$$

Cap Torque Test

Cap torque can be measured as application torque, removal torque, or give torque. There are slight differences in the tests, which are accounted for in the test procedures blow. The torque tests are designed to indicate the ability of the container to accept the required amount of application torque, and to withstand the torque required to remove a properly sealed cap. If, for whatever reason, the container cannot be capped properly, the design or manufacturing run may be producing leakers.

Apparatus

O.I. Torque Tester, model 25 (appropriate for most sizes).

Procedures

1. *Application Torque.* This is generally a specified condition for testing such things as sealing surfaces of containers or sealing characteristics of caps. Place an uncapped container in the adjustable jaws of the torque tester, position a cap and tighten it until the specified torque is shown on the indicator.

2. *Removal Torque.* This is the torque required to remove a cap from the container. The removal torque is used to evaluate the performance of the capping equipment or to evaluate the various design factors involved in cap and bottle finishes.

A. Place a capped container in the jaws of the torque tester and apply torque very gradually to loosen the cap.
B. Record the maximum torque reached on the indicator. If the removal torque is below the established minimum, the container finish and/or the caps should be inspected and tested further.

3. *Give Torque.* The give torque is the amount of force required to further tighten a cap already applied to a container until the container of the cap fails ("gives").

A. Place the capped container between the jaws of the torque tester and tighten the cap slowly.
B. Record the torque reading on the indicator at the exact instant when the cap tightens on the bottle or "gives".

If heat-sealing closures are used on containers prior to the capping operation, removal torque tests will be invalid inasmuch as torque is no longer a critical factor and because heat-sealing tends to diminish the bottle and cap's ability to retain specified torque, due to the sealing material and the application of heat.

Container Label Panel Distortion

This test is required only for containers that will require external labeling. A distorted label panel on a container will usually result in the label being applied with wrinkles or with areas that are not adhered properly to the container. A container label panel is designed to be straight in the vertical dimension. This test determines the amount of inward curvature (concavity) since that is what is detrimental to a labeling operation. However, certain blow molding conditions may cause an inward curvature to develop.

Apparatus

1. Straightedge equal in length to the label used with the test bottle.
2. Metal gauges 1/4 in. wide × 3 in. (6.35 mm × 7.5 cm) long, of various thicknesses: 1/32, 1/16, 3/32, 1/8 in. (0.8, 1.6, 2.4, 3.2 mm) and others.

Procedure

1. Place the straightedge against the label panel in the area where the label would be applied. Keep the straightedge parallel with the vertical axis of the bottle. Move it back and forth across the panel to visually detect the point of maximum concavity.
2. Determine the measure of concavity by finding the largest gauge that will fit under the straightedge.
3. If the clearance is 1/16 in. or more, repeat the test with bottles filled with water at 73 ± 4°F (23 ± 2°C). Test the filled bottles in a vertical position to determine if the liquid pressure will cause the label panel to assume its desirable position.
4. Report the curvature as the largest gauge size that can be fitted under the straightedge at the point of maximum clearance. Specific tolerances will have to be determined by the bottle design and suggested end use.

Leaker-Seeper Test

The purpose of this test is to determine two specific things:

1. Will a properly applied cap effectively seal the container?
2. Are containers from the packaging line effectively sealed?

This test should be performed in a lab environment approximately every four hours. It should be performed in the production area at least every two hours, and more often if problems are realized. The two-hour test intervals should be observed by the blow mold production operation and the final packaging operation. Some of the factors that cause failures during this test are:

Uneven container sealing surfaces
Neck dimensions out of specification
Caps out of specification
Insufficient cap torque

Not all containers rely on the external sealing surface to provide sealing. Some containers are designed with a flange or rim at some point along the inside of the neck finish. This flange or rim may be designed to interact with the cap design to provide adequate sealing. Be aware of the container and cap designs of the samples being tested.

Apparatus

Torque tester with torque rating necessary for container(s) to be tested (e.g., O.I. Torque Tester, model 25, 50 lb torque rating).

Procedure in the Blowmold Department

1. Fill the container to the nominal fill level with water or, if available, the intended liquid product.
2. Cap the container to minimum specified removal torque.
3. Place the container on its side, with dark-colored paper towels or cardboard directly beneath the neck of the container.
4. Check for leakage/seepage after thirty minutes, then again after one hour.
5. For any wetting of the cardboard or paper towel classify the container as a *leaker*. For formation of a liquid droplet between the cap bead and the finish classify the container as a *seeper*.

6. In recording test results, be sure to indicate from which extruder the failure came, as well as the cavity identification.

A Quick Leaker-Seeper Test

Cap an empty container and, while holding the nozzle under water, exert pressure via a firm, constant squeeze. Observe whether any air bubbles escape within ten seconds or less. (Filling a container with water or product and capping, then exerting pressure to see if liquid seeps out is *not* an acceptable method of defect identification.)

Procedure at the Packaging Line

1. Obtain a sample at any point after the capping operation.
2. Place the sample on its side and follow steps 3–6 above.

The procedures described above should also be followed for containers having heat-sealed caps, although there is no need to be concerned with the cap torque.

Sealing Surface Evaluation

The test described here is a simple, quick method of determining the quality of container sealing surfaces. It is not intended to be a means to determine acceptance or rejection, but can serve as an indication of developing problems. If suspect containers are identified by use of this test, then the preceding leaker-seeper test should be performed immediately.

Apparatus

1. A clipboard or other paper support with a smooth, flat finish
2. Bond photocopier paper
3. A well-inked stamp pad

Procedure

1. Put five sheets of the photocopier paper on the clipboard. The top sheet is to accept the impressions that will be made; the others are for padding.
2. Using both hands, press the container sealing surface onto the stamp pad and cover well with ink.

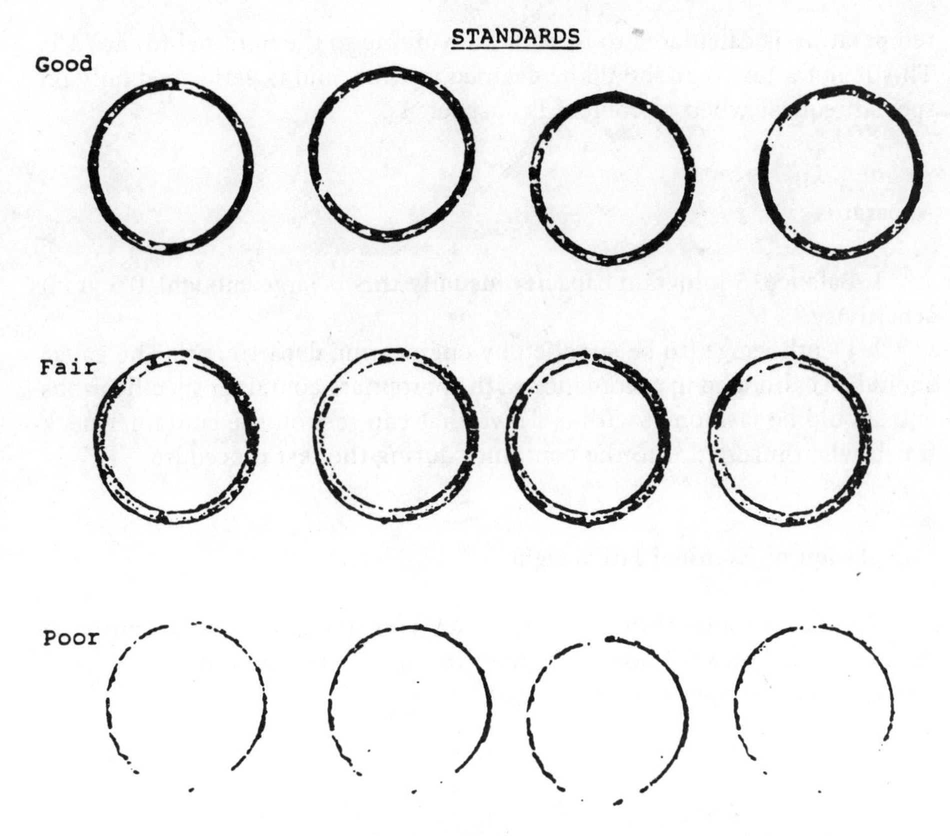

Figure 14-2 Comparison impressions for container sealing surface evaluation.

3. Guide the inked surface of the container down onto the paper; hold the container at the neck and at the base to assure smooth, even pressure and application. Press *lightly,* being sure not to twist the container or smear the ink.

4. Repeat steps 2 and 3 at least three times.

5. Label the impressions with container mold and cavity numbers.

6. Compare the impressions with the samples shown in Fig. 14-2.

7. Record findings as *good, fair,* or *poor* and take corrective actions as needed.

8. The containers may be cleaned using a towel dampened with denatured alcohol and returned to stock.

Determining Outage (Head Space) at a Nominal Fill

This test determines the amount of linear head space in a container filled to nominal volume. Water is used as the fill material for the test and the temperatures must be 23 ± 2°C. The amount of water to be used at this temperature is calculated to be equal in volume to the nominal fill at 73°F. This is not a test that should be deemed routine, and is performed only on special request when a problem is suspected.

Apparatus

1. Balance, 5 kilogram capacity (usually this is large enough), 0.5 gram sensitivity.

2. Depth gauge (to be supplied by engineering department). The gauge should be calibrated in accordance with appropriate container specifications and should be fashioned with a sleeve that can rest on the container neck finish when inserted into the container during the test procedure.

Calculation of Nominal Fill Weight

Table 14-1 shows gram-fl oz equivalents for standard nominal container volumes. For other volumes, calculate the weight of water at 23°C by one of the following equations [3]:

$$\text{Grams} = \frac{\text{Nominal fill U.S. fl oz} \times 3772.16 \text{ g/gal water/23°C}}{128 \text{ fl oz/gal}} = \text{Nominal fill U.S. fl oz} \times 29.470$$

$$\text{Grams} = \frac{\text{Nominal fill Brit. fl oz} \times 3772.16 \text{ g/gal water/23°C}}{128 \text{ oz/U.S. gal} \times 1.0409 \text{ Brit. fl oz/U.S. fl oz}} = \text{Nominal fill Brit. fl oz} \times 28.312$$

Table 14-1 Water weight at 23°C for nominal fill in various containers

U.S. Fl Oz	*Grams Water*	*U.S. Fl Oz*	*Grams Water*
6	177	48	1415
12	354	64	1886
16	472	96	2829
22	648	128	3772
32	943	160	4715

Procedure

1. Weigh the container to the nearest gram.
2. Fill the test container almost to nominal fill with tap water at 23 ± 2°C. Hold the container at an angle to avoid forming air bubbles that will cling to the inside walls of the container and cause inaccurate test results.
3. Cover the container opening and slowly invert the container four times to displace any air bubbles.
4. Dry the outside of the container and place it on the balance. Add additional water with a dropper until the net water weight equals the nominal weight determined from Table 14-1 or by calculation.
5. Lift the container by its neck several times, then place it on a flat surface.
6. Insert the gauge into the container with the sleeve resting on the neck finish. Keep the gauge in a vertical position and push it slowly into the container. Remove the gauge and read the outage.
7. Repeat the measurement after drying the gauge tips and record the average of the two readings to the nearest 1/32 in. (0.79 mm).
8. *Caution:* Avoid squeezing or pressing down on the container while taking measurements.

H Dimension Test

The *H* dimension of a container is defined as the height of the nozzle—the distance between the sealing surface and the shoulder (base of the nozzle) of the container (see Fig. 14-3).

If *H* is below the minimum specification, the cap skirt will ride the shoulder of the container and prevent an effective seal. Variation on individual containers may indicate a slanted or uneven sealing surface, which

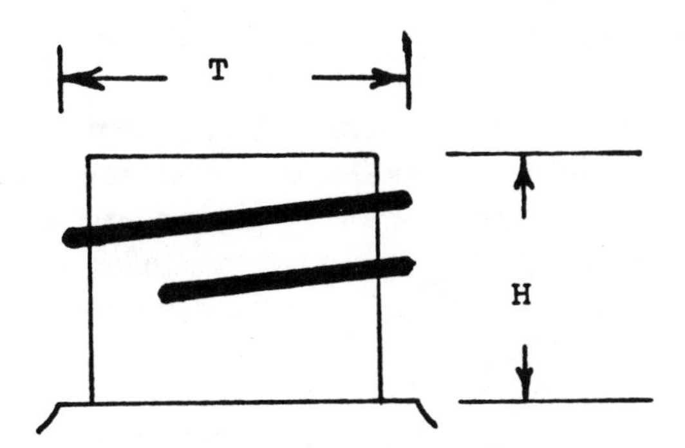

Figure 14-3 Container nozzle dimensions. *H*, height from shoulder to sealing surface. *T*, outside diameter of threads.

will result in poor seals at normal torque or excessive pressure concentrated at one point on the cap; such pressure will result in stress cracking.

If *H* is above maximum specification, or varies greatly from container to container, it will create problems during the labeling operation and the capping operation. It will also create a problem with overall container appearance.

H dimension measurements should not include variations due to ridges, nicks, grooves, saddling areas, or other finish defects that are in themselves reason for rejection. Variations due to mold misalignment must also be addressed separately and are not included in the *H* measurements.

Apparatus

Dial caliper, 6 in. range with dial calibrated to 0.001 in. (15 cm range, 0.025 mm calibration).

Procedure

1. Open the caliper so that the slide is extended approximately 1 in. (2.5 cm) beyond the butt of the instrument.
2. Place the end of the slide against the shoulder of the container and with a smooth, even motion bring the stock of the instrument down so that the butt of the instrument is in parallel contact with the sealing surface; read the dimension to the nearest 0.001 in.
3. Repeat the above procedure so that at least six measurements are taken, evenly spaced around the container.
4. Record the minimum and maximum measurements found.

Slanted Finishes

On a container nozzle, the difference between the minimum and maximum *H* dimensions is defined as *sealing surface slant.* This slant should not exceed the specified tolerance for the given container.

Figure 14-4 shows the effect of mold misalignment on *H* dimension measurements. Mold misalignment may be determined as the difference between *H* dimension measurements made adjacent to each side of the container parting line. Subtract this value from the difference between the minimum and maximum *H* dimension to determine the *sealing surface slope.*

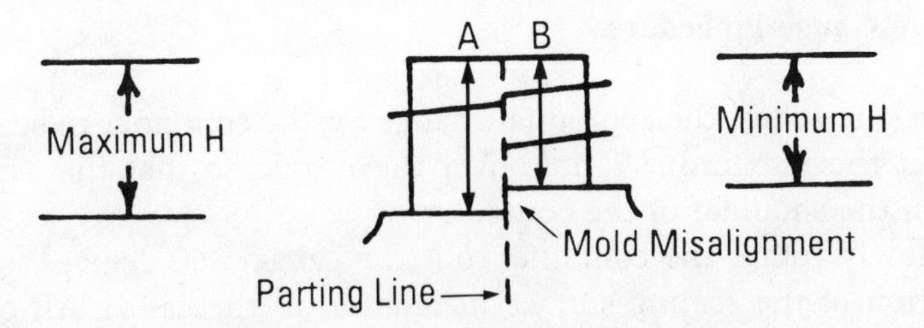

Figure 14-4 Measurements at various points around nozzle circumference will indicate minimum and maximum *H*. Measurements adjacent to each side of parting line, *A*, *B*, will reveal mold misalignment.

Using a Go-No Go Gauge

Combination *H* dimension and *T* dimension Go-No Go gauges can be fashioned easily to serve as very useful tools for quick checks, especially in the production areas. Figure 14-5 shows gauge construction. The *T* dimension is the overall thread diameter of a container, shown in Fig. 14-3. Its measurement is discussed in a separate test procedure, below.

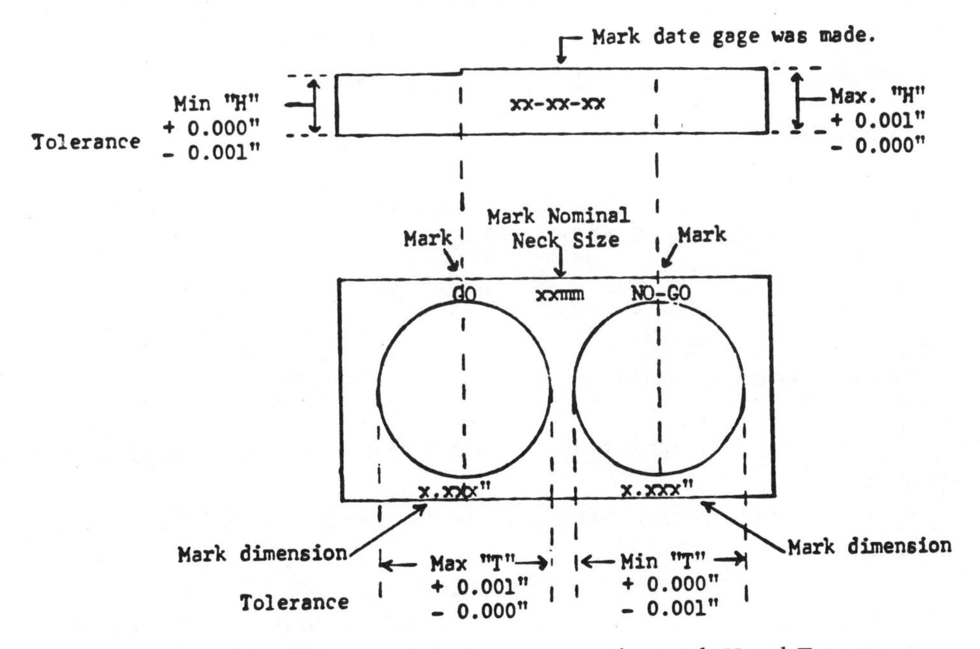

Figure 14-5 A stainless-steel Go-No Go gauge for quick *H* and *T* tests.

Go-No Go Gauge Procedure

1. Be sure to use the appropriate gauge for the container to be checked.
2. Set the maximum *T* hole over the nozzle so that the gauge rests evenly on the shoulder of the container.
3. Slowly rotate the container (not the gauge) 360 degrees, observing the position of the sealing surface in relation to the gauge. All points on the sealing surface should fall within the minimum and maximum *H* limits of the gauge.
4. If the limits do not appear to have been met, or if oversizing or ovalization of the nozzle *T* dimension prevents use of the gauge, then the dial caliper must be used.

Note: The height of a container may play a critical role in how the container will process in a capping and labeling operation. If there is a height specification for a container, then it is easy to fashion a device with which to measure overall height in a consistent manner. A simpler rule of predetermined calibration, depending on the allowable tolerance, may be used. Do not refer to the *height* requirement of a container as the *H* dimension; they are separate and distinct dimensions.

T Dimension Test

The *T* dimension is the outside diameter of the thread on the container finish (Fig. 14-3). It must be maintained within established specifications in order to assure proper cap application and retention.

Apparatus

1. Dial caliper as specified for the preceding *H* dimension test.
2. A combination Go-No Go Gauge, as described in the *H* dimension test, may be useful for quick checks.

Procedure with Dial Caliper

1. The *T* dimension is to be interpreted as the average of two measurements.
2. Take one measurement near the parting line at full thread cross section.
3. Take one measurement 90 degrees to the parting line.
4. Average the two measurements. The average must meet the specified dimensional limits.

Interpretation of Test Results

A. When mold misalignment is suspected, two crisscrossed readings at full thread cross section near the parting line should be taken and averaged for the determination of the *T* dimension. If the difference between these crisscrossed readings divided by two is significant (e.g., > 0.004 in. or 0.10 mm) then the mold alignment should be checked immediately.

B. The difference between the two readings (taken at the parting line and 90 degrees to the parting line) is interpreted as the ovality of the container neck finish. The recorded measurements should be identified as P/L and 90-P/L, respectively. The difference between the two readings should not exceed the difference between the specified maximum and minimum *T* dimensions.

C. It is not permissible to use a closure (cap) on the container for acceptance/rejection determinations relative to undersized *T* dimensions. It is permissible to decide whether or not a finish with one *T* reading over maximum can be properly capped. It is imperative, however, that caps used to make this decision be dimensionally within specification.

Go-No Go Gauges

The use of Go-No Go gauges is limited to determinations where:

- *T* dimension readings are not required.
- Ovalized finishes or mold misalignment is not a critical factor.
- Historical data indicate that the gauge can be used for acceptance/rejection decisions with a high measure of confidence.

1. Use the appropriate gauge for the finish being checked.
2. Check the finish threads with the maximum and minimum diameter holes.
3. Finishes that do not fit into the minimum hole (No Go) but that do fit into the maximum hole (Go) without forcing the gauge have a maximum *T* dimension within the specified range.
4. The gauge does not measure the minimum or average diameters.
5. When the No-Go hole can be forced over a finish with a minimal (very slight) amount of pressure, the finish has a low average diameter (undersized circumference) and should be measured with the dial calipers.

Ash Content of Plastics and Color Concentrate

Titanium dioxide is frequently used as an opacifying agent in plastic containers. Although dyes are also present, their amounts are insignificant compared to the TiO_2 content. Therefore when we burn a sample from a plastic container, the percent ash residue is approximately equal to the percent TiO_2. Following is the prescribed test method.

Apparatus

1. Crucibles (bottoms only), 50 mL, Coors porcelain, high form, Fisher no. 07-965G.
2. Triangles, 2 1/2 in. iron with flanged clay pipe stems.
3. Fisher burner, a single jet gas burner.
4. Muffle furnace, with chamber 9 in. wide, 4 3/8 in. deep, 4 in. high. A Jelrus no. 365 furnace is recommended because of its shallow chamber. Cut a piece of 1/4 in. thick high-temperature board to fit the bottom of the chamber as a support for the crucibles.
5. Crucible tongs.
6. Desiccator.
7. Scale which will weigh to nearest .1 gram.

Procedure

1. Before each use, clean the crucibles and dry them by heating to redness for a few seconds with the burner. Allow to cool for a few minutes then transfer to the desiccator. When the crucibles have cooled to room temperature, weigh each to 0.1 mg. The empty crucible is weighed for tare purposes only.
2. To get a representative sample from a given container, cut a 4 gram ring from the circumference for each crucible. Then cut the ring into 1/2 in. squares.
3. Weigh 4.0 grams of sample to 0.1 mg into the preweighed crucible.
4. Place not more than four crucibles with samples in them on the high-temperature board in the muffle furnace. The furnace should be cold or below 600°F (315°C). Leave the door fully open. (Four crucibles may be used to test four different containers.)
5. Turn on the furnace and adjust the temperature controller to a predetermined setting. The correct setting is one that will slowly raise the temperature so that the samples will ignite and, without requiring further attention, burn gently until all of the plastic is consumed. Because of the large amount of smoke produced, the furnace must be used in a ventilation hood.
6. When the samples have stopped burning, close the furnace door to a point at which an approximate 1/2 in. opening remains for air circulation. Turn the temperature controller to the highest setting.
7. When all carbon black has been burned from the crucibles, remove

them from the furnace with tongs. Allow the crucibles to cool several minutes on triangles, then transfer them to a desiccator and allow them to cool completely to room temperature.

8. Reweigh to the nearest 0.1 mg and calculate as follows:

$$\text{Ash, \%} = \frac{\text{Weight of ash} \times 100}{\text{Sample weight}}$$

Drop-Impact Test

The purpose of the drop-impact test is to determine the resistance of the container to splitting or rupturing during processing, warehousing, or use by the consumer.

Apparatus

Drop test tank. The tank, shown in Fig. 14-6, is merely a vessel that will contain the water in the test container should it rupture during the drop test. Drop test release heights are marked on the guide chute that

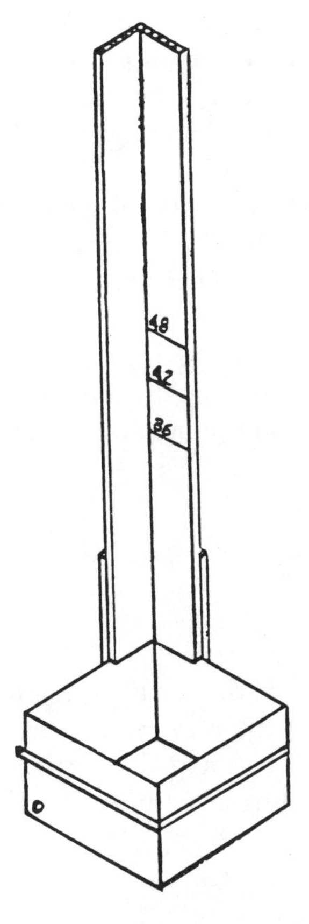

Figure 14-6 Drop test tank with release heights marked on guide chute.

stands vertical to the tank so that tests may be performed consistently. A small port at the base of the tank allows the water to be drained out when a container fails the drop test.

Procedure

1. Fill the container(s) to the nominal fill level with tap water at approximately 73 ± 10°F (23 ± 5.5°C), and cap tightly.
2. Before performing the drop test, check all welds for pinholes by placing the container on its side with parting lines (seams) into a horizontal plane. Compress center of container with the palm of hand and check all pinch-off points and welds for any signs of water seepage or leakage.
3. Using the drop test tank, hold the container upright in the guide chute at the maximum specified height and release it. The height will be depend on the specific container size and use. A rule of thumb is to choose a drop height equal to the height at which the container will most often be handled by the end user, and from which it may be subjected to accidental dropping.
4. Drop the container from the primary height three times. Ideally the container should pass (not break or split at any point) all three drops. However passing on the first drop but failing on either the second or third drop is generally deemed acceptable, although the failure on drop two or three is an indication of weakening container characteristics and appropriate adjustments in container production should be made.

Many containers will have a secondary drop height specified. When this is the case, failure on the first drop at the maximum height warrants immediate attention, but for acceptance criteria a second container may be dropped from the minimum (secondary) drop height. If it passes this drop, no formal rejection is necessary.

In drop testing containers it is sometimes desirable to drop the container flat on the bottom the first and third drops, but flat on the handle side on the second drop.

Wall Thickness Test

The wall thickness is an extremely important factor in the overall strength and rigidity of a container. If the thickness is below the minimum specification, the container will very likely fail the drop test and the vertical strength (CYP) test, and create problems during labeling and filling operations.

Apparatus

Any good micrometer dial caliper capable of measuring 0.001 in.

Procedure

1. Feel the container to determine the thinnest area.
2. Cut the bottle in the thinnest area. Make a clean smooth cut that is free of burrs—they can interfere with the micrometer readings.
3. Use the caliper to measure the area until the minimum thickness is determined.
4. Record the minimum wall thickness to the nearest thousandth of an inch. It is also useful to note in what area of the container the thin wall was found.
5. Measure the sample again to identify the thickest wall area.
6. Even if minimum wall thickness is maintained, a container may be weakened if there is a comparatively thick wall opposite the thin area.
7. To measure the variation, make two measurements 180 degrees apart on the same horizontal plane.
8. If the difference (maximum minus minimum) between these two measurements exceeds the value for the minimum measurement, the cause should be identified and corrected immediately.

Weld Strength Test

Good weld strength is of extreme importance as a safeguard against container failure during processing, warehousing, or use due to low impact or stress crack resistance. The following test can be used to determine weld quality. While it is not an exact procedure, with knowledge and experience in blow molding it will serve as a useful tool for the quality inspector and manufacturer alike.

Procedure

1. Retrieve samples from the blow molding machine before the deflashing operation, and allow to cool for 10–15 min before hand deflashing.
2. Air cool the samples for an additional 30 min after deflashing.
3. Cut out a section of the container containing at least 1 in. of weld area.

A. In all bottom welds and on handled containers, a 1 in. sample section will be indicative of the quality of the entire weld.
B. The entire weld on the top of nonhandled containers is easily tested by cutting 90 degrees from the parting line.

4. Bend the weld 180 degrees outward and rate as follows:

Poor: Weld snaps, resulting in a complete, or nearly complete, break on both sides of bend
Fair: Weld snaps but skin on inside of bend does not break
Good: Weld does not snap; skin on outside of bend shows cracking but there is significant adhesion
Excellent: No breaking of skin on either side; outside of bend shows only signs of stretching

Container Crush-Yield Point Test

The load-bearing strength of a plastic container varies according to size and design, and the type and distribution of plastic resin(s) used. The following test indicates a method to determine the maximum load a container will support under given test conditions. This load is defined as the crush-yield point (CYP) of the container. A container found to be below the required CYP measurement will potentially collapse under prolonged warehouse storage, and may collapse during the filling operation.

Apparatus

The test apparatus is easily constructed from readily available parts (see Fig. 14-7) and is easily operated. Basic components of the assembly are:

1. Compression shaft and head that transmits a load to the sample container.
2. Speed reduction motor and gears designed to move the shaft at a constant rate of one inch per minute.
3. Reversing drum switch that controls the direction that the shaft travels.
4. Scale (bathroom type) with an accuracy of ±1 pound, that measures the load being applied to the sample. Seca Corp. model 760 has a dial and pointer that can be removed, inverted, and reinstalled to improve readability.
5. A clutch to disengage the gears when manual adjustment of the shaft is desired.

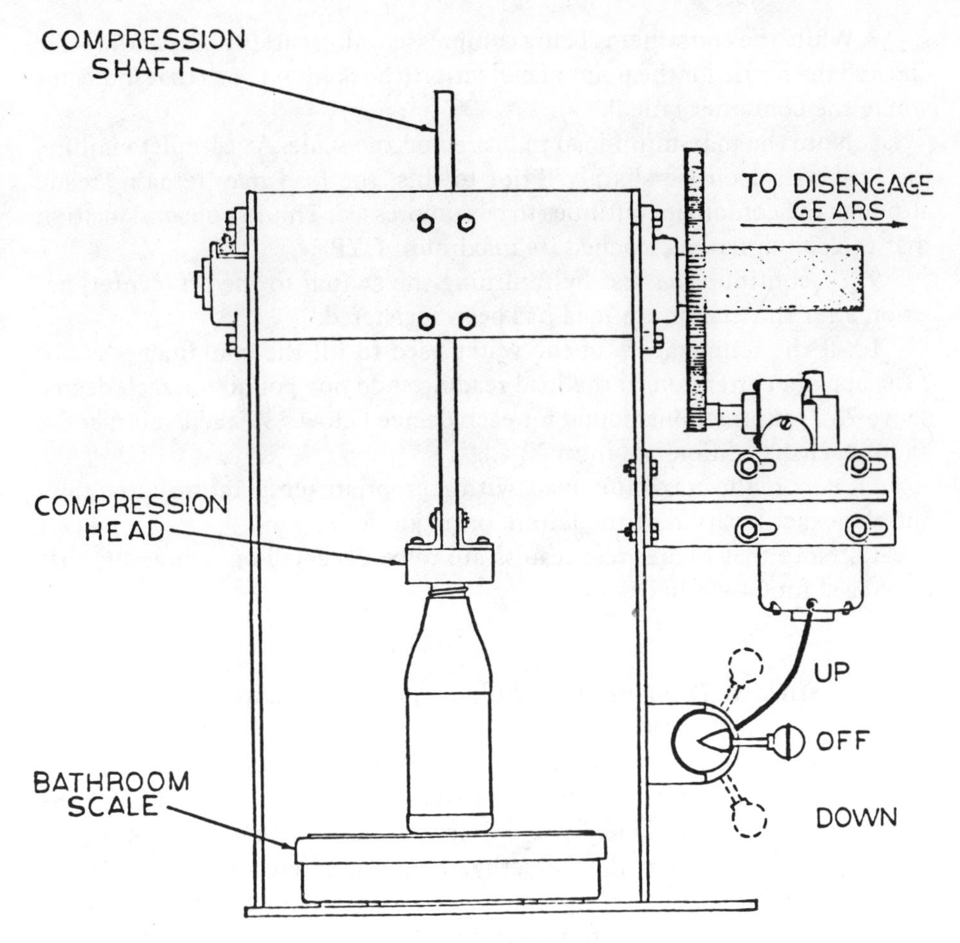

Figure 14-7 Apparatus to test container crush-yield point (CYP).

Procedure

1. Take containers from blow mold machine at lest 30 min prior to testing. Make certain that none has pinholes or windows.

2. Fill the containers with water at 73 ± 7°F (23 ± 4°C) to the optimum outage at standard fill. This should be indicated in the container specification. To do this effectively, add excess water and then reduce the level to the desired outage with an aspirator tube set to the proper length.

3. Cap the container with flat-top caps, using enough torque to assure a positive seal.

4. Raise the compression shaft and place the filled container on the scale, centering it under the compression head.

5. Set the scale to read zero pounds by adjusting the zero adjust control.

6. Push the reversing drum switch handle to the down position. This activates the motor and drives the compression head downward.

7. While the container is being compressed, alternately observe the scale dial and the bottle for the point of deflection (the point on both that indicates where the container failed).

8. Note the maximum load indicated on the scale. At complete failure, the load will decrease sharply. Prior to this, the load may remain steady although the container continues to be compressed. This is a clear indication that the container has reached its maximum CYP.

9. Discontinue the test by returning the switch to the off (center) position after the maximum load has been registered.

10. If the temperature of the water used to fill the container was *not* 73°F, apply a correction to the load reading. Add one pound for each degree above 73°F, subtract one pound for each degree below 73°F (add/subtract 0.8 kg for each 1°C difference from 23°C).

11. Record the maximum load (with appropriate temperature correction) and the exact location of the failure or major deflection.

12. Note that higher test results are to be expected on containers that have aged for 24–48 hours.

Test for Surface Treatment on Polyethylene Containers

In order to achieve good label adhesion, flame treatment of polyethylene containers is necessary. The degree of treatment is directly related to the surface tension of the flamed polyethylene surface. Label adhesion is generally satisfactory when the polyethylene surface tension is equal to or greater than the surface tension of water, as determined in the following water test. The carbol fuchsin stain test has been found useful for estimating the degree of flame treatment (surface tension) above the water wet level, as well as determining uniformity of treatment on various parts of the container. Better treatment is shown by a more intense dye stain. The water test is useful as a routine control test, while the stain test is better suited to determining optimum flaming conditions.

These tests are appropriate only when attempting to determine how labels with a pressure-sensitive adhesive will adhere to the container. In-mold labeling or glue lap labels do not require that a container be flame treated.

Reagents Needed

1. Kinyoun carbol fuchsin stain solution
2. Sodium chloride (salt)

Stain Solution. The stain solution should be well mixed before diluting. Shake the container thoroughly and let stand several minutes to release entrapped air. Dilute the stain solution as received with distilled water in a 1:19 ratio vol/vol in a suitable container, such as a widemouth gallon jar. Keep this solution in a closed container and shake well before each use.

Water Test Procedure

1. Hold the container to be tested under running cold tap water.
2. Remove the container from the water and observe its surface. If the water spreads evenly over the entire label panel area, and holds for at lest 20 sec, the surface is properly flame treated. If the water runs off or forms into droplets in less than 20 sec, the flame treatment is deficient.

Stain Test Procedure

1. Immerse the flame treated part of the container instantaneously into the 1:19 diluted carbol fuchsin solution for 4 sec.
2. Drain off the excess stain solution for approximately 4 sec, holding the part in a manner that leaves a uniform layer of stain on the treated area.
3. Immediately wash off all excess stain solution in a gentle flow of cold tap water. Do not use excessive water pressure.
4. Position the container to air dry so that the water will drain off rather than dry on the treated surface.
5. If desired, nonpigmented containers may be filled with clean, dry salt before evaluation. This provides a white background and prevents the dye on one side of the test container from interfering with the evaluation of the other side.
6. Rate the flame treatment according to the stain conditions as follows:

STAIN INTENSITY	
Excellent	A very intense color
Good	A definite, predominant color
Fair	A moderate degree of color
Poor	A slight color or no color
STAIN UNIFORMITY	
Excellent	Uniform intensity of stain
Good	A major area shows uniform stain intensity
Fair	A minor area shows uniform stain intensity
Poor	A few patches of stain are observed
Very poor	A very slight trace of color or no color at all is observed

Environmental Stress Crack Test

This is an accelerated test used to determine the relative performance of containers over prolonged periods of warehouse storage. The test containers are filled with product and held in an oven at 140°F (60°C) for five to seven days or until cracks develop. The test determines the effect on stress crack susceptibility of any change in molding conditions, container design, or resin type. It is a suggested test for quality assurance purposes. Each twenty-four hour period in the oven simulates approximately one week of normal warehouse storage time.

Apparatus

1. Cap wrenches of various sizes
2. Oven shelves, flat
3. Spring torque tester
4. Ratchet wrench
5. Forced air oven, 140°F
6. Tote boxes, thermoplastic ABS, with stacking corners
7. Thermometer 0–220°F (−17–104°C), maximum registering, armored
8. Balance
9. Siphon for fill adjustment

Bore a hole in a No. 8 neoprene stopper to accept a 6 in. length of rigid plastic tubing (1/4 in. OD × 1/8 in. ID at bottom). Cut a shallow slot across the bottom of the stopper for air circulation. Connect the top of the rigid tube to an aspirator using clear Tygon tubing.

When working with materials such as liquid detergents, connect a one liter polypropylene (Nalgene) filtering flask as a trap in the vacuum line.

Oven Operation

1. Install and adjust oven according to manufacturer's instructions. Omit all water connections and remove the humidity control to avoid its deterioration.

2. Remove the top water pan and the top and bottom baffles to provide additional sample space.

3. Set the control switch to the "on" position to give the high heating rate at all times.

4. Install three flat shelves in the oven. Each should hold three tote boxes and three more can be fitted on the bottom of the oven.

5. Keep a maximum registering thermometer in the oven and use it to check the oven temperature daily. Adjust the oven thermostat as required to maintain a temperature of 140°F.

Procedure

1. Select test containers of uniform weight and quality. If possible use containers from only one mold. Identify each container with extruder number and cavity identification. Take the samples at random from regular production.

2. Record the empty weight and cavity number of each container. Be certain that every sealing surface is in good condition. A tight seal is essential for this test.

3. Fill the containers with the appropriate product type.

4. Add an excess of product to each container, then use the siphon to reduce the level to the outage shown in each container specification. Ordinarily it will be necessary to set the siphon tube length about 1/16 in. less than the specified outage. Avoid using excessive vacuum.

5. Cap all containers to 15 in.-lb torque with standard nonventing caps.

6. Place the containers in the tote boxes. *Caution:* Keep common product types in the tote boxes together. Do not mix different products in the boxes, inasmuch as a test failure will result in product leakage and many products contain chemical components that are not compatible.

7. Place the tote boxes with the sample containers in the forced air oven at 140°F and record the date and time. Be sure that the tote boxes do not jam the fan when the oven door is closed.

8. Check for stress crack failures every eight hours by squeezing each container carefully. Remove any that are found to be leaking and record the date, time, and location of failure. To check a container that is soft yet has no apparent leak, turn the sample on its side and squeeze again. This will detect leaks near the top.

9. Remove all containers from the oven and discontinue tests after one week. If desired, the test may be conducted for a longer period of time, but one week is generally considered ample time and three weeks is deemed maximum.

10. Include the following information in the oven test report:

Container style
Size
Color
Weight
Mold number
Number of days to failure (to nearest one-third day)
Failure location
Machine type
Machine number
Special molding conditions
Plastic resin supplier
Plastic resin number
Color concentrate supplier

Color concentrate number
Percent regrind in use

Soot Chamber Antistatic Test

Plastic containers attract and hold static electric charges produced by friction. Dirt and dust particles are then attracted to the charged article. Antistatic materials may be added to the plastic resin to reduce or eliminate the problem by attracting moisture from the air to dissipate the charge. The following test is designed to determine the effectiveness of the antistatic additives.

The samples to be tested are preconditioned at low humidity and then given a static charge. After allowing a charge bleed-off period, the sample(s) are subjected to filtered smoke. The smoke is attracted to and causes discoloration on those samples that have retained the static charge because of inferior antistatic properties. Conversely, lack of discoloration indicates that the antistatic material is providing effective protection.

Apparatus

1. "Suit Case" soot chamber per Celanese Polymer Co. Drawing E-00057-191 (Reference: Celanese # JGF-64-191).
2. Cotton cheesecloth, 40 × 40 string count.
3. Paper towels, Scott C-fold 150 brand.
4. Filter paper, Whatman #1, 9 cm.
5. Relative humidity indicator, Serdex model 201.
6. Laboratory tongs.

Reagents

1. Toluene
2. Calcium chloride, anhydrous, 4 mesh from acceptable vendor.

Procedure

1. Clean the chamber with a dry cloth to remove any loose soot from previous tests. Prior to each test replace the cheesecloth in the filter holder with the specified material. Check that the cloth is free of wrinkles and provides a good seal.

2. Check the condition of the calcium chloride drying agent and replace if necessary. Latch the drying and combustion drawers tightly.

3. Check that the samples are clean, dry, and coded for proper identification. Place the samples in the soot chamber, evenly spaced, and include a separate paper towel for each sample. Spread the towels to provide air circulation around them. If desired the towels may be predried for a few minutes in an oven set at 140°F (60°C).

4. Place the humidity indicator in the chamber and close the door. Open the damper between the drying and combustion portions of the chamber and turn on the blower. When the relative humidity has dropped to 15 ± 5 percent, turn off the blower and begin timing the *two-hour precondition period.* Operate the blower as required during this period to maintain the specified relative humidity.

5. After preconditioning, remove the samples from the chamber and charge each by rubbing with one of the conditioned paper towels. Use a new towel for each sample. To charge a container, grasp the neck with one hand and hold the container on a 45-degree angle with the bottom edge resting on the table. Cup the towel in the other hand and rub the container rapidly ten times in one direction from neck to bottom applying considerable force. Rotate the container 180 degrees and repeat the operation on the opposite side.

6. To charge a cap, grasp the edge of the cap skirt with the thumb and forefinger of one hand. With the other hand, wrap the paper towel around the skirt, apply pressure and rotate the towel through 180 degrees ten times.

7. Immediately return the charged samples to the chamber in random order with the greatest possible spacing. A two-inch spacing is the absolute minimum. Close the chamber door, turn on the blower and begin timing the *two-hour static charge bleed-off period.* Turn off the blower when the relative humidity drops to 15 ± 5 percent (this will usually require less than thirty minutes).

Note: Bleed-off periods longer than two hours will only be necessary for special studies.

8. After the bleed-off period, remove the humidity indicator and immediately reclose the chamber door.

9. Close the damper and turn off the blower.

10. Using tongs, dip a quarter section of filter paper in toluene and allow the excess to drip off.

11. Place the filter paper in the combustion drawer, ignite it and close the drawer quickly. Wait thirty seconds for the paper to burn then turn on the blower *for ten seconds only.*

Caution: Toluene is flammable and toxic. Store it in a safety can only and remove the bulk of the solvent to a safe distance before igniting the filter paper. Avoid contact with the liquid or inhaling the vapor.

12. After fifteen minutes, remove the samples from the chamber.

13. Observe the soot patterns. To preserve the soot patterns, spray each sample with a fast-drying clear acrylic lacquer.

14. Estimate to the nearest 10 percent the soot-free surface area of each

Table 14-2 Quality classifications and acceptability

*Classification**	*Description*	*AQL†*
Critical	Will render the product unfit for use and may cause injury or illness to consumer. Fails to conform to all applicable specifications and regulations.	0.1%
Major	May render the product unfit for use and may meet with consumer rejection. Very unlikely to cause personal injury or illness.	1.5%
Moderate	Major effect on neatness and appearance and will make the product difficult to use. Will not pose a threat to user, but will meet with negative reaction.	4.0%
Minor	Appearance and use are affected slightly. Unlikely that consumer will notice any problem. Conforms fully to specifications.	6.5%

*According to acceptance sampling plan
†Acceptable quality level

sample. This figure is considered the percent effectiveness of the antistatic treatment.

QUALITY CONTROL ACCEPTANCE SAMPLING PLAN

The example of a quality control acceptance sampling plan given here will aid in establishing a consistent policy for determining acceptance/rejection of blow-molded containers. An acceptance plan tailored to a specific operation is certainly warranted.

When presented with a suspect lot of containers (a lot is any designated segment of production; e.g., one hour of container production, one shift, one day, etc.), the quality control inspector must make a logical determination as to how many individual units will be inspected, inasmuch as 100 percent inspection is usually not practical due to time and manpower constraints. A good basis for normal inspection is Mil-Std 105 D, Inspection Level I, for Normal Inspection. This standard provides a table indicating the number of random samples that should be taken from any given size lot, and the number of defects per sample group that are considered tolerable.

The following acceptance sampling plan will assist in classifying the various defects that you may be encountered. The effects on product quality and the acceptance level in each classification are given in Table 14-2.

ACCEPTANCE SAMPLING PLAN FOR BLOW MOLDED CONTAINERS

Refer to the appropriate acceptable quality levels (AQL) in Table 14-2.

Critical

1. Holes or tears of any size that prevent the bottle from holding the intended product.
2. Defects in the neck finish that prevent capping: neck OD, including threads.
3. Incomplete or improper neck finish causing functional defect (leakage or cap stripping) or packaging line malfunction. This includes defects in ID, height, trim characteristics, and concentricity of milling.

Major

1. Low overflow capacity, below minimum specification.
2. Incomplete or improper neck finish. This includes defects in ID, height, trim characteristics, and concentricity of milling.
3. Severe rocker bottoms (prevents use on filling line).
4. Chip contamination larger than 1/16 in. (0.0625 in., 1.59 mm)
5. Container deformed (dented or bent) to the extent that it is unusable.
6. Stress cracks.
7. Drop-impact test failures.
8. Flash in excess of 0.031 in. (0.79 mm) in neck area, bearing surface area, or sharp flash in handle area. Flash is defined as the excess plastic created during the manufacture of the container and intended to be trimmed from the finished product. The neck dome and container bottom tail are considered flash, and are also referred to as the pinchoff areas.
9. Cocked neck finishes greater than 5 degrees from a plane perpendicular to the base of the container.
10. Excessive concavity or convexity, out of specification ±0.015 in. (0.38 mm).
11. Containers that do not meet top load (CYP) requirements.
12. Blocked nozzles (drooped necks).

Moderate

1. Poor mold aligment, greater than 0.005 in. (0.127 mm).
2. Improper container height.
3. Wall thickness below minimum.
4. Flash in excess of 0.031 in. (0.79 mm) in the body of the container.
5. Improper container finish: insufficient opacity in colored containers so that product fill level is visible, color outside of established tolerances, and antistatic test failures.
6. Deep spirals in neck from reaming operation.
7. Resin contamination; three or more inclusions (black specks or gels) exceeding 1/32 in. (0.031 in., 0.79 mm) in greatest dimension, with none exceeding 1/16 in. (0.0625 in., 1.59 mm).

(continued)

ACCEPTANCE SAMPLING PLAN (*continued*)

Minor

1. Excessive grease or dirt—subject to limit standards.
2. Water marks (orange peel)—subject to limit standards.
3. Scuffs and scratches—subject to limit standards.
4. High overflow capacity.

Other Quality Control Records

An effective quality control program is one that is understood by the participants, supported by the persons actually manufacturing the product, and sensitive to the needs of the blow mold operation. There are two additional tools that will be a valuable part of a comprehensive quality control program, the extruder operating profile, and the troubleshooting check sheet. These will be developed by the quality control personnel and blow mold operators as they jointly gain knowledge and experience in the art of blow molding plastic containers:

Extruder Operating Profile

This form (Fig. 14-8) should indicate the normal operating parameters of the extruder. Each area that may be monitored should be listed, and the optimum setting. These areas include:

- Pyrometers
- Timers
- Parison controllers
- Air pressures
- Hydraulic pressures and oil temperatures
- Metered areas
- Water pressure and temperatures
- Target cycle time
- Target container weight

If a quality problem is realized, this operating profile is a good place to start checks to see if any extruder conditions have been changed or have gone out of control due to mechanical malfunction.

I. PYROMETERS

1. Feed section ____________
2. Transition section ________
3. Metering section 1 _______
4. Metering section 2 _______
5. Head __________________
6. Stock temperature _______
7. Variacs _______________

II. TIMERS

1. Charge delay ___________
2. Exhaust time ___________
3. Blow delay _____________
4. Blow time ______________
5. Blow pin delay _________

III. METERED AREAS

1. Mold open time _________
2. Extruder load __________
3. Extruder speed _________

IV. AIR

1. Low pressure ____________
2. High pressure ___________

V. HYDRAULIC

1. Low pressure ____________
2. High pressure (RAM) _______
3. Fill pressure ____________
4. Oil temperature __________

VI. WATER

1. Chilled H_2O temperature ___
2. Temperature _____________

VII. RECOMMENDED CYCLE TIME _______________

VIII. TARGET BOTTLE WEIGHT _______________

Figure 14-8 Extruder operating profile sheet.

BLOW MOLDING DEPARTMENT

PROBLEM

PROBABLE CAUSE

SOURCE CHECK

Steps
1. ______________________________
2. ______________________________
3. ______________________________
4. ______________________________
5. ______________________________
6. ______________________________

COMMENTS

Figure 14-9 Troubleshooting check sheet.

Troubleshooting Check Sheet

This record (Fig. 14-9) is developed when problems occur, to be used for future reference. It indicates:

- What the problem was
- What caused the problem
- What was done to solve the problem
- Applicable comments

The growing collection of check sheets constitutes a plant-specific or product- or process-specific troubleshooting guide that will be especially useful when a problem that is out of the ordinary or relatively unusual is encountered. For that reason, it is valuable to keep the check sheets updated and easily accessible.

ACKNOWLEDGMENT

I wish to recognize the many packaging engineers and quality control managers who, through the years, have worked so diligently to research and refine the test methods contained in this chapter. It is the dedicated efforts of these professional men and women that allow this instructional material to have practical value.

REFERENCES

1. Juran, Joseph, ed. *Quality Control Handbook.* American Soc. Quality Control.
2. *Handbook of Chemistry and Physics.* 44th ed. p. 2191.
3. Ibid.

SELECTED READINGS

Acceptance Sampling Plans. No. 3, Plastic Bottles. Phoenix, AZ: Dial Corporation.
Manufacturing Standards Manual. Phoenix, AZ: Dial Corporation. Secs. 830–832.

CHAPTER

15

Safety

THOMAS E. DOUGLAS

The dictionary defines safety as freedom from danger or injury. While this definition is simple and short, the ramifications of safety are not. This chapter identifies a number of the voluntary and mandatory safety and safety related programs that can apply to a blow molding operation.

Often an injury is blamed on the carelessness of an individual or group, with little being done to prevent a recurrence of the injury. Owing in many cases to management's inaction in predicting, preparing for, and reacting to unsafe conditions, many laws have been enacted to establish and maintain safety programs. These laws, which carry severe penalties for violations, are also briefly described in this chapter.

RESPONSIBILITY

Businesses have a moral and legal responsibility to provide a safe place for employees to work. They are also held responsible for the safety of the environment to the extent that it is affected by business or manufacturing activities. Thus the matter of safety should not and cannot be taken lightly at any level, especially in a manufacturing organization.

Aside from the moral issues, failure to meet legal requirements can result in enormous fines being levied against both corporations and individuals found guilty of neglect, ignorance, or disobedience of the laws. All companies should designate a competent department to be specifically responsible for monitoring and assuring compliance with all applicable laws and regulations, federal, state, and local.

Machinery and Equipment Manufacturers

It is strongly recommended that all requests for quotations and all purchase orders have a section explicitly stating the manufacturer's responsibility to design and build all equipment to meet existing standards and legal requirements of all related laws and regulations. The company legal counsel should review and approve this statement prior to use. The Occupational Safety and Health ACT (OSHA) has become a standard for machine building and should be specifically mentioned along with any other applicable laws known to the purchaser.

Before any piece of equipment is accepted or used it should be thoroughly inspected and tested to assure that it meets all codes and regulations and that all safety devices are in place and functional.

A safety check list is suggested to assure that nothing is missed on this inspection. The completed list should be retained to provide a permanent record of the condition of the machine when received. An example list is given as an appendix to this chapter. This example is not intended to be all-inclusive, but to be used as a guide in preparing a list tailored to a specific situation.

Machinery User

A blow molding company has an obligation to its employees to provide adequate training and instruction in the proper maintenance and safe use of all equipment, simple as well as complex. Machinery suppliers normally provide service and instruction manuals that are an excellent starting point for preparation of a training program. Too often, however, the manuals are stuck in a drawer somewhere or are lost or discarded and the employees are left to their own devices (testing, trial and error, etc.) to try to operate the equipment in a safe and efficient manner.

Training responsibilities must be assigned and those responsible for training held accountable for proper execution of the program. This program should not be limited to the initial start-up of the equipment, but should provide planned and scheduled follow-up sessions to refresh safety practices in the minds of all involved with the equipment. The program should include both those who operate the equipment and those who must service and maintain it.

IDENTIFYING HAZARDS

Employees at all levels of the company must be trained in and alert to safety hazards. This includes both the ability to identify hazards and the knowledge of how to report them and assure that prompt action is taken to correct them.

It is strongly recommended that every company have an official safety committee headed by a person of high authority within the company. This committee should consist of a cross section of members from all areas in the company and should have absolute authority to assure that all safety related problems are addressed and corrected in a timely manner.

A number of laws provide guidelines for identifying hazards and dealing with them. They include Environmental Protection Agency (EPA) regulations, the Right-to-Know laws (29 CFR 1910.1200), the Superfund Amendments and Reauthorization Act (SARA Title III), state right-to-know laws, and local ordinances throughout the country.

SAFETY RULES

Every company should establish a list of safety rules. It should be included in the employee's handbook, should be gone over point by point during employee orientation and training sessions, and should be posted conspicuously throughout the plant and offices as a constant reminder to all. An example list of safety rules follows.

SAFETY RULES

To guide each employee in carrying out his/her work in a safe manner, certain rules and regulations have been adopted and are listed below. Violators of these rules will be subject to disciplinary action.

1. Required Safety Apparel
   a. Safety glasses should be worn by all employees throughout the plant.
   b. Face shields should be worn when grinding metal, filling propane tanks, and servicing batteries of forklift trucks.
   c. A welding helmet must be worn when welding.
   d. Color tinted safety glasses must be worn when using a cutting torch.
   e. All employees must wear shoes with rubber soles and heels in the plant. Maintenance personnel must wear a leather work shoe, not a canvas shoe.
   f. Employees with hair longer than shoulder length must wear their hair up while working around machinery.

(continued)

SAFETY RULES *(continued)*

- g. Shorts can be worn in plant but must be within 6 inches of the knee. No halters or open-back tops are to be worn. No open shirts are to be worn in the plant.
- h. All employees when operating machines and/or when using a knife, must wear armbands and chest protectors.
- i. Particle masks must be worn in any area where plastic dust or other airborne contaminants are heavy.
- j. Ear protection must be worn in the Production Department, Material Handling Department, Compressor Room, and the Pump Room.
- k. All operators and relief operators must wear safety goggles or wraparound safety glasses whenever using saw cutter finishing equipment.

2. Forklift Truck Safety

Any employee operating a forklift truck or golf cart will observe the following rules:

- a. Be certain that brakes and horn are working properly at all times.
- b. Always park with forks lowered.
- c. Observe safe speed at all times and be extremely cautious in congested areas.
- d. Never back up without being sure the way is clear.
- e. Always sound horn when passing other employees and when entering blind areas, such as when moving through doorways.
- f. When moving with forks raised, watch for and avoid overhead obstacles such as pipes, electrical wires, shelf and walkway bracing, and other items.
- g. Never ram machines, walls, doors, etc.
- h. Never drive at a speed that would cause the vehicle to skid should a sudden stop be required.
- i. Only authorized valid licensed operators should operate fork lift trucks.
- j. If the load is large enough to prevent the operator from seeing clearly, the forklift will be backed rather than driven forward.
- k. Always come to a complete stop before changing from backward to forward movement, or from forward to backward movement.
- l. Always carry loads with forks as low as possible.
- m. Safety bucket should always be used when raising or lifting someone with forklift.
- n. No fast takeoff on forklift.
- o. Always stop before entering aisleway.
- p. Forklifts in maintenance area for repair should not be used.
- q. Employees operating a golf cart must sound horn and use caution lights when entering aisleways and blind areas.

Remember: the attitude of the driver toward the safety of himself and his fellow employees, as well as his attitude toward company property, is the key to safe driving.

3. General Plant Safety

- a. Never throw plastic, tools, or any other objects within the plant or on company premises.

(continued)

SAFETY RULES *(continued)*

b. There will be no horseplay at any time.
c. No running in the plant.
d. No pushing or shoving in the plant.
e. Before any equipment is used, always make certain that adequate guards are in proper position. If guards have been removed in order to make repairs, make certain they have been replaced or other adequate protection has been provided. Tag a machine as inoperable if a guard is off.
f. Smoking is permitted only in the lunchroom and restrooms in the plant and designated smoking areas in the administrative offices.
g. Never place hands into any moving machine unless all safety devices are operating properly and it is safe to do so. Only authorized mechanics and maintenance personnel may reach around or othewise bypass a safety guard when working on machinery or equipment.
h. Electrical covers, insulation, etc., must be in proper position at all times when power is on, unless repairs are being done by experienced maintenance personnel. Maintenance personnel must never leave consoles, control panels, and similar items open and unattended while power is on.

Violation of safety rules has long been an accepted reason for disciplinary action or even termination of employment and should remain so. Safety must remain the number one responsibility if a business is to survive.

LEGAL REQUIREMENTS

Over the last three decades, federal, state, and local governments have come to recognize the need to protect both the individual and the environment. As a result they have enacted many laws in an attempt to provide a cleaner, safer place in which to live. In many cases the laws are extensive in their coverage and involve common data that relate to more than one law or ordinance. No one should consider these regulations lightly, as the penalties can be severe.

Because of the complexity, extensive coverage, and local variations in the many laws, regulations, and ordinances it is not possible to cover here all of the specific areas that apply to all blow molding operations. The following is a general listing and brief explanation of basic laws and regulations. It is intended to provide an initial awareness of them and indicate possible sources of additional information.

Occupational Safety and Health Act (OSHA)

This act was one of the first attempts to provide safety guidelines relating to workplaces. Many specific standards have been established by OSHA regarding electrical safety, hazard guarding, color coding, and many other concerns. Both federal and state agencies exist and can be contacted for detailed information regarding specific needs. Training courses in occupational safety practice and administration are available from many agencies, colleges, and universities. They are strongly recommended both for initiating a program and for maintaining an awareness of changes to and extensions of the laws.

Federal Register 29 CFR 1910.1200

Known as the Right-to-Know law, this law covers employees' right to know about the chemical hazards to which they are exposed in the manufacturing sector. This law applies to blow molders and requires the employer to provide training and Material Safety Data Sheets (MSDS) on all hazardous materials with which employees may come in contact.

The nature of a specific molding operation, the amount of other secondary operations provided, and other operations within the facility will dictate the extent to which training must be provided and reports made to the government.

State Health Hazard Communication Standards

Many states have adopted standards known as state right-to-know laws. These laws differ from the federal standards in that they specifically address the community's right to know about the hazards in a local workplace. These laws usually involve local authorities such as the fire marshall, the Emergency Response Committee, and other designated agencies.

Labeling standards, employee training, and reporting are requirements of these laws. Formal training of supervisors and those responsible for in-company safety instruction, through seminars and other available resources, is strongly recommended. There must also be continual follow-up to assure compliance with changes and additions to these laws.

Environmental Protection Agency

The United States Environmental Protection Agency, better known as the EPA, has broad authority over any emissions to the environment—land, air or water. The requirements and restrictions are massive and should be studied carefully by the assigned department to assure compliance. New and expanded EPA rulings are being enacted continually, which necessitates ongoing action and review, not a one-time compliance program.

Complying with EPA restrictions can seem overwhelming at first. It is suggested that a competent consultant be retained for the initial evaluation; that can save hours of laborsome reading and interpretation of the rulings. In addition a knowledgeable consultant can offer alternatives to meeting the various requirements without massive expenditures. Many times the mere substitution of one chemical for another can accomplish compliance.

Proper testing and record keeping are essential. Duplicate records are suggested in case of fire or loss of the originals.

Superfund Amendments and Reauthorization Act of 1986 (SARA)

One of the latest acts, this law builds upon the EPA Chemical Emergency Preparedness Program (CEPP) and various state and local programs aimed at helping communities to better meet their responsibilities in regard to potential chemical emergencies.

This law affects all manufacturing facilities in some way and periodic reports must be filed with local, state, and federal agencies. The most prominent section of the law is known as SARA Title III, the Emergency Planning and Community Right-to-Know Act of 1986.

Title III has four major sections that must be addressed and with which compliance must be assured. These are:

- Emergency planning
- Emergency notification
- Community right-to-know reporting requirements
- Toxic chemical release reporting emissions inventory

Food and Drug Administration

The United States Food and Drug Administration (FDA) controls and approves those items contained in or relating to human consumption. Whenever a product is to be used in conjunction with consumable items the ingredients in that item must meet FDA specifications.

Blow molding is used primarily to produce containers and is thus often involved with meeting FDA requirements. The suppliers of base materials such as resin, additives, and colors should be notified if a product in which their material is used will be subject to FDA regulations and should be required to submit written certification that their materials are approved by the FDA for the specific blow molding application.

Proper records of these certifications are essential and a method of tracking the flow of the materials they cover through a plant into the end product is required. Detailed requirements can be obtained from the Food and Drug Administration.

Department of Agriculture

The United States Department of Agriculture (USDA) controls items used in producing foods and that come in contact with foods as they are produced. All articles produced for use in this area require approvals and certifications.

Compliance

Because the various laws and regulations affecting the blow molding industry are so diverse and extensive, and are changed or added to so frequently, the problem of compliance must be an ongoing concern. Many of the laws and agencies use common data and lists. Each law should be dealt with and recorded separately, with review dates planned and scheduled to assure continuing compliance.

Regulatory Agencies to Contact

1. U.S. Department of Labor Occupational Safety and Health Administration (OSHA), Washington, DC 20210
2. Federal right-to-know laws. Same as # 1
3. State Health Hazard Communication Standards. Contact your state Department of Labor to determine if your state has such a standard.
4. United States Environmental Protection Agency (EPA), Washington, D.C. 20460
5. Superfund Amendments and Reauthorization Act of 1986 (SARA Title III) United States Environmental Protection Agency WH-548A, Washington, DC 20460
6. United States Food and Drug Administration (FDA) 5600 Fishers, Lane Rockville, MD 20857
7. United States Department of Agriculture (USDA) 14th St. and Independence Avenue S.W., Washington, DC 20250

SAFETY CHECK LIST

	OK	NEEDS REPAIR
1. OSHA CODES		
a. All guards and gates in place.		
b. Color codes correct.		
c. Electrical grounding.		
d. Safety devices functional. (Test)		
e. Safety signs in place.		
2. Mechanical Functions		
a. All bolts tight.		
b. Proper machine mounting.		
c. Proper instructions.		
d. Instruction manual on file.		
e. All welds solid		
f. Safety bar.		
3. Hydraulics		
a. No leaks.		
b. Pressure properly set.		
c. All hoses in good condition.		
d. All connections tight.		
e. Safety precautions taken if accumulator bottles present.		
f. Proper support and protection of all hydraulic lines.		
g. Hydraulic schematic available.		
4. Electrical		
a. All wires and connections covered.		
b. All junction boxes covered.		
c. All controls clearly marked.		
d. All safety devices functional.		
e. Emergency stop provided at all locations necessary.		
f. Proper fuses or circuit breakers.		
g. Lock-out provision on main disconnect.		
h. Electrical circuit diagram available and accurate.		
i. Electrical circuit meets applicable codes.		
j. Interfaces with other equipment.		
5. Pneumatics		
a. No air leaks.		
b. Proper filters, etc. in place.		
c. Pressures properly set.		
d. All connections tight.		

(continued)

SAFETY CHECK LIST *(continued)*

	OK	NEEDS REPAIR
e. Hoses and pipes of proper size, type, and mounting.	_____	_________
f. Pneumatic diagram available.	_____	_________
g. Disconnects and shutoff valves properly positioned and functional.	_____	_________
6. Operator		
a. Operating manual available.	_____	_________
b. Operator training session prior to release to production.	_____	_________
c. Safety review with operator.	_____	_________
d. Housekeeping requirements reviewed with operator.	_____	_________
e. Emergency procedures reviewed with operator.	_____	_________

SECTION IV

PLASTIC BLOW MOLDING MATERIALS

CHAPTER

16

Materials Review

JAMES P. PARR AND ECKARD F. H. RADDATZ

POLYMER PRINCIPLES

A polymer is a material with a molecular structure that consists of many repeating subunits called monomers, or, simply, mers. The polymer molecule can be thousands of times as massive as the monomer. Many naturally occurring materials are polymeric. Cellulose, the predominant constituent of wood and cotton, is one of these materials. Starch, one of our most valuable foodstuffs, is simply a stereoisomer, or mirror image, of cellulose.

The twentieth century saw the era of man-made polymers emerge. We encounter many of these materials every day. The clothes we wear, the carpets in our homes, and many other items we buy, sell, or use are made from various types of these polymers. The plastics used in blow molding are all members of this class of compounds. A clear understanding of the chemical and physical nature of these materials is invaluable for the conscientious molder.

Structure and Synthesis of Thermoplastics

The long chain structure of polymers used in blow molding is their most immediately striking aspect. All of these compounds are based on the

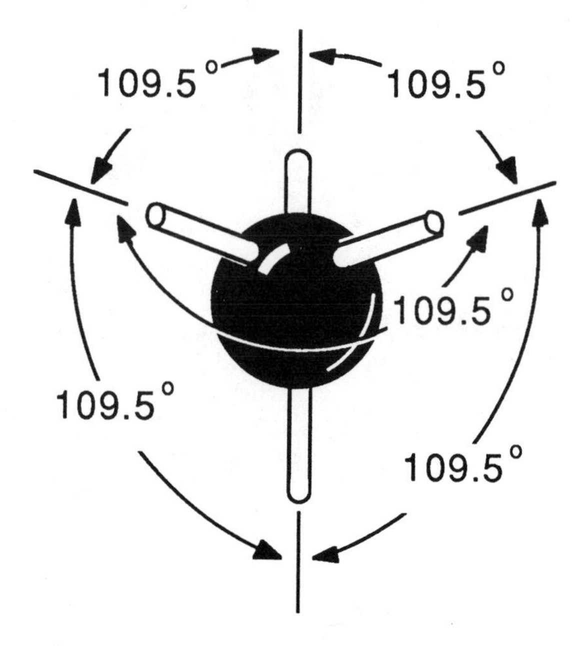

Figure 16-1 Tetrahedral bond angles of carbon.

carbon atom and its ability to form stable chemical bonds. From the study of organic chemistry one finds that the carbon atom has the ability to form four bonds with other atoms. These bonds form tetrahedral angles of 109.5 degrees, as shown in Fig. 16-1. Often these bonds form between adjacent carbon atoms creating a chain formation.

The class of compounds called hydrocarbons is such a group of materials exhibiting carbon-carbon structure. Typically, these molecules are formed from chains of carbon atoms that have three hydrogen atoms attached to each end, or terminal, carbon atom, and two to each interior carbon atom. These compounds are called alkanes or paraffins.

For example, the gas propane, a common home heating fuel, consists of three carbon atoms bonded together. The two end carbons each have three bound hydrogen atoms, while the other carbon has two. This is because each carbon must have four bonds. The end carbons must use only one bond to attach to the middle carbon, therefore they have three remaining bonds for hydrogen atoms. Propane is schematically depicted in chemical terms as follows:

```
    H   H   H
    |   |   |
H — C — C — C — H
    |   |   |
    H   H   H
```

or $CH_3—CH_2—CH_3$ or C_3H_8

⟨ Propane ⟩

A shorthand structural formula method simply ignores the hydrogen atoms:

$$C\text{—}C\text{—}C$$

Propane

Sometimes a hydrocarbon molecule can exist in a different form in which two or more carbon atoms use two of their bonds to bond each other. These atoms are thus double-bonded. Double bonds are not as stable as single bonds and are prone to chemical reaction. These compounds are called alkenes or olefins. The most simple olefin, ethylene, consists of two carbon atoms and four carbon atoms. The structural formula for this molecule is

$$\begin{array}{ccc} H & & H \\ & \diagdown \; \diagup & \\ & C{=}C & \\ & \diagup \; \diagdown & \\ H & & H \end{array} \quad \text{or} \quad C{=}C$$

Ethylene

There are only two hydrogens with each carbon because the doubly bonded carbons each use an extra bond for their own attachment. The ethylene molecule is the building block for the most basic of the large-volume commercial polymers, polyethylene (PE). Polyethylene is a so-called addition polymer, so named because it forms through the linking together, one after another, of ethylene molecules. The addition process continues until a molecule of immense length (up to 200,000 carbon atoms in length) is formed. The polymerization reaction occurs because the double bonds of adjacent molecules can be made to open up and attach to one another. The basic formula for polyethylene is

$$CH_3\text{—}\{\text{—}CH_2\text{—}CH_2\text{—}\}_n\text{—}CH_3$$

Polyethylene

Where n is one less than the number of ethylene molecules that are ultimately joined together.

The other addition polymers of greatest commercial interest in blow molding include polypropylene (PP), polystyrene (PS), polyvinylchloride (PVC), polyacrylonitrile (PAN), and polyacetal (POM). All of these polymers utilize precursor materials that include double bonds to form their long chainlike molecules.

Another valuable class of thermoplastics are formed from two different reactants in a condensation reaction. Nylon, or polyamide (PA) is a good example of these materials. The two starting materials are, for 6-6 nylon, hexamethylenediamine and adipic acid. Their structures and the reaction look like this:

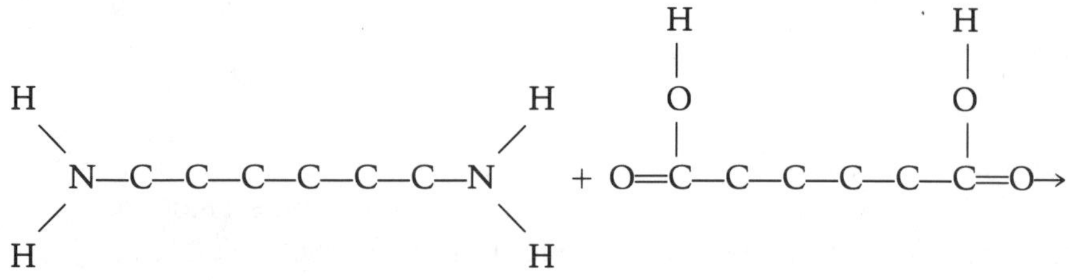

```
                                      H              H
                                      |              |
H                    H                O              O
 \                  /                 |              |
  N—C—C—C—C—C—C—N       + O=C—C—C—C—C—C=O→
 /                  \
H                    H
```

Hexamethylenediamine　　　　　　Adipic acid

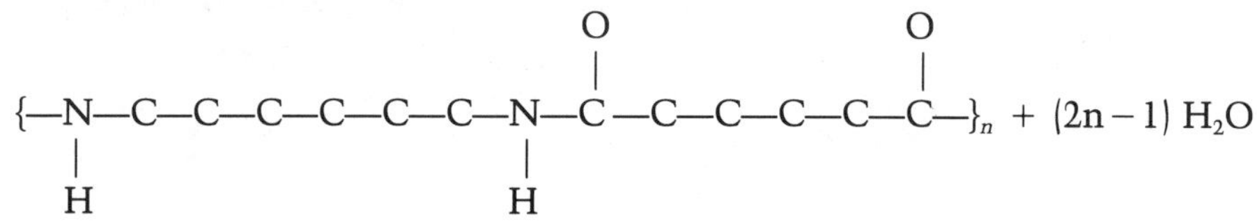

```
                         O                   O
                         |                   |
{—N—C—C—C—C—C—C—N—C—C—C—C—C—C—}n + (2n − 1) H2O
   |                 |
   H                 H
```

Nylon 6 = 6

As the materials react, a molecule of water is liberated for each amide linkage (NH—CO) that is formed. Other valuable blow molding resins formed in a condensation reaction include polycarbonate and polyesters. One might observe that condensation polymers could be produced from a series of different reactants having different carbon chain lengths. This allows a great deal of flexibility in the choice of physical properties.

These types of polymers are called linear polymers. Linear polymers can be melted and solidified many times. They are therefore part of the class of materials called thermoplastics. A different class of polymers have a structure similar to a three-dimensional network and are called, accordingly, network polymers. Both of these structures are shown in Fig. 16-2. Network polymers cannot be remelted and are called thermosetting polymers. These polymers are not used for blow molding. Vulcanized rubber and cross-linked polyethylene are examples of these types of polymers.

The nature of the side groups that are attached to the main chain helps determine the properties of the polymer. For example, polyethylene and polystyrene both have a backbone of carbon atoms. However the presence of the attached phenyl groups (benzene rings) of the PS molecules (Fig. 16-3) leads to a considerable difference in their properties. Foremost among these differences are stiffness and clarity [1].

Another aspect of molecular structure arises when there is a possibility for different arrangements of the groups that are attached to the main polymer chain. These various arrangements result in spatial differences that result eventually in physical property differences. These different structures are called isomers. Isomers can be best understood by looking at your own hands. One is a left hand while the other is a right hand. Although they are

- A - A - A - A - A - A - A -

- A - A - A - A - A - A - A -

- A - A - A - A - A - A - A -

LINEAR POLYMER

- A - A - A - A - A - A - A -

- A - A - A - A - A - A - A -

- A - A - A - A - A - A - A -

NETWORK POLYMER

Figure 16-2 Bonds between adjoining chains restrict molecular motion in network polymers. **A,** atom in polymer chain; **-**, bond between chain atoms.

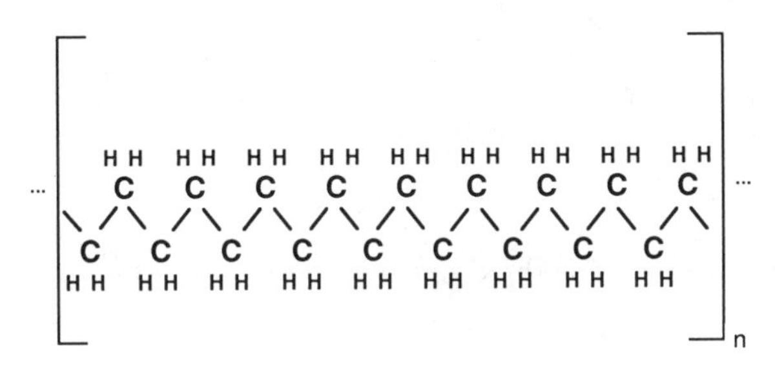

POLYETHYLENE

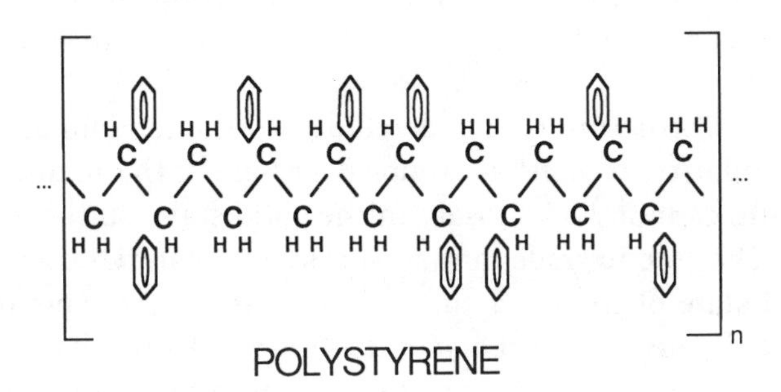

Figure 16-3 Benzene rings in the structure of polystyrene impede crystallization. **C,** carbon atom; **H,** hydrogen atom; ⬯ benzene ring.

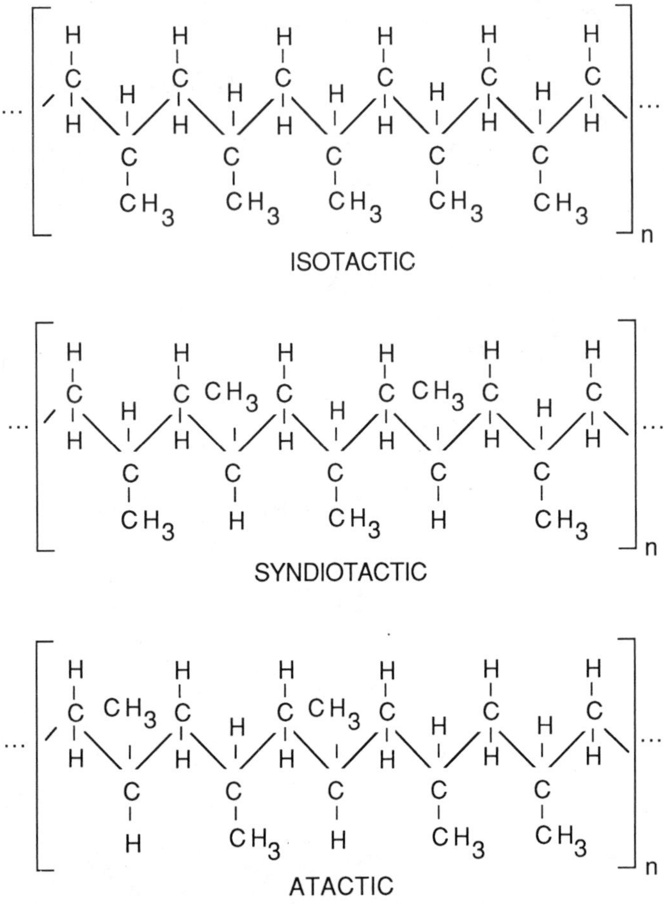

ISOTACTIC

SYNDIOTACTIC

ATACTIC

Figure 16-4 The three stereoisomeric forms of polypropylene.

constructed of the same basic elements (thumb, ring finger, etc.) they are essentially different in their spatial arrangement. This can be verified by putting a left glove on your right hand.

As an example, it is useful to consider polypropylene. There are three possible configurations for a polypropylene molecule, as shown in Fig. 16-4.

The atactic structure is disordered due to the random arrangement, or nonstereoregularity, of methyl groups (CH_3) along the main carbon chain. This structure cannot pack closely in the solid state, so the polymer is not crystalline. The atactic grade forms a soft sticky material used in adhesives. The ordered state of the isotactic arrangement is quite able to crystallize, however. The higher degree of crystallization results in a harder, stiffer solid state. The isotactic form is the type most commonly encountered in molded parts.

Another aspect of polymer synthesis is the possibility of using more than one starting monomer, in the case of addition polymers, or two starting monomers for condensation polymers. The material produced under these circumstances is called a copolymer. High-density polyethylene is most often produced as a copolymer, since the pure ethylene polymer, or homopolymer, is much more brittle. A commonly used comonomer is hexene ($CH_2{=}CH_2{-}CH_2{-}CH_2{-}CH_2{-}CH_3$). The copolymer has the following appearance:

```
{—C—C—C—C—C—C—C—C—C—C—}n
              |
              C
              |
              C
              |
              C
              |
              C
```

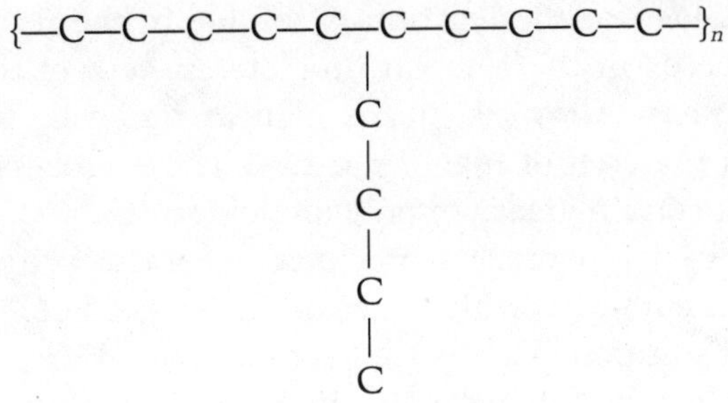

There are typically 5 to 30 branches per thousand carbon atoms. Only four carbons extend from the main chain backbone because two carbons become part of the backbone. This type of copolymerization is called graft copolymerization. Sometimes the polymer can consist of three monomers and is called a terpolymer.

Two other copolymerization structures are seen. These are the block and random copolymers. If we label the two starting materials *A* and *B*, these copolymers take on the following appearance:

$$(\text{—A—A—B—A—B—B—B—A—B—A—A—A—A—B— A—})_n \text{ Random}$$

$$(\text{—A—A—A—A—B—B—B—A—A—A—A—B—B—B—})_n \text{ Block}$$

Polymerization reactions can occur because the end products of the reaction are in a lower energy state than the reactants. This results in a large liberation of energy during the reaction. This amount of heat must be removed from the reaction chamber to allow proper control of the process.

There are several main types of commercial reaction processes used in the production of polymers. These include one- and two-phase polymerization and precipitation polymerization [1].

One-phase polymerization can be separated into two types, mass and solution processes. In the mass polymerization process, a pure monomer is allowed to polymerize. The rate of heat transfer out of the system defines the rate of reaction. This process is used to produce low-density polyethylene (LDPE). In the solution process, the monomer and polymer are both dissolved in a solvent, which facilitates heat removal.

Two-phase processes include emulsion and suspension technologies.

One type of two-phase process utilizes a fluidized bed of polymer particles, supported by a gaseous reactant.

Emulsion polymerization utilizes a reaction between the monomer, which comprises one phase, and the catalyst, which is dissolved in a second liquid phase, near the interface between the two. The reaction utilizes emulsifiers such as soaps, which promote contact between the phases.

The suspension process is similar but does not utilize emulsifiers. Instead, fine particles of materials are used which help promote a fine suspension. The catalyst for the reaction is soluble in the reactant phase, so the reaction proceeds in the reactant droplets. In both of these processes, the nonreactive phase allows a means of heat removal. The suspension process produces material of higher purities. These processes are used in the production of certain grades of polyvinylchloride.

The third main type of process, precipitation or slurry, involves reaction in a suspending medium in which the monomer is soluble. The polymer precipitates from solution. This is the technique used for the Phillips and Ziegler processes for the production of high-density polyethylene (HDPE). Again, the carrier liquid, or diluent, is utilized to facilitate heat transfer.

After the raw polymer is produced, the material is usually pelletized. The molten polymer is forced through a die and cut into the pellets. Additives are typically added in this step. The pelletization process helps to homogenize the material and provide a good dispersion of the additives. Some products of high molecular weight are not able to be pelletized without a great degree of degradation, so they are sold as powders. The greater exposed surface area of powders makes these grades more susceptible to oxidative attack and moisture absorption.

MOLECULAR WEIGHT

The molecular chains can be produced with very high molecular weights. In general, it can be seen that all of the physical properties of the material are enhanced by an increase in the molecular weight. The molecular weight is simply the mass of all the monomer molecules that were combined to produce the polymer molecule, minus any portions that were liberated.

When considering the molecular weight of a long chain polymer, one realizes that in all probability it is unlikely that all the molecules are of the same absolute weight [2]. More likely, the polymer molecules are of varying weights that cluster around some average value. This is in fact the case with all polymers in commercial use. This average value is taken to be the molecular weight of the polymer.

This distribution function of the molecular weights for a resin is commonly referred to as its molecular weight distribution (MWD). The distribution of the individual values can be described by standard statistical terms. The distribution for a given polymer can take on a number of different forms,

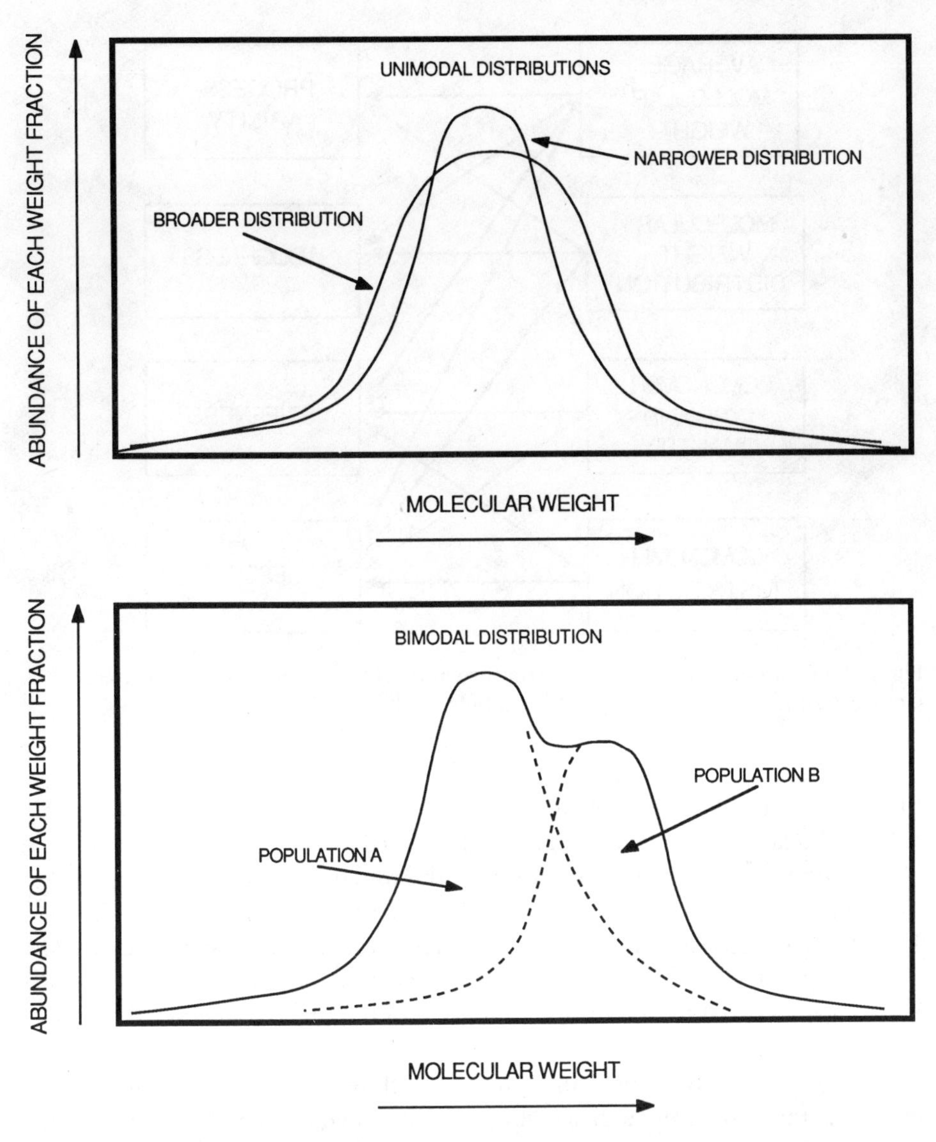

Figure 16-5 Differences in molecular weight distributions affect physical properties of a polymer.

including unimodal and bimodal, as shown in Fig. 16-5. A bimodal distribution is comprised of two separate distributions that are combined.

Many of the properties of polymers, both morphological and rheological, can be related to the character of the molecular weight distribution. Figure 16-6 shows the relationship of various properties to the molecular weight and molecular weight distribution of a polymer.

Molecular weight is measured by a number of techniques that vary in cost and complexity. The melt index is a quick, simple indication of molecular weight. It helps indicate the expected processability of a resin. Since

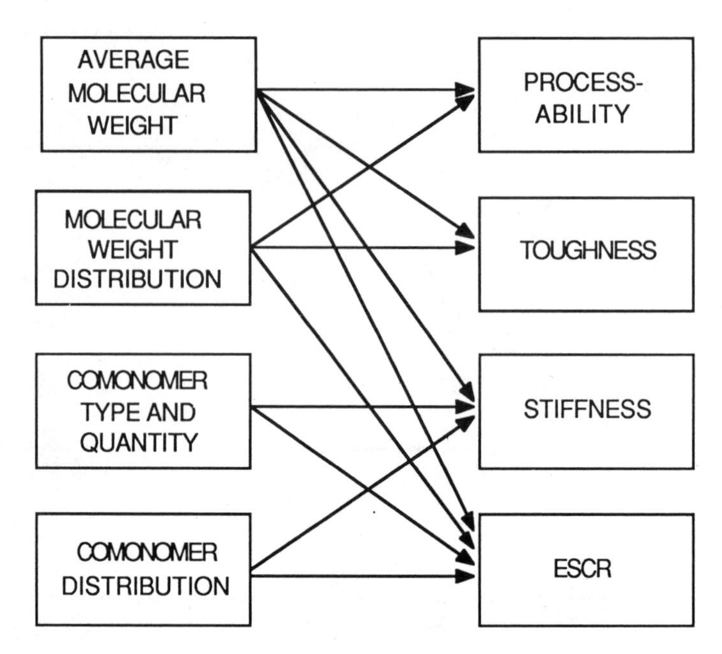

Figure 16-6 Effects of molecular structure and properties on resin performance. Courtesy of Eckard Raddatz and Armin Schwaighofer.

the melt index is inversely proportional to the molecular weight, lower values indicate higher molecular weights.

Another technique for determining the molecular weight of a polymer involves viscosity measurements of solutions of the polymer and a solvent. The intrinsic viscosity, K value, and reduced solution viscosity tests are all of this type. They are based on the principle that the viscosity of a liquid is strongly affected by the number and size of polymer molecules dissolved in it.

Gel permeation chromatography is a different type of technique. It is used to determine molecular weight and also molecular weight distribution. The principle is to dissolve the polymer in some solvent. The solution is then passed through a column that is packed with a material which has pores that are on the same order of size as the polyethylene molecules. As the solution passes through, the molecules are retained in the pores for various times depending on their size. The higher molecular weight materials tend to pass through faster since they are retained in fewer of the pores than the smaller molecules.

MOLECULAR ARRANGEMENT

When the individual polymer molecules are looked at in terms of how they interact with other molecules, some revealing observations can be drawn. The forces that hold molecules together in the solid state are weaker than the forces that bond the individual atoms together in the molecular structure. These intermolecular forces include van der Waals forces, dipole attractions, and hydrogen bonds. These forces, if high enough, can cause a polymer with a regular structure to crystallize [1].

The degree of crystallinity of a polymer is a very basic determinant of its physical properties. Crystallinity is a state of order in which the various molecules pack together in a regular pattern. Crystalline polymers have a precise melting point. Properties associated with high crystallinity include high stiffness, density, tensile strength, hardness, solvent resistance, and impermeability. Environmental stress crack resistance, notched impact strength, and transparency are reduced at high crystallinity levels.

Because the melting point results in the molecules breaking out of their dense crystalline structure, there is a large decrease in density at this point. Therefore there is a greater tendency for crystalline materials to shrink and warp as they are cooled in a mold.

Amorphous polymers have a disordered solid state. This results in a high degree of elasticity in the solid state. Amorphous polymers do not have a definite melting point, instead they soften over a broad temperature range. These materials typically have a structure with either a number of bulky groups attached to the main backbone, or a random stereoregularity.

Polymers with relatively simple structures usually have a greater tendency to crystalline. Crystallinity can be increased by stretching in either the solid or molten state. This phenomenon is called orientation. Orientation yields a structure with good properties in the axis of the stretching force, but the properties are not as good in the transverse direction.

The degree of crystallinity of a resin is influenced also by the rate at which the molten polymer is cooled. This is utilized in fabrication operations to help control the degree of crystallinity. The balance of properties can be slightly altered in this manner, allowing some control over such parameters as container volume, stiffness, warpage, and brittleness. Nucleating agents are available that can promote more rapid crystallization, resulting in faster cycle times.

An important transition occurs in the structure of both crystalline and noncrystalline polymers. This is the point at which they transition out of the so-called glassy state. The glassy state is characterized by rigidity and brittleness. This is because the molecules are too close together to allow extensive slipping motion between each other. When the glass transition temperature, T_g, is above the range of the normal temperatures to which the part is expected to be subjected, it is possible to blend in materials that can reduce the T_g of the blend. This yields more flexible, tougher materials.

ROLE OF RHEOLOGY IN PROCESSING

Rheology means the study of flow. Flow is an important property to study for a polymer. Polymers exhibit flow characteristics that are termed *vis-*

coelastic. Viscosity is a measure of the relationship of shear stress to shear rate for a flowing material. Shear stress is analogous to the pressure, or driving force, that causes the material to flow. Shear rate is a measure of the response of the fluid to the driving force. Viscosity is the relationship of these two conditions. A low-viscosity material has a greater response to an imposed shear stress, resulting in higher flow. A higher viscosity material will have a lower flow response to the same shear stress.

A normal, or Newtonian fluid has a viscosity that is constant at varying shear stresses. Water is like that. Polymers, however, exhibit pseudoplastic flow. This means that the shear rate increases exponentially with shear stress. Shear-rate–shear-stress curves are plotted on logarithmic axes for this reason.

Elasticity is the other phenomenon, in addition to viscosity, that is seen in a viscoelastic fluid. Viscosity may be thought of as a resistance to flow, the phenomenon that dissipates the energy of, for example, an object falling into a pool. Elasticity is instead a storage mechanism. That means that during flow, some of the energy input is stored in an elastic manner, similar to a spring. This behavior is perhaps most simply seen whenever a molten polymer is forced through a die under pressure. Invariably, the extrudate will be larger than the die through which it passed. This is because energy was stored in the polymer due to the pressure gradient that forced the flow. After the material exited the die, the energy was released, causing the expansion or swelling to occur. This phenomenon is commonly called die swell.

These flow properties have a very basic effect on all polymer processing techniques. It is important to have a fundamental understanding of them. The most basic flow phenomena encountered in the blow molding process include flow in the extruder during plastication, flow through the head and die, parison sag while hanging from the mold, and the stretching of the parison during the blowing cycle.

The melt index (MI), a simple flow value, is measured according to ASTM D1238, under which specification several different conditions are shown. The test requires that the amount of material extruded at a standardized temperature through a die under pressure from a set mass over a period of ten minutes be measured. The testing apparatus is shown in Fig. 16-7. This weight of the extruded material is then reported as the MI. The most commonly used condition for polyethylene is that which employs a 2.16 kg weight at 190°C (4.76 lb at 374°F). When the melt index is run using a higher load of 21.6 kg (47.62 lb) at 190°C, it is often referred to as the high load melt index or the HLMI (condition F according to ASTM D1238.) The various weights and temperatures used are shown in parentheses with the abbreviation MFI; e.g., the MI is shown as MFI (190/2.16), and the HLMI as MFI (190/21.6).

While the basic flow measure is the melt flow index, other measurements can be made on a rheometer that has the capability to easily change

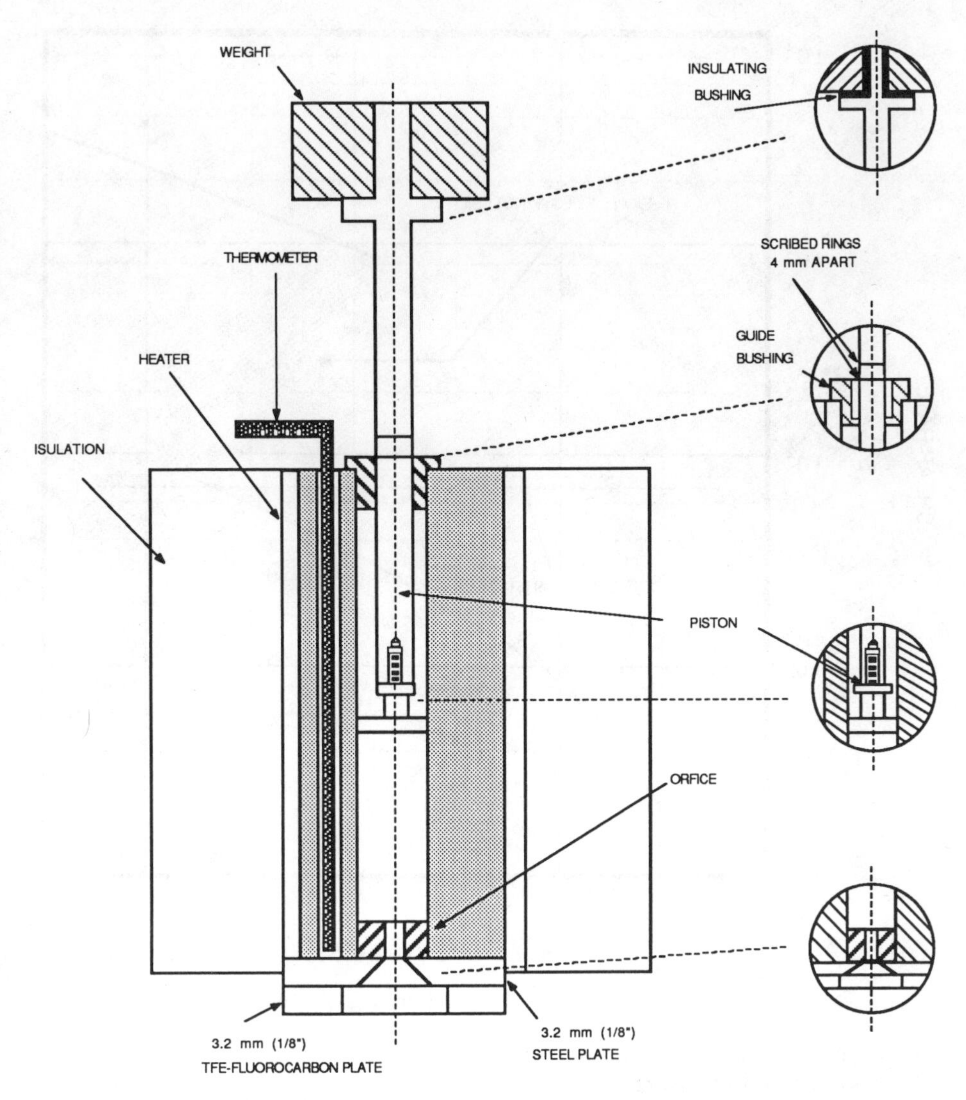

Figure 16-7 Extrusion plastometer. Courtesy of Armin Schwaighofer.

the shear stress. It is important to make measurements at different conditions, because of the exponential nature of polymer flow. The viscosity of the polymer at a given temperature is dependent not only on the polymer itself, but also on the shear rate. This has a practical significance in that the different flow conditions encountered in the process present a range of conditions. For example, the sag behavior is dependent on the low shear flow, while the high shear flow is more determinative of the flow through the die lips. A flow cure for a resin is shown in Fig. 16-8.

The effect of temperature should also be examined. If the viscosity or elasticity of the material changes too strongly with temperature, then the processing range will be somewhat restricted.

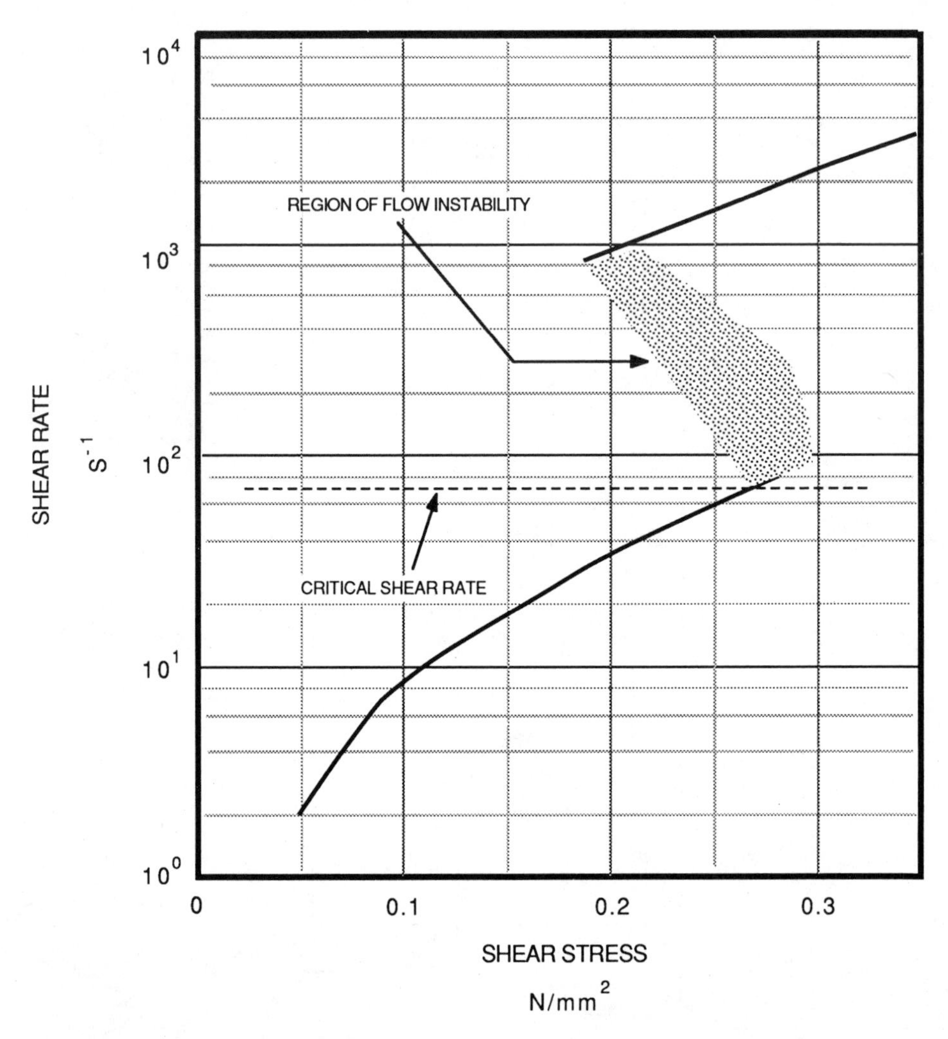

Figure 16-8 Shear diagram for high-molecular-weight–HDPE blow molding grade. Courtesy of Eckard Raddatz.

The first flow is seen in the extruder. The screw should be designed with the flow data for the material that is expected to be processed on it. This will insure the best possible performance. It is especially important if the material is extremely viscous and sensitive to overheating.

The flow through the head/die tooling results in the formation of the parison. Phenomena involved here include the flow around the central mandrel, or through a more complicated routing arrangement, and the reunification of the resulting flow front boundaries. The material must reunite in a homogeneous manner without defect. The pressure must be equilibrated within the radial entrance section leading to the lands of the die or the wallthickness of the parison will not be even. Stagnation points are to be avoided to reduce contamination and color change problems.

The parison formation step can be either continuous or intermittent. It is possible to encounter the critical shear stress of the resin during this step, especially with intermittent extrusion (Fig. 16-8). If this occurs, melt fracture can be encountered. This is a surface waviness in the transverse direction to the extrusion. The spacing of the ridges and their amplitude can vary for different tooling geometries and resin flow rates. To relieve this condition it is necessary to either reduce or increase the shear rate by changing either die size, extrusion rate, or melt temperature.

As previously mentioned, the extrudate will swell as it exits the die. The swelling behavior is a manifestation of the elastic component of the resin's flow. It is possible to measure this increase and obtain values to describe it numerically. The values are typically expressed as a ratio comparing either the weight or the diameter of the extrudate to a standard value. These data are valuable in the design of dies. They can also be used to compare different resins to determine if they can be processed in the same tooling.

Because the parison is shaped like a tube, the swelling effect is manifested in two different ways, as seen in Fig. 16-9. The overall diameter of the parison will be some value in relation to the die diameter. This is

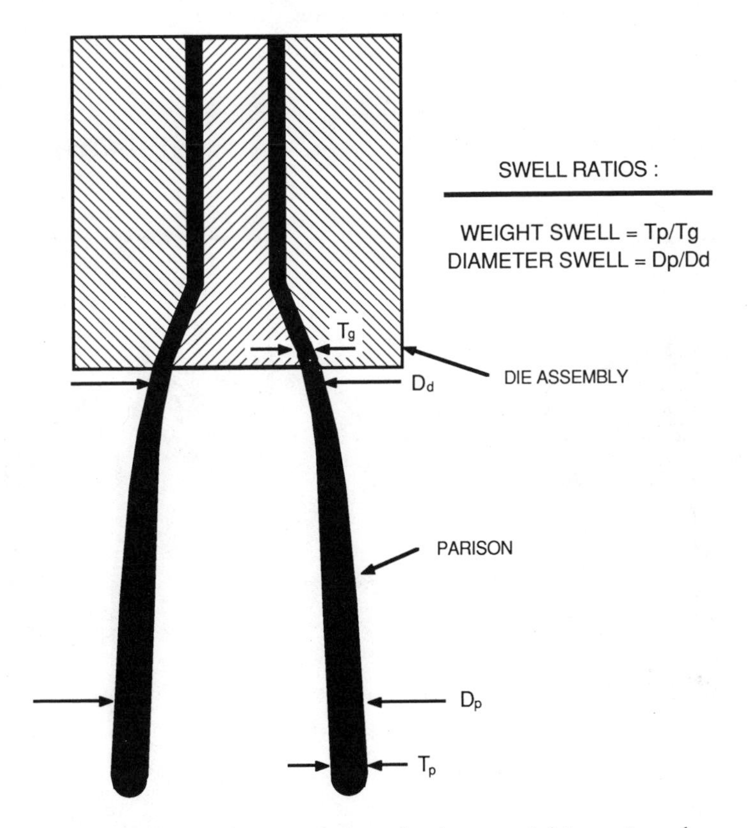

Figure 16-9 Swelling behavior of viscoelastic material in parison formation.

sometimes referred to as the diameter swell. It is important especially when handleware is to be produced. The wall of the parison must fall within the handle portion of the mold if the handle is to form properly.

The other effect is the relationship of the wall thickness of the parison to the die gap. This effect is often called the weight swell. This is because the weight of the container is set by changing the die gap. If a higher swell resin is processed in a given tooling at a standard weight, the die gap must be decreased to provide the same part weight. Sometimes this will result in the critical shear rate of the resin being exceeded. Other times it will result in higher pressure drops across the die and possible overheating. Two resins with widely different swell values can sometimes not be run on the same tooling.

The swelling behavior of the parison is a time-related effect. If the material is extremely viscous, the time for the parison to fully swell may be longer than the time for the molds to close. In this case the parison will appear to be shrinking just as the molds close. This is because the increase in the wall thickness requires a compensating decrease in the length of the parison.

After the formation of the parison is completed, gravity immediately begins to exert a force on it. The effect of this force is an increase in the length of the parison, and a corresponding thinning. The effect is greatest near the die, but is also experienced all along the parison as shown in Fig.

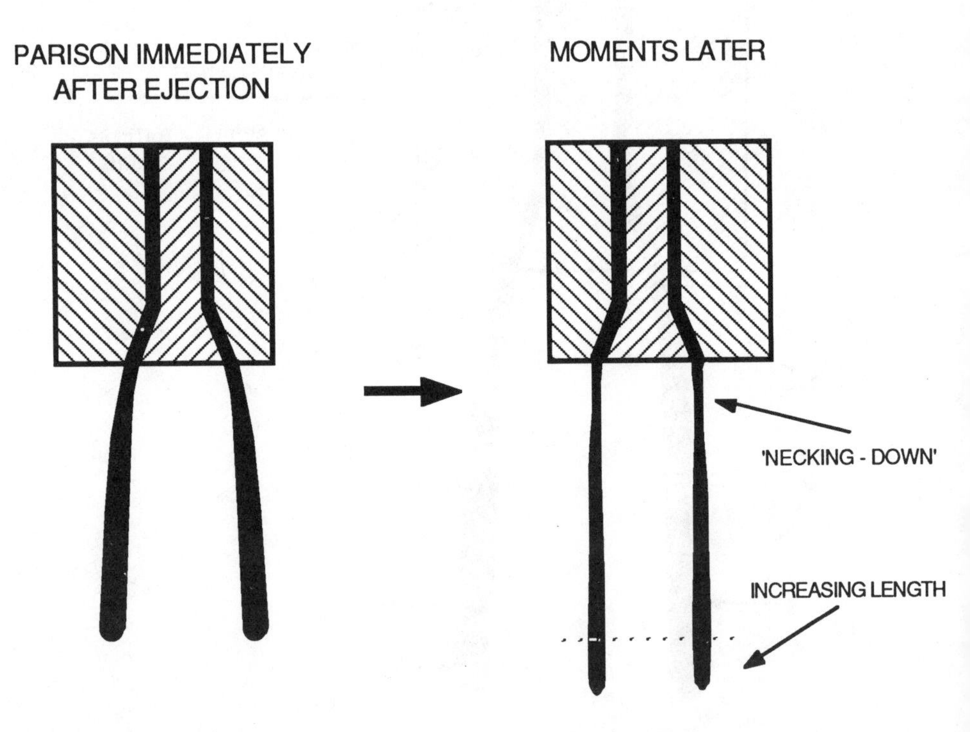

Figure 16-10 Sagging behavior of parison.

16-10. In continuous extrusion, the parison is subjected to gravity throughout the entire cycle. Intermittent extrusion is the preferred method for production of large parts so that the length of time in which gravity can act is reduced.

The sag flow takes place under conditions of very low shear rate. For this reason, the lower shear rate flow measurements are valuable tools for analyzing the sag behavior. It has been stated that a typical viscosity for a good large part resin falls in a range of around 5–10,000 Pa·s at a shear rate of $10s^{-1}$ [3].

The next step is the blow stage. During this step the parison is inflated rapidly. The viscous and elastic components both play key roles in this step. If a material has too high a viscosity, the rate of stretch will be faster than can be accommodated, perhaps resulting in a blowout. The same can be true if the material does not have enough elasticity. These problems are seen mostly on molds of complicated geometry or high blow ratios. The behavior of this step is more properly defined by use of a concept called elongational viscosity, which concerns the elongation of a particular section of a material in response to a force acting along the vector of elongation. The reader is urged to consult more extensive treatments of this subject [10].

The flow rate of the material also has an effect on the way the parison is pinched and welded. Designs for lower flow materials have more obtuse angles to reduce the tendency to cut through the parison.

ADDITIVES

The minute a polymer molecule is produced, the possibility exists that it can be involved in a chemical reaction that can weaken or break it. Special additives are thus included to help the resin retain its properties. Other materials are included to provide it with physical property variations.

Antioxidants and Stabilizers

One of the primary causes of degradation of a polymer is oxidative attack. Oxygen molecules react with free radicals to form unstable compounds that can further degrade the polymer chain. Degradation is manifested as a loss in physical properties, change in rheological parameters, and change in color. The high temperatures encountered during the processing stage facilitate this possibility. Protection requires an oxidation inhibitor. Also important is a safeguard against longer term attack.

Often the antioxidant (AO) system will have two or more ingredients. This is due to the fact that two different types of antioxidants can work synergistically to provide better protection at lower concentrations and cost than either one by itself. Commonly encountered primary antioxidants such as BHT, hindered penolics, and secondary amines protect by absorbing destructive molecular fragments called free radicals. Secondary AOs include phosphites, and thioethers [4]. They protect against unstable oxygen-rich molecules called peroxides. Phosphites provide exceptional resistance to yellowing.

Polyvinylchloride is noted for its instability at higher temperatures, since chlorine atoms easily separate from the molecule. This requires that it be stabilized for heat. Organotin compounds such as dialkyl tin mercaptide are the leading materials in this service. Ca and Zn soaps are also used, as well as epoxides, phosphites, beta diketones, and polyols [5].

Another important consideration is that the polymer be resistant to the effects of ultraviolet radiation if the finished article is to be used in the sun. Common UV stabilizers include hindered amines, benzophenones, and carbon black. Of these, carbon black is usually considered to be the most effective, although this presupposes black is an acceptable color. If another color is desired, it is important to take into account the effects the color can have on the stabilizer system. The order of resistance for some common colors in combination with a given UV stabilizer in high-density polyethylene is shown below.

Black	Best protection
White, Blue	
Red (inorganic)	
Yellow (inorganic)	
Green	
Red, Yellow (organic)	
Natural (nonpigmented)	Least protection

Lubricants

Lubricants serve the function of helping to improve the flow characteristics of a resin, and of helping to improve the surface finish of the finished part. Lubricants can be of the internal or the external variety.

Internal lubricants reduce molecular interaction, reducing the viscosity of the melt. Rigid PVC materials are carefully formulated with these types of materials. External lubricants help reduce adhesion to the die, mold, or other processing equipment. A commonly encountered lubricant in the injection blow molding area is zinc stearate. It is included in HDPE, PP, PVC, and PS injection blow molding grades as a pin release. It is added at the level of from 0.1 to 0.5 percent.

Colorants

Colorants for plastics used in blow molding are usually in a compounded form that is added to the resin in a proportion of 1 to 8 percent of total composition. The colorant typically consists of 50–60 percent carrier resin, which is usually the same type as the resin to which it is added. The carrier typically has a higher flow rate to facilitate easy dispersion. If the carrier resin is not of the right type, the physical properties of the resultant product can be severely affected.

The remaining 40–50 percent of the colorant concentrate consists of a pigment and perhaps a stabilization system. The concentrate has the advantage, compared to the pigment, of easier dispersion in a wider variety of extruders and no dust or associated health hazard. The pigments can be either organic or inorganic. Organic pigments are much more vivid and available in a wider range of colors than the inorganic pigments. They tend to be higher in cost, however. Inorganic pigments have higher opacity and are more stable [6].

Colorants often have additional effects that must be considered. The phthalo blues can increase warpage for polyolefins. Carbon black is known to be the best UV stabilizer for polyolefins. Titanium dioxide, the leading white pigment, is abrasive and can cause increased wear in the extruder.

Modifiers, Fillers, and Other Substances

Many grades of resin require an additional additive that helps improve their performance. Polypropylene (PP) is often produced with a small percentage of rubber to improve low-temperature impact resistance. Polystyrene (PS) is also modified in this way. Rigid PVC is modified with various grades of other polymers including ABS, EVA, and some acrylics.

Plasticizers are an important class of additives that are frequently encountered in PVC compounds. These materials are added to improve flexibility. Other properties, including hardness and tensile strength, are reduced. The materials serve to lower the glass transition point of the compound. They also reduce the viscosity of the melt in a manner similar to internal lubricants.

A plasticizer is similar in action to a high molecular weight solvent. Plasticizers are more soluble in an amorphous polymer. Some of the most common plasticizer types include pthalic, phosphoric, and carboxylic acid esters, among others [4].

Another way to obtain better resin performance is to include some type of filler that serves to increase stiffness, heat resistance, and can perhaps reduce the cost of the overall compound. Mica is used for the reinforcement of HDPE to provide a much stiffer material. Glass fibers are used in nylon, not only to improve stiffness but also to help improve the melt strength.

Because plastics are poor conductors of electricity, it is frequently a problem that static electricity buildup on a product causes an attraction for dust particles in the air. Antistatic materials are used to impart a slight conductivity to the surface of the product to allow these charges to dissipate. Most commonly used in blow molded containers are agents that are provided dispersed in the polymer substrate [7].

Antistatic agents have separate chemical groups in their structures that are attracted to different things. One type attracts water, while another attracts hydrocarbons. This is similar to the action of a soap. Some of these materials include alkyl ammonium, phosphonium, or sulfonium salts, sodium alkyl sulphonates, ethoxylated fatty amines, fatty acid esters, and polyethylene glycol esters or ethers. The materials are effective because they migrate to the surface of the polymer where they attract water molecules from the atmosphere. The film of moisture allows a major decrease in the material's surface resistivity. Should the antistatic concentration be too high, the surface of the part can become coated and feel greasy. This can lead to printing problems.

REQUIREMENTS FOR BLOW MOLDING MATERIALS

From a processing technology standpoint, all materials with melts that have the following properties are suitable for blow molding applications [8]:

- Sufficient thermal stability for the processing temperature range and, if necessary, for repeated processing
- Sufficient flowability of the homogeneous, plasticated melt
- Sufficient stretchability of the tube (parison) even at high stretching speeds
- Excellent repeatability of parison weight and length
- A smooth parison surface
- Compatibility with additives such as master batches, pigments, etc.
- A sufficiently wide processing range for the required finished part properties
- Excellent lot to lot consistency

Process-Specific Requirements

The special requirements of the various blow molding technologies must be added to the above general list.

Extrusion Blow Molding

The various melt streams formed in the flow channels must be capable of reuniting into a consistent parison. The vertically hanging extruded parison must have sufficient melt strength to allow time for the mold to close. Good welding in the pinchoff area of the blow mold is important.

Injection Blow Molding/Stretch Blow Molding

Because the parison is supported on a core in this process, it is possible to use materials with a melt strength that would be too low for extrusion blow molding. Multiple cavity molds, often with hot runners, dictate that a material with easy flow and low adhesion to the core be utilized.

Coextrusion Blow Molding

Low melt strength materials may be used if they are supported on a strong structural layer. The performance of the structural layer is typically put on a cost basis.

Dip Blow Molding

Sufficient thermal stability, good separation from the core, and clean separation without melt stringing are the most important criteria for this process.

End-Use Criteria

In addition to the general technological requirements, the requirements of specialized end uses must be taken into account. There are as many of these as there are applications, but some of the more common include:

- Chemical resistance
- Permeation characteristics
- Environmental stress crack resistance (ESCR)
- Mechanical properties (e.g., cold impact resistance)
- Physiological properties (e.g., for food applications)
- Optical properties
- Weather resistance
- Long-term properties

Chemical Resistance

A good guideline for this important property can be obtained from the data sheets of the polymer manufacturers or the general technical literature. The resistance depends most of all on the chemical composition of the specific polymer type. When using a blend of materials, it is essential in most cases to perform a storage test under operating conditions.

Permeation

In addition to chemical resistance, most containers must be tested for gas and moisture permeation. Permeation is defined as the penetration of vapors or solvents through the walls. Permeation is a molecular process. It is increased when the polymer is highly soluble in the permeant, a high difference in concentration and/or pressure exists across the wall of the container, or if the article has defects such as pores, tears, or extreme wall thickness variation.

Environmental Stress Crack Resistance (ESCR)

When in contact with certain polar liquids, such as detergents, polyethylene tends to stress crack. Molded-in and additional external mechanical stress, such as stacking force, tend to facilitate this crack formation. The resistance to stress cracking increases with increasing molecular weight and decreasing density. Copolymers tend to stress crack less than homopolymers. Molecular weight distribution also plays a key role in the ESCR of an HDPE.

A well-known testing method for ESCR is the so-called Bell test (ASTM D1693). However, more practical test results can be obtained by bottle testing (ASTM 2561). Bottle testing can also be performed with small modifications. In order to compare values from different ESCR tests, it is required that all testing parameters be held constant.

Other polymers also exhibit stress cracks when exposed to the right type of liquid. One of these combinations is polycarbonate and chloroform. Polystyrene can be stress cracked by vegetable oils and white spirits [9].

MATERIALS FOR BLOW MOLDING

Blow molding has emerged as one of the leading fabrication processes in the world today. A large part of the success of this industry is due to the broad range of materials that can be employed, and the varied balance of properties they offer. The range of thermoplastics available to the blow molding designer is increasing so rapidly that it is almost impossible for one person to keep fully up to date with current technology.

Blow molding currently consumes about 10 percent of all resins produced in this country, approximately five billion pounds a year, making it the third largest processing technique in the plastics products industry. The most commonly used resins in 1987, along with their percentage shares of the U.S. blow molding market, are shown as follows [11]:

RESIN	PERCENTAGE
High-density polyethylene (HDPE)	65
Polyethylene terephthalate (PET)	22
Polyvinylchloride (PVC)	6
Polyproplyene (PP)	4
Low-density polyethylene (LDPE)	2

Material Types

The thermoplastics that have obtained a greater or lesser significance in blow molding may be placed in two groups according to their polymerization type:

ADDITION	CONDENSATION
Polyethylene	Polyamide
Polypropylene	Polycarbonate
Polyvinylchloride	Polyester
Polyacrylonitrile	
Polystyrene	
Polyacetal	
Fluorpolymers	

Materials for Packaging

In addition to the three preceding end-use requirements, the following must be added if the product will serve as a packaging material:

- Stiffness under load
- Dimensional stability at high filling temperatures
- Good surface quality

- Good printability
- Good drop impact resistance even at low temperatures

Materials for Technical Blow-Molded Parts

For technical parts, good mechanical properties are of prime importance. However, excellent surface quality must also be obtained if, for example, the part will subsequently be painted.

Resin Selection

Depending on the size of the container, a resin may or may not be a good choice. The following list reflects many practical applications.

Small Parts, Under 5 Liters

Polyolefins (HDPE/LDPE/PP)
Polyvinylchloride (PVC)
Polyacrylonitrile (PAN)
Polystyrene (PS)
Styrene-based polymers (SAN)
Polyester (saturated; linear: PET)
Polyvinylidene fluoride (PVDF)

Large Parts, over 5 Liters, Including Technical Parts

Polyolefins (HDPE/PP)
Polycarbonate (PC)
Polyacetal (POM)
Polyamide (PA)
Polyvinylidene fluoride (PVDF)

REFERENCES

1. Domininghaus, D.I. H. "Introduction to the Technology of Plastics." *Schweizer Maschinenmarkt,* vols. 68, 70. Fachpresse, Goldach, Switzerland.
2. Daniels, F., and R. Alberty. *Physical Chemistry.* 4th ed. New York: John Wiley and Sons, 1975.
3. Subramanian, P., and J. Feathers. "Blow Molding of 6,6 Nylon Compositions." *Proceedings, ANTEC.* Soc. Plastics. Eng., 1988.
4. Gaechter, R., and H. Mueller. *Plastics Additives.* Hanser, New York, 1983.

5. Wypart, R. W., and J. W. Summers. "New Generation of Low Taste and Odor, High Clarity, Low Blush Vinyl Compounds." *Proceedings, ANTEC.* Soc. Plastics Eng., 1986.
6. Goldring, T. "Colorants." In *Modern Plastics Encyclopedia.* New York: McGraw-Hill, 1988.
7. Rogers, J. L. "Antistatic Agents." In *Modern Plastics Encyclopedia.* New York: McGraw-Hill, 1988.
8. Raddatz, E. "Rohstoffauswahl fuer die Herstellung von Grosshohlkoerpern und technischen Blasformteilen." *Fachtagung Fortschritte beim Blasformen von Thermoplasten.* Industrieverband Verpackung und Folien aus Kunststoffe e.V.
9. Powell, P. *Engineering with Polymers.* Chapman and Hall, 1983.
10. Lodge, A. S. *Elastic Liquids.* Academic Press, New York, 1964.
11. Rosato, D.V. "Blow Molding Expanding Technologywise and Marketingwise." *Proceedings, ANTEC.* Soc. Plastics Eng., 1987.

ACKNOWLEDGMENT

The material in this chapter contributed by Eckard Raddatz is adapted from his "Materials for Blow Molding of Packaging and Technical Articles," presented at the Sixth Plastics Symposium, German Engineers' Society, Brazilian Group, 1987.

CHAPTER

17

Polyethylenes and Polypropylenes

JAMES P. PARR AND ECKARD F. H. RADDATZ

POLYETHYLENE

The discovery of polyethylene was made in 1931 by Imperial Chemical Industries (ICI) while researching the effect of high pressure on chemical reactions. Commerical production of low-density polyethylene began in 1939, using a high-pressure, high-temperature system.

Low-density polyethylene (LDPE), the first variety of polyethylene available, is produced by a free radical chain reaction process. Generally the conditions are in the range of 1000 to 3000 atm and 80 to 300°C (175–570°F). The polymerization technique includes the use of a free radical initiator, usually a peroxide, which is passed with the reactants through narrow-bore tubes or stirred reactors or is batch processed in an autoclave. The polymer produced in this type of reaction exhibits a highly branched structure due to the random nature of the reaction. Figure 17-1 shows the structures of low-density, high-density, and linear low-density polyethylenes.

The reaction is extremely exothermic. The use of effective cooling in the reactor design keeps the reaction under control. After polymerization has been completed, the polymer is separated and homogenized. The mixing and compounding equipment used has the effect of reducing the very high molecular weight species that can be present [1].

Development of the low-pressure polymerization technique by Phillips,

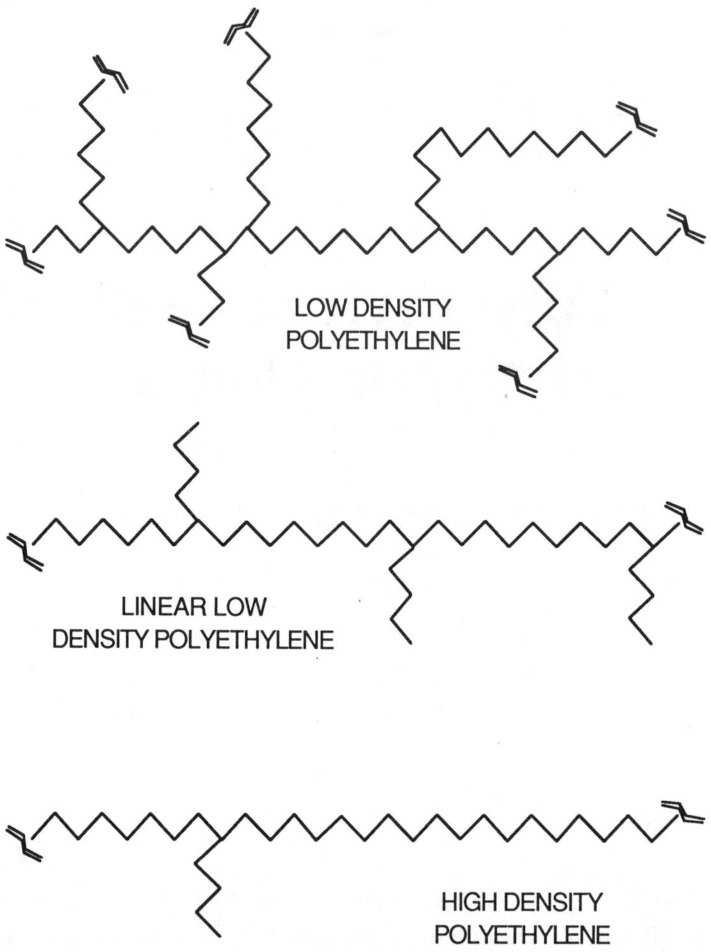

Figure 17-1 Structures of the three types of polyethylene.

Standard Oil of Indiana, and Hoechst (Ziegler process) in West Germany followed in 1953. Full-scale commerical manufacturing began in 1955. The noteworthy characteristic of these processes is that they produce a very linear molecule with controlled, short branches. Due to the closer molecular packing than with LDPE this type of polymer has a higher density, hence its designation as high-density polyethylene (HDPE).

In the Ziegler process, a catalyst complex is first prepared from an aluminum/titanium system. This mixture is then fed with the reaction components into a reaction vessel at controlled temperatures and pressures. An agitator is used to keep the suspension well mixed. As the polymer is formed, it precipitates from the solution to form a slurry, which becomes progressively thicker as the reaction continues. The polymer is discharged to a secondary vessel where the catalyst residues and diluent are neutralized.

The Phillips process employs a chromium oxide catalyst on a silica alumina base that is activated by elevation to high temperatures. The ethylene is dissolved in a liquid hydrocarbon solvent that also suspends the polymer as it is formed and acts as a heat transfer medium. The reactor design is similar to a loop. The reaction mixture is pumped continuously around through the loop to effect the suspending and heat exchanging functions. The molecular weight of the system is very dependent on process temperature.

Typically, the resins produced in both processes have densities up to 0.960 g/cc, with molecular weights in a corresponding range. These slurry processes utilize secondary separation steps to purify the reacted polyethylene from the unreacted monomer, diluent, and any impurities.

The Unipol process was developed by Union Carbide in the 1970s. It features gas phase operation and a fluidized bed. The monomeric ethylene performs the suspending and cooling functions. The process has the advantage of producing the powder in a very pure form with no diluent, which eliminates the need for liquid separation.

Also during the 1970s, linear low-density polyethylene (LLDPE) resins were developed. These resins utilized the same basic processes used for high-density products. The main innovation was the ability to control the spacing of side branches, allowing the synthesis of linear molecules at densities in the range of the old high-pressure reaction products.

Properties

There are three basic polyethylene groups based on density, low, medium, and high. These groups are defined within ASTM D1248 by density as follows, the homopolymer being a separate class of high density:

TYPE	DENSITY GROUP	DENSITY, g/cc
1	low	0.910–0.925
2	medium	0.926–0.940
3	high	0.941–0.959
4	homopolymer	0.960–higher

Some basic properties of typical grades from each density range are shown in Table 17-1. Polyethylene is a long chain aliphatic hydrocarbon with a structure which is illustrated as follows:

$$—CH_2—CH_2—CH_2—CH_2—CH_2—$$

The material has a low glass transition temperature of about −110°C (−80°F) as a result of the flexibility of the C—C bond. The density is reduced by

Table 17–1 Typical physical properties of selected blow molding grades of polyethylene

		Resin Type				
Property	*ASTM Test Method*	*LDPE*	*LLDPE*	*General Purpose HDPE Copolymer*	*HDPE Dairy Grade*	*High MW HDPE Copolymer*
Density, g/cc	D792	0.923	0.930	0.955	0.960	0.950
Melt index, g/10 min	D1238	2.0	0.80	0.30	0.65	<0.10
Tensile strength at yield, psi	D638	1200	1900	2600	3800	2800
Flexural modulus, psi	D790	35,000	40,000	200,000	240,000	175,000
Melt point, °C	—	98	122	128	130	127

the addition of comonomer side chains during the reaction. Propylene, butene, hexene or octene are all used as comonomers, although the more commonly used are butene and hexene. The polymer is highly regular, and high crystallinity exists. In low-density polyethylene the maximum crystallinity is found to be of the order of 60 percent; the high-density product has a crystalline region of up to 90 percent. The remainder of the material is amorphous.

Polyethylene is milky white and opaque in color and waxy in feel. As would be expected from what is in reality a high molecular weight paraffin, it has exceptional resistance to chemical attack and therefore finds application in all varieties of chemical packaging.

HDPE Resin Types and Characteristics

There are a variety of HDPE grades available for the blow molder. Usually all these products are offered as pellets, although powdered grades are often available. One of the largest volume types is the homopolymer grade used for the production of one-gallon milk containers.

The conversion of this material is almost exclusively performed on intermittent extrusion blow molding machines designed for high production rates of approximately 60 bottles/min. The typical melt index (MI) and density values for these grades, available from most of the major U.S. suppliers, are about 0.6–0.7 g/10 min and 0.960 g/cc, respectively. The high density is necessary to provide maximum stiffness at the lowest possible weight. The high flow is for ease of processing. A variation of this type is a lower flow material for use on the wheel type blow molding machines. Here the lower flow is utilized to provide more melt strength since the parison is pulled continuously by the wheel.

The grades used for the broadest number of applications are commonly called the general-purpose grades. The standard type usually has a density of 0.955 g/cc and an MI of around 0.25–0.35 g/10 min. A lower density version is also available with a density of about 0.950 for use in application requiring more flexibility or environmental stress cracking resistance (ESCR). Some typical application areas include bleach, detergent, and motor oil containers. These materials are readily processable on a wide variety of machines and have good physical properties. Parts weighing up to five pounds are easily produced with them. They are often supplied in antistatic formulations. The same type of material is used for injection blow molding with a zinc stearate concentration of from 0.1 to 0.5 percent.

Other grades are offered for more specialized needs. High ESCR grades utilize different molecular weight distributions to provide higher ESCR than is available in the standard 0.950 grade. High-gloss resins are used in some injection blow molding applications when a glossier container is desired for, perhaps, a cosmetic container.

For large part blow molding, high molecular weight HDPE is normally used. The higher molecular weight helps provide the melt strength needed for the production of such heavy parts as 55-gallon drums, which can weigh up to twenty-two pounds. The high molecular weight also provides more toughness to meet the demanding performance requirements of such parts as automobile fuel tanks. These grades are usually supplied in a density range of 0.945–0.955 g/cc, and a melt flow range of from 2.0–15.0 g/10 min (melt flow index, MFI, 190°C/21.6 kg). Applications include shipping drums, bulk shipping containers, 90-gallon refuse carts, gasoline tanks, lawn and garden spray tanks, and a variety of other tanks.

Low and Linear Low-density Polyethylene

LDPE is distinguished from HDPE most of all by its lower degree of crystallinity. Its density range is from 0.918 to 0.930 g/cc. LDPE is, therefore, significantly softer and less stiff than HDPE. It is often used for mixing with HDPE. Processing is possible in extrusion blow molding as well as injection blow molding and dip blow molding. Squeeze bottles, transfusion tubes and containers, and also large storage containers are produced from LDPE.

LLDPE is also used for parts requiring high flexibility, but slightly more toughness than conventional LDPE, such as road channelizing barriers, and liners for steel and fiber drums. Low-density materials with melt indices of less than 2.0 are suitable.

Processing

All types of polyethylene are suitable for processing on both extrusion and injection blow molding machines. Single-screw extruders are used most often. Smooth barrel extruders are used with good success for all but the highest molecular weight materials. For these materials, a grooved barrel extruder is necessary to achieve the highest outputs at the lowest possible temperatures. Melt temperatures should fall in the following ranges [2]:

LDPE	160–170°C (320–338°F)
LLDPE	160–180°C (320–356°F)
HDPE	160–210°C (320–410°F)

The higher temperatures can be used for LDPE if a more transparent part is desired. The mold temperature for all types of polyethylene should be between 5 and 40°C (41 and 104°F). Molds should be sandblasted for good venting.

POLYPROPYLENE

Polypropylene is produced from propylene with Ziegler-Natta catalysts, which produce a stereospecific molecular arrangement (see Fig. 16-4). Reactor technologies include slurries using either unreacted monomer or another aliphatic hydrocarbon liquid carrier, and monomeric gas phase fluidized beds.

Rapid improvements in catalyst technologies have been experienced. These newer catalysts produce higher yields and have been displacing the previously existing standard catalyst systems. Higher yields lead to better color, taste, and odor properties. The catalysts can be either supported or unsupported.

A typical process utilizes a primary catalyst that contains a transition metal halide. If supported, it will be deposited on an inert carrier such as an alkaline metal earth halide. An organometallic or organometallic halide cocatalyst is also used. Electron donors and hydrogen are used to control, respectively, stereoregularity (crystallinity) and molecular weight [3].

Properties

Some of the desirable properties for which designers look to polypropylene are:

- Very low density
- High stiffness
- High surface hardness
- Very good chemical resistance
- Contact clarity and gloss
- Good resistance to high temperatures
- Very low water absorption and transmission
- Very good processing
- Very good toughness at low temperatures, especially when alloyed with elastomers
- A favorable price/property relationship, which makes PP an ideal base material for reinforced and elastomer-modified grades

In addition to homopolymers, ethylene copolymers are available that contain up to 6.5 percent ethylene. The copolymer has more clarity and impact resistance, both strong advantages for a blow molding product. The maximum ethylene content is about 6.5 percent. The melt flow of these grades is about 1 to 3 g/10 min at MFI 230°C/2.16 kg. Some properties of these resins are listed in Table 17-2.

The most common grade of polypropylene sold in the U.S. blow molding market is random ethylene copolymer, which has about a 70 percent market

Table 17-2 Typical physical properties of selected blow molding grades of polypropylene

		Resin Type		
Property	*ASTM Test Method*	*Homopolymer**	*Copolymer*	*Clarity Copolymer**
Density, g/cc	D792	0.905	0.900	0.900
Flow rate, 230/2.16, g/10 min	D1238	1.6	2.3	2.0
Tensile strength at yield, psi	D638	5000	4250	4100
Flexural modulus, psi	D790	210,000	130,000	130,000
Melt point, °C	—	165	147	143
Heat deflection at 66 psi, °C	D648	107	92	82

*Also used in injection blow molding

share. In terms of process type, about 60 percent of the resin is used in extrusion blow molding. Of the remainder, all but about 5 percent is claimed by injection blow molding for use in the stretch blow process. This is the newest of the processes and it can result in better properties, especially clarity and impact strength.

Polypropylene needs a moderate level of antioxidant to have good melt flow and color stability. The development of specially optimized polypropylene blow molding grades has made it possible today to produce large surface area parts with excellent wall thickness distribution and good surface finish.

When the material is expected to be exposed to multiple heat histories, a secondary antioxidant is also included. Other additives can be included to:

- Improve melt flow stability
- Improve mold release
- Impart antistatic properties
- Enhance stiffness and reduce cycle times (nucleating agents)
- Reduce haze

In the injection blow molding grades of polypropylene 0.05–0.5 percent calcium stearate is used as a pin release. Fillers are also used to improve stiffness or reduce resin cost. Glass fibers or talc can increase hardness, stiffness, heat stability, and other properties. Adding elastomers can influence toughness and flexibility, even at low temperatures.

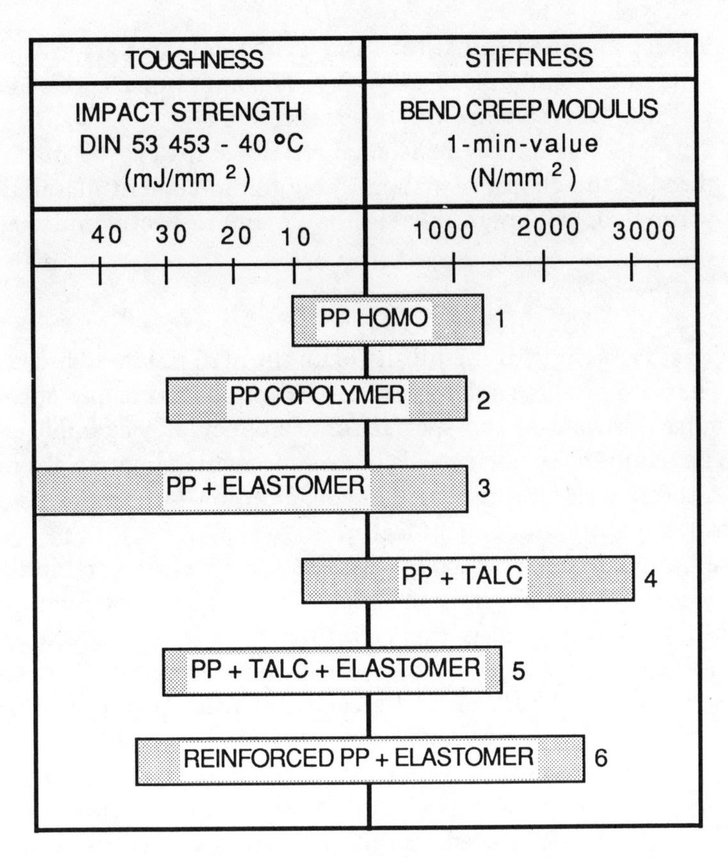

Figure 17-2 Toughness-stiffness relationships of various polypropylene resin types. Courtesy of Eckard Raddatz and Armin Schwaighofer.

A good description of mechanical properties of the different types of PP is the so-called stiffness-toughness relation. The creep modulus is used as a measurement of the stiffness. Toughness is measured using the impact strength at −40°C (−40°F) (Figure 17-2). From this we can determine (as numbered in the figure):

1. Homopolymer types have a higher stiffness than copolymer types.
2. Copolymer types have a higher (cold) impact resistance than homopolymers and achieve most of their stiffness.
3. An improvement of the (cold) impact resistance can be achieved through compounding with elastomers (EPDM). However, with these a small loss of stiffness must be expected.
4. A reinforcement with, for example, 20 percent talc (TV20) leads to considerable improvement in stiffness and certainly, in comparison with a homopolymer type, will also lead to a loss of impact resistance.

5. Impact-modified and at the same time reinforced grades (as with 20 percent talc and EPDM) allow a good compromise between stiffness and impact resistance improvement.
6. Stiffness and impact resistance relationships can be made flexible through the choice of an applicable reinforcement material for polypropylene, an applicable elastomer, and respective mixing proportions.

Moreover, it is possible to chemically foam the material by adding a blowing agent. That can produce a density from 500 to 600 grams per cubic meter, better noise absorption, and special surface effects (for example, pearlescence). The reduction of noise levels in cars—for which lightweight materials are essential in order to improve fuel consumption—is currently an area of work at many automobile manufacturers.

The annual volume of polypropylene used for blow molding in the United States is about 150 million pounds. The annual growth rate is around 10 percent at present and is expected to remain at that rate for the next three to five years. One of the most important recent applications for PP was the multilayered bottle used for catsup and similar types of food products, such as syrup and spices. The medical product area includes bottles for cough syrup, sterile fluids, antacids, cold medicines and mouthwash. Cosmetic applications include cream, lotion, and eyeliner containers. Although it is most often used in the food, drug, and cosmetic packaging markets, the most exciting growth opportunities for polypropylene could lie in the area of technical parts.

Processing

Polypropylene presents no extreme difficulties for the blow molder. The screw should have a depth compression ratio of about 3.5:1. The *L/D* ratio should be a minimum of 20, with longer screws providing better homogeneity.

Temperature settings on the barrel should be about 170–190°C (338–374°F) at the feed throat and increase in steps up to about 200°C (392°F) at the die. The melt temperature should lie between 180 and 240°C (356–464°F).

Polypropylene has about 70 percent of the die swell of HDPE. It can be processed on polished molds to accentuate its glossy appearance. Mold temperature should be about 10–40°C (50–104°F) [2].

ACKNOWLEDGMENTS

The material in this chapter contributed by Eckard Raddatz is from his presentation "Materials for Blow Molding of Packaging and Technical Articles," presented at the Sixth Plastics Symposium, German Engineers' Society, Brazilian Group, 1987.

The authors express special thanks to Joe Cox, Fina Chemical, Houston, TX for his assistance.

REFERENCES

1. Brydson, J. A. *Plastics Materials.* Illiffe Books Ltd., London, England, 1969.
2. *Blow Molding of Thermoplastics.* Hoechst AG., Frankfurt, W. Germany, 1976.
3. Personal communication from J. Cox, Fina Chemical, September, 1988.

CHAPTER

18

Polyesters

JAMES P. PARR AND ECKARD F. H. RADDATZ

POLYESTERS

When considering blow molding polyester, the material of specific concern is saturated linear thermoplastic polyester [1]. This thermoplastic polyester is manufactured by a poly-condensation reaction of dimethyl terephthalate and ethylene glycol, or terephthalic acid and ethylene glycol. The structure of this molecule, polyethylene terephthalate (PET), is shown in Fig. 18-1. Methanol is yielded as a by-product of the reaction. Medium molecular weight materials are produced from the melted reaction mass. Higher molecular weights are obtained by allowing further condensation in the solid phase.

The solid phase polymerization step utilizes chips of material that have been produced in the melt phase. Residual ethylene glycol, water, and acetaldehyde diffuse from the chips at temperatures of between 190 and 230°C (375–445°F). The lower temperatures required in this type of polymerization help reduce the production of acetaldehyde as a by-product. Yellowness is also reduced. The physical strength properties increase in proportion to the degree of condensation, which can be in the range of 100–200. However, the production costs also increase.

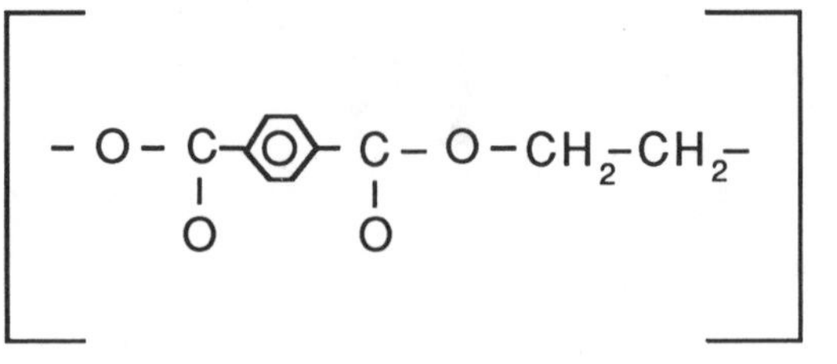

$$\left[-O-\underset{\underset{O}{|}}{C}-C_6H_4-\underset{\underset{O}{|}}{C}-O-CH_2-CH_2- \right]_n$$

Figure 18-1 Structural formula of polyethylene terephthalate (PET).

Properties

The key properties of PET include the following:

- Excellent gloss and clarity
- Very tough and impact resistant
- Good chemical resistance
- Low permeability to CO_2
- Good processability
- Good dimensional stability

PET has a melt point of 247°C (477.6°F). It has a density of 1.4 g/cc. The possible material decomposition and the expected finished part properties are determined most quickly by measurement of the intrinsic viscosity (IV). The IV is a viscosity that is extrapolated from a standardized solution of polymer. At this time, production of stretch-blown bottles uses a raw material with an IV value of 0.7 to 0.85 deciliter/gram (dL/g). Some of the properties of PET resins, along with PETG (see below), are shown in Table 18-1.

Also to be considered is the crystalline behavior. PET is very different from other partially crystalline plastics, such as polypropylene. According to heat history the crystalline portion can be from less than 5 percent to about 35 percent.

PET is offered in the market in the form of chips or pellets, either dried or not. At high processing temperatures, a moisture content of greater than 0.005 weight percent can lead to hydrolytic decomposition and thereby a reduction in mechanical properties. Drying in air cabinets should proceed at 140–180°C (284–356°F) for a minimum of three hours. The dew point of the air should be less than −40°C (40°F).

The significance of PET in the bottle or packaging market was achieved through stretch blow molding technology. The major application for PET is in containers for carbonated beverages. Sizes range from $\frac{1}{2}$ to 3 liters. The application examples today include drink and pharmacy containers with volumes of from a few milliliters up to about 30 liters.

Table 18-1 Typical physical properties of blow molding grades PET and PETG

Property	*ASTM Test Method*	*PET*		*PETG*
		Homopolymer	*Copolymer*	
Intrinsic viscosity, dL/g	D2857	0.72	0.83	—
Density, g/cc	D792	1.4	1.4	1.27
Tensile strength at yield, psi	D638	8000*	8000*	7100*
	—	24,000†	27,000†	—
Flexural modulus, psi	D790	350,000*	350,000*	290,000*
	—	700,000†	750,000†	—
Melt point, °C	—	254	246	—
Vicat softening point, °C	D1525	—	—	82

*Unoriented
†Oriented

Processing

PET is processed almost exclusively in blow molding using a stretch blow process. The stretching orients the molecules, increasing the strength and barrier properties of the container. There are two main processes, one-step and two-step.

The one-step process includes the injection and blowing stations on the same machine, whereas the two-step method utilizes separate machines. The key factor in stretch blowing of PET is to have a well-equilibrated temperature distribution throughout the preform. After initial injection molding, the preform is cooled below its glass transition temperature, T_g, then reheated to the thermoelastic range.

Figure 18-2 shows stretch-blow temperature plots for PET and PP. Pearlesence will be seen in the side walls if the temperature is too low during the blow phase. A two-stage blow pressure sequence of about 200–230 psi low pressure, and 400 psi high pressure, is sufficient. The mold temperature should be about 35–40°F (1.5–4.5°C).

Acetaldehyde is a volatile gas (BP −21°C [−5.8°F]) that is produced in the manufacture and subsequent processing of polyester. This gas has a fruity aroma that can change the flavor of drinks, even at low concentrations. The development of acetaldehyde depends on the melt temperature and the residence time in the melt flow system.

Figure 18-3 shows the influence of melt temperature (measured with a probe pyrometer) on the content of acetaldehyde in the preform. The content

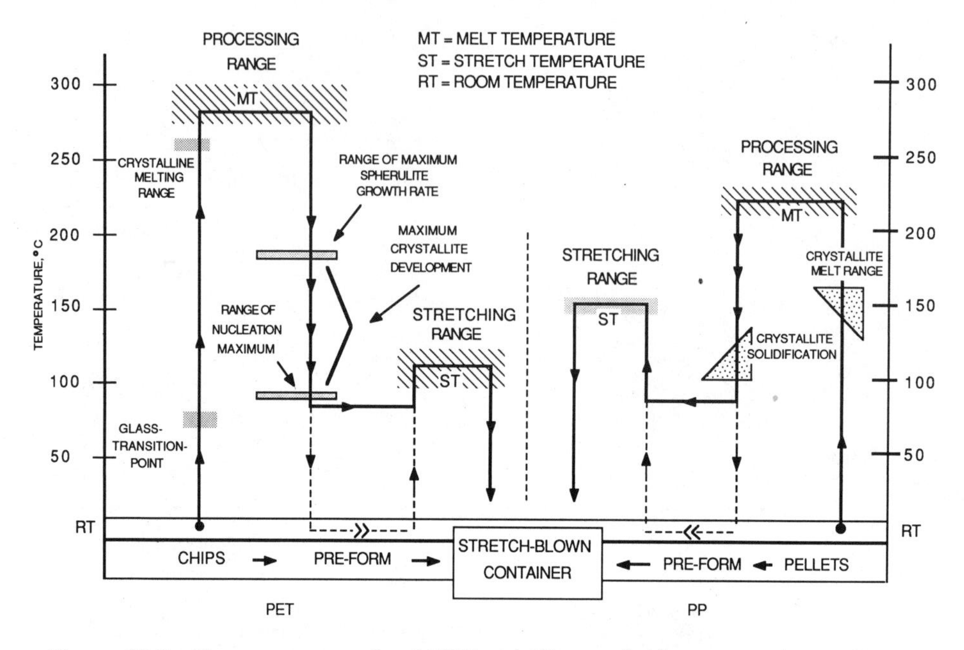

Figure 18-2 Temperature cycle of PET and PP stretch-blown containers. Courtesy of Eckard Raddatz and Armin Schwaighofer.

exponentially increases with increasing melt temperature in the range which was investigated. The content also increases when the runner system temperature is higher. It was determined that the acetaldehyde content in the preform side wall increased 2 ppm in response to an increase of 35°C (from 265 to 300°C [509–572°F]) in the runner temperature. The influence of the

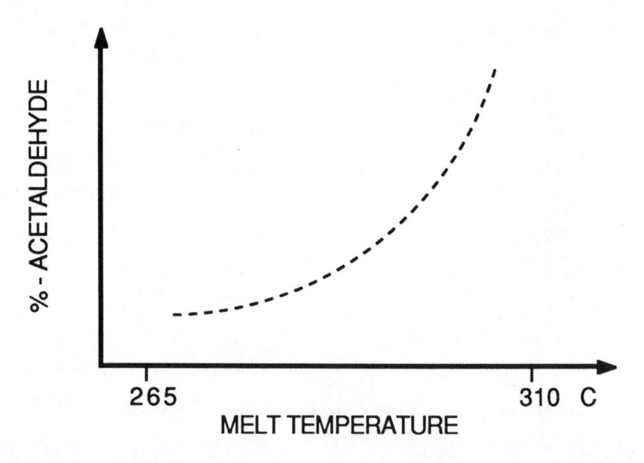

Figure 18-3 Influence of melt temperature on acetaldehyde content of a PET preform. Courtesy of Eckard Raddatz and Armin Schwaighofer.

gate heaters is relatively small. At longer cycle times, a higher gas development does not occur, but material degradation, evidenced by yellowing, does [2].

PETG Copolyester

Another type of polyester in the PET family is called PETG. The G stands for the other glycol used in the copolymerization, 1,4-cyclohexanedimethanol.

PETG is an amorphous polymer with a glass transition temperature of about 178°F (81°C). Some of its properties are shown in Table 18-1. This polymer has good resistance to dilute mineral acids, bases, salts, and soaps, and to aliphatic hydrocarbons, alcohols, and some oils. It is subject to attack by halogenated hydrocarbons, ketones, and aromatic hydrocarbons.

PETG can be used for conventional extrusion blow molding, and produces clear, glossy containers. Useful applications include shampoo, soap, detergent, and oil containers. It is not suggested for use with carbonated beverages.

This polyester must be dried at a maximum temperature of 150°F (66°C) for four hours before processing in a manner similar to other polyesters. If the temperature is too high, there is a danger that the material will fuse together. Hopper driers are recommended. Exact drier requirements should be obtained from the vendor to insure optimum performance.

Because of its high viscosity, PETG is best processed using low shear screws. A depth compression ratio of 2.5:1 for a 2.5 or 3.0 in. screw is suitable. In general, it is usually found to be processable on machines suitable for PVC and PC. Melt temperatures in the range of 415–440°F (213–227°C) are typical. Coolant to the barrel should not be cooler than 250°F (121°C).

The head and die designs should be nonrestrictive. Designs suitable for PVC have been found to be generally suitable. The adapter and manifold sections should be equipped with heaters to prevent cold spots and high head pressures. The material has a low parison swell, which requires relatively larger die diameters than for PVC.

The surface of the mold can either be polished or matte. The matte finish is used to help avoid problems with venting. PETG is noncorrosive to steel, aluminum, and beryllium-copper alloys. Mold temperature should be approximately 40–70°F (4.5–21°C). The shrinkage of PETG is 0.4–0.6 percent [3].

ACKNOWLEDGMENTS

The material in this chapter contributed by Eckard Raddatz is from his presentation, "Materials for Blow Molding of Packaging and Technical Articles," presented at the Sixth Plastics Symposium, German Engineers' Society, Brazilian Group, 1987.

The authors express special thanks to Bob Thomas, Hoechst Celanese, Spartanburg, SC for his assistance.

REFERENCES

1. "Kann PET die PVC-Flasche in West-Europa substitutieren." *Verpackungs-Rundschau.* No. 2 (1986).
2. "Polyester Hoechst." *Werkschrift.* Hoechst AG.
3. Publication No. TR-60D. Eastman Chemical Co., July, 1987.

CHAPTER

19

Polyvinylchloride

JAMES P. PARR AND ECKARD F. H. RADDATZ

The first of the major modern blow molding materials to be developed was polyvinylchloride, PVC. In 1912, Klatte in Germany successfully polymerized vinyl chloride, although commercialization of the polymer was not achieved until the 1930s.

Polymerization is achieved by the use of peroxide catalysts. The type of polymerization employed, called free-radical polymerization, results in a polymer chain structure that is largely amorphous, although short segments of the chain have a syndiotactic arrangement. The structure of PVC is shown in Figure 19-1. Maximum crystallinity is around 10 to 15 percent [1]. The high rigidity of the material is due to the strong dipole linkage forces between the chains of the polymer.

There are three basic polymerization techniques used commercially for the production of PVC: emulsion, suspension, and mass polymerization. During the course of the reaction a certain amount of exothermic heat is produced which must be carefully monitored and adjusted.

The suspension and mass systems are those used to produce materials most suitable for the blow molding manufacturing technique. The mass polymerization process yields resins with better processing characteristics due to their higher weight to surface area ratio, and the absence of any residual emulsifiers or protective colloids. They are of exceptionally high

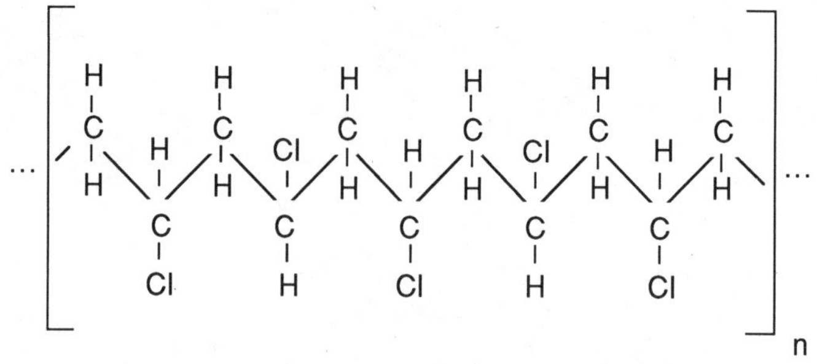

Figure 19-1 Structural formula of polyvinylchloride.

clarity, with very low moisture absorption. Both of these characteristics are significant for blow molded articles used in the packaging industry.

PROPERTIES

The K value is a measurement of the average degree of polymerization and therefore represents a measure for the flow properties. The measurement is made according to DIN 53726. For container applications, the suspension and mass polymerization products with K values between 58 and 65 are used. The density for clear polymers, according to level of toughness, can be between 1.39 and 1.31 g/cc at room temperature.

PVC is a brittle, rigid material in the unplasticized state, with a glass

Table 19-1 Typical properties of some PVC compounds for blow molding

Property	ASTM Test Method	*Grade* General Purpose	Food Grade	General Purpose Injection Blow	General Purpose Orientation
Density, g/cc	D792	1.32	1.32	1.33	1.35
Tensile strength at yield, psi	D638	5900	6300	7200	7100
Flexural strength, psi	D790	10,700	11,300	13,000	12,300
Tensile impact, ft lb/in.	D1822	85	75	55	60
Thermal deflection temp. at 264 psi, °C	D648	64	62	56	67

transition temperature, T_g, around 80°C (176°F). Unplasticized PVC is sold in powder form or as a ready-to-process pelletized material for a variety of applications, such as pressure pipe and window frames. The impact resistance can be improved by the addition of modifiers such as ABS, chlorinated polyethylene, or rubber. In this state the material is resistant to acids, alkali solutions, alcohols and light naptha, and oils.

PVC is also often modified by the processor through special compounding to provide special proprietary properties in given ranges. The modifiers include plasticizers, lubricants, and antistatic agents. The properties of some PVC bottle grades are listed in Table 19-1.

The addition of plasticizers imparts flexibility to the resin, which opens up another range of possible applications. The plasticized materials do, however, lose some of their chemical resistance. Whatever the amount of plasticizer required, PVC is not processable without the addition of heat stablizers. Heat stabilization allows processing the resin on conventional equipment.

The properties that make PVC a large-volume material for the packaging industry are, among others:

- Glass clarity
- Smooth surface
- Low gas and aroma permeation
- High stiffness with low wall thickness
- High chemical resistance
- Easy labeling and printing

PVC bottles are preferred for the packaging of edible oil, vinegar, marinades, cosmetics, detergents, and mineral water.

PROCESSING

The different processing conditions of PVC are due to different temperature-dependent states. Below the glass transition temperature, bottles are hard and tough. Above the glass transition temperature (80°C, 176°F), in the so-called softening range, the yield strength increases, reaching a maximum between 90 and 100°C (194–212°F). In the so-called thermoelastic range from 90 to 110°C (194–230°F) a further deformation of the blown container is possible. This is the basis for stretch blow molding, and also partly for injection blow molding. Above 180°C (356°F), the thermoplastic range is reached where extrusion blow molding may occur up to about 210°C (410°F) (Figure 19-2) [2].

Blow molding of PVC is best accomplished on machines designed with its needs in mind. PVC processing can lead to material degradation and

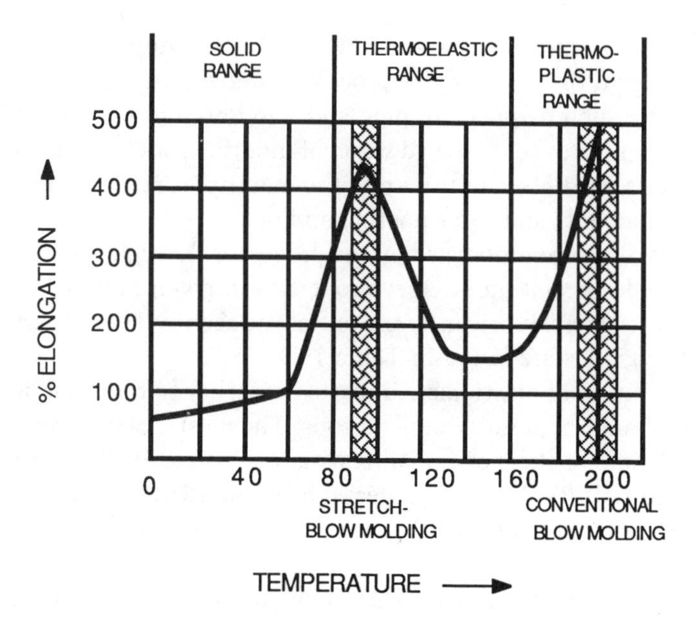

Figure 19-2 Temperature-elongation curve for polyvinylchloride. Courtesy of Eckard Raddatz and Armin Schwaighofer.

evolution of corrosive vapors. Extruder barrels and screws should be made of a corrosion-resistant metal such as nitrided steel. The screw should be ground and polished after nitriding. The screw is often hard faced to improve resistance to abrasive pigments or other additives. The screw should be of a tapered design with a depth compression ratio of about 1.8 or 2.3 to 1. A barrel cooling system is important.

The extruder die head design must be chosen to eliminate any areas of possible flow stagnation in order to reduce degradation potential. The use of accumulator heads is not common due to the longer residence times they involve. Corrosion-resistant steels should be used for the construction.

During processing, consistency of operation is important, and the extruder should not be left off with the heat zones still connected. The temperature ranges on the machine should range from 140–150°C (284–302°F) near the feed section to 180–210°C (356–410°F) at the die. The melt temperature should range between 190 and 215°C (374–419°F). These temperatures are listed only as a general guide, and may not be applicable to a specific machine or compound.

Due to the nice gloss and clarity of PVC, the molds are usually polished to provide the highest gloss. This is possible with PVC because of its low shrinkage, which does not permit it to form the orange peel surface seen in polyethylene with similarly polished molds.

ACKNOWLEDGMENT

The material in this chapter contributed by Eckard Raddatz is from his presentation, "Materials for Blow Molding of Packaging and Technical Articles," presented at the Sixth Plastics Symposium, German Engineers' Society, Brazilian Group, 1987.
The authors express special thanks to Paul Zorzi, Georgia Gulf, Plaquemine, LA for his assistance.

REFERENCES

1. *Hostalit.* Hoechst AG.
2. "Solvic (R) fuer Hohlkoerper aus Hart—PVC." *Werkschrift.* Solvay.

CHAPTER

20

Engineering Plastics

MICHAEL L. KERN, STANLEY E. EPPERT, JR., JAMES P. PARR AND ECKARD F. H. RADDATZ

ABS: ACRYLONITRILE-BUTADIENE-STYRENE

ABS designates a unique family of engineering polymers. The acronym is derived from the names of the three monomers used to produce the polymers: acrylonitrile, butadiene, and styrene. The three-monomer system can be tailored to end-product needs by varying the ratios in which they are combined. Acrylonitrile contributes heat stability, chemical resistance, and aging resistance; butadiene imparts low-temperature property retention, toughness, and impact strength; and styrene adds luster (gloss), rigidity, and processing ease. Tailoring of these three ingredients, plus modifications to the rigid phase molecular weight and additive package, can be achieved to improve melt strength characteristics as well. Due to these attributes, ABS has been used in blow molding applications for over twenty years and is now finding much success in engineering applications requiring more structural integrity and rigidity than was typically associated with blow molding.

Properties

Overall chemical resistance is quite good for ABS products. In general, they are very resistant to weak acids as well as weak and strong bases; somewhat resistant to strong acids; and are soluble in polar solvents such

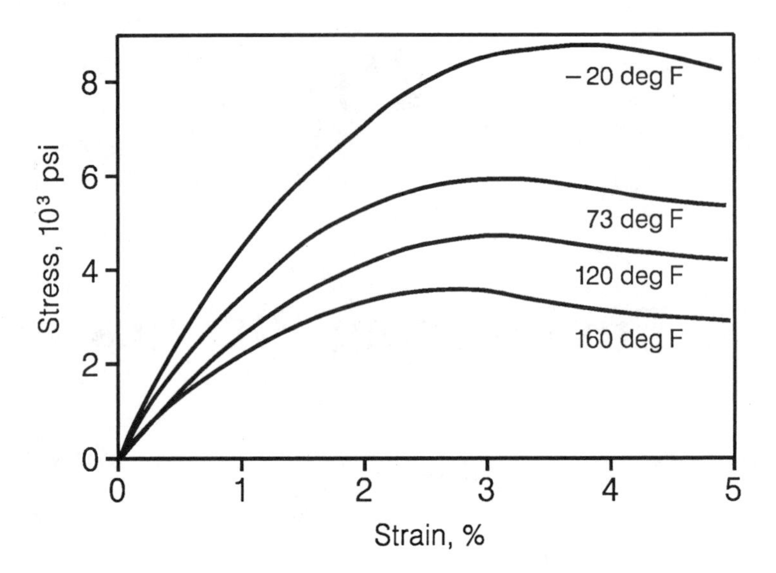

Figure 20-1 Tensile stress/strain curves for general-purpose, high-impact ABS.

as esters, ketones, and some chlorinated hydrocarbons. Resistance to specific chemical media that will come in contact with the ABS should be evaluated, or the effects determined from the supplier prior to use in a particular application.

Like many other thermoplastics, ABS has poor resistance to ultraviolet radiation. Direct sunlight will cause significant changes in appearance and properties. ABS must be protected with UV-resistant additives, laminates, or paint coatings to retard color, gloss, and property value changes.

Figure 20-1 shows tensile stress/strain curves for a typical high-impact blow molding grade. ABS is characterized by the capability to deform in a ductile manner at low temperatures. Tensile strength at yield ranges from approximately 3000 to 9000 psi. Strain at yield typically is 2 to 5 percent. Typical design limits are set at 0.75 percent strain for continuous load-bearing applications. For one-time, snap-fit designs, strains up to 8 percent may be permissible.

The key physical properties of four commercial ABS blow molding materials are compared in Table 20-1. ABS 1 has the highest room-temperature (73°F) impact with good low-temperature impact performance; ABS 2 has the highest low-temperature impact; ABS 4 has the highest tensile and flexural strength and modulus (rigidity). One very important physical property characteristic of ABS is its specific gravity of 1.02 to 1.04, which will allow either heavier walls at the same weight or the use of less material at the same wall thickness than is possible with the majority of other engineering thermoplastics.

Table 20-1 Key physical properties of four commercial ABS blow molding materials

Property	*Test Conditions*	*ABS 1*	*ABS 2*	*ABS 3*	*ABS 4*	*Units U.S. (metric)*	*Test Method*
Izod impact strength	73°F (23°C)	8.6 (459)	8.0 (427)	8.2 (438)	8.0 (427)	ft lb/in(J/m)	ASTM D-256 method A
	−40°F (−40°C)	4.5 (240)	5.0 (267)	2.5 (133)	2.9 (155)		
Test specimen: 0.125″ bar, notched							
Tensile strength	73°F (23°C)	4600 (32)	5000 (34)	4950 (34)	6400 (44)	PSI (MPa)	ASTM D-638 0.2″ (5mm)/min
Tensile modulus	73°F (23°C)	2.2 (1.5)	2.5 (1.7)	2.5 (1.7)	3.0 (2.1)	$\times 10^5$ PSI(GPa)	
Test specimen: Type 1, 0.125″ thickness							
Flexural strength	73°F (23°C)	7250 (50)	8600 (59)	8000 (55)	9800 (68)	PSI(MPa)	ASTM D-790 0.05″(1.3mm)/min
Flexural modulus	73°F (23°C)	2.3 (1.6)	2.6 (1.8)	2.4 (1.7)	3.0 (2.1)	$\times 10^5$ PSI(GPa)	
Test specimen: 0.125″ × 0.5″ × 5″ bar							
Deflection temperature*							
Under load, annealed	264 psi (1820 kPa)	218 (103)	214 (101)	219 (104)	220 (104)	°F(°C)	ASTM D-648
unannealed		192 (89)	195 (91)	193 (89)	194 (90)	°F(°C)	
Test specimen: 0.5″ × 0.5″ × 5″ bar							
Rockwell hardness*	73°F (23°C)	86R	90R	90R	105R	R (R)	ASTM D-785
Specific gravity*	23°/23°C	1.02	1.02	1.02	1.04	— —	ASTM D-792 method A

Source: Kern 1988.
*Compression molded; all others from injection-molded specimens.

Processing

Blow molding ABS is quite different from processing polyolefins. Key processing and molding factors are discussed below. It is important to contact the raw material supplier to obtain specific safety data sheets and processing information, and the machinery manufacturer to confirm equipment capabilities prior to molding a new material.

Drying

ABS is slightly hygroscopic and should be thoroughly dried prior to processing. A minimum of four hours at 190–195°F (88–91°C) is recommended, with enough air flow to minimize any temperature differential through the dryer. Moisture is readily noticeable as splay, rough chicken-track type lines on the surface, or bubbles, any of which can have an adverse affect on the integrity (strength) of the molded part. Moisture can also reduce melt strength. A moisture level of 0.02–0.03 percent or below is recommended for blow molding. To obtain optimum drying it is recommended that a dehumidifying dryer capable of circulating a continuous supply of warm, dry air to the pellets at a controlled temperature be used. A dessicant-type unit is preferred as it can provide air with the recommended dew point of less than 0°F (−18°C) for faster and more complete drying in humid weather.

Machine Downtime

General-purpose ABS blow molding grades exhibit good heat stability at the recommended stock temperatures of 380–430°F (193–221°C) and can withstand downtime of short durations. If downtime is to be extended, then the machine should be purged with polySAN or acrylic and the temperatures reduced. When the machine is to be shut down, it should be run dry of ABS and followed with polySAN or an acrylic purge. Since accumulator heads can retain heat for a long period of time and have extended heat-up periods, care must be exercised to protect against material degradation.

Purging

When using ABS to replace another material, or when switching from one grade or color of ABS to another, these steps should be taken into consideration:

1. Mold in an order that will allow moving from less viscous to more viscous materials

2. Mold from lighter colored to darker grades
3. If the order of molding requires processing of a higher viscosity or darker material first, use either polycarbonate or polySAN to expedite purging

Surface contamination can cause problems with appearance, pinch-point bonding, and paint adhesion. For these reasons, it is very important to thoroughly purge the processing equipment before saving ABS parts.

Screw Design

ABS tends to be more shear sensitive than polyolefins, and does not require high shear to produce a homogeneous extrudate for a good parison. Because of this, a high compression ratio, high-shear screw with long, shallow metering sections, or high-shear barrier screws are not recommended. Also because of their greater shear sensitivity, ABS materials will increase in melt temperature as the degree of shear on the material increases; a higher screw speed will cause an increase in the stock temperature.

In order to control the stock temperature within the optimum processing window, a low shear screw with a 2.0:1 to 2.5:1 compression ratio and a length to diameter ratio of 20:1 to 24:1 is recommended. Figure 20-2 presents the recommended screw characteristics.

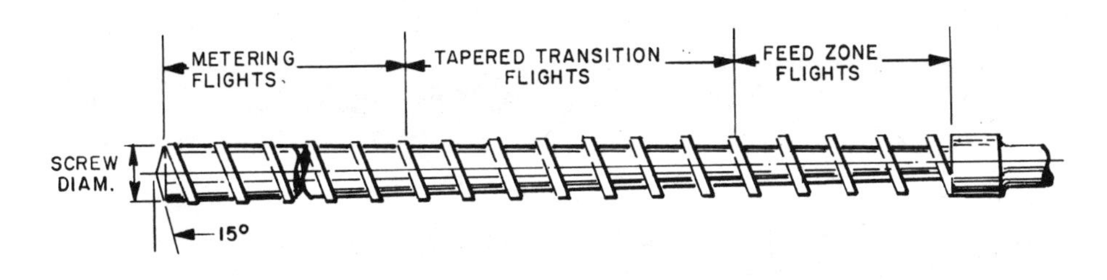

Compression ratio (by flight depth): 2.0:1–2.5:1
Helix angle (approx.): 17.7°

Length:Diameter ratio:	20:1	24:1
Feed section:	4–6 flights	4–6 flights
Tapered transition section:	6–10 flights	8–12 flights
Metering section:	6–8 flights	8–10 flights

Adapted from Borg Warner Chemicals, Inc., Tech. Bulletin PB117A, 5.

Figure 20-2 Design and general characteristics of screws recommended for use with ABS.

Due to recent developments by some machinery manufacturers, barrier-type screws with reduced shear have been used successfully for ABS. Even with these, attention must be given to keep shear heating to a minimum.

Processing Temperatures

Temperature settings on the barrel and accumulator should be adjusted to obtain a stock temperature of 380–430°F (193–221°C) for most ABS materials. Table 20-2 lists the typical temperature profile range for ABS. The optimum temperature will, however, depend on the equipment and the specific material being processed. It is recommended that the stock temperature and material residence time be controlled on the low or minimum condition in order to reduce the possibility of thermal degradation and to improve the parison melt strength.

Die Swell

Like most engineering thermoplastics, ABS materials do not have as much die swell as polyolefins. As a result, the die tooling diameter required to produce a parison of a given diameter and lay-flat will be larger with ABS than with most polyolefins. Many of the newer machines that are designed for use with engineering thermoplastics are equipped to accommodate larger head tooling. ABS die swell varies with material grade, stock temperature, and shear rate (the rate of accumulator discharge). Die swell of ABS blow molding materials varies from 5 to 30 percent, while the die swell of olefins is considerably higher. Figure 20-3 displays swell ratio versus temperature

Table 20-2 Typical temperature profile for ABS

	Range *°F (°C)*
Barrel, Zone 1	350–380 (177–193)
Barrel, Zone 2	360–390 (182–199)
Barrel, Zone 3	370–400 (188–204)
Barrel, Zone 4	380–410 (193–210)
Barrel, Zone 5	390–420 (199–216)
Accumulator	
Upper zone	390–420 (199–216)
Lower zone	390–420 (199–216)
Die	390–430 (199–221)

Source: Kern 1988

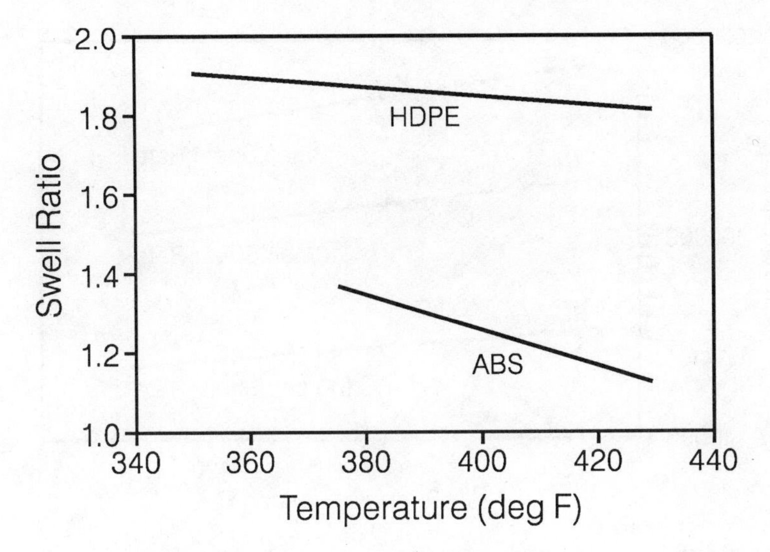

Figure 20-3 Die swell ratio as a function of temperature for typical blow molding grade ABS and HDPE at 100/sec shear rate.

data for typical blow molding grades of ABS and HDPE. With ABS, die swell decreases with increasing stock temperature, and increases with increasing shear rate through the die.

Viscosity

Like die swell, the viscosity of a material is also dependent on material grade, stock temperature, and shear rate. Many of the ABS grades have a higher viscosity than polyolefins within the processing window. The higher viscosity (stiffer) materials require more hydraulic pressure on the accumulator ram and the mandrel, especially if larger head tooling is used. Some of the accumulator heads designed for use with engineering thermoplastics have increased hydraulic capabilities. In general, the viscosity of an ABS material decreases as the shear rate and/or melt temperature increases, as displayed in Figure 20-4.

Generally, for a particular grade of ABS, melt strength is proportional to viscosity. The viscosities of typical blow molding grades of ABS and HDPE are shown at a shear rate of 100 reciprocal seconds in Figure 20-5. The greater rate of change in ABS viscosity with temperature calls attention to the fact that, even though ABS is somewhat forgiving, temperature has more effect on the ABS melt strength than is typically noticed with HDPE. For ABS, the lower the temperature, the better its melt strength.

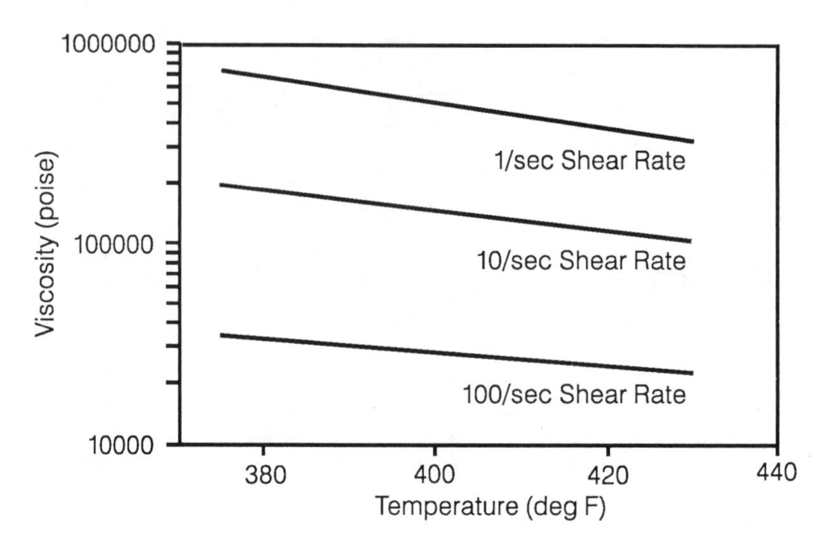

Figure 20-4 Viscosity of a typical blow molding grade ABS as a function of temperature and shear rate.

Accumulator Head

The head designed for use with ABS should include the following:

1. Capability for larger diameter head tooling
2. Improved streamlining and reknit features
3. Tooling with reduced land length to reduce back pressure
4. Increased hydraulic pressure to operate the ram at the required speed, and to operate the larger diameter programmed tooling

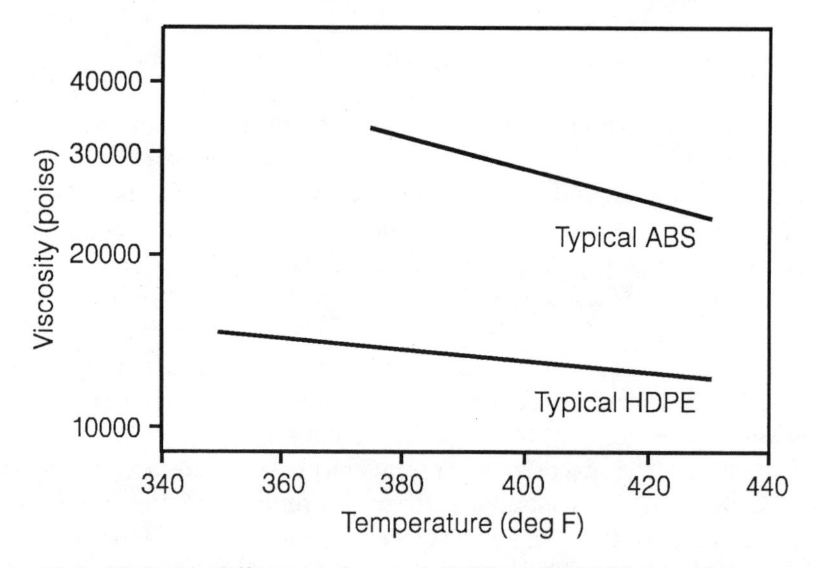

Figure 20-5 Viscosity difference of typical HDPE and ABS blow molding materials as a function of temperature at 100/sec shear rate.

Increased hydraulic pressure on the ram is also needed to insure a fast parison drop speed with ABS materials, which have marginal hot melt strength. The head should be streamlined to reduce areas of stagnation and to facilitate color changes and purging. Special finishes, such as nickel, are being evaluated on the tooling to reduce material sticking and die line formation.

Several equipment suppliers are currently developing a head design that will allow multilayer blow molding. That would be very advantageous in the use of ABS where UV resistance or added chemical resistance is required.

Press

Presses are being designed with larger platens to produce larger parts. Increased clamping force is being specified in cases where larger pinch and compression details are anticipated and where higher blow pressures may be used. Faster platen closing speeds are being recommended for use with engineering thermoplastics with marginal hot melt strength to reduce part thinning due to parison sag.

Mold Temperature

The best surface finish can be obtained with warm mold temperatures. A range of 160 to 190°F (71–88°C) is recommended for use with ABS. A warm mold also reduces molded-in stress to insure good dimensional stability. A warm mold will help provide strong weld points.

Shrinkage

Properly molded ABS materials of the type listed in Table 19-1 exhibit a mold shrinkage of 0.005–0.008 inches per inch. However, parison melt temperature, mold temperature, blow time, blow pressure, and wall thickness can play an important role in determining the actual shrinkage. Since the shrinkage of ABS is less than the polyolefins, draft angles and undercuts must be modified.

Regrind Use

Clean scrap and trim may be ground and blended with virgin material of the same grade. Extreme care must be taken to insure that no cross-contamination or foreign material is present in the scrap or trim before grinding. Metal inserts must be cut out of scrap parts before grinding. The regrind must be kept clean and dried if it is not used immediately. Cross-contamination can cause holes, thin spots, delamination, and weld line weakness. The maximum acceptable regrind use level will depend on a number of factors, such as extrusion melt temperature, shear history, and residence time at extrusion temperature.

If there is a significant change in color, viscosity, melt strength, or odor, the polymer may be exhibiting signs of degradation. Parts containing regrind should be tested carefully to insure that the physical properties remain within the expected limits. The processor should contact the raw material supplier to determine optimum regrind level usage for each code.

Conclusions

The use of ABS for blow molding applications is growing. There are significant differences between the processing and molding of HDPE and ABS thermoplastics. Material and equipment advances have increased the viability of blow molding ABS. If these differences between HDPE and ABS are taken into account and the guidelines presented above are adhered to, the processor should find ease in blow molding this engineering thermoplastic.

PPE: POLYPHENYLENE ETHER

Modified polyphenylene ether alloys are a family of engineering thermoplastics that offer long-term heat resistance and a wide range of physical properties. They possess excellent processing stability and the melt strength required to blow mold large parts having complex designs. These polymers are an excellent choice for applications requiring structural stability at elevated temperatures.

Polyphenylene ethers are produced by the oxidative coupling of substituted phenols. The miscibility of PPE with polystyrene at all proportions provides the capability of producing blow molding products with heat distortion temperatures (HDT) ranging from 180 to 270°F (82–132°C). (This range reflects products that are commercially available at the present time. PPE alloys can be formulated with HDTs as high as 350°F [177°C].) PPE alloys are available in both general-purpose and flame-retardant grades with flammabilities ranging from UL 94 HB to V–O at 0.060 in. minimum wall thickness. Most polyphenylene ether alloys have low moisture absorption levels, which results in good electrical properties over a broad range of temperature and humidity conditions. The material experiences minimal chemical attack from water, acids, bases, and most salt solutions. Exposure to some organic chemicals may cause the material to crack or soften.

Physical Properties

PPE alloys can be made flame retardant by the addition of thermally stable phosphorus-based additives. Rubber impact modifiers as well as a variety of fillers can be added to the polymer blends to improve toughness, strength, and rigidity. It is therefore possible to develop blow molding products that have a wide range of physical properties. Typical physical property ranges for blow molding grades of modified PPE are given in Table 20-3.

Processing

PPE alloys can be blow molded on various types of equipment. They perform especially well on machines equipped with accumulator heads. As with other engineering thermoplastics, PPE alloys need to be processed in a slightly different manner than polyolefins. Modified polyphenylene ether does not experience the same swell, memory, and temperature effects as seen in the processing of polyolefins such as HDPE: therefore certain considerations must be made in equipment selection, material handling, and operating parameters.

Extruder Screws

PPE alloys have been successfully processed on several styles of extruder screws, including general-purpose screws, barrier screws, polyethylene screws, and polystyrene screws. Engineering thermoplastics are more thermally sensitive to shear than polyolefins. Thus, increasing the shear on modified PPE results in an increase in its stock temperature and a decrease in viscosity. Excessive screw speeds that heat the material beyond the recommended processing window should be avoided in order to maintain proper melt strength of the parison. When processing PPE alloys on high-shear screws, screw speeds should be maintained within a range that allows for good control of the material stock temperature. A general-purpose, low-shear screw with a compression ratio in the range of 2.5:1 and a length to diameter ratio ranging from 20:1 to 24:1 provides good control over the material's melt temperature and is normally recommended for PPE alloys. There are also barrier type screws available that have reduced shear designs for use with engineering thermoplastics.

Screen packs or breaker plates are not recommended for use with PPE alloys because they tend to increase stock temperatures and complicate color and product changes. Shear pins and high-intensity mixing sections are not recommended because they generate high shear, which may result in excessive stock temperatures.

Table 20-3 Typical property data ranges for blow molding grades of modified PPE

	Test Method	*Test Conditions*	*Unit*	*General Purpose*	*Flame Retardant*
Izod impact strength	ASTM D 256	73 °F	ft lb/in	5.0–6.0	5.0–5.5
0.125″ (3.2mm) bar, notched	method A	23 °C	J/m	270–320	270–290
Gardner impact strength	ASTM D 3029	73 °F	ft lb	18–20	15–22
0.125″ (3.2mm) specimen	method G	23 °C	J	24–27	20–30
Tensile strength	ASTM D 638	73 °F	psi	5000–9000	6000–8000
0.125″ (3.2mm) thickness	2.0″/min	23 °C	MPa	34.0–62.0	41.0–55.0
Tensile modulus	ASTM D 638	73 °F	10^5 psi	3.2–3.3	3.2–3.6
0.125″ (3.2mm) thickness	2.0″/min	23 °C	GPa	2.2–2.3	2.2–2.5
Flexural strength	ASTM D 790	73 °F	psi	8000–13000	8500–14000
0.125 × 0.5 × 5.0″ bar	0.05″/min	23 °C	MPa	55–90	59–96
Flexural modulus	ASTM D 790	73 °F	10^5 psi	2.5–3.5	3.2–3.7
0.125 × 0.5 × 5.0″ bar	0.05″/min	23 °C	GPa	1.7–2.4	2.2–2.5
Rockwell hardness	ASTM D 785	73°F/23°C	R	104–114	105–120
Deflection temp. under load	ASTM D 648	unannealed			
0.5 × 0.5 × 5.0″ bar		264 psi	°F	180–270	180–270
		1820 kPa	°C	82–132	82–132
Flame class rating	UL STD 94	HB		0.062″	—
		V0		—	0.062–0.090″
		V1		—	0.062–0.090″
		5V		—	0.125–0.150″
Specific gravity	ASTM D 792 method A	73°F/23°C		1.04–1.10	1.04–1.10

Source: Stanley E. Eppert, Jr.

Accumulator Heads and Tooling

Streamlined first-in/first-out accumulator heads equipped with programmable parison control should be used with PPE alloys. Streamlining aids in color changes and material purge by reducing dead space, which is responsible for material stagnation in the head. PPE alloys can be run on tooling sized for either engineering thermoplastics or polyolefins. However, the modified PPE does not experience the same degree of swelling that is sometimes seen with materials such as HDPE. In some instances it may be desirable to process modified PPE using head tooling that is 20–25 percent larger in diameter than is typically used for polyolefins. The larger die size will help minimize the amount of preblow stretching and cooling of the parison.

Molds

Molds for PPE alloys are normally operated in the 170–200°F (77–93°C) range. Therefore it is necessary to have molds constructed of metals that have the capability of withstanding high processing temperatures and temperature cycling. Use of inferior metallurgy can result in the cracking of a mold. Pinch-offs should be designed to ensure a proper seal of the material and facilitate trimming operations. Flash areas should be designed with adequate clearance so that the polymers, which are more viscous than polyolefins, do not hold the mold open.

Drying

It is suggested that PPE alloys be dried before processing in order to ensure cycle consistency and good part appearance. Excessive moisture can cause splaying, bubbles, or roughness to appear on part surfaces, as well as a reduction of the parison's melt strength. Sufficient drying can be achieved through the use of hopper or tray dryers. It is recommended that the material be dried for two to four hours. Drying temperatures generally range from 180 to 220°F (82–104°C), depending upon the material's heat distortion temperature. Temperatures above the heat distortion temperature of the material should be avoided as this may result in the fusing of pellets.

Regrind

PPE alloy regrind, when blended with virgin pellets, displays excellent processing stability with little change in product performance. Maximum regrind levels will depend on several factors, such as melt temperature and shear history. It is good practice to test finished parts containing regrind to ensure that important physical properties are within specifications. When handling regrind, care should be taken to prevent the material from becoming contaminated by other polymers or foreign matter that may be present. Contamination, especially from less stable polymers, can result in the reduction of both processing and product performance.

Processing Temperatures

PPE alloys are more heat stable than many other blow molding polymers and generally require higher processing temperatures. Depending on the composition of the material, optimum processing temperatures generally fall within the range of 400–500°F (204–260°C). Materials with higher heat distortion temperatures tend to require higher processing temperatures. The viscosity of modified PPE, which directly affects the material's melt strength, is much more temperature-sensitive than HDPE (see Figure 20-6). As with other polymers, the viscosity of engineering thermoplastics decreases as the material's shear rate or melt temperature increases. It is important to control each material's stock temperature within the proper processing window in

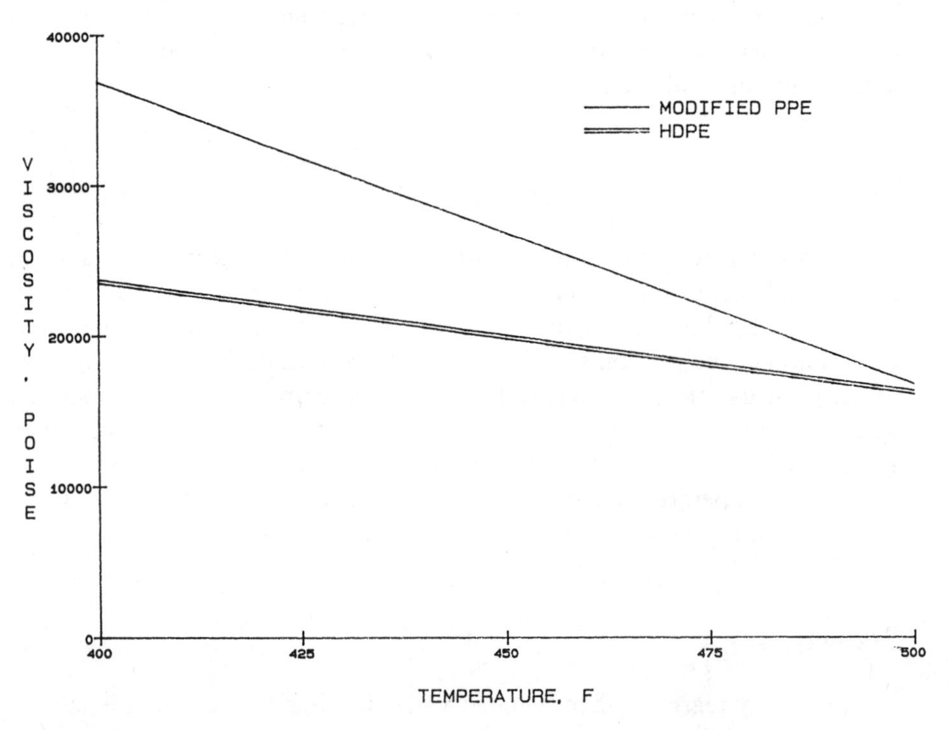

Figure 20-6 Viscosity vs. temperature for typical blow molding grades of modified PPE and HDPE. Courtesy of Stanley E. Eppert, Jr.

order to form good parisons and weld lines. Stock temperatures can be controlled by means of machine set temperatures and extruder screw speeds.

Purging

Success in purging to and from PPE alloys is primarily determined by technique and machine design. Purging is performed much more easily on machines having heads with streamlined flow channels. A continuous purge, which is accomplished by continuously extruding material at low speeds through an open accumulator head that is set for a narrow die gap opening, works very well for PPE alloys. This method, when supplemented with occasional full-purge shots to clear any loose material that may get hung up in the die opening, saves both time and material.

When possible, PPE alloys should be sequenced according to color and heat distortion temperature. Light colors should be processed before dark colors and grades having lower heat distortion temperatures should be processed before those with higher HDTs. The high processing temperatures used to process PPE alloys may degrade polyolefins and other polymers. Materials that are processed prior to PPE alloys must be completely purged from the machine in order to prevent cross-contamination, which greatly hinders modified PPE performance. Surface contamination from these polymers on the inside of the parison results in weak bonding at the pinch points of the mold. If desired, an intermediate purging material may be used when purging to and from another type of polymer. Purging materials for PPE alloys include polystyrene, high-impact polystyrene, wet polycarbonate, mica-filled polyethylene, and a few commercially available purging compounds.

Mold Temperatures

For good surface appearance, mold temperatures should be maintained in the 170 to 200°F (77–93°C) range. Higher mold temperatures also improve weld line strength and reduce the chance of premature freeze-offs, which can prevent the material from completely filling the mold.

Shrinkage

Properly molded PPE alloys can be expected to experience 5–7 mil/in. of shrinkage in a blow-molded part. Part shrinkage can be influenced by mold temperature, blow pressure, cooling time, and wall thickness. Higher mold temperatures and greater wall thicknesses will increase shrinkage, while higher blow pressures and longer cooling times tend to reduce shrinkage.

Material Selection

When selecting a blow molding material it is important to review items such as end-use applications, physical property requirements, and processing requirements. Assistance with material selection, part design, and processing is readily available from manufacturers of PPE alloys.

OTHER MATERIALS

In recent years, the applications of the traditional blow molding markets have matured. A wealth of new blow molding equipment developments has accompanied this maturity, as would be expected for such a young industry.

Faced with declining growth rates and increasing rationalization in the traditional markets, and an array of newly developed materials with excellent physical properties, the blow molding industry began seeking opportunities in areas where high-performance products were being manufactured by alternate processes with competing materials such as steel. The result has been the rise of the use of engineering resins in blow molding operations. These materials include ABS, polyacetal, polycarbonate, polyamide, acrylonitrile copolymers, polyphenylene ether, polyetherimide, polysulfone, ionomers, and a wide range of blends and alloys [6].

The products from these operations are expected to perform in a variety of ways not typically associated with blow molding. For example, many of these parts are used in high-temperature service or in load bearing functions. Some properties of selected resins are listed in Table 20-4.

Polyacetal

Polyacetal is a polymer with a repeating unit of ($—CH_2—OH—$). This unit is called an oxymethylene unit, so polyacetal is also called polyoxymethylene (POM).

As a result of the linear structure of its molecular chain, polyacetal has a solid state with a highly crystalline structure. Because of this, its physical properties have a high hardness, stiffness, and dimensional stability over a wide temperature range.

The highest service temperature, according to the stress level and duration, lies between 100 and 150°C (212–302°F). Polyacetal is resistant to almost all organic solvents, as well as caustic. It has low gas and moisture permeability.

The blow molding applications are in automobile and appliance construction and in packaging material. Some uses include pressure containers and fluid reservoirs.

Polyacetal resin can be processed on all of the recognized blow molding

Table 20-4 Typical properties of some various blow molding resins*

Property	ASTM Method	Poly-acetal	Polyamide		Poly-carbonate	Thermoplastic Polyester Elastomer	Modified PBT & PC	Modified PPO
			Amorphous	*6-6*				
Density, g/cc	D792	1.41	1.11	1.09	1.20	1.16	1.19	1.06
Tensile strength at yield, psi	D638	9800	9000	8400	9000	5200	7000	9000
Flexural modulus, psi	D790	440,000	290,000	270,000	340,000	18,000	290,000	325,000
Heat deflection temp. at 265 psi, °C	D648	125	115	63	132	—	96	129
Water absorption 24 hr, %	D570	0.2	—	1.03	0.15	0.4	—	0.10
Melting point, °C	—	164	—	260	—	196	—	—
Vicat softening point, °C	D1525	163	—	—	154–160	155	—	—

*All values dry-as-molded, unless otherwise specified, and 23°C

machines used for the processing of the polyolefins. The extruder can be equipped with either a progressive compression screw or a short compression screw. The melt temperature should be between 170 and 185°C (330–365°F). The extruded parison shows an essentially small swell ratio as it exits the die compared, for example, to the polyolefins. The types with high melt strength are especially useful for extrusion blow molding [2].

Polyamides

Polyamide (PA), commonly called nylon, is used in technical blow molded parts and as a barrier layer in coextruded containers. Amorphous grades are used as well as some polycaprolactam and filled and unfilled 6-6 nylon [3]. The material has good high-temperature properties, with a service temperature limit of about 200°C (392°F). It also has excellent solvent and chemical resistance. It may be modified to provide excellent impact resistance.

Because of its water absorption characteristics, it is necessary to verify that the material is absolutely dry before processing. Wet material gives a parison with a streaky, speckled surface.

For processing, extruders such as those designed for PVC have proven successful. Processing with an accumulator head is also possible. Melt temperatures fall in a range from 430 to 600°F (221–315°C). The material has a shrinkage of about 1 percent. Parison lengths of up to two meters have been obtained [4].

Applications of blow-molded PA include radiator surge tank and pressure vessels, air ducts, and power steering and brake fluid reservoirs.

Polycarbonate

Polycarbonate (PC) for extrusion blow molding must be a special high melt strength grade. Containers of polycarbonate distinguish themselves through the following properties:

- Very good transparency
- Excellent impact strength to −150°C (−238°F)
- Light weight
- Dimensional stability
- Heat resistance to 135°C (275°F)
- Smooth surfaces that are easy to clean
- No taste or aroma
- Surfaces that are printable without special preparation

Polycarbonate can be used for returnable containers due to its break resistance. However, because of its low gas and moisture permeation resistance, it is not suitable for packaging oxidation-sensitive or carbonated products. PC also has low chemical resistance. Current applications include:

- 20-liter milk containers
- 20-liter drinking water containers
- 5-gallon containers for drinking water dispensers
- Returnable milk bottles

Blends, Alloys, and Specialty Resins

While it is not possible to list all the thermoplastics and their applications, some are extremely interesting in certain applications:

- Blends and alloys
- Modified elastomers
- Acrylonitrile copolymers
- Fluorine-containing materials

Future developments in extrusion blow molding technical parts from polymer mixtures and alloys remain promising [3]. Further possibilities of ever-increasing significance for the combination of properties of two different materials will be explored through coextrusion.

A variety of blended and alloyed materials are now being utilized in applications where high stiffness, high heat resistance, or excellent dimensional control are desired. Often the parts have a flat or double-wall geometry. Some of the combinations encountered in this area include polybutylene terephthalate (PBT) with PC, PS with polyphenylene sulfide, and polyamide with polyphenylene oxide.

The blend components are chosen for the properties each provides. Often one material is chosen to provide the needed toughness or temperature performance while the other component is chosen to improve processability.

Thermoplastic elastomers are materials that emulate certain properties of rubber. Their difference lies in their ability to be remelted. There is a broad range of polymer types from which to select. One of the application areas for these products is tough, flexible automotive components such as the bellows covers for the axles on front wheel drive automobiles.

Acrylonitrile copolymers are noted for their resistance to liquid hydrocarbons, ketones, and chlorinated solvents. Comonomers include styrene, butadiene, and nitrile rubber. Applications include automotive and agricultural chemical containers, food packaging, and correction fluid containers [6].

Small containers from five to twenty-five liters are produced by extrusion blow molding processing of polyvinylidene fluoride (PVDF). This fluorine-containing thermoplastic is recognized for its outstanding chemical and mechanical properties [7]:

- Great chemical resistance to strong acids, aliphatic and aromatic hydrocarbons
- Good thermal stability up to 165°C (329°F) (melt point 177°C [351°F])
- Low residual monomer content
- Resistance to water
- Good aging characteristics in warm storage conditions

The density of the extrusion blow molding type is about 1.78 g/cc and the melt flow index at 230°C/5 kg is about 23 g/10 min.

Polystyrene

Polystyrene (PS) is one of the largest volume thermoplastics. Production began in Germany in 1936. The monomer is styrene, which is simply ethylene with a phenyl group (benzene ring) substituted for one of its hydrogens. Catalytic polymerization proceeds in either a mass or a suspension process. The structure of this molecule is shown in Figure 16-3.

The resulting polymer, known as crystal PS, is a clear, hard, stiff, amorphous polymer. It is resistant to acids, alkali solutions, alcohol, and mineral oils. It is not resistant to most aliphatic and aromatic solvents. It is amorphous because the phenyl groups attached to the carbon backbone impede crystallization. Unmodified crystal grades are brittle [1]. Some properties of this material are shown in Table 20-5.

Table 20-5 Typical properties of polystyrene grades used in blow molding

		Grade	
Property	*ASTM Method*	*Crystal*	*Impact Modified*
Density, g/cc	D792	1.06	1.05
Melt flow, 200/5.0	D1238	3.0	4.0
Tensile strength at yield, psi	D638	7000	4000
Flexural modulus, psi	D790	453,000	280,000
Flexural strength, psi	D790	11,000	6000
Vicat softening point, °C	D1525	104	101
Heat deflection temp. at 264 psi, °C	D648	200	190

Impact-modified grades of polystyrene typically utilize polybutadiene elastomers to increase impact strength. Both polymerized and compounded grades are produced. Modification yields materials that are not as clear or stiff. Acrylonitrile-butadiene-styrene (ABS) resins, discussed in the previous section, are closely related to impact-modified PS grades.

For blow molding, the higher molecular weight grades are used. The largest application area is in injection blow molded parts, although PS is also used in extrusion blow molding. Stabilization against oxidation is necessary, and some release agent is incorporated. PS is used as an alloying agent in several of the new engineering alloys and blends now appearing on the market.

ACKNOWLEDGMENT

The material in this chapter contributed by Eckard Raddatz is from his presentation, "Materials for Blow Molding of Packaging and Technical Articles," presented at the Sixth Plastics Symposium, German Engineers' Society, Brazilian Group, 1987. The section on ABS: Acrylonitrile-Butadiene-Styrene is contributed by Michael L. Kem; the section on PPE: Polyphenylene Ether is contributed by Stanley E. Eppert, Jr.

The authors express special thanks to:
Katrina Branting, DuPont, Wilmington, DE
Rubi Deslorieux, General Electric, Pittsfield, MA
Richard Westphal, Huntsman Chemical Co., Chesapeake, VA.

SELECTED READINGS

Borg Warner Chemicals, Inc., Technical Bulletin PB117A
Borg Warner Chemicals, Inc., Technical Publication PR 149.
Brewer, G. "Polyphenylene Ether, Modified." *Modern Plastics Encyclopedia.* New York: McGraw-Hill, 1988.
Ferguson, L. E. "Blow Molding ABS and PPE/Styrene Alloys." *RETEC Paper* 1987.

REFERENCES

1. Swett, R. M. "Polystyrene." In *Modern Plastics Encyclopedia.* New York: McGraw-Hill, 1986-1987.
2. Brandes, R. *The Engineered Blow Molding Opportunity.* Pittsfield, MA: General Electric.
3. Subramanian, P., and J. Feathers. "Blow Molding of Nylon 6,6 Compositions." *Proceedings, ANTEC.* Soc. Pastics Eng., 1988.

4. "Amorophous 'Zytel' Nylon Resins Are Tailored to Large Part Production." *Tech Topics.* No. 87/2. Wilmington, DE: DuPont,1987.
5. "Schlagzaehe Polymerlegierung Aufbau und mechanische Eigenschaften." *Plasterverabeiter* 38 (1987). Heft 1.
6. Lund, P. "Enhanced Impact-High Nitrile Resin for Blowmolding." *Proceedings, ANTEC.* Soc. Plastics. Eng., 1988.
7. "Das Polyvinylfluroid von Solvan." *Werkschrift.* Solvay.

SECTION V

PRODUCT DESIGN

CHAPTER

21

Designing Products for Blow Molding

JAMES P. PARR AND JOHN A. SZAJNA

BASIC DESIGN FACTORS

The following sequence of steps is suggested for the design of a plastic part:

1. Initial product conception
2. Selection of material
3. Preliminary mechanical design
4. Prototyping
5. Adjustment of design

Steps 4 and 5 are repeated until a satisfactory solution is achieved. This chapter will discuss the considerations involved in steps 1 and 3. Steps 2 and 4 are dealt with in greater detail in other portions of this book.

The most important aspect of designing a blow molded plastic object is to carefully define all the requirements of the application, and match those with the proper combination of part design and material properties [1]. Some of the important material properties include: elastic modulus, impact resistance, fatigue resistance, environmental stress crack resistance, temperature response, chemical resistance, abrasion resistance, optical and electrical properties, and weatherability.

Generally, when a part is being designed, several competing requirements make their demands simultaneously. This often requires overdesign of one area to meet the minimum requirements in another, or limits the material selection to a single resin, or even grade of resin. The part should ideally offer the maximum space and material utilization.

Whenever more than one material is capable of performing the function required of the molded item, a cost analysis should be the final criterion used to make the material selection. This comparison can only be made after prototype parts are produced and subjected to testing, unless experience is sufficient to make a preliminary judgment.

The design process requires many judgments, and trial and error modifications are a necessary part of the process. Because so many facets of the design process are subjective, a high degree of imagination is an indispensable asset for a designer.

Characteristics of Plastics

Plastics are by their nature different from such traditional materials as metal, wood, and glass. As such, they offer distinct advantages and disadvantages that are relevant to product design. The broad range of materials that can be used in blow molding offers the designer the option of trying various property combinations while using the same production method.

The container market has been the traditional stronghold of blow molded plastics. There are some strong reasons for this success. Plastics resist rust and corrosion, breakage, and moisture, factors that plague metals, glass, wood, and paper. Plastics also usually offer savings in weight, which is especially important in shipping costs. They are also good insulators of heat and electricity. A grade can usually be found that is chemically resistant to a proposed lading.

The momentum of plastics in the packaging sector has been augmented by recent breakthroughs in coextrusion technology that have overcome problems with permeation of certain gases. This subject is discussed in greater detail in chapter 5. New grades of resin and new processing modes have opened other areas.

In addition to the many advantages, the designer must be aware of and account for some inherent weaknesses of plastics. Most resins are more or less notch sensitive, especially in low-temperature impact strength situations. Environmental stress cracking can lead to premature failure if the part is subjected to stress in an aggressive medium, typically a detergent in aqueous solution. Notches also facilitate this type of failure. A UV stabilization formula is necessary if the article will be exposed to the sun for a significant period of time, since the ultraviolet rays can break the intermolecular bonds of most types of polymers. Also, the service temperature range of most plastics is narrower than for some competing materials [2].

The viscoelastic nature of plastics must be considered one of their design difficulties. Viscoelasticity means that a material exhibits behavior that is a combination of elastic deformation and viscous dissipation in both the solid and liquid states. Plastics cannot be treated mathematically as an elastic solid material like metal or glass, because the material creeps when under stress. That is, the strain increases with time for a given load. Creep (cold flow, or nonrecoverable deformation) under load is a continual process in a plastic product under any stress. This requires that creep properties be examined for structural elements in both tensile and compressive loadings [3].

The rate of creep is dependent upon the stress or strain levels, the service temperature, and the rate of loading. After short-term loadings, a portion of the deformation is recovered, but some is permanently retained. Due to these uncertainties, plastic structural elements have a certain indeterminability that has impeded their broad acceptance. Overdesign is the normal method of dealing with the situation in the absence of detailed creep data [2].

Creep data are typically determined by testing specimens at constant values of temperature and load. While the tests are usually of a tensile nature, results for compressive creep are practically the same. The creep modulus is the value of stress/strain at any given point in the duration of the test for the given conditions of time and temperature. The data can be graphed in a variety of ways, usually showing a family of curves with one parameter (stress, time, temperature, or modulus) held constant. Flexural creep data are also often determined.

To calculate a loaded member using elastic theory equations, one must determine the maximum time and temperature for which the product will be required to perform at the highest level of stress. The appropriate creep modulus is then obtained for these conditions and substituted for Young's modulus in the calculation [3]. The reader is urged to consult more extensive treatments of this subject should mechanical performance be a key aspect of a given design.

Blow Molding Design Parameters

Blow-molded objects must be designed to take into account the specific physical phenomena involved in and resulting from the blow molding process. By their nature, blow molded parts are basically hollow. Most traditional parts have tended to be generally cubical or cylindrical, although many newer applications utilize broad flat sections. The hollow interior can be injected with foam or some other substance for improved insulation or stiffness. The interior can also be used as a form of inner ducting or conduit for wiring. Sometimes the object can be cut in two to yield two parts from the same production cycle.

Blow molding offers some attractive benefits in comparison to other types of fabrication options [4]. Perhaps the foremost of these is its simple one-piece, one-step nature. Assembly operations are usually minimal, especially when replacing hollow metal containers. For an object such as an automobile gas tank, the ability to easily fill all the limited space to provide the highest possible storage space is a definite design strength. Larger parts may be produced with lighter weight molds and lower clamping pressures than are required for injection molding [5].

Where the parison enters and exits the cavity, it must be cut and welded. This section is heavy due to the welding and lack of stretching in that area. It distorts more than the rest of the container due to retained heat and also produces a surface visual defect. These seams are a source of notch defects and stress concentrations [6].

The necessity of blowing the part requires an opening into the interior, which may or may not be used as a type of spout or neck. If not, it may be plugged or sealed in some manner. In a few cases it is possible to seal the opening with a special attachment or by using a small tangentially inserted needle, before the parison has completely solidified. This is the basis for aseptically filled containers.

As in most plastics forming processes—whether injection molding, extrusion, or thermoforming—the ideal is a uniform wall thickness in the part. Wall thickness in a blown part must be expected to vary due to the nature of the process. An example of this is shown in Figure 21-1. The part should be designed with this variability in mind. The wall uniformity depends on

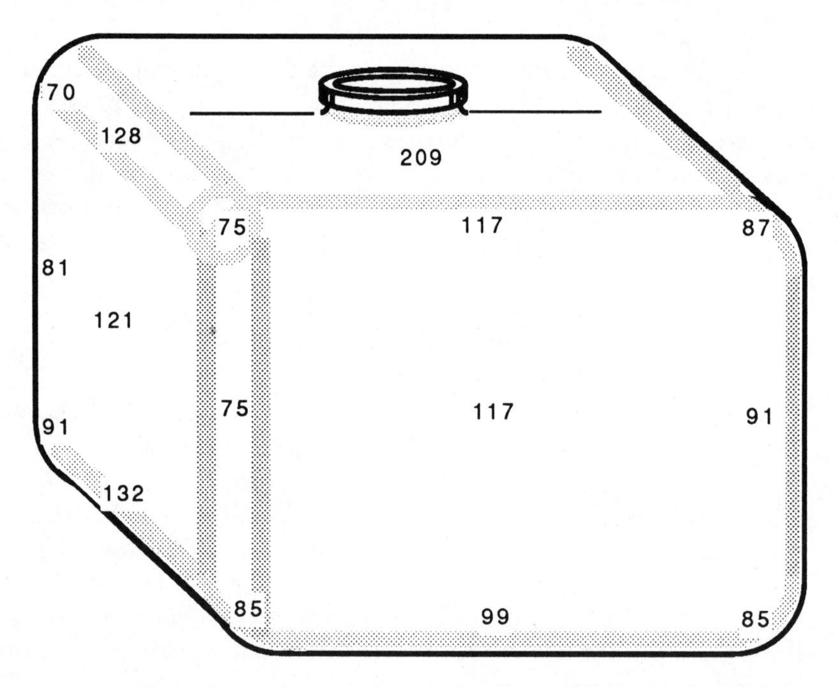

Figure 21-1 Wall-thickness diagram for a 275-gallon cubical container (all values × 0.001 in.).

the configuration of the mold cavity. Especially troublesome is the extreme difference in thickness, near each pinch-off, of sections 90 degrees apart at the same axial position. In extremely large parts, drawdown of the parison can create extreme thinning at the top of the mold, especially with lower melt strength materials. Parison programming, both axial and radial, can help alleviate these problems to a certain extent. Tooling can be shaped for extra control.

The blow molded product should be designed in such a manner as to minimize any points of extreme stretching or of too deep a draw, because of the associated stretch orientation and thinning. Surfaces should be as smooth as possible and all corners should have generous radii. The ideal blow molded part design would be perfectly symmetrical [7]. This would then allow as even a wall thickness distribution as possible. However, other factors, such as aesthetics and function, play a part in design.

The amount of stretching a parison is subjected to is a function of the part size and configuration in relation to the parison size and orientation. In general, this can be expressed as follows:

$$\text{Average part thickness} = \frac{\text{parison surface area}}{\text{part surface area}} \times \text{parison thickness}$$

Localized areas will be lower than this nominal value if the parison contacts the wall sooner in other areas of the mold. Different materials, or grades of materials, will exhibit differing stretch behavior. The blow ratio of a molding is a way of representing the amount of stretch involved for a given combination of parison size and part size. For cylindrical containers, this can be expressed as

$$\text{Blow ratio} = \frac{\text{mold diameter}}{\text{parison diameter}}$$

Generally, this value is between 1.5 and 3, but it can be up to 7 in unusual cases.

Because molding is executed with a melt that is then solidified, shrinkage and warpage are experienced with most materials. Higher crystallinity polymers have higher shrinkage values (see Table 21-1). Shrinkage is dependent upon the wall thickness, due to the different cooling rates. The cycle time to cure the bottle will be what it takes to cool the thickest wall section.

Containers

The overwhelming majority of blow molded objects are some form of container. Since a container is a type of package, it is important for the

Table 21-1 Shrinkage and other properties of some common blow molding materials

Polymer	*Shrinkage** (*%*)	*Linear Coefficient of Thermal Expansion* $K^{-1} \cdot 10^{-4}$	*Specific Volume* V *at 20° C* (cm^3/g)
LDPE	1.2–2	2.3 (20°C)	1.09
HDPE	1.5–3	2.0 (20°C)	1.05
Polyacetal	1 –3	1.3	0.7–0.71
Polypropylene	1.2–2.2	1.6	1.10
Polystyrene	0.5–0.7	0.7–0.8	0.89–0.95
Polyvinylchloride	0.5–0.7	0.8	0.81

*Measured on an axially symmetrical test bottle with an average wall thickness of 0.7–1 mm, by method of R. Holzmann, Kautex-Werke, Hangelar.

designer to understand the basic functions of a package. The following are some suggested functions [8]:

- To allow transport
- To protect product integrity
- As a marketing tool
- To protect the environment from a spill

The ability to withstand the demands of the various modes of transport is obviously a very basic requirement of a package. Long bumpy rides in a trailer or railcar require a tough material and design. Handling must be facilitated, especially in large-volume containers that must be moved with forklifts. Stacking of the containers in hot warehouses must be expected unless an overpack will serve this function.

Containers may be rigid, semirigid, or flexible, and of either a simple or complex nature. A milk container must stand alone, be handled, and transport the milk as an entity. A 275-gallon bulk transport bottle, however, is packaged in a steel or cardboard overpack, which not only greatly enhances the liner's physical stability but also provides load-bearing and transport properties. Container types that can be blow molded include tubes, boxes, bottles, jars, vials, pails, and barrels. Some of these containers are shown in Figures 21-2 and 21-3.

The most common container type encountered in blow molding is the bottle. The bottle design includes several elements. The design should be aesthetically pleasing so as to enhance product sales. Ease of dispensing is a vital factor. Perhaps the economics of the product require a least-cost approach. The bottle design in most cases will need to survive filling, cap-

Figure 21-2 Blow-molded products serve a wide variety of applications. Courtesy of Johnson Controls, Plastics Machinery Division.

ping, label application, warehousing, retail outlet distribution, the store shelf and, ultimately, the consumer.

The variety of considerations that must be addressed by the engineer is mainly functional. However, marketing parameters may enter into and influence some of the mechanical considerations. Basically, the bottle is intended as a package for a product. The following concerns, as well as others, must be addressed:

- Type of product to be contained
- Compatibility of resin and product
- Labeling requirements
- Size, shape, and appearance
- Size and type of closure

When packaging an item, it is important to consider whether the package and product are compatible on a chemical and physical basis. In dangerous goods transport, it is required that the packages pass stringent permeation and chemical resistance tests with each proposed loading before being allowed in service [1]. Containers for tomato-based products must minimize penetration by oxygen, which rapidly destroys them. A detergent packaged in homopolymer HDPE would quickly stress-crack the container.

A plastic bottle is compatible with the product it contains if it meets the following following criteria:

1. There is no chemical reaction between resin and product.
2. Permeation loss of the ingredients of the product is below an acceptable minimum.
3. A sufficient barrier protects the product from attack by the exterior environment.

Figure 21-3 Heavy-duty containers provide safe shipping for hazardous chemicals. Courtesy of Hoechst Celanese Corporation.

Figure 21-4 Hot foil stamping provides an attractive decorating option. Courtesy of Gladen Corporation.

Selection of the resin to be used for the bottle is often determined by compatibility considerations. Most manufacturers of plastic resins have published data that indicate whether their resin is applicable for a specific application. The supplier may provide more than one material, or information regarding the possibilities for surface treatments or options for multilayer usage of several materials.

Another key consideration in relation to product integrity is the effect of storage on the odor and taste properties of the contained product. Taste can be absorbed by the container from the contents or vice versa. Tensile strength of the container is important in carbonated beverage service due to the internal pressure. It is also required to have a leakproof seal between cap and neck finish.

The marketing aspect of advertising the contents by means of the package's shape and decoration is a large factor in the design. The things that affect marketability are more subjective than the purely mechanical aspects. Sometimes a clear stripe that reveals the amount of product is used as a marketing feature. Fruit drink containers may look like little animals to attract a child's attention. Oil containers captured their packaging market largely because of the convenience of their funnel-top design and resealability.

Requirements of transparency, translucency, or opaqueness can affect the choice of plastic to be used for a particular application. For example, if the bottle must be transparent to show the product color or a see-through label, then polyethylene would be excluded. Various materials have different degrees of translucency and can be chosen according to the need to partly reveal or conceal the contents. An opaque bottle can be practically any material with the addition of color. Opaque bottles are one means of masking the fill level.

Color, like styling, will be determined by those responsible for marketing the product. Blow-molded containers may be decorated in a variety of ways. Silk screening or other printing methods can apply a design directly to the container's surface. Labels can be applied, or the container can be stamped using special pigment-coated transfer tapes and a heated stamping tool such as that shown in Figure 21-4 [7]. These processes are discussed in more detail in chapter 11.

DESIGN CONSIDERATIONS

Corner and Edge Rounding

Care must be taken when designing the part to adequately round its corners and edges for two reasons. The first is the typical wall thinning effect in corner regions. Surface area/volume relationships show that a sphere always has thicker walls than a cube of the same volume and part weight. Compounding this situation, as the parison is being inflated the areas that contact the walls first tend to solidify. As the inflation continues, the corners are the last part of the parison to come into contact with the mold, with relatively thinner corners the result.

Adequate rounding of the container corners helps to alleviate this thinning problem. An examination of a modern one-gallon dairy container (Figure 21-5) reveals a highly advanced approach. Originally more rectangular, this container now is almost spherical, with flat faces. This allows the most efficient use of resin possible.

A practical guideline for designing a rectangular container is to allow a minimum corner or edge radius of at least one-third the depth of the mold half. For cylindrical containers the radius should be at least one-tenth the container diameter.

The second reason to avoid sharp corners or notches is the notch sensitivity of most plastics. This is very important when designing heavy-duty containers that must meet impact testing criteria. Fracture can be initiated in the roots of threads, the intersection of the neck and body, or near the pinch-off. Typically these notches are caused by projections on the mold that are easy to grind to a more desirable rounded form.

Figure 21-5 The design of the standard one-gallon dairy container minimizes resin usage. Courtesy of Johnson Controls, Plastics Machinery Division.

Volume

When designing the cavity of a mold, one must determine the final volume of the container based on the working drawing [9,10]. The contributions to that overall cavity volume, V_{oc}, can be described as

$$V_{oc} = V_c + V_b - V_s - V_r$$

where:

V_c = usable volume of container

V_b = volume increase due to sidewall deformation (bulging)

V_s = volume loss due to shrinkage of container

V_r = volume of resin in part

Approximations may be made of the volume of a proposed container design by using standard geometrical formulas. Some standard formulas for selected geometrical shapes are shown in Figures 21-12 and 21-7.

The difficulty in using such formulas, however, is that a container is rarely a pure geometric solid. More commonly the container has rounded edges, and is composed of various shapes that have been combined. In these cases, there are two key points to keep in mind.

First, divide the container into the various contributing component shapes, and calculate the contribution of each.

Second, calculate the volume lost through corner and edge rounding. As an example, look at Figure 21-6. Here we see a cross section through a rectangular container with rounded edges. The volume of the portion of the container with the cross section shown (e.g., the body of the container) is calculated by multiplying the area shown (47.1416 in.^{2}) times the height of the constant cross section.

Now suppose that in this example the top and bottom of the container are also rounded. Here we calculate in a similar fashion the volume that is lost due to rounding. Only now we calculate the difference in volume of a sphere with radius equal to the radius of the container edges, and a cube of edge length equal to two times the radius.

As an example, suppose we have a container that has the cross section shown in Figure 21-6 and is 10 in. tall. All corners and edges have a radius of 1 in. The volume is calculated as follows:

Basic formulas

$$\text{Volume of cube } 8 \times 6 \times 10 = 480 \text{ in.}^3$$

$$\text{Volume of cube } 2 \times 2 \times 2 = 8 \text{ in.}^3$$

$$\text{Volume of sphere 1 in radius (Fig. 21-12)} = 4.189 \text{ in.}^3$$

$$\text{Area of cross section (Fig. 21-6)} = 47.1416 \text{ in.}^2$$

Contributions to volume

1. Rounded cubic solid (top and bottom portions of the container, above and below points where corner radius begins)

   Volume with square edges $= 2(\text{height}) \times 8 \times 6 = 96$ in.^{3}

   Correction for corner and edge rounding

   corners:

   8(volume of $2 \times 2 \times 2$ cube) $-$ 4.189 (sphere volume)

   $= 3.811$ in.^{2}

edges:

difference in area of circle with 1 in. radius and square

with sides of 2 in. (Fig. 21-6) = 0.8584 in.^{2}

edge length = (6 − 2) in. + (8 − 2) in. = 4 + 6 in. = 10 in.

10 in. × 0.8584 in. = 8.584 in.^{3}

corrected volume = 96 − 3.811 −8.584 = 83.605 in.^{3}

2. Body section (section of constant cross section)

Cross-sectional area = 47.1416 in.^{2} (Fig. 21-6)

Height of section of constant cross section (height of container minus top and bottom radii) = 10 − 2 = 8 in.

Volume = height × cross-sectional area = 8 × 47.1416
= 377.133 in.^{2}

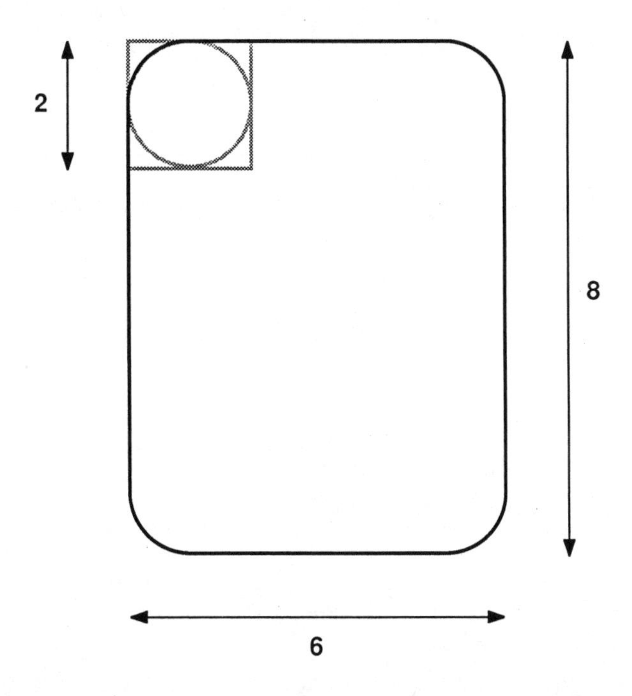

Area of rectangle 8 in. x 6 in. = 48 square in.
Radius of corner = 1 in.
Area of circle of 1 in. radius = 3.1416 square in.
Area of square with 2 in. sides = 4 square in.
Difference = 0.8584 square in.
Area of rounded rectangle = 47.1416 square in.

Figure 21-6 Illustration of method for determining area of rounded rectangle.

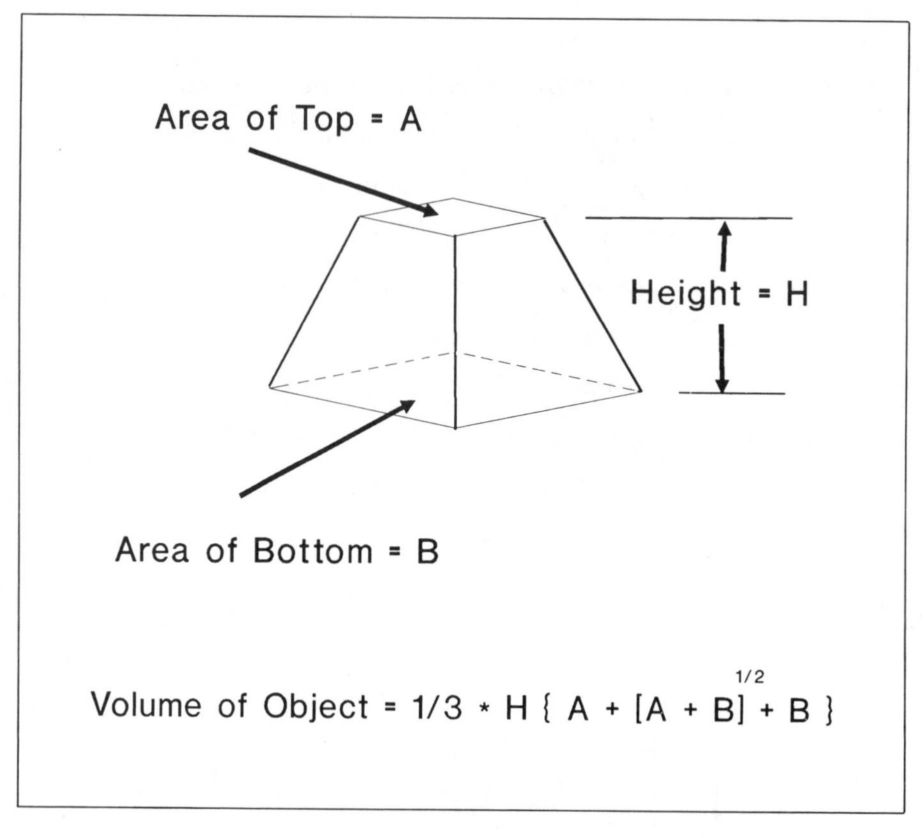

Figure 21-7 Illustration of method for determining volume of cubic section.

Total volume

$$\begin{aligned} \text{Volume} &= \text{volume of top and bottom} + \text{volume of body} \\ &= 83.605 \text{ in.}^3 + 377.133 \text{ in.}^3 \\ &= 460.738 \text{ in.}^3 \end{aligned}$$

Note that the total volume lost to the corner and edge rounding is $480 - 460.738 = 19.262$ in.^{3}, or about 4.0 percent of the volume of the nonrounded cube.

It is often not possible to exactly determine the volume in this way, if the cross section is not a standard shape. In these cases, the cross-sectional area may be estimated with graphical techniques or by using a planimeter.

When producing a series of containers of the same overall shape in a range of volumes, it is possible to produce the mold in sections. In this way, different molds are not needed: volume can be changed by adding or removing sections. This is usually done with simple shapes such as cylinders.

Of course, manual methods for calculating cavity volume are rapidly

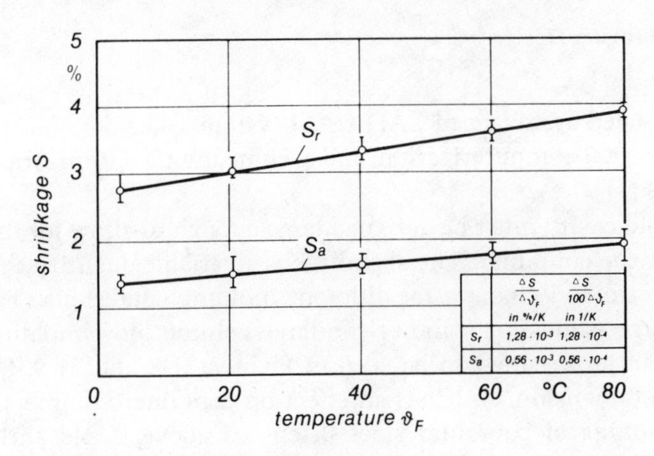

resin density = 0.946 g/cm; MFI 190°C/5 kg = 1.6 g/10 min
average wall thickness = 0.66 mm
extrudate temperature = 190°C
ϑ_F = temperature of blow mold; blowing pressure, 9 bar
inflation ratio = 2:1
cooling time = 10S

Figure 21-8 Axial shrinkage, S_a, and radial shrinkage, S_r, of an axially symmetrical high-shouldered HDPE bottle as a function of mold cavity temperature (measured after 14 days storage at room temperature). Courtesy of Hoechst AG.

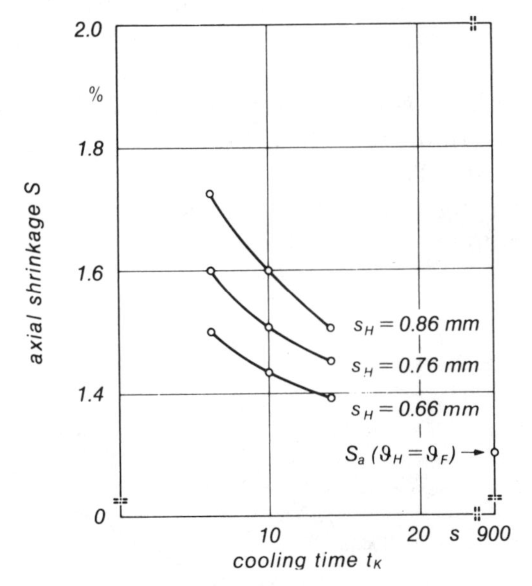

melt temperature, 190°C
mold temperature, 14.5°C
inflation ratio, 2.1:1
blowing pressure, 9 bar
S_H = average wall thickness of a bottle

Figure 21-9 Axial shrinkage, S_a, of rotationally symmetrical high-shouldered HDPE bottles with various wall thicknesses as a function of cooling time in the blow mold under blowing pressure with compressed air convection cooling. Courtesy of Hoechst AG.

being superseded by advanced CAD/CAM systems. The next few years will probably see total computerization, and accompanying optimization of such calculations [11].

The mold cavity must be designed large enough to allow for the overall shrinkage of the container. The shrinkage is a variable quantity that differs with each resin and changes for different molding conditions. Table 20-1 lists shrinkage values for some of the large-volume blow molding resins. The effects of processing can be seen in Figures 21-8 and 21-9. Shrinkage acts in each dimension, so it has an effect on container volume similar to the proportioning of container sizes discussed above. Mold shrinkage is generally higher in the neck region of bottles than in the body due to the higher thicknesses.

Container bulging is due to the hydrostatic pressure of the fluid in the container, and the deflection of the walls as a result of this pressure. The shape of the bottle can be altered to take this factor into account. Bulging can vary with wall thickness distribution, temperature, and density of lading, so it is difficult to exactly determine in advance. The volume increase can be up to 5 percent. Figure 21-10 shows the effects of some of these variables.

Bulging is usually greatest in flat-sided containers. This can be reduced by using recesses, ridges, or grooves in order to improve the section modulus of the part. Bulging of label panels can result in wrinkling of labels, especially with flat-sided bottles.

If a bottle, similar in shape to an existing bottle, is being considered, it can be helpful to determine the extent of the bulge in the existing container. This can be done easily. First, fill the container with water. Then lower the full container into a larger vessel filled with water and deep enough to accept the test container up to its neck. The "bulge" volume of water in the test

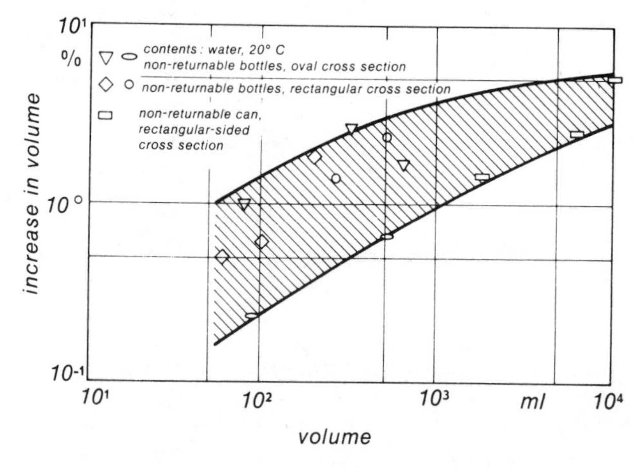

Figure 21-10 Increase in volume of nonreturnable bottles and cans made of HDPE caused by hydrostatic pressure. Courtesy of Hoechst AG.

container will overflow since the internal hydrostatic pressure due to the force of gravity has been eliminated. When pulled back out of the water, the bulge will reappear, but a portion of the container will now be empty. The bulge volume can be determined by carefully refilling the container from a graduated cylinder [9].

The total volume of the container can be determined by carefully weighing it while both full and empty. The weight of water it contains can be converted to an accurate volume if the temperature of the water is known. At 25°C (77°F), for example, water's density is 0.99567 g/ml. Tables of the

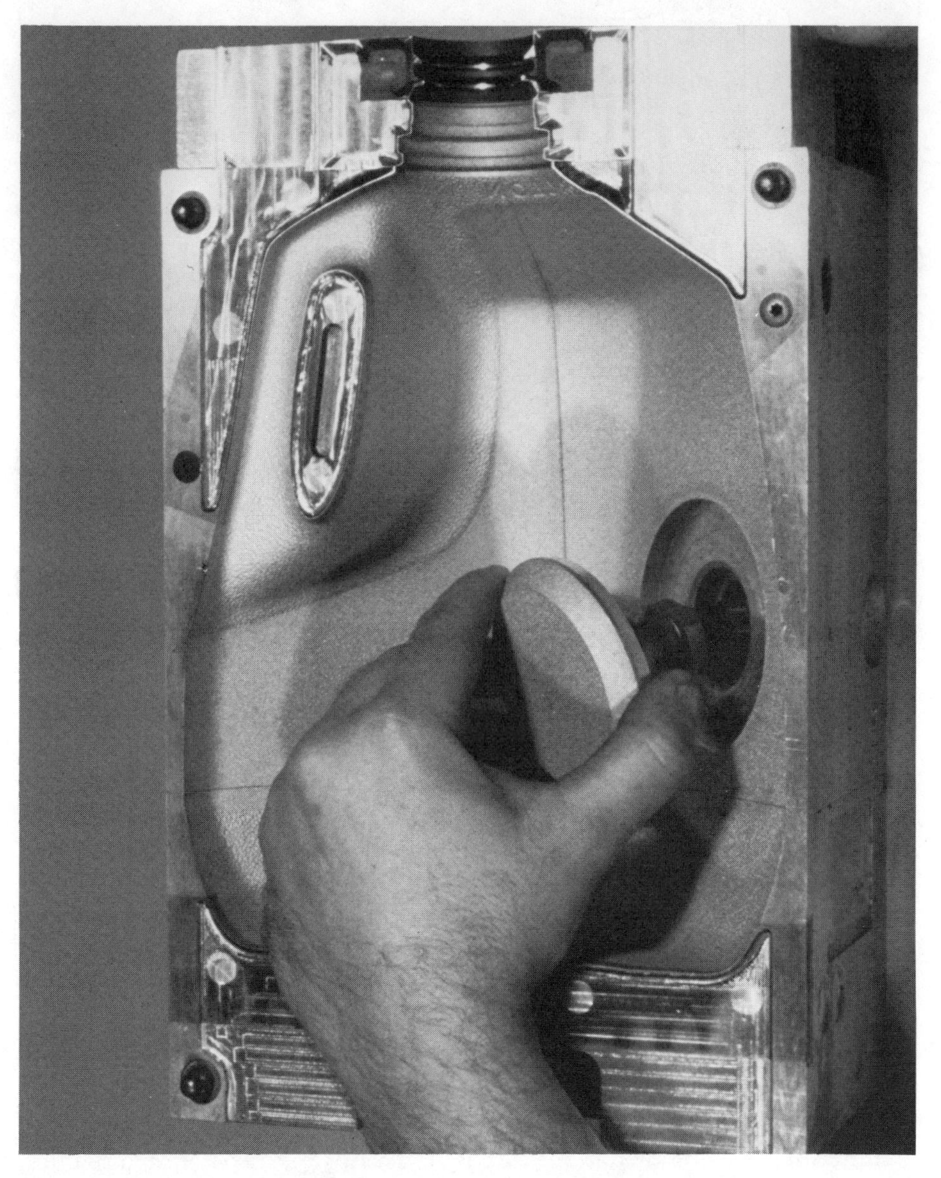

Figure 21-11 Interchangeable inserts can be used to adjust container volume. Courtesy of Johnson Controls, Plastics Machinery Division.

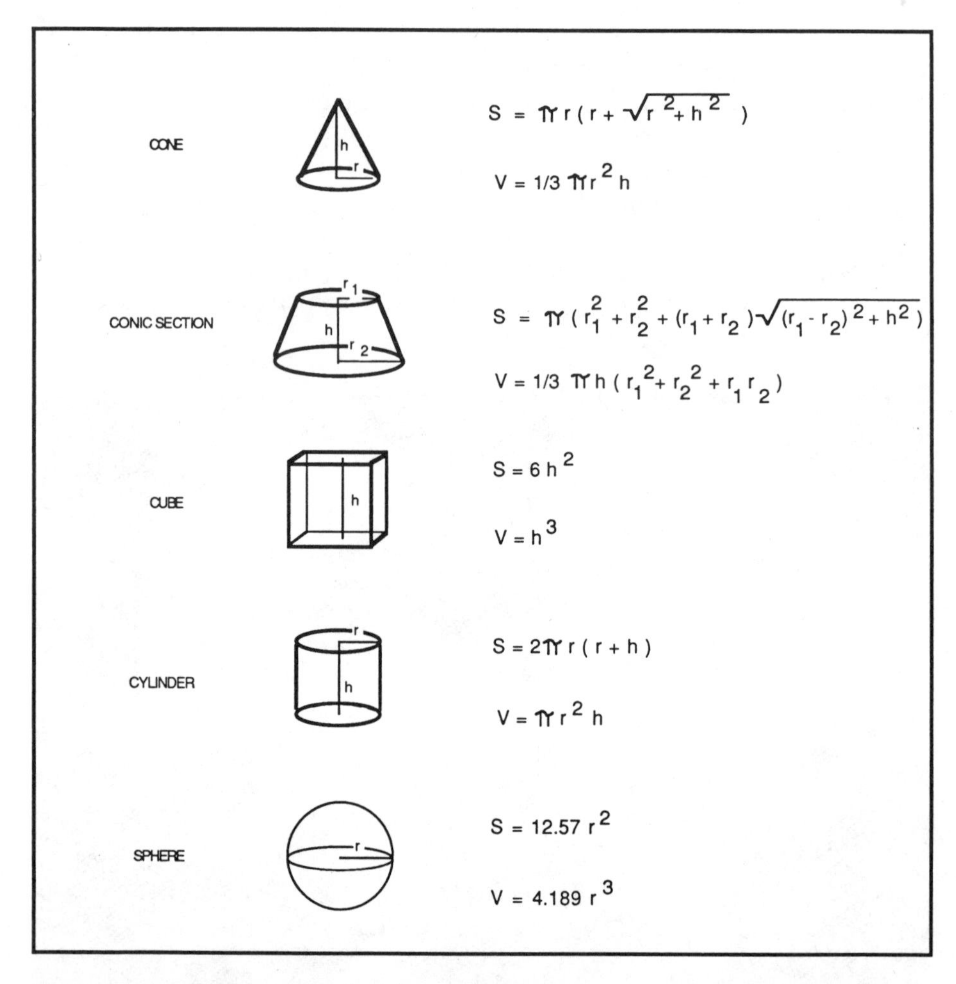

Figure 21-12 Relationship of surface area, *S*, to volume, *V*, in common geometric solids. H = height; r = radius.

density of water are available in most good chemistry, physics, or engineering handbooks, such as the *Handbook of Chemistry and Physics,* 58th ed., CRC Press Inc., West Palm Beach, FL, 1978.

After the mold has been built, slight adjustments must usually be made if the volume is desired to be exact. When adjusting the volume, it is easy to reduce it by grinding the parting surfaces of the mold. Increasing the volume requires a more extensive and expensive machining process.

Another approach to volume adjustment is seen in the dairy industry. It is important for sales purposes that the container look full, and also that it contain exactly one gallon. Here, the containers are produced with inserts in the side walls. The depth of these inserts can be changed to adjust the volume slightly, as shown in Figure 21-11. The inserts also help add rigidity to the sides of the container. This approach can be helpful when reduced weights are attempted in a mold which has already been in production at a heavier weight, since the reduced volume of material in the bottle wall translates into higher volume in the container.

It is desirable for economic reasons to maximize the container volume and thickness for a given weight of raw material. The relationship of surface area to volume for several different geometrical shapes is shown in Figure 21-12.

The amount of space a container will occupy in a box or on a shelf should be considered. Larger containers are designed with the dimensions of standard pallets or truck beds in mind. Square shapes can be packed more efficiently than round shapes. Novel interlocking shapes can also lead to better packing efficiency in the shipping container.

One should remember, however, that style and labeling considerations often play more of a role in container design than pure volume/shape efficiency since not many containers are spherical. This is due to the need for the container to be distinctive and stylish to attract the attention of the consumer.

Necks, Spouts, and Other Openings

Each part must be designed with an opening that is utilized to blow it, with only a few exceptions. Often the opening is utilized as a neck or spout. In the neck section, it is possible to control the thickness to a greater degree than in the rest of the part through careful matching of the blow pin and neck cavity dimensions. This area can be treated as more of a compression

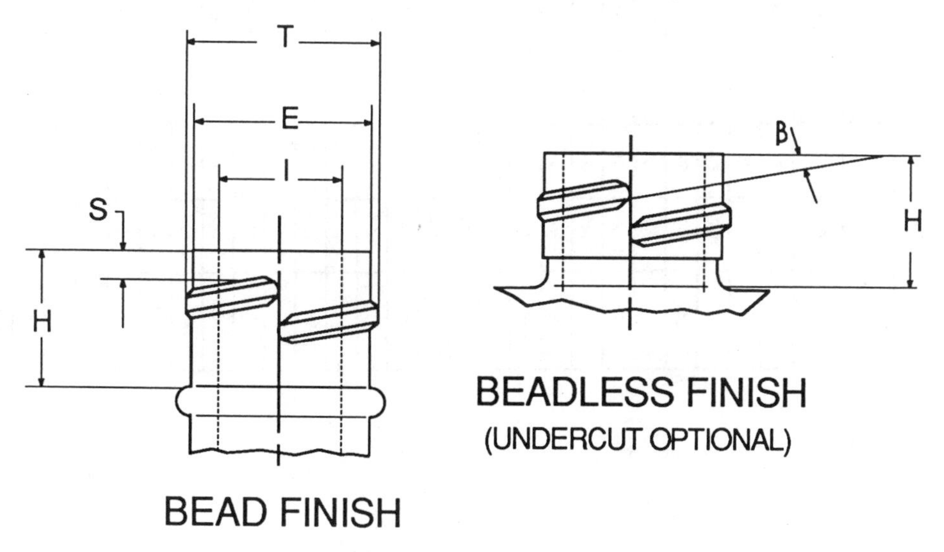

Figure 21-13 Standard dimensions for plastic bottle neck finishes. Courtesy of Armin Schwaighofer.

Table 21-2 Selected recommended voluntary neck finish specifications of the plastic bottle institute

SP - 400 FINISH

mm	T 6/		E 6/		H 2/		S		I 4&5/	HELIX ANGLE β	CUTTER DIA.	THD'S. PER INCH
	MAX.	MIN.	MAX.	MIN.	MAX.	MIN.	MAX.	MIN.	MIN.			
18	.704	.688	.620	.604	.386	.356	.052	.022	.325	3°30'	.375	8
20	.783	.767	.699	.683	.386	.356	.052	.022	.404	3°7'	.375	8
22	.862	.846	.778	.762	.386	.356	.052	.022	.483	2°49'	.375	8
24	.940	.924	.856	.840	.415	.385	.061	.031	.516	2°34'	.375	8
28	1.088	1.068	.994	.974	.415	.385	.061	.031	.614	2°57'	.500	6
30	1.127	1.107	1.033	1.013	.418	.388	.061	.031	.653	2°51'	.500	6
33	1.265	1.241	1.171	1.147	.418	.388	.061	.031	.791	2°31'	.500	6
35	1.364	1.340	1.270	1.246	.418	.388	.061	.031	.875	2°21'	.500	6
38	1.476	1.452	1.382	1.358	.418	.388	.061	.031	.987	2°9'	.500	6
40	1.580	1.550	1.486	1.456	.418	.388	.061	.031	1.091	2°0'	.500	6
43	1.654	1.624	1.560	1.530	.418	.388	.061	.031	1.165	1°55'	.500	6
45	1.740	1.710	1.646	1.616	.418	.388	.061	.031	1.251	1°49'	.500	6
48	1.870	1.840	1.776	1.746	.418	.388	.061	.031	1.381	1°41'	.500	6
51	1.968	1.933	1.874	1.839	.423	.393	.061	.031	1.479	1°36'	.500	6
53	2.067	2.032	1.973	1.938	.423	.393	.061	.031	1.578	1°31'	.500	6
58	2.224	2.189	2.130	2.095	.423	.393	.061	.031	1.735	1°25'	.500	6
60	2.342	2.307	2.248	2.213	.423	.393	.061	.031	1.853	1°20'	.500	6
63	2.461	2.426	2.367	2.332	.423	.393	.061	.031	1.972	1°16'	.500	6
66	2.579	2.544	2.485	2.450	.423	.393	.061	.031	2.090	1°13'	.500	6
70	2.736	2.701	2.642	2.607	.423	.393	.061	.031	2.247	1°8'	.500	6
75	2.913	2.878	2.819	2.784	.423	.393	.061	.031	2.424	1°4'	.500	6
77	3.035	3.000	2.941	2.906	.502	.472	.075	.045	2.546	1°1'	.500	6
83	3.268	3.233	3.148	3.113	.502	.472	.075	.045	2.753	1°9'	.500	5
89	3.511	3.476	3.391	3.356	.550	.520	.075	.045	2.918	1°4'	.500	5
100	3.937	3.902	3.817	3.782	.612	.582	.075	.045	3.344	0°57'	.500	5
110	4.331	4.296	4.211	4.176	.612	.582	.075	.045	3.737	0°51'	.500	5

SP - 415 FINISH

mm	T 6/		E 6/		H 1/		L	S		I 4&5/	W 3/	HELIX ANGLE β	CUTTER DIA.	THD'S PER INCH
	MAX.	MIN.	MAX.	MIN.	MAX.	MIN.	MIN.	MAX.	MIN.	MIN.	MAX.			
13	.514	.502	.454	.442	.467	.437	.306	.052	.022	.218	.045	3°11'	.375	12
15	.581	.569	.521	.509	.572	.542	.348	.052	.022	.258	.045	2°48'	.375	12
18	.704	.688	.620	.604	.632	.602	.429	.052	.022	.325	.084	3°30'	.375	8
20	.783	.767	.699	.683	.757	.727	.456	.052	.022	.404	.084	3°7'	.375	8
22	.862	.846	.778	.762	.852	.822	.546	.052	.022	.483	.084	2°49'	.375	8
24	.940	.924	.856	.840	.972	.942	.561	.061	.031	.516	.084	2°34'	.375	8
28	1.088	1.068	.994	.994	1.097	1.067	.655	.061	.031	.614	.094	2°57'	.500	6

SP - 410 FINISH

mm	T 6/		E 6/		H 1/		L	S		I 4&5/	W 3/	HELIX ANGLE β	CUTTER DIA.	THD'S PER INCH
	MAX.	MIN.	MAX.	MIN.	MAX.	MIN.	MIN.	MAX.	MIN.	MIN.	MAX.			
18	.704	.688	.620	.604	.538	.508	.361	.052	.022	.325	.084	3°30'	.375	8
20	.783	.767	.699	.683	.569	.539	.361	.052	.022	.404	.084	3°7'	.375	8
22	.862	.846	.778	.762	.600	.570	.376	.052	.022	.483	.084	2°49'	.375	8
24	.940	.924	.856	.840	.661	.631	.437	.061	.031	.516	.084	2°34'	.375	8
28	1.088	1.068	.994	.974	.723	.693	.421	.061	.031	.614	.094	2°57'	.500	6

1. Dimension "H" is measured from the top of the finish to the point where diameter T, extended parallel to the centerline, intersects the top of the shoulder. 2. A MINIMUM OF 1½—full turns of thread shall be maintained. 3. Use of bead is optional. If bead is used, bead dia. and "L" MINIMUM must be maintained. 4. Hole dia. "I" to be measured through full length of finish unless otherwise specified. 5. Concentricity of "I" Min. with respect to diameters "T" and "E" is not included. "I" Min. is specified for filler tube only. 6. "T" and "E" dimensions are the average of two measurements taken 90° apart. The limits of ovality will be determined by the container supplier and container customer, as necessary. 7. All dimensions are in inches unless otherwise indicated. *To the best of our knowledge the information contained herein is accurate. However, The Society of the Plastics Industry, Inc., assumes no liability whatsoever for the accuracy or completeness of the information contained herein. Final determination of the suitability of any information or material for the use contemplated, the manner of use and whether there is any infringement of patents is the sole responsibility of the user.*

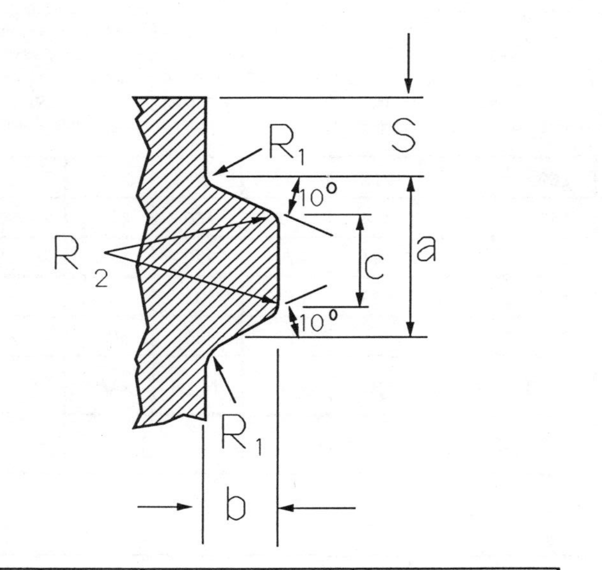

TPI	a	b	c	R1	R2
5	.120	.060	.051	.020	---
6	.094	.047	.040	.020	.020
8	.084	.042	.036	.020	.020
12	.045	.030	.011	.015	.005

"L" STYLE

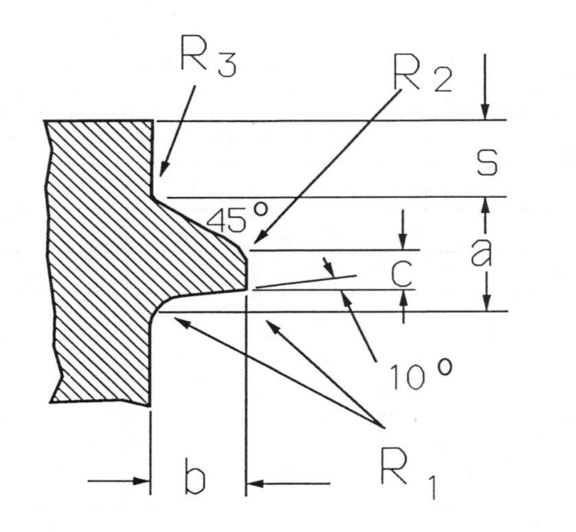

TPI	a	b	c	R1	R2	R3
5	.120	.060	.049	.010	.03 0	.030
6	.094	.047	.039	.010	.03 0	.030
8	.084	.042	.035	.010	.03 0	.030
12	.051	.030	.016	.010	.008	.020

"M" STYLE

Figure 21-14 Two standard thread profiles and suggested voluntary dimensions. Courtesy of Armin Schwaighofer.

molding. Some type of plug or attachment must be used to fill the hole if the item is not to be opened for use.

Since the overwhelming majority of blow-molded items are used in packaging, it is important to study threaded openings, which are closed with some type of cap or plug. Obviously, the container is only as good as its sealed opening [12]. The stripping torque for a bottle neck design is often run as a quality control test.

The important dimensions of a threaded neck finish are shown in Figure 21-13. Bottle finish (neck) sizes are specified in millimeters, type of thread design, and series number, which specifies the helix angle and number of turns of thread. The recommended voluntary guidelines for these dimensions, as per the Plastic Bottle Division of the Society of the Plastics Industry (SPI), are shown in Table 21-2. This is not a complete listing, which is obtainable from the Plastic Bottle Institute (PBI) (Technical Bulletin PBI2, revision 2). The two main thread shapes, L and M, are shown in Figure 21-14. The L style is similar to that found on glass bottles, while the M style is more typical for plastic containers. Matching closure dimensions are shown in Figure 21-15. It is important to match the closure thread types to the container closure type to form a good seal, and to have efficient capping on filling lines.

Other neck openings are designed for snap caps, or for spray attachments. Some typical designs for these types of neck configurations are shown in Figures 21-16 and 21-17.

Widemouthed containers are blown through a chamber type cavity that adjoins the neck. This is illustrated in Figure 21-18. The chamber is subsequently trimmed off with a rotating knife or saw blade. Twin cavity molds for these containers utilize a small section between the cavities, which are oriented mouth to mouth. A side-mounted blow needle is required in this case.

Bung openings on Department of Transportation (DOT) specification

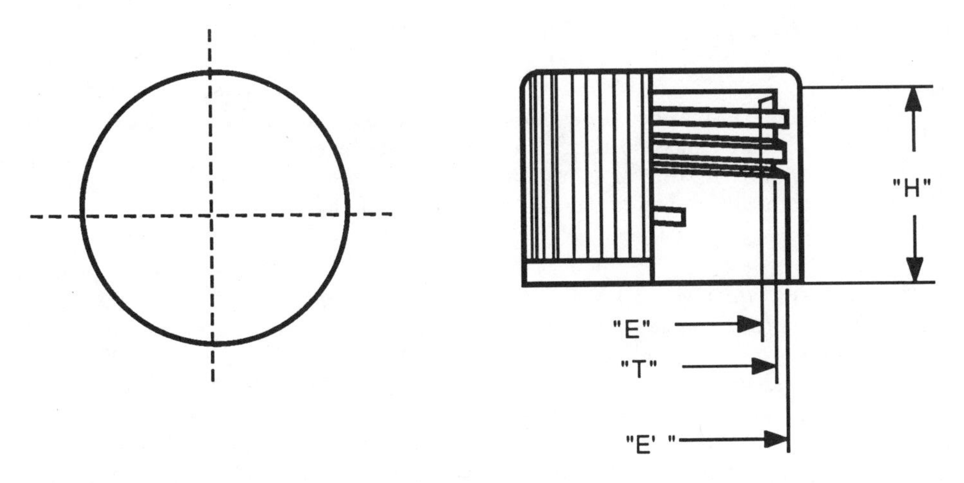

Figure 21-15 Key dimensions in screw cap design, M style thread. Courtesy of Armin Schwaighofer.

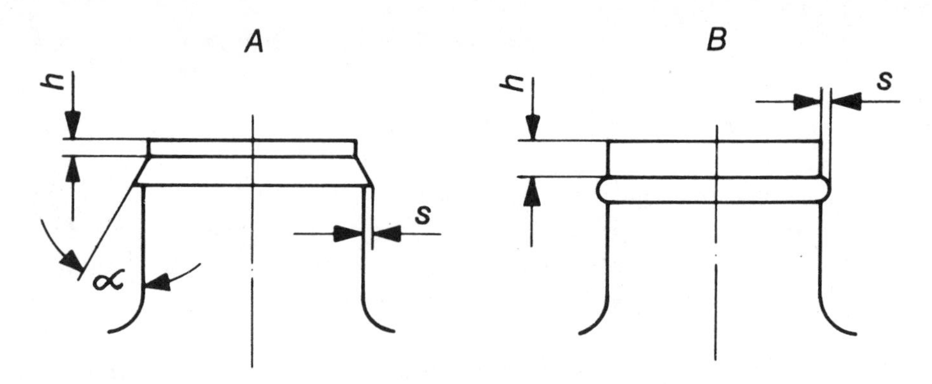

Figure 21-16 Pouring neck with click stop for snap closure. A. Design with conical torus. h = 1–3 mm, α = 30–45°, s = 1–2 mm. B. design with annular torus, dimensions as in A. Courtesy of Hoechst AG.

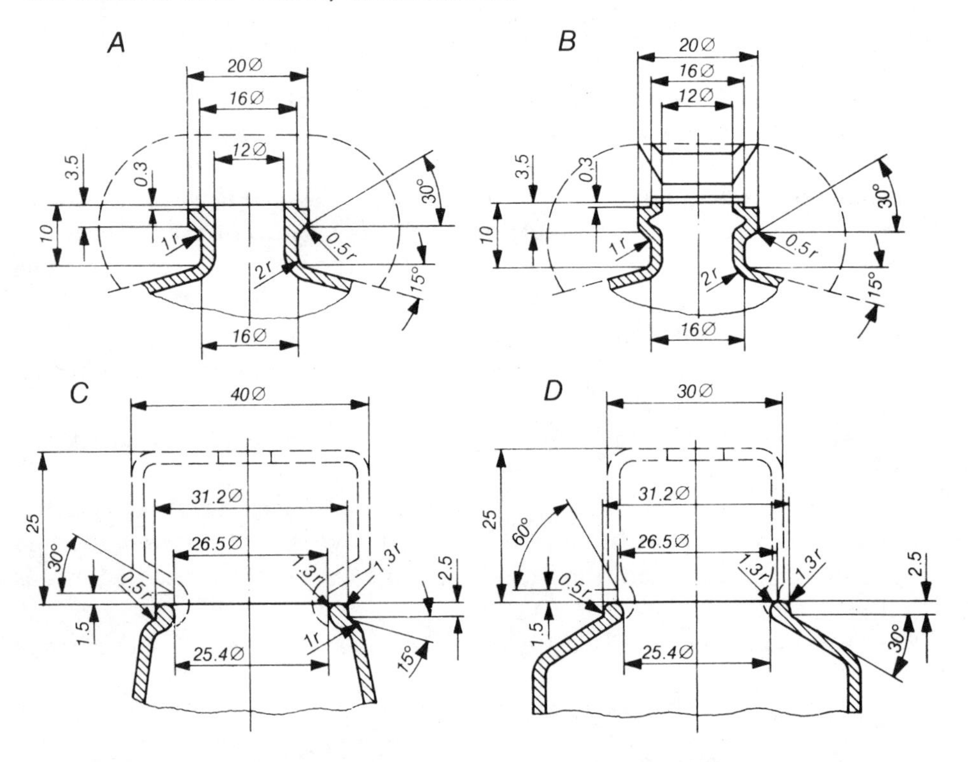

Figure 21-17 Design recommendations for the top of an aerosol container. A. Internally calibrated glass bottle type top. B. Externally calibrated glass bottle type top. C. Top for 1 in. lift valve on slimline aerosol container. D. Top of 1 in. lift valve on aerosol shouldered container. Courtesy of Hoechst AG.

Figure 21-18 *(facing page)* Mold design with "lost feedhead" blowing chamber for widemouthed containers. A. Neck with beveled edge to increase sealing surface, and centrally inserted blowing needle. *a* dome-shaped lost feedhead blowing chamber, *b* channel blowing needle, *c* bore for ejector pin, *d* cutting edge and pinch-off pocket, *e* knife insert to mark parting line. B. Mold construction with laterally inserted blowing needle. *a* preset breaking point with venting channel, *b* cutting edge and pinch-off pocket, *c* lost feedhead blowing chamber, *d* bore for ejector pin, *e* threaded groove for needle holder and channel for blowing needle. Courtesy of Hoechst AG.

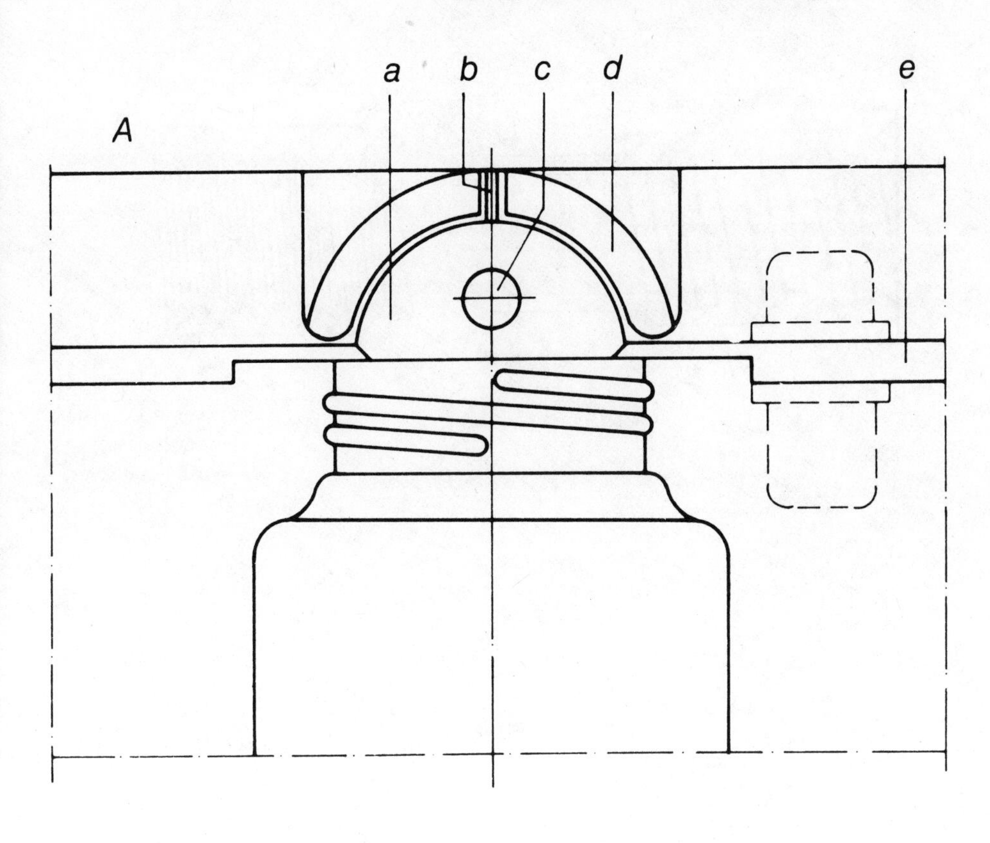
A
a
b
c
d
e

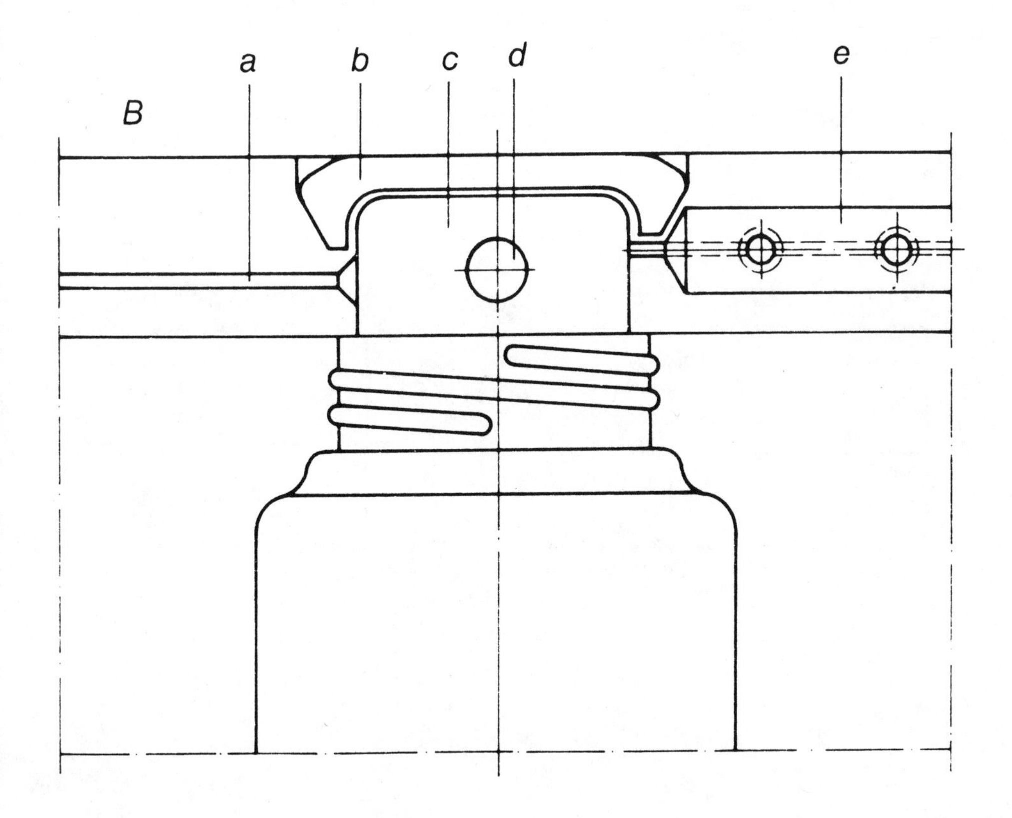
B
a
b
c
d
e

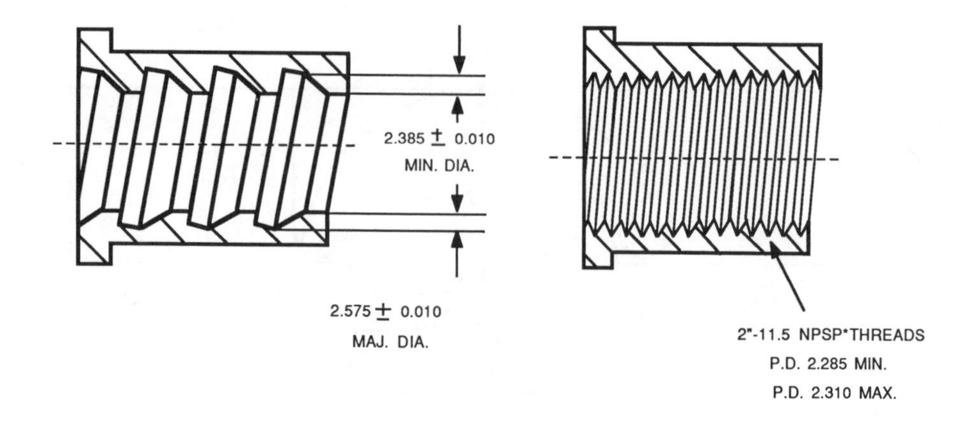

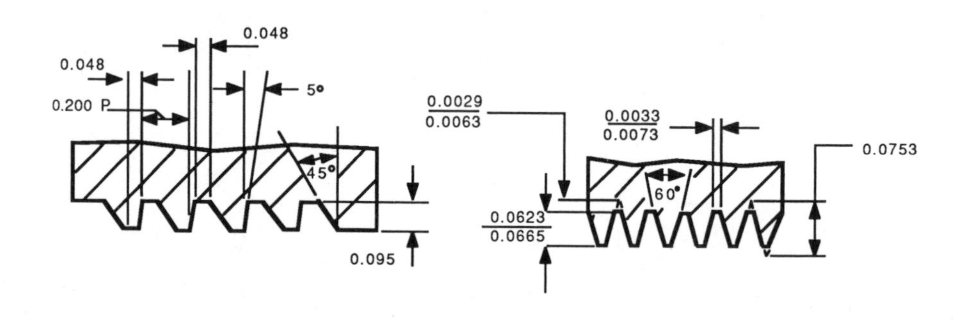

BUTTRESS INTERNAL THREADS

NOM. SIZE	THD'S PER INCH	DIA	
		MAJ	MINOR
2"	5	2.575 ± 0.010	2.385 ± 0.010

NPSP* INTERNAL THREADS

NOM SIZE	THD'S PER INCH	PITCH	PITCH DIA	
			MIN	MAX
2"	11.5	0.0869	2.285	2.310

* - AMERICAN NATIONAL PIPE THREADS STRAIGHT FOR PLASTIC DRUM CLOSURES AND OPENINGS

Figure 21-19 Suggested voluntary thread specification for plastic drums. Courtesy of Plastic Drum Institute, SPI.

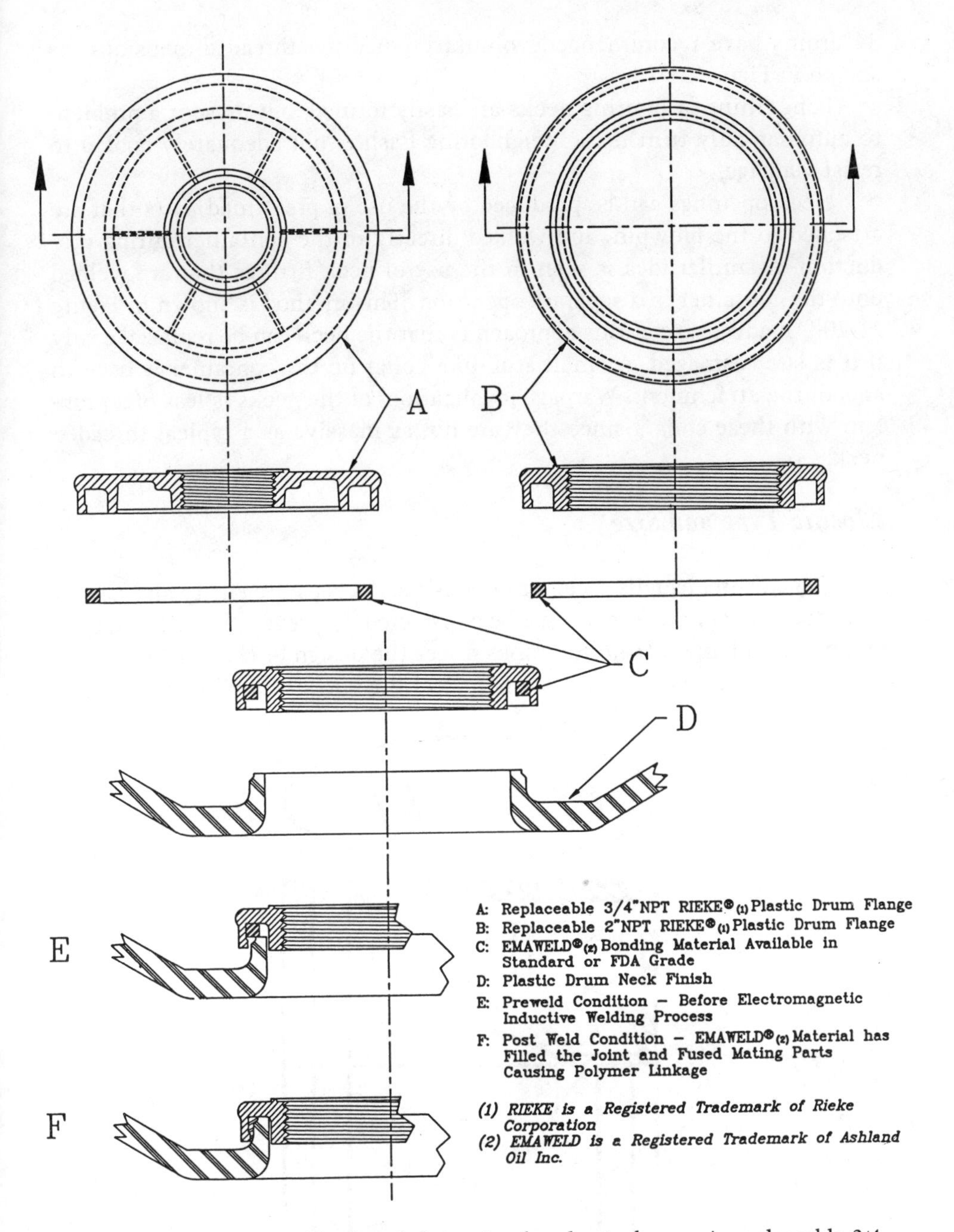

Figure 21-20 Replaceable threaded opening for plastic drums. A, replaceable 3/4 in. NPT plastic drum flange. B, replaceable 2 in. NPT plastic drum flange. C, bonding material, standard or FDA grade. D, plastic drum neck finish. E, preweld condition, before electromagnetic inductive welding process. F, postweld condition: bonding material has filled the joint and fused mating parts, causing polymer linkage. Courtesy of Rieke Corp. and Ashland Oil Inc., Emabond Systems.

34 drums have recommended voluntary guideline thread dimensions, as shown in Figure 21-19.

Long filling or pouring necks are easily formed, but present a problem to automatically trim if the neighboring flash is not adequately cooled to resist warpage.

Neck openings can be produced by the use of preformed parts that are attached to the blowpin, and welded directly to the container during production. A similar idea is seen in the use of neck fittings that are welded onto the container in a separate operation. This method is shown in Figure 21-20. An advantage of this approach is that the neck can be replaced easily if it is later damaged. A small stub-like collar on the container is used to anchor the attachment. Warpage (ovalization) of the necks is less of a problem with these collars since they are not as massive as a typical threaded neck.

Closure Type and Size

The closure, usually a cap or plug, is itself a separate entity and has its own design parameters that must be considered. However, the size and type of closure and any feature functions affect the design of the bottle.

The basic purpose of the cap is to seal the bottle and allow dispensing

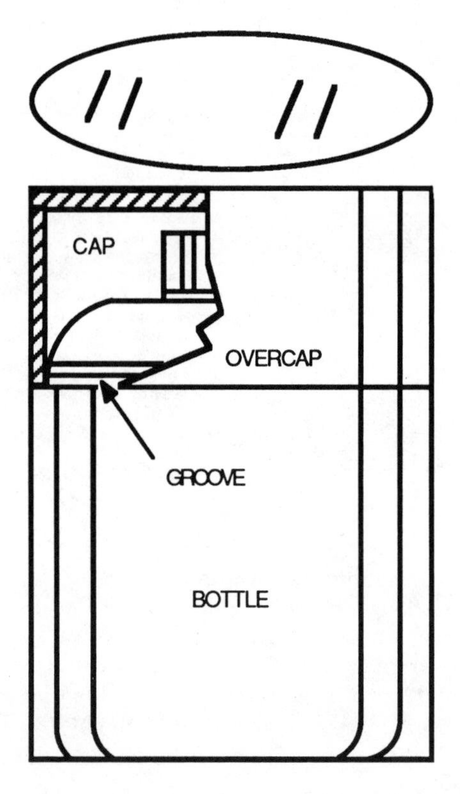

Figure 21-21 An overcap can improve the appearance of a container. Courtesy of Continental Plastic Containers.

of the contents. Many closures are multipurpose and function as safeguards against tampering or use by children. The cap may also fit into the overall image of the package. For example, the cap may consist of two functional elements, one of which is the threaded section and sealing surface common to all caps, while the others is an outer contour that blends into the bottle shape. This is depicted in Figure 21-21.

Cap orientation is desirable for many dispenser type closures. With screw caps, this orientation is subject to an angular tolerance, which in some cases can be substantial, as shown in Figure 21-22. This tolerance should be negotiated with the bottle/cap suppliers.

Closure size can be a marketing tool. Figure 21-23 shows how a large-diameter closure presents a more massive appearance. A tall closure can have a similar effect. Small-diameter, tall closures tend to be popular in cosmetic lines. Different sizes of caps for bottles of different volumes can eliminate the possibility of switching caps with different price markings.

The neck openings for containers of various sizes can be sized to provide higher filling speeds. Pint and quart containers may have a 28 mm diameter neck, while half- and the one-gallon containers might have a 33 or 38 mm diameter neck. Extremely viscous products may use larger necks to facilitate pouring.

Child-resistant closures vary in construction and operation. The press-and-turn types require a longer neck for extra vertical clearance since the outer shell of the closure must be depressed for removal. The squeeze-and-turn varieties need stops like those in Figure 21-24 to prevent the removal of the cap if it is not squeezed at the proper points. These stops will vary in design for the specific cap style.

Base Design

Due to warpage, container bases are seldom likely to be flat. The thicker weld seam shrinks more than the other sections of the base. This leads to

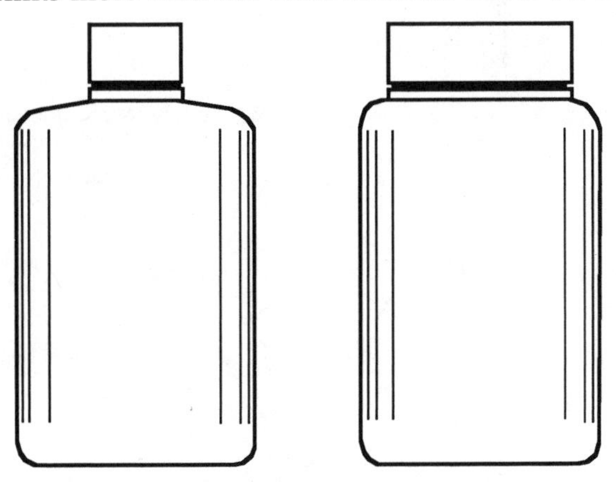

Figure 21-22 A larger cap creates a different visual impression. Courtesy of Continental Plastic Containers.

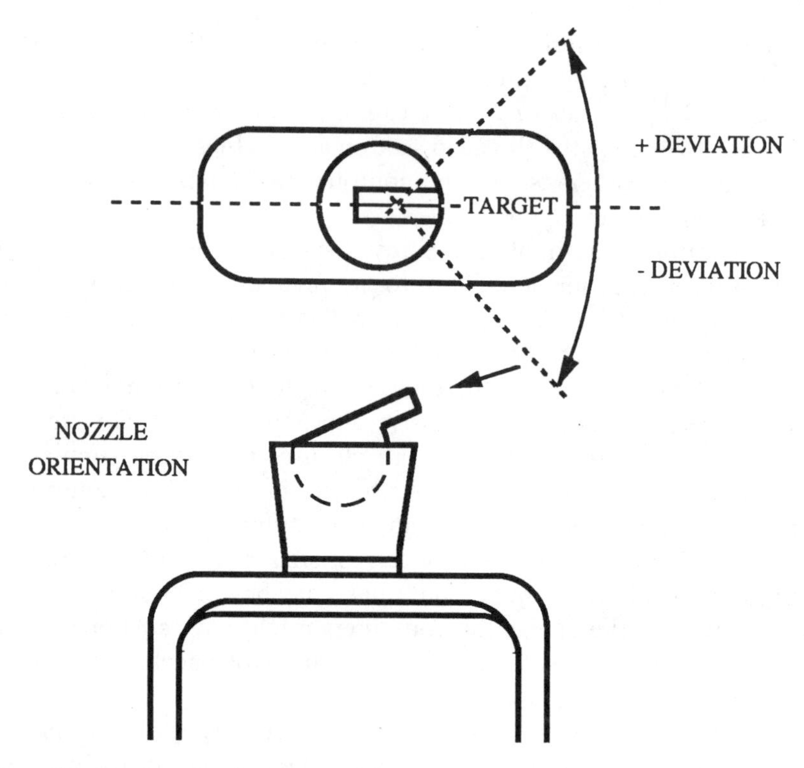

Figure 21-23 Angular deviation of dispenser screw caps is an important consideration. Courtesy of Continental Plastic Containers.

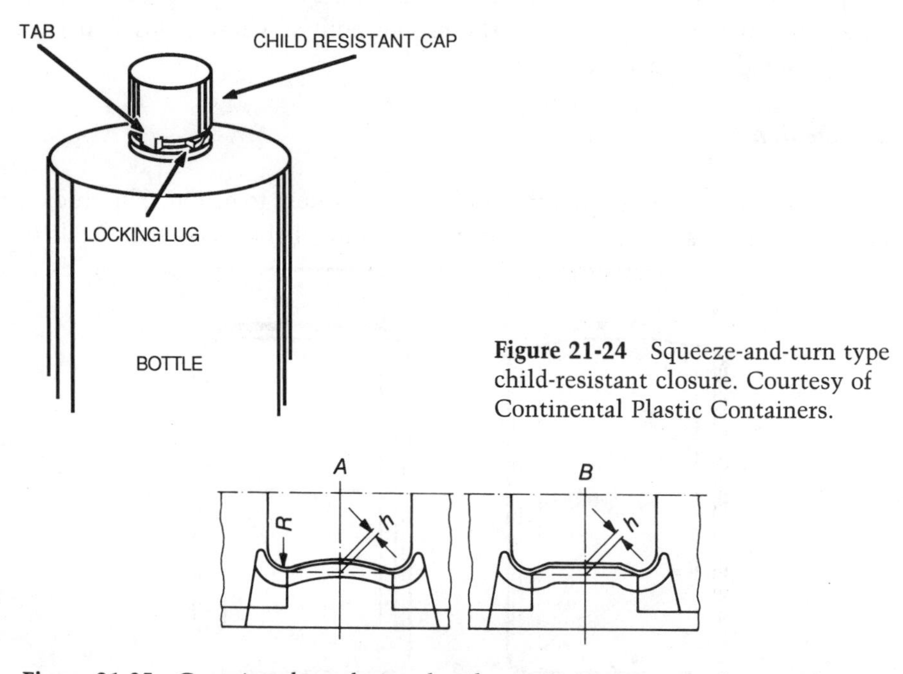

Figure 21-24 Squeeze-and-turn type child-resistant closure. Courtesy of Continental Plastic Containers.

Figure 21-25 Container base design details. A. Convex-concave base with rounded edges. B. Hip-roof shaped base for square and rectangular cross section with peripheral base edge. h = height of undercut. Courtesy of Hoechst AG.

the "rocker bottom" phenomenon. The typical solution to this problem is the pushup, a domed recess in the base, shown in Figure 21-25. This recess can be up to 0.060 in. in depth to avoid mold release difficulties. Greater depths can result in problems during demolding.

The rim around the pushup is also often recessed at the parting line so that the warpage associated with the seam does not prevent the bottle from standing flat. This also helps reduce stress on the seam, which can lead to early stress cracking failures.

On stretch blown PET containers for carbonated beverages, the container base must be spherical due to the internal pressure. Injection molded base cups are attached separately so that the container can stand upright. Another design, the petaloid base, provides a self-standing container with several egg-shaped feet on which it balances.

Attachments

Handle, valves, brackets, and other types of attachments can be produced in a separate fabrication process and attached to the molding by insertion into the mold between cycles. Baffles can even be secured inside the part by ingeniously using the blow pin or a separate attachment arm to

Figure 21-26 Insert-welded lifting bars can be seen on this 90-gallon automated waste cart. Courtesy of Zarn, Inc.

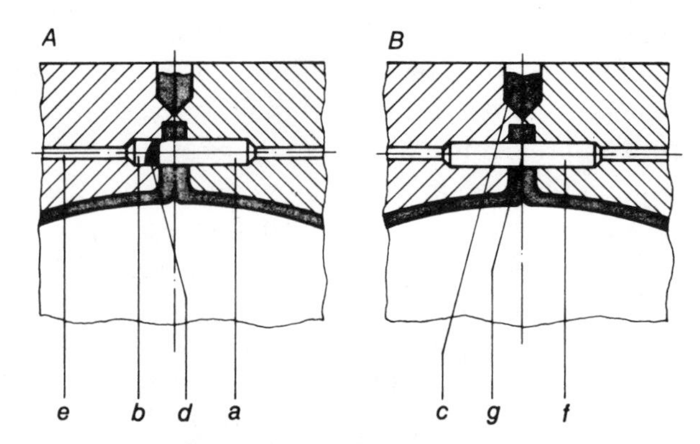

Figure 21-27 Mold design for eyelets. A. With punch bolts and blind holes. B. With two squeeze bolts. *a,* punch bolt; *b,* blind hole; *c,* pinch-off; *d,* pinchoff to be removed or torn off when the handle is inserted; *e,* duct for compressed air; *f,* squeeze bolt; *g,* pinch-off film to be torn off when the handle is inserted. Courtesy of Hoechst AG.

hold the insert as the parison is ejected. The arm is then retracted outside the parison before mold closing.

Figure 21-26 shows a new design of 90-gallon automated refuse cart. The lifting bars are blow molded in a separate operation. They are then inserted into the mold and welded to the cart body during the blow stage. The welds produced in this manner are strong enough to last for up to ten years in service.

Eyelets can be pressed into a part in a flange extension that can later be drilled or pressed out to provide an attachment site for a pinned or inserted fitting, as shown in Figure 21-27. Inserts with barbed feet can be pushed through the opening and anchored by the barbs.

Double-Wall Construction

A common design concept for blow molding is used in the production of packaging or casing for objects such as tools and appliances. These cases are formed from two halves, each half having two walls and being hollow between. One of the mold halves is like a male projection rather than the typical cavity form. These types of parts are called double-wall cases. An example is shown in Figure 21-28.

The geometry of the double-wall construction provides some benefits. Because the shell consists of two walls tacked together at their edges, greater stiffness is achieved than if a single panel of twice the thickness were utilized. This is due to an increased moment of inertia in the direction of the bending moment. The corners are extremely stiff for the same reasons [13].

The design also has the advantage of providing an inherent cushioning effect. Impacts striking the outer shell of the cases are transmitted to the

Figure 21-28 Blow molded double-wall cases provide rugged protection for valuable items. Courtesy of Phillips Chemical Company.

object inside indirectly by a spring action of the case exterior. A portion of the shock is thus reduced.

The hollow interior of the case provides a space that can be foamed for greater stiffness or insulation. The double-wall design provides a case that resists corrosion and is highly impact resistant. The two halves of the case can be formed in the same operation. A hinge can be molded between them, or a metal hinge pin inserted later.

SPECIAL CONSIDERATIONS FOR BOTTLES

The most important structural or mechanical considerations in a bottle include:

- Vertical strength
- Wall thickness uniformity
- Highlight definition
- Pushup strength
- Label considerations
- Rigidity
- Shape
- Hot fill capability

A bottle is subjected to various operations before it reaches the consumer. Among these is the filling operation. One filling method involves a spring-

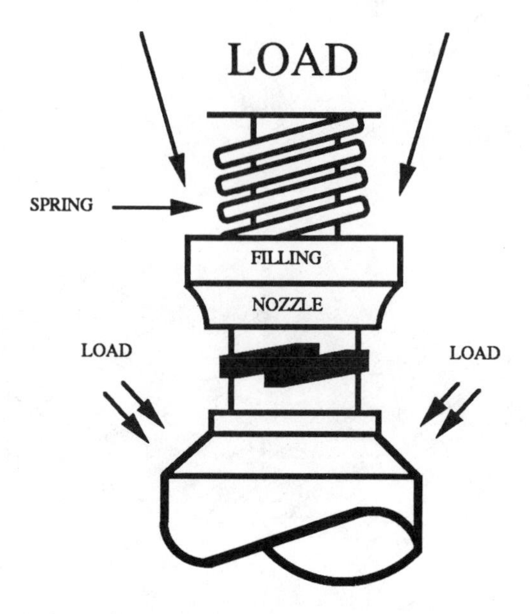

Figure 21-29 Loads from filling operations require bottle stiffness. Courtesy of Continental Plastic Containers.

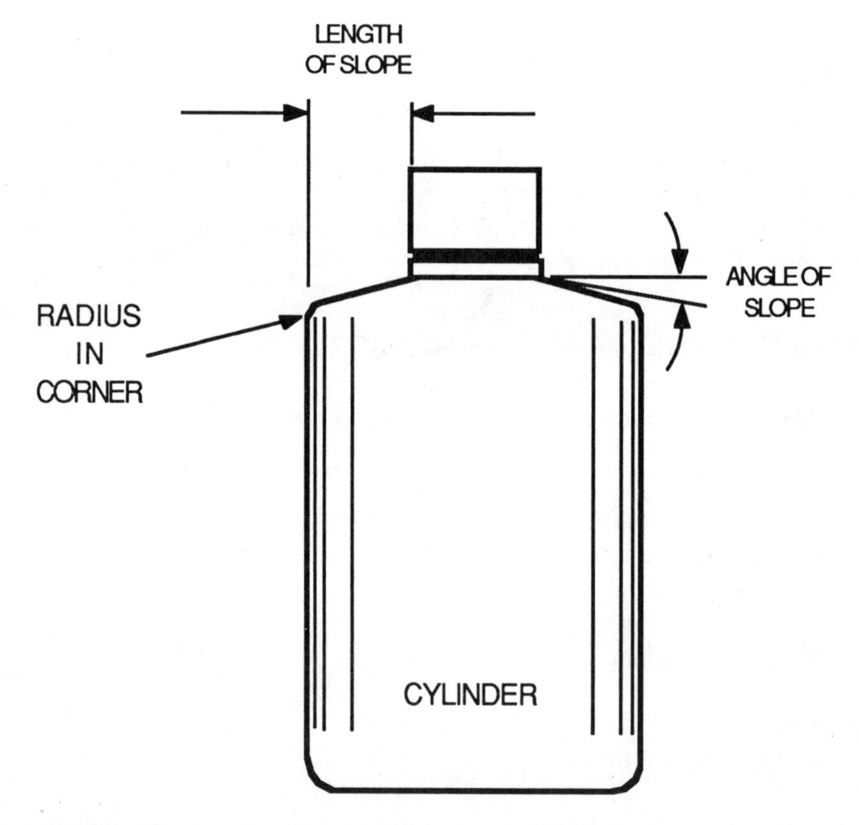

Figure 21-30 The shoulder design is critical to top-load resistance. Courtesy of Continental Plastic Containers.

loaded nozzle as shown in Figure 21-29. The nozzle is actuated by the neck of the bottle as it is extended into the bottle. The load in this operation can exceed 25 pounds.

On a cylindrical bottle with a sloped shoulder, the angle of the slope and the length of the slope are prime considerations. If this shoulder is rather flat, it may collapse. Bottles with a shoulder slope angle of 12 degrees, and length of 1/2 in., or an angle of 30 degrees and a 2 in. length should be considered adequate for filling line pressures. This is shown in Figure 21-30. Larger radii at the junction of the side wall and the slope decrease the hinge action and add to vertical strength.

If the bottle is expected to be subjected to heavy vertical loadings, horizontal corrugations and bellows should be avoided. These features may both reduce the loading capacity, and lead to stress concentrations which can likely lead to stress cracking. Examples of this are shown in Figures 21-31 and 21-32.

On F-style bottles where the top of the bottle is not sufficiently arched

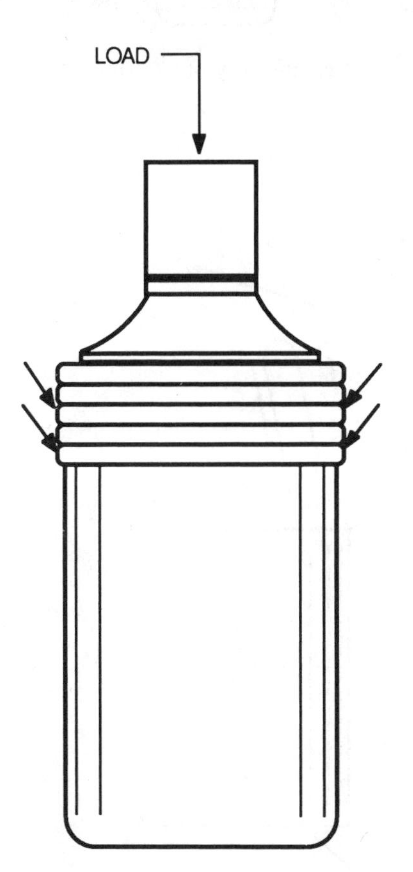

Figure 21-31 Corrugations can lead to reduced stacking capability. Courtesy of Continental Plastic Containers.

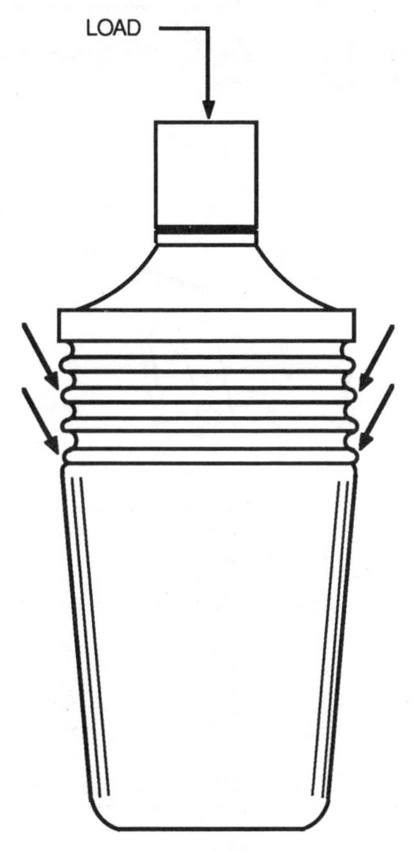

Figure 21-32 Bellows can lead to stress concentrations or stress cracking. Courtesy of Continental Plastic Containers.

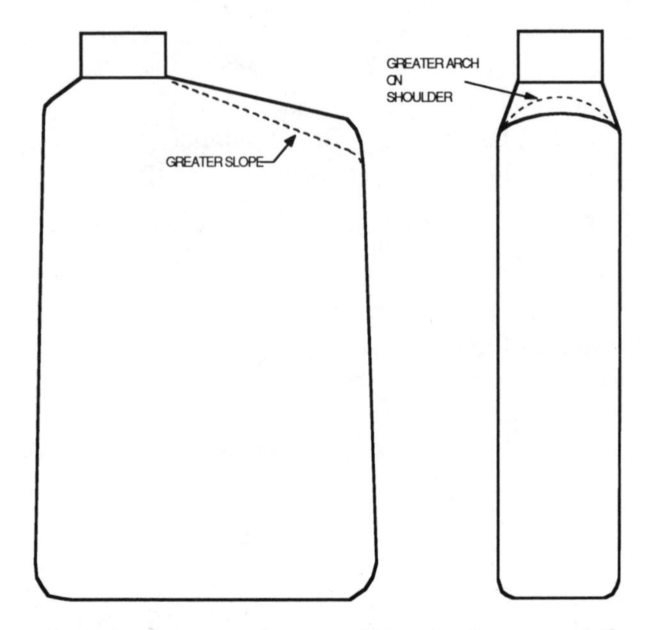

Figure 21-33 Key design features of an F-style bottle. Courtesy of Continental Plastic Containers.

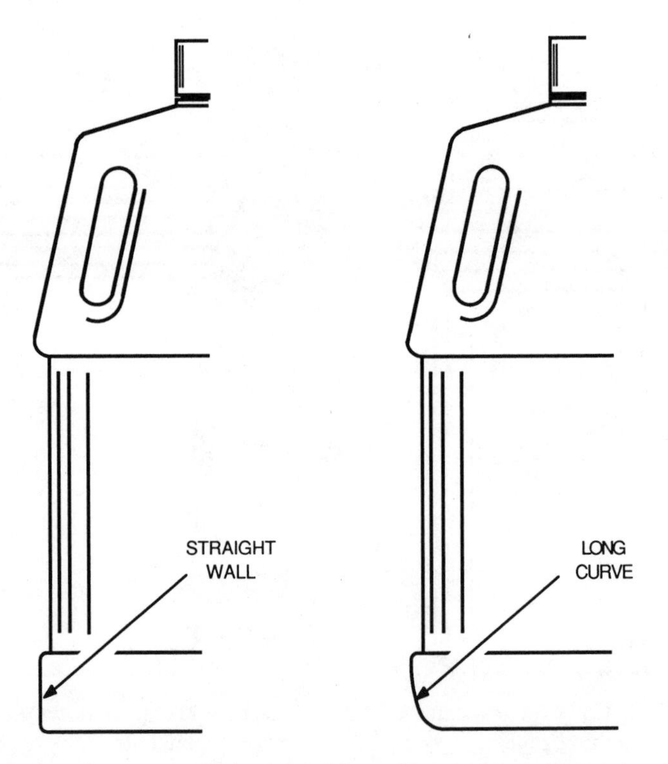

Figure 21-34 The base curve radius should not be too sharp. Courtesy of Continental Plastic Containers.

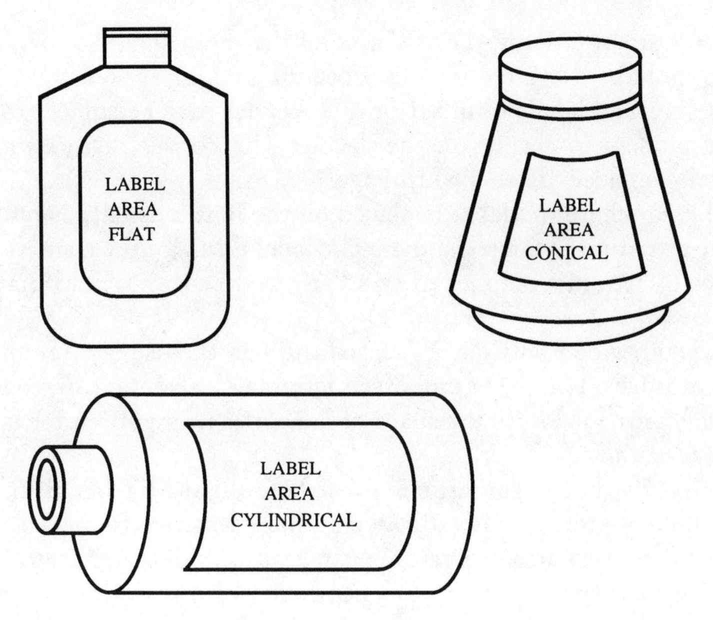

Figure 21-35 Typical label surfaces. Courtesy of Continental Plastic Containers.

or drafted, the finish will tend to tip backward when a load is applied. This weakness will interfere in filling, capping, and stacking. As in cylinder bottles, the slope angle helps determine the vertical strength in this area. This is seen in Figure 21-33.

The base chime design can contribute to poor vertical loading if the base corner radius is too sharp, as shown in Figure 21-34. An adequately rounded chime helps eliminate the extreme thinning that can lead to poor strength, as discussed previously. Uneven wall thickness distribution wastes time and material. The bottle should be designed in a way to avoid this.

Highlights are intersections of two plane surfaces that add definition to the bottle. Unfortunately these edges also trap air and cause thinning. Rounded edges reduce these problems.

Although labeling decisions and design may be a marketing function, the type and size of the label will affect the mechanisms of the bottle design. For paper labels, the label area should be free of compound curves (that is, curvature in more than one axis), as shown in Figure 21-35. The method of application may restrict the type of "framing" of the label, as with a depressed label area. Generally, adjustment to volume for regular glued paper labels is not required.

Heat transfer (decal) labels may require a margin for the web that carries the label. Due to the method of application (heat), the bottle is subject to additional shrinkage after label application. This must be considered in the mold design for prelabeled bottles. Thin wall spots are to be avoided due to the possibility of melting through the container.

In-mold labeling (IML) is often used to increase productivity. This tends to induce a slight bulging. The volume of the mold should be reduced if a standard mold is converted to IML processing.

Silk-screened labels or offset-printed bottles may be subject to heat in a drying process to accelerate the drying. The bottles in this method of labeling are subject to added shrinkage.

The area where the label is placed on the bottle usually is outlined to focus the attention of the consumer. The label break, or the transition edges should be treated in a manner to avoid any sharp edges or radii that are too small.

The rigidity of a bottle is increased through the use of circumferential grooves or ridges. Raised, or embossed letters can also increase the rigidity of the label area. Of course the shape of the bottle is important for the image it conveys to the customer.

Finally, the bottle sometimes must be designed for hot filling. Temperatures encountered in hot fill operations can range from 100 to 190°F (38–88°C). The higher temperatures can lead to bulging and wrinkling of labels. Hot decal prelabeled bottles can help avoid this problem. Paneling can be encountered if the caps are not vented. Stacking force is also reduced until the bottle has had an opportunity to cool. Labeling is discussed in detail in chapter 11.

SHAPE-STIFFNESS RELATIONSHIPS

One of the difficulties when designing with plastics is the lack of an accurate method for determining the long-term behavior of the object in a load-bearing condition. Most applications in current use have been developed on a trial and error basis because of this. Much theoretical knowledge is available regarding the complex time-dependent property behavior of plastics. What is missing, however, are detailed basic data on most grades of resins. This is because of the infinite range of time frames and loading rates necessary to fully define the values. An additional hindrance is the fact that all traditional design theory is based on elastic materials [2].

Plastics have much lower Young's moduli than competing materials (e.g., glass, paper, metal), and generally lose strength rapidly as temperature increases. This further complicates design in load-bearing applications. Whenever possible, it is best to use previously designed parts as the basis for new ones.

Naturally, the greater the wall thickness of a part, the higher the impact strength and stiffness will be. For a container of a given weight and volume, the shape that has the minimum surface area will provide the maximum thickness. This condition occurs in a cylindrical container when the height is equal to the diameter. Some other ideas to keep in mind are [1]:

- Convex containers resist buckling better than purely cylindrical ones.
- Rectangular container corner areas support most of their load, so this is where selective thickness increases are most effective.

- Studs, ribs, or other means of locating stacked containers in a perfectly vertical orientation eliminate load shifting and buckling.

The relationship of similar containers of different weights has been found to be approximately

$$\frac{F_1}{F_2} = \frac{(E_1 W_1)^2}{(E_2 W_2)^2}$$

where:

F = stacking force

E = Young's modulus of the as-molded material

W = the weight of the container

When the material in each container is the same, the relationship of the stacking forces is simply the ratio of the squares of the container weights [14].

There exist, however, options that can increase the stiffness of an object purely through the use of geometric elements in the structure of the object. Various types of surface features such as waists, reinforcing bands, grooves, and beads, shown in Figure 21-36 are used. These features of an object improve the stiffness and flexure by increasing the moment of inertia in relation to the expected orientation of the bending force. The important thing here is that in blow molding, the areas are usually variably thick in a pattern that is a function of the depth of the wall. If a pattern that creates localized sections of extreme thinning is used, the purpose of the surface feature can be defeated.

Whenever bands are present around the circumference of the container, they should be well rounded and not too deep or raised. They should not be too close to the shoulder or base of the container, to avoid stress concentrations or poor top loading capacity [15].

A highly successful approach to stiffness in blow molding is to use the parison in a manner that makes a flat, thin section with a double wall. The corners of double-walled parts are very strong due to the section modulus of the corner and to the higher thickness of the typical pinchoff seam weld relative to the adjacent wall sections.

The double walls can be made to support each other for more stiffness. Projections on either mold half are used to mold buttress-shaped projections. If the projection is deep enough into the cavity, the two sides actually touch and weld together.

When designing squeeze bottles, the stiffness of the selected resin determines the shape the bottle will be. HDPE, PP, and other similarly stiff materials must use rectangular or oval cross sections, while LDPE and other softer materials are suitable for round cross sections. This is because the dents caused when the bottle is squeezed pop out of the more resilient materials more easily.

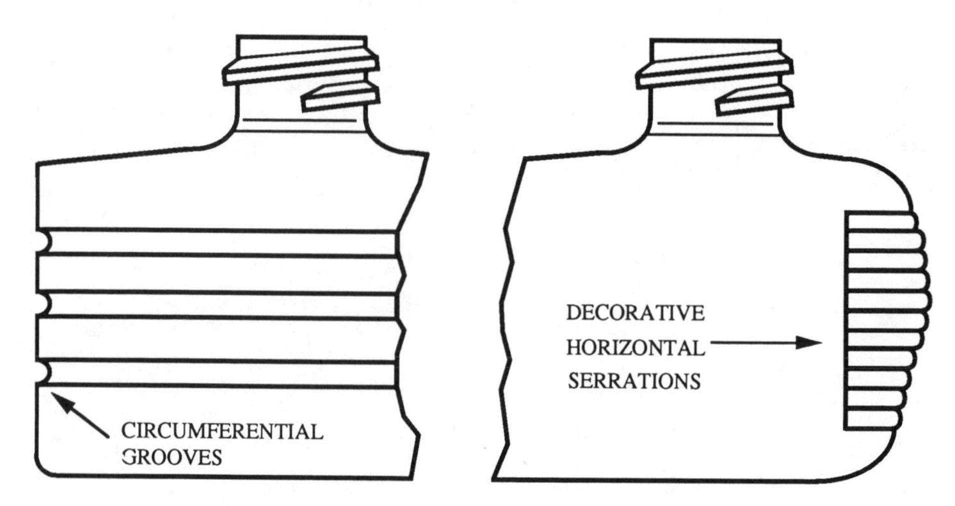

Figure 21-36 Surface features can increase side-load capability. Courtesy of Continental Plastic Containers.

Absorption of the fluid by the container walls can lead to a phenomenon due to the swelling of the material called "paneling." This causes the container wall to collapse inward. This is because the inner surface expands. Bottles can also collapse when they are hot filled, allowed to cool, and are not properly vented. This type of problem is not as noticeable in a rectangular or oval container as in a round container. Reinforcing bands and gradation of cross section can reduce paneling or collapse.

PREDICTING PERFORMANCE

A vital step in a product design project is to determine if the product is capable of performing the task for which it was designed, and what level of safety factor is available. This requires analysis, testing, or both.

The key area on which most analysis is focused is the mechanical load-bearing function for both tensile and compressive stresses. Valuable design equations available in standard texts on the mechanics of materials can often be applied, based on the product geometry. They can yield excellent predictions of short-term (dynamic) loading capabilities, as well as long-term (creep-related) approximations. The key factor here is to anticipate the extremes of temperature that can be encountered, especially high temperatures. Generous safety margins should be allowed in order to compensate for a variety of factors that can reduce the allowable load under extreme conditions. Off-center loadings can result in premature buckling and collapse in 55-gallon drums.

Top-load tests are usually either dynamic or static. A dynamic test can give a good instantaneous reading of the maximum load the container can bear in a short-term situation such as a filling operation. Longer term static load tests are useful in evaluating the effects of creep that might come to play in long-term storage, such as would be encountered in an ocean shipment.

Prediction of impact resistance requires a more empirical approach due to the varying modes of failure which can be encountered. Puncture can occur from either a sharp or a blunt object. The orientation of the impact can change.

Because impact conditions can vary so greatly, predictions of impact behavior can only be approximate. For that reason most standards are written in terms of some sort of empirical testing criterion, most often a water-filled drop test. One approach to drop testing is to determine the mean failure height for the container. Another approach is to determine the percentage of containers that fail when dropped from a given height. These procedures are standardized under ASTM D2463. They require approximately twenty containers each.

When testing larger items, such as 55-gallon drums, it is not practical to test so many containers. Instead, a pass/fail drop test is performed. This is performed at low temperatures, when the material is most brittle, to simulate use under the conditions of extreme winter. The tests are typically from the maximum height at which one could expect an accidental drop to occur. United States Department of Transportation specification 34 (DOT 34) currently requires a 4 ft, 0°F test for drums used in hazardous materials service.

Environmental stress crack resistance (ESCR) is an area that must be considered. Many blow-molded items are used as a container for materials that are stress cracking agents. When this is the case, correlations of shelf life in the field to values obtained in laboratory tests are invaluable in allowing estimation of expected performance. When the performance of a given container material/loading combination has been found to be acceptable, new materials or loadings can be compared with the standard material or loading, using such easily performed tests as ASTM D1693 (bent strip ESCR test), or D2561 (blown container ESCR test).

ESCR has been shown to follow an Arrhenius type relation to temperature [14]. That is, the life of a given container is related exponentially to the inverse of temperature. The relationship may be described as:

$$\text{Log } t = C_0 + C_1 \cdot \log P + \frac{C_2}{T}$$

Where t is time, P is internal pressure, T is temperature, and C_0, C_1, and C_2 are constants. Through selection of three different sets of temperature and pressure combinations, it is possible to evaluate the constants with as little as three different tests. These constants can then be used to predict shelf lives for actual field conditions. Of course the tests must be performed in similar containers produced from like materials, with the same loading.

In containers that are expected to perform a significant amount of service outdoors, resistance to the effects of ultraviolet radiation is a concern. Accelerated exposure conditions can be employed to allow determination of the suitability of the pigment-stabilizer combination. Tensile or impact properties can be obtained for both exposed and unexposed test samples, and compared to establish the effects of the exposure.

ACKNOWLEDGMENTS

This project would not have been possible without the Hoechst blow molding handbook, which provided the foundation upon which this chapter was written, and the men who were responsible for it: Eckard Raddatz and Christian Gondro, Hoechst AG, Frankfurt, West Germany.

John Szajna, Continental Can Co., contributed the sections titled Closure Type and Size and Special Considerations for Bottles in their entirety.

The authors give special thanks to:

H. R. Glover, Chevron Chemicals, Orange, TX

Dieter Wunderlich, FGH Systems, Denville, NJ

Steve Stretch

Christopher Phillips, Johnson Controls, Manchester, MI

Philip F. Hartung, Hartung Product Design Service, E. Greenwich, RI

Don Peters, Phillips Chemical Company, Bartlesville, OK.

REFERENCES

1. Gondro, C., and E. Raddatz. *Large Blow Molded Hollow Articles from High Molecular HDPE.* Hoechst AG.
2. "A Systematic Approach to Plastics Material Selection and Design." *Modern Plastics Encyclopedia.* New York: McGraw-Hill, 1986–1987. Pp. 400–402.
3. Powell, P. C. *Engineering with Polymers.* Chapman and Hall, 1983.
4. Junk, P. "Blow Molding of Technical Parts—Development Trends." *Proceedings, ANTEC.* Soc. Plastics Eng., 1988.
5. E. Raddatz. "Materials for Blow Molding of Packaging and Technical Articles." 6th Plastic Symposium, German Engineers' Society, Brazilian Group, 1987.
6. Levy, S., and H. DuBois. *Plastic Product Design Engineering Handbook.* New York: Van Nostrand Reinhold, 1977.
7. "Blow Molding Marlex." Phillips Chemical Co. V.2(2, 3, 5, 6), v.3(2, 3, 4), v.4(1, 4, 6) (March 1961–December 1963).
8. Sacharow, S., and R. C. Griffin, Jr. *Basic Guide to Plastics in Packaging.* Boston: Cahners Pub. Co., 1973.
9. Hartung, P. *Plastic Bottle Designing.* E. Greenwich, RI: Hartung Product Design Service, 1986.
10. Gut, I. H. "Formenberechnung fuer hohlkoerper aus kunststoffen." *Plastverarbeiter* 29(10) (1978): 553–555.
11. Szajna, J. L. "Traditional and CAD/CAM Plastic Bottle and Tool Design." *Proceedings, ANTEC.* Soc. Plastics Eng., 1985.
12. House, R. "Plastic Threaded Closures: Basic Specifications and How They Relate to Containers." *J. Packaging Technology* 1(4) (August 1987).
13. Rathgeber, J. "Blow Molding: The Evolution of an Imaginative Technology." *Proceedings, ANTEC.* Soc. Plastics Eng., 1987.
14. Herrman, F., H. Hofler, O. Hollricher, and G. Kienitz. "Qualitaetskriterien und qualitaetssicherung von pe-hohlkoerpern."*Kunststoffe* 73 (1982):2.
15. Szajna, S. L. "What Every Designer of Plastic Bottles Should Know About Structure." *Plastics Engineering* (March 1981): 83–87.

CHAPTER

22

Computer Container/Preform Design

TOD F. EBERLE

There are significant new market opportunities for reheat-stretch blow molded containers, beyond the traditional beverage applications. Such package sectors as food powders, dressings and sauces, edible oils, brine-packed vegetables, potable spirits, health and personal care products, and household chemicals all have experienced inroads made by oriented plastics containers. These have displaced long entrenched packing materials, such as metals or glass, but also have won out over certain nonoriented plastics packaging. Most reasons for switching to oriented plastics packaging revolve around cost as it relates to material savings, production capacity, and operating efficiency. But packaging decisions must ultimately relate to consumer acceptance of and preference for a new package design. Indeed, often the aesthetic or ergonomic aspects of the container are most important in the decision to go to a new package design.

It is quite common for package designs to be as unique to a branded product as the brand name. The original shape may have been created by shaping a model to fit a person's hand, or embellishing it with features to make it stand out on the store shelf. The drafted documentation for the mold cavity that manufactures this package may be incomplete or may not exist, because the actual cavities were traced from a sculptured, handcrafted master model. The cavities may have been carefully adjusted by hand until the final product met manufacturer's performance specifications.

When plastics customers examine the possibility of switching to the clear, oriented container, they are very thorough in evaluating its merits. In order to begin any evaluation, they must have samples, and these must be provided quickly, with short lead times, according to the performance specifications. The specifications may include final weight, minimum wall thickness, maximum height or diameter, overfill capacity, visual fill level, closure removal torque, column strength, gas permeability, elevated temperature creep, scuff resistance, vacuum paneling resistance, and many other factors. The quality of the prototype must be as good or better than the ultimate production samples, since it is the prototype that serves as the basis for the production investment decision. It becomes increasingly difficult to meet all performance, quality, and aesthetic criteria in shorter design cycles without the use of additional productivity-enhancing engineering tools. Computer-aided design and engineering can provide:

- Full documentation of complex contours
- Volumetric analysis without building models
- Material flow, rigidity, and heat or gas transfer analysis
- Interactive redesign to meet specifications with less prototyping
- Solid shaded images or models for marketing presentation and approval
- Databases that can be used for programming computer numerical control (CNC) machine tools, or to speed design through modular or parametric construction

Technology is emerging that can analytically eliminate the art in the design, documentation, and manufacturing of containers and their molds, and replace it with hard engineering. A proposed design can be quickly evaluated for producibility using reheat-strength blow molding manufacturing methods. This is not to say that designs produced by such methods cannot be artistic; rather, the techniques used are based on the advanced capabilities of computers, acting as extensions of the creative human mind. The advantages of using such tools are discussed in this chapter. Engineers and designers who take advantage of these emerging techniques will be able to create better designs, ones that can be predicted to meet performance specifications. Art and experience are not replaced, but augmented. With guesswork eliminated, the overall efficiency of design is often drastically improved. (Table 22-1).

The plastics container industry now operates in a highly competitive world market. The companies that will thrive in such a market will be those that offer exceptional service at a fair price to their customers. The United States and Western Europe, with their high wage scales and living standards, can best compete on the basis of superior technology, service, and quality. It is the purpose of this chapter to describe a means for achieving that competitive edge: CAD/CAE, computer-aided design and computer-aided engineering.

Table 22-1 Software for computer-aided engineering products installable on a variety of platforms

Company	*Product*	*Utility*
Autodesk, Inc.	AutoCad	PC-based CAD, low cost
Cadam, Inc.	CADAM	IBM-based CAD
CAMAX Systems, Inc.	CAMMAND	Sophisticated 3D CAD/CAM
ICAD, Inc.	ICAD	2½D CAD
Dassault Systems, Inc.	CATIA	IBM-based 3D surface CAD/CAM
PDA Engineering	PATRAN	Modeling, imaging, analysis
EMRC	NISA	Modeling, imaging, analysis

SYSTEMS AND METHODS

One of the greatest advancements in production engineering has been the advent of the analytical personal computer, which has brought CAD and CAE to the desktop (Figure 22-1). In the early 1980s investment in CAE technology often meant purchasing expensive mainframe and minicomputers, and hiring specialists to maintain them. Continuous improvements in processing power, speed, and the "friendliness" of the user interface now have put computers within the economic grasp of even the smallest shops. Systems are being created for the IBM PC-AT and compatible machines which, at an increasingly reasonable cost, are weaning designers from the manual drafting board forever.

A step up from the personal microcomputer is the minicomputer workstation. Such stations are most commonly based upon fast 32-bit microprocessors and the Unix operating system. They generally feature very high resolution 19-inch monitors with the capability to display thousands of colors simultaneously. They can function as general-purpose computers that not only perform the primary task, such as computer-aided drafting, but also design evaluation exercises. These can be purchased as a complete package, supported by a single organization, they can be gathered from a group of vendors, or they can be developed by the individual user to perform tasks specific or proprietary to an organization. Importantly, individual stations can be interconnected (networked) in order to share printing, plotting, library, and mass storage facilities. See Table 22-1 and the following list of manufacturers of stand-alone or networked workstations for computer-aided engineering:

Adra Systems Inc.
Apollo Computer, Inc.
Apple Computer
Applicon
Calma Co.
CimLinc, Inc.
Compaq
Computervision Corp.
Data General Corp.
Digital Equipment Corp.

Evans and Sutherland Computer
Hewlett-Packard Co.
IBM Corp.
Intergraph Corp.
McDonnell Douglas
NEC Information Systems, Inc.
Prime Computer, Inc.
Silicon Graphics, Inc.
Sun Microsystems, Inc.
Tektronix, Inc.
Zenith Information Systems

Large organizations with many ongoing engineering projects and the need to install dozens of workstations may decide to install mainframe environment programs, which can support very powerful software programs with exceptionally fast response. However, these large computer facilities are very expensive to acquire and to maintain, and can rapidly become overutilized. When a mainframe system becomes overutilized, the incremental cost of adding additional capacity can be very high, forcing an organization to make do with the present, perhaps obsolete, system. For this reason many organizations are choosing the networked workstation approach for their CAD resources. Each additional seat added can function autonomously and thus does not affect the productivity of the others.

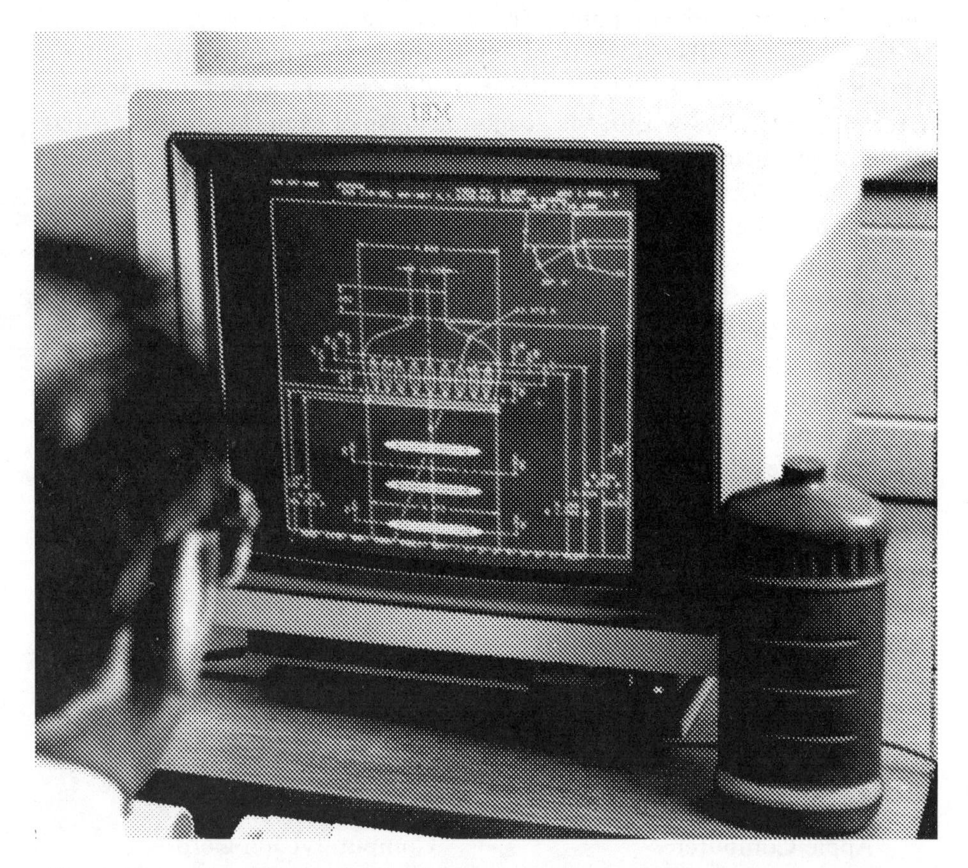

Figure 22-1 Cincinnati Milacron design specialist analyzing bottle parameters.

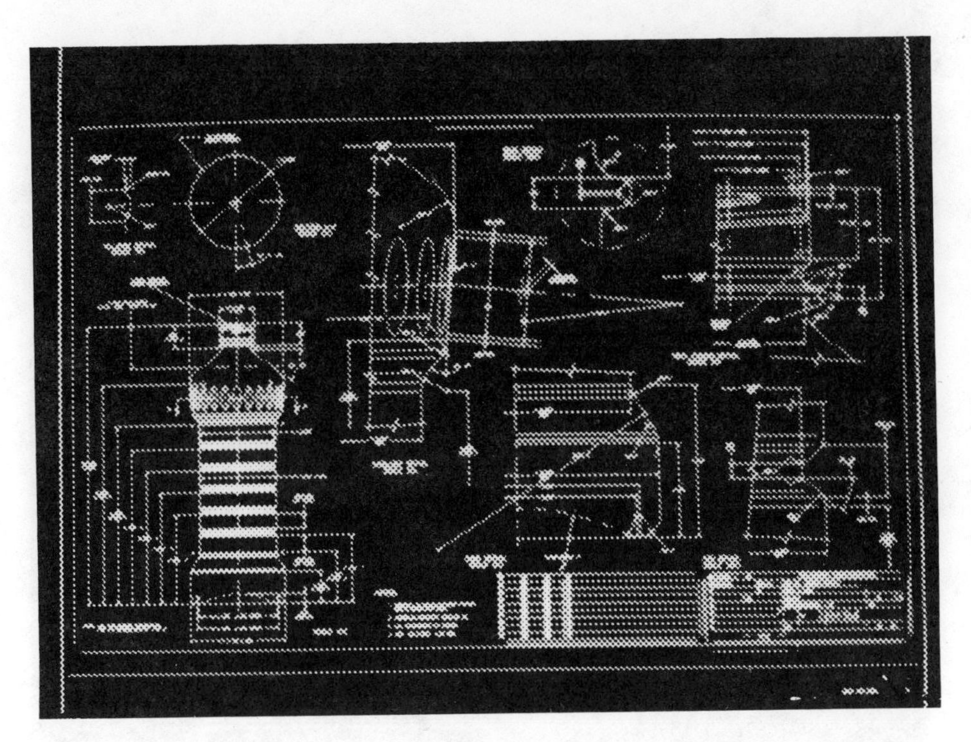

Figure 22-2 Detail drawing of blow mold cavity, generated using CADAM software, running on IBM Host-dependent Graphics Workstation.

Companies doing good business with experienced engineers and all-manual design and drafting techniques may consider computers a superfluous expense. Only recently has the cost begun to drop within the reach of most companies. Even so, many organizations feel that they may continue to do just fine without the headaches of additional high tech equipment. Any organization is only as good as the skills and expertise of its people—a fleeting advantage subject to job changes and mortality—unless, of course, some of those skills can be retained for others that follow.

Saving the accumulated experience of good engineers is a recent, and invaluable, characteristic of computers and their ability to store and manage vast amounts of information. Any competent draftsman can develop and fully document a drawing for a container that can be converted into a mold, and eventually to some finished part (Figure 22–2). This has been done without computers for decades, and with just conventional drafting tools and hand calculators for fifteen years. But that process typically relied upon experience, upon approximation, upon trial and error.

There are computer systems available that solely perform computer-aided drafting. Changes to a design are facilitated without erasure, and the quality of the documents is certainly improved, but no additional integrated capability for making engineering decisions is provided. This is the status of computer assistance in many companies today.

A step up from drafting is computer-aided design (CAD). This is im-

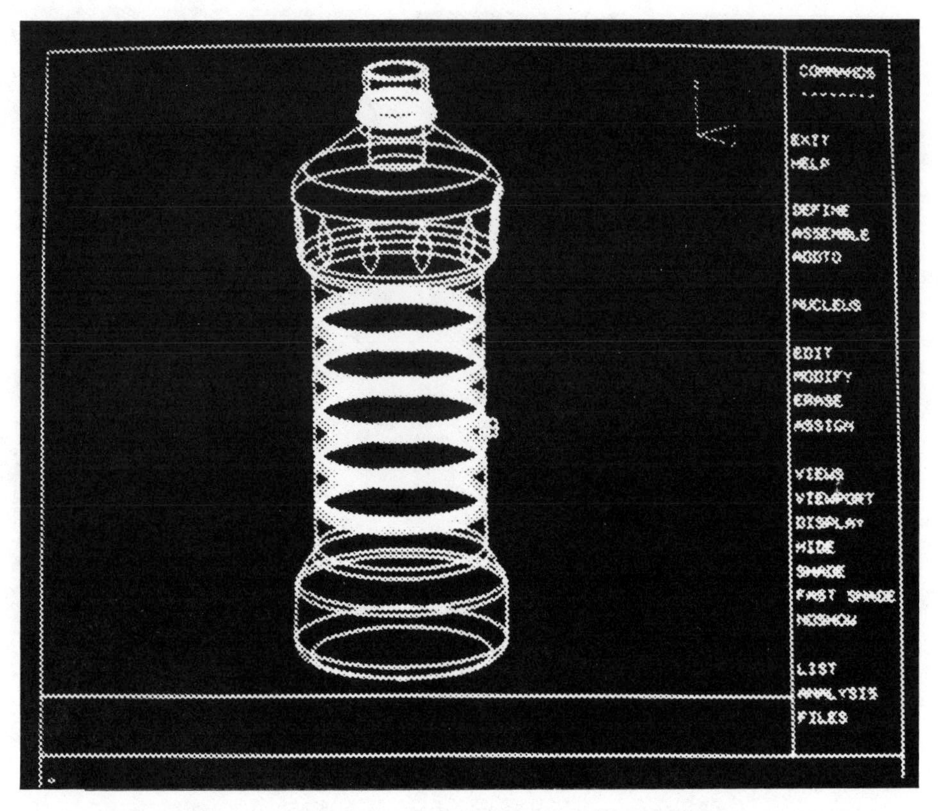

Figure 22-3 Wire frame drawing for sectional property analysis.

plemented to a limited extent in many available systems. In the simplest case, there is the ability to analyze the geometric characteristics of a drawing, for example sectional areas and inertial characteristics, from which further, manual calculations may reveal the volume or some other feature of the eventual solid (Figure 22-3).

The highest level is computer-aided engineering (CAE). At this level computer resources can provide real answers and provide solutions for informed decisions. Often, separate analysis software packages act upon a single geometrical database to provide truly integrated modeling of the intended characteristics of the ultimate design. Components of a computer-aided engineering system may include:

- Parametric design sessions to build figures from libraries of standard components
- Solid modeling of an object for imaging and physical properties analysis

Physical properties analysis may include:

- Orientation ratios optimization
- Shrinkage, weight, and final volume prediction
- Injection mold fill optimization
- Process-induced stress minimization
- Pressurization resistance and column strength determination
- Expert system database searches, decisions, and recommendations based on known phenomena or material behavior

Most of the true CAE applications software for the plastics industry is resident on mainframe or superminicomputers, and thus is less accessible to the smaller organizations that form the majority of companies participating in the blow molding industry as primary processors. However, small companies are working feverishly to develop programs that meet the need at a justifiable cost, for example, one-year return on capital employed.

If a company is in the business of designing reheat-stretch blow molded (RHB) containers, or is considering converting an existing package to an RHB-derived oriented container, what analytical resources will best assist in arriving at an informed decision?

For any given RHB container, the shape of the preform to be designed will be determined by the geometry of the container. Very often the desired closure has already been determined, so that the dimensions of the thread finish and those of the container will fully determine the basic lengths and diameters of the preform. This must be compared with what standard production machinery and processing technology require before all the relevant parameter values are established.

The container itself must be designed to meet performance specifications and evaluated for ease of manufacture by the stretch blow molding method. Material distribution, in particular, will determine such critical features as column strength, paneling resistance, internal volume, and the ability to fully replicate mold contours.

A system that can speed the determination of accurate figures for the mass of material needed to meet performance specifications, as well as the area of certain critical sections for the determination of internal volume, is the least that should be available to do the job.

IMPLEMENTED CAE DESIGN SYSTEMS

A significant implemention of CAE is illustrated by the system used by Cincinnati Milacron. The capabilities of existing resources at Corporate R&D were evaluated for applicability to the RHB process and compared to the capabilities of emerging technologies.

The corporate CAD system included CADAM, an IBM mainframe-based $2\frac{1}{2}$D system with sectional area analysis and limited surfacing capability, and CimLinc, an integrated system bringing together CAD, solids imaging and modeling, and machine tool postprocessing. This, and systems like it, features the Unix operating system, open architecture, and icon-based on-screen menus for ease of use. Finite-element analysis packages such as NISA display and Patran, and data display and analysis packages such as RS/1 run on a DEC Vax 11/780 computer.

Cincinnati Milacron has been active for several years in providing container and preform design technical services to customers of its line of two-stage RHB machinery. It also offers this service for experimental and exploratory studies, where the technology might be applied to a process or market not previously envisioned. This phase of support has been carried out under the direction of Corporate R&D, with the bulk of conventional applications, such as the more mature two-piece beverage container technology, being handled by the Plastics Machinery Division. Analytical and technical services include:

- Container design and evaluation for producibility by RHB
- Design of preform injection molds and container blowmolds
- Technical analysis and preproduction quality assurance
- Prototyping and market sampling
- Manufacturing plant layout and equipment specification

The R&D environment often tests the limitations of a technology, and thus has been a source of inquiries as to the solutions to problems beyond the realm of conventional existing CAD technology. Analytical resources are available in Research that might not be justifiable in production manufacturing departments.

A database was complied into which the following parameters were inserted:

1. Maximum and minimum orientation ratios
2. Reference dimensions for the common thread finish designs
3. Maximum and minimum dimensions of containers for each type of RHB equipment
4. Oven configuration for each type (infrared or RF, lamp spacing, etc.)

An expert system, called *RHB,* was then developed to search the database during a development session, in order to establish the relevant preform design parameters for a proposed container design. A report from this program enables quick rendering of the design on the separate CADAM drafting system.

These data were also used to develop a geometric model of the proposed container using Patran. The output of this full 3D representation was di-

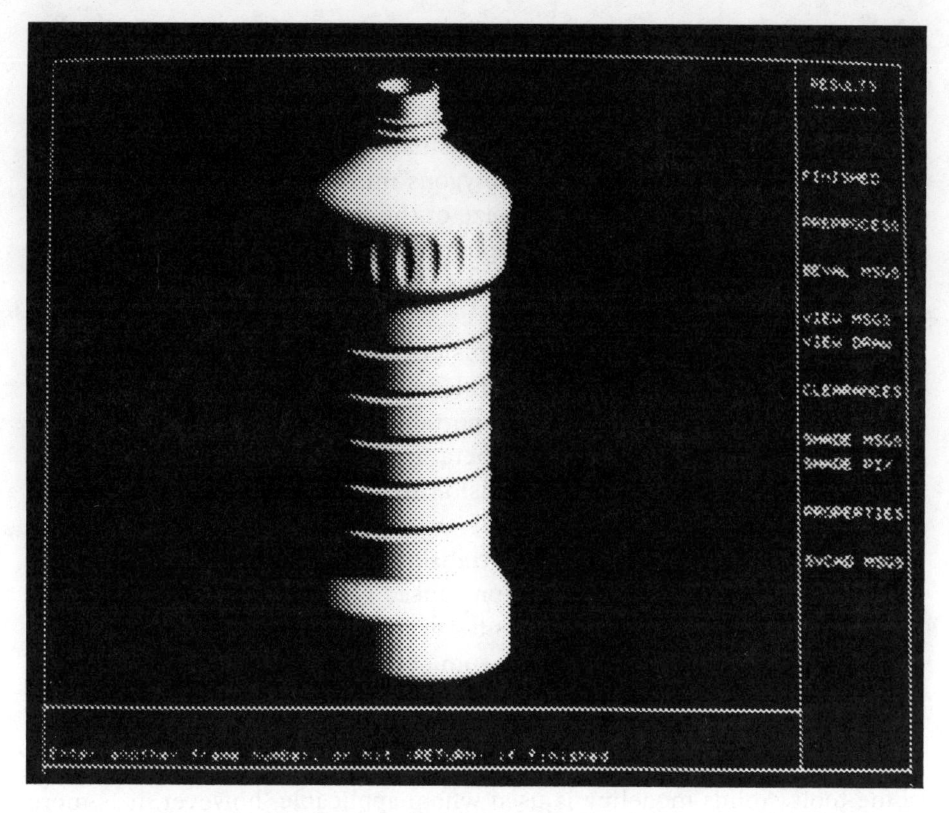

Figure 22-4 Final 3D representation.

rected to graphics devices for examination of shape and contour, (Figure 22-4), as well as to NISA, a finite-element modeling package, which enabled examination of the anticipated container performance under various load conditions.

Two of the container shapes to which these techniques were applied are shown in Figures 22-3 and 22-4. Test results and further details are provided in the presentation.

COMPUTER-AIDED ENGINEERING AT THE MERCHANT CONVERTORS

The two major practitioners of PET orientation blow molding technology in the United States are the merchant convertors Johnson Controls, Inc., and Constar International. Both use sophisticated, integrated design systems to provide fast, high-quality responses to their beverage and custom container customers.

Constar uses Euklid by Matra Datavision, implemented on Digital Equipment Corporation VaxStation II workstations. A significant feature of this software is the use of solids modeling. This modeling method can generate shapes faster than wireframe methods can, although at some expense in model size and accuracy (current solids modeling technology approximates curved surfaces with many small polygons; in order to achieve high accuracy, the subdivision of polygons must be eight or more per arc, which can drastically increase the size of the model database).

Johnson Controls, Inc., has implemented Dassault Systemes' CATIA software on IBM graphics workstations that communicate with a central 3090 mainframe located in Milwaukee WI, where the software is professionally managed by a team of systems engineers.

The advantage for Johnson Controls in establishing a national standard for CAD/CAM software is, in a word, portability. Models can be interrogated at any site in the United States where a workstation is located. Design groups at the Plastic Container Division may transfer all models for container and cavity geometries *directly* to their moldmaker's design group with *no* loss of information due to translation between different software programs, nor errors in interpretation caused by misreading prints.

CATIA offers many advantages for designers of PET containers. It has both 2D and 3D model spaces, with some of the most sophisticated surface modeling capabilities available. Essentially *any* sculptured surface can be created and precisely defined geometrically in CATIA. These surfaces are connected to form bounded volumes, as well as cutter paths for CNC machine tools. Solids modeling is used where applicable; however, it is more accurate to avoid the polygonal approximation of solids geometry when surfaces with substantial continuous curvatures are developed.

Any surface, solid, or volume element can be displayed as a solid, shaded image for contour visualization or marketing presentation. Any image displayed on screen can be reproduced as hard copy by a variety of pen, electrostatic, or color-raster plotters.

The software provides for precise calculation of container or preform volume, surface area, mass, and inertial properties, regardless of geometric intricacy.

Once each preform or container model is created and approved, expansion of the container geometry by appropriate scaling factors yields the dimension and contours of the blow or injection molds. The cavity, or "steel" dimensions, are sent to mold design groups and to CNC programmers at Johnson Controls' sister divisions, who convert the containers surface information to mold cutter paths.

It is possible to generate a container surface contour, develop a CNC cutter path, and produce a model cavity component ready for polishing and further finishing within four hours.

In summary, the major container manufacturers, who compete by offering fast response to their customers, are achieving competitive advantage by investing in sophisticated tools for engineering analysis and design, and in the engineering personnel to make those tools productive. The goals to be achieved are direct-to-production development projects, drastically reduced lead times, fast response to observed quality evaluations, and the ability to offer more quality products to customers.

CONCLUSIONS

Computers continue to revolutionize the plastics industry in general and reheat-stretch blow molding in particular.

Analytical techniques now automate and make precise such formerly tedious tasks as volume and shrinkage analysis, neck finish detailing and specification, and prediction of material behavior. Empirical data are correlated with modeling predictions to bring the whole process under control. And knowledge gained by the experience of one individual is accessible to many.

Systems available on small, 32-bit workstations offer by the early 1990s performance equivalent to mainframe software systems at a substantially more affordable cost. All of the necessary engineering resources can be combined and made accessible at each designer's desk. Such integration is clearly superior to the use of distributed resources running on different, independent platforms, such as were evaluated for this chapter.

These tools will significantly improve the quality and efficiency of design for builders of machinery, molds, and dies, for the primary processor, and ultimately for their many customers in the packaging industry.

SECTION VI

MOLD DESIGN AND ENGINEERING

CHAPTER

23

Extrusion Blow Molds

JAMES P. PARR

The success of any blow molding operation is dependent upon the molds which are used in this process. These molds must be accurate, efficient, durable and, most of all, cost effective. The material and method of construction are chosen for their suitability for the desired end product production quantity, and for the resin type.

The detail of the mold is important because multiple molds must produce interchangeable parts of the same volumetric and dimensional accuracy. Without extremely good duplication, processing can be a problem, especially with remote trimming in multiple-mold operations.

Efficiency in processing results from a mold that cools the molten extrudate into a form as free from warpage as possible in as short a time as possible. An efficient mold also reduces the number of parts that are rejected due to design imperfections.

Durability is needed for long mold life. The molds are closed under pressure, which can eventually wear down the mating surfaces. The pinch-off sections also can be abraded by the material it cuts and welds. Corrosion is sometimes a key factor in mold material selection when, for example, processing PVC or acrylonitrile.

Of course, the selection of the mold material and construction must be based overall on the expected sales volume of the product. Alternative types can then be evaluated based on economic criteria, in addition to the practical

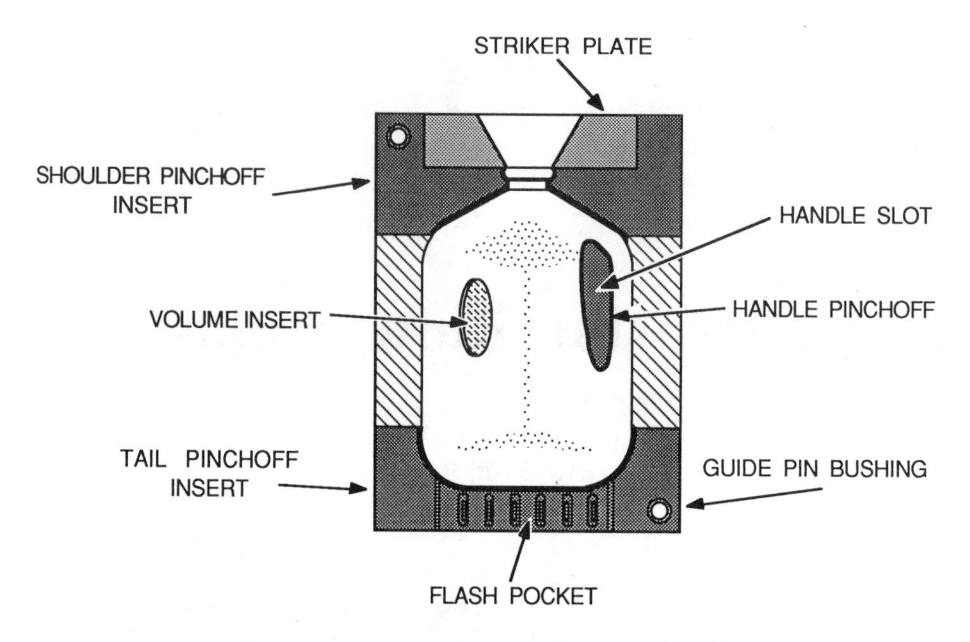

Figure 23-1a Basic features of a typical mold.

Figure 23-1b Molds for 48-ounce widemouthed square containers. Courtesy of Wentworth Mould & Die Ltd.

in-plant concerns. The choice of production method depends upon the expected run-length of the product, the number of cavities necessary, and the surface irregularity of the product. Large numbers of cavities or extremely irregular surfaces favor casting over machining.

Castings are typically not as durable or as thermally efficient as molds cut from solid blocks. They are often used for large-capacity parts when the mold size exceeds the range of the available slab stock. Castings also tend to be cheaper, a factor to consider when production quantities are small.

A typical mold is shown in Figure 23-1a with some of the commonly discussed components labeled for identification. Actual molds are shown in Figure 23-1b.

MATERIALS FOR CONSTRUCTION

Aluminum Alloy

The aircraft grade aluminum alloy 7075-T6, which contains zinc, magnesium, copper, and chromium, is currently the most popular choice for blow molds [1]. This is attributable to the metal's high thermal conductivity, machinability, light weight, and resistance to corrosion. Its main drawback is that it is soft and easily damaged. Some of the important properties for this material, and the other materials to be discussed are listed in Table 23-1. The properties of the metals can vary tremendously depending on the temper.

The aluminum alloy is obtained in heat-treated blocks up to about 8 in. thick. It is then usually cut with modern machine tools. Molds with dimensions greater than the available stock sizes can also be produced. Smaller subunits can be bolted together to form large assemblies, which can then be machine cut just like a smaller piece.

A related material, aluminum alloy 356-T6, is usually cast directly into the desired configuration, especially when producing large molds. This method, however, can result in high porosity with a resultant loss of cooling efficiency. The cast material also has lower toughness, which mandates the use of steel or beryllium-copper pinch-off inserts.

Beryllium-Copper Alloy

Beryllium-copper alloys are used extensively in modern blow molds. These alloys have excellent thermal conductivity, corrosion resistance, and mechanical toughness. Their chief drawbacks include high cost and poorer machinability than aluminum. Beryllium-copper alloys are about three times as dense as aluminum alloys, which further increases the cost.

Table 23-1 Properties of common mold materials

	Thermal Conductivity (BTU in/ft² H°F)	*Density (lb/in³)*	*Hardness**	*Tensile Strength (PSI)*	*Composition*
Beryllium-copper alloy					
CA 172	770	0.298	37 RC	150,000	1.8% Be
CA 824	750	0.304	34 RC	135,000	1.65% Be
Aluminum alloy					
7075 T6	900	0.101	150 BR	73,000	1.6% Cu, 0.23% Cr 5.6% Zn, 2.5% Mg
AISI P-20 steel	200	0.282	30–36 RC 290–330 BR	120,000	0.35% C, 1.0% Ni and 1.0% Cr
AISI 420 stainless steel	166	0.280	50 RC	213,000	0.38% C, 0.8% Si 13.6% Cr, 0.5% Mn 0.3% V
Kirksite A	640	0.250	100 BR	35,000	Contains Cu, Al, Mg

*RC, Rockwell C; BR, Brinell

Alloy CA 172 contains 1.8 percent beryllium and is the harder of the main two alloys used. This material is machine cut. Molds of fine detail can be cast with alloy CA 824 if a highly irregular mold is desired. This alloy contains 1.65 percent beryllium. These materials can be heat treated to improve their hardness [2,3].

In most instances, beryllium-copper is used in pinch-off inserts to utilize its extra hardness as compared to aluminum. By limiting its use to inserted areas, the overall cost of the mold can be reduced while providing extra durability [4]. Sometimes, especially when processing corrosive materials, the entire mold is produced from beryllium-copper to take advantage of its excellent corrosion resistance.

An additional note must be made concerning the health hazards of beryllium. Inhalation of beryllium dust can cause lung damage. There is also a question as to whether it is a carcinogen. These issues should be investigated if machining or casting operations are to be performed with this material.

Steel

Steel is most often used in blow molds for use with PVC or engineering resins. This is based on its corrosion resistance and extreme toughness. Excellent surface texture is obtainable through etching processes. Its major disadvantage is its low thermal conductivity.

AISI P-20 is a widely used prehardened mold steel. This alloy can be nitrided and photoetched. It is an alloy containing less than 1 percent carbon, manganese, silicon, and molybdenum, and less than 2 percent chrome.

For corrosive resins, AISI 420 stainless is used. This through-hardening alloy contains 13.6 percent chromium. It must be heat treated for best wear characteristics. This material can also be photoetched using a different technique. Cooling channels drilled into this alloy resist corrosion, a plus for cooling.

Steel molds can be fabricated by machining, casting, hobbing, or even welding. Machining is difficult, of course, because of the high hardness of steel. Service lives of steel molds can be up to ten million cycles [5].

Steel is also used in areas of the mold or its mounting hardware where high wear is expected. These areas include the mounting platens and bars, guide elements, and pinch-off inserts. Steels are also used in the auxiliary parts such as the blow pins, needles, and shear surfaces. Standard tool steel grades are used for these applications.

Miscellaneous Materials

Zinc alloy (Kirksite) can be used for casting large molds or large quantities of smaller molds. This alloy has good thermal conductivity [1]. The molds need steel or Be-Cu inserts for pinch-off areas. Zinc can also be alloyed with Al and Cu for better dimensional stability. The material is somewhat susceptible to corrosion, however.

Another mold material in use is a copper-nickel alloy. Depending on its composition, this alloy offers thermal conductivity values in the range of beryllium-copper and aluminum alloys. Its hardness is somewhat less than beryllium-copper at comparable thermal conductivity, however. It can be formed using all of the previously mentioned techniques.

Synthetic plastics such as acrylates, polyesters, and epoxies can be cast to produce low-cost molds and prototypes or extremely short production runs. They can be filled with metal powder to improve their dimensional stability and thermal conductivity.

MOLD COOLING

Cooling of a plastic part consists of three separate transfer mechanisms:

1. Conduction of heat in wall of part
2. Conduction of heat in mold wall
3. Convective transfer of heat in cooling fluid

Step 1 is dependent upon resin type, temperature, and wall thickness. Step 2 depends upon the mold material's thermal properties, porosity, and mold/cooling layout geometry. Some thermal properties of selected resins are shown in Table 23-2. Step 3 can be optimized with regard to temperature, fluid flow rate, and prevention of scale formation on the liquid side. The cooling rate of most processes is limited more by the rate of conduction within the plastic than by the rate of conduction in the mold. The cyclic time of a part is usually strongly dependent on its wall thickness.

Table 23-2 Thermal conductivity values for some thermoplastics vs. temperature (kcal/m · h · °C)

	Temperature		
Polymer	*50°C*	*100°C*	*150°C*
HDPE	0.34	0.29	0.22
LDPE	0.28	0.22	0.21
Polyacetal	0.35	0.32	—
Polycarbonate	0.21	0.22	0.22
Polypropylene	0.18	0.16	—
Polystyrene	0.14	0.14	—
Polyvinylchloride	0.14	0.14	—

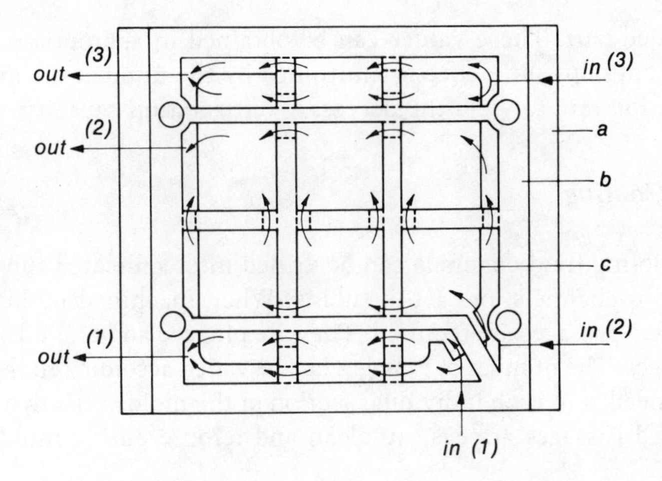

Figure 23-2 Cooling a blow mold by means of three cooling circuits and cooling chambers. *a*, mounting plate; *b*, blow mold; *c*, visible contours of mold cavity for a canister with a filler connection at one corner. Courtesy of Hoechst AG.

Consideration should be paid to the fact that wall thickness differences lead to cooling rate differences, and in turn, warpage and molded in stresses. The rate of cooling sometimes cannot proceed as rapidly as possible but must be retarded, especially when thick-walled parts might warp if cooled too fast. This is due to the high gradient of temperature across the part section that can occur under those circumstances. Slower cooling allows more stress relaxation (annealing). Sometimes molds that are too cold can cause unwanted condensation, which spoils the surface finish and retards cooling. Dehumidification of the blow molding area can help eliminate this problem.

Cooling of the flash is important, in order to help efficiently trim the part. In addition, the shoulders and bottoms of bottles are usually thicker than the walls, and require more cooling lines. Since the flash and the shoulder and bottom sections are contiguous, the mold can be split into three cooling zones, as shown in Figure 23-2. These can then be provided with different flow or temperature values. The number of cooling channels can be greater in the high heat load areas.

A supply of cooling water to the blow pin is necessary to produce well-formed threads and eliminate warpage that could ovalize the openings. Flash grippers and other part removal equipment need a cooling water supply to prevent sticking.

The molding can be released sooner if a postmold cooling apparatus is provided. Various types of jigs allow the part to maintain its shape with an increase in production rate. These stations can be equipped to trim the part simultaneously.

The cooling load necessary for the production of an item can easily be calculated from the difference in enthalpy values of the molten feed and the demolded part. These values can be obtained in appropriate reference texts. The enthalpy difference is multiplied by the number of cavities and the production rate to yield the necessary refrigeration capacity.

External Cooling

The cooling fluid channels can be drilled interconnected tubelike passages, cast-in channels, or cast-in tubing. When machined molds are produced, drilled passages are required. They are plugged and cross-drilled with other passages. The number of passages can be varied according to the amount of cooling needed in each individual section of the mold, as shown in Figure 23-3. Drilled passages are easy to clean and reroute during mold reconditioning [1,6].

The typical size of drilled cooling lines is 7/16 in. (11.1 mm). The spacing of the lines center to center should be a minimum of twice this distance. The distance from the mold surface should also be two diameters so as not to create cold spots on the cavity surface, which could lead to warpage.

Cooling channels that are produced during the casting process must usually be impregnated with an epoxy or silicate material after fabrication to seal leaks. A mazelike barrier system of baffles is patterned to force turbulent flow. This is depicted in Figure 23-4. After the rear of the mold is squared off, a backing plate is attached with an appropriate gasket. The backing plate can be removed for easy cleaning.

Tubing can be laid into the casting chamber to provide a leakfree, easily fabricated cooling network. Although easily installed, this method does not lend itself to cleaning.

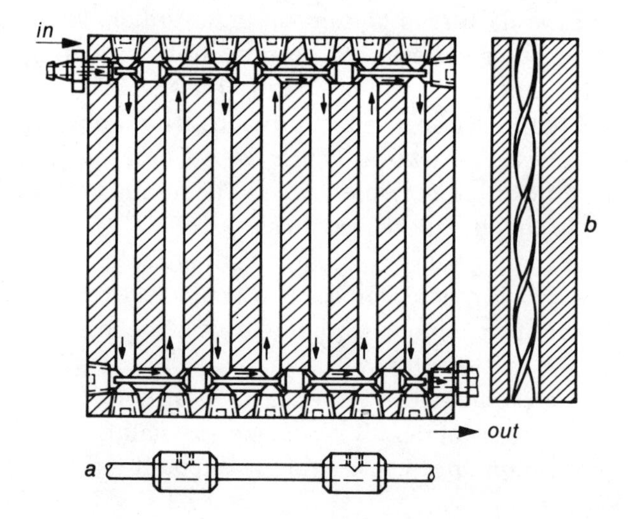

Figure 23-3 Cooling channels with labyrinth-type water flow. *a,* continuous rod with stoppers; *b,* inserted copper spiral. Courtesy of Hoechst AG.

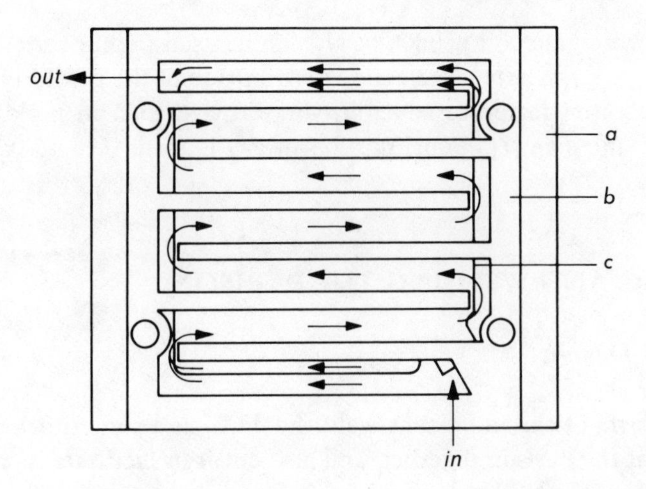

Figure 23-4 Channel system with labyrinth-type water flow produced by baffles. *a*, mounting plate; *b*, blow mold; *c*, visible contours of mold cavity for a canister with a filler connection at one corner. Courtesy of Hoechst AG.

For sections of the mold that extend deeply into the interior of a part, such as the "eye" pinch-off section of a bottle, lack of sufficient cooling can lead to excessive cycle times. This is because the slower cooling rate near the hot spot requires longer cooling times. The heat buildup can also lead to product deficiencies through increased stress and brittleness. Flood cooling can be provided to these areas in cut molds in the form of hollow backed inserts. O-rings around the mating surface boundary provide a watertight seal.

It is important to locate the cooling fluid entrances near the bottom of the mold and the exits at a higher level. This helps eliminate any air being retained, which would reduce flow or heat transfer capability.

Internal Cooling

Sometimes an additional means of cooling is chosen. It involves cooling the inside of the part in one of three ways:

1. Venting of blow air to create turbulence inside the part
2. Blowing with a cryogenic liquified gas to quickly cool the inside of the part
3. Blowing with a fine mist of water and/or ice

These internal blowing methods yield parts in a faster cooling time and often help to eliminate some of the stresses seen in molded parts that were only externally cooled. Methods 2 and 3 utilize the latent heat of a phase change to increase the heat absorption.

Cryogenic gas cooling (CO_2 or N_2) can be used within certain parts that have a long central axis for effective distribution of the gas. Internal cooling allows a greater degree of control over warpage than in moldings cooled from only one side [7]. Consumption of gas is from 0.25 to 0.50 kg/kg of plastic [8].

CUTTING AND WELDING THE PARISON

Pinch-off Design

The parts of the mold that weld the ends, and sometimes the interior portions, of the parison together, and also cut it or facilitate its removal are called the pinch-offs. Pinch-off design has an important effect on the success of a blow molding product because the weld seams are usually the weakest

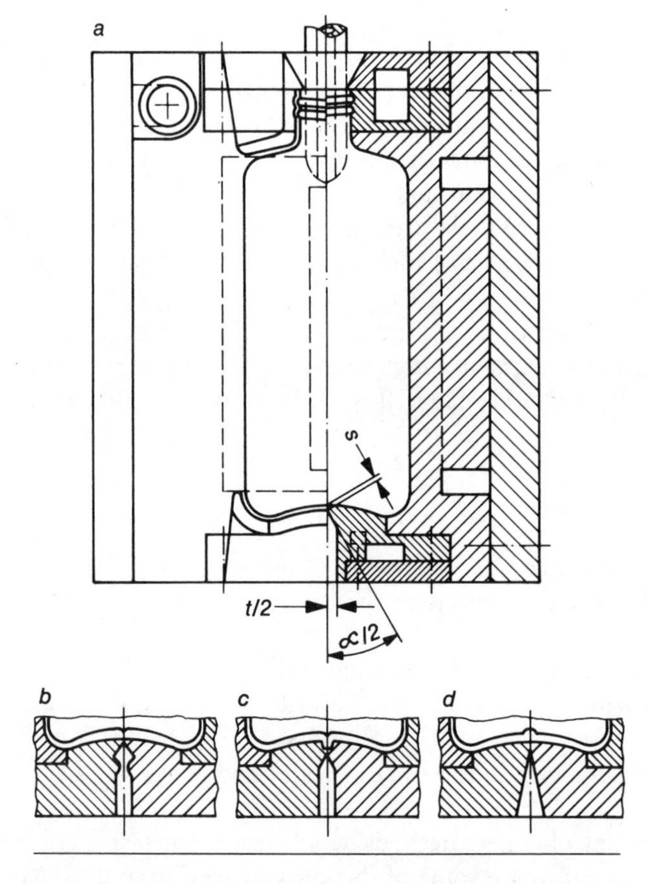

Figure 23-5 Design of welding edges and pinch-off pockets. *s*, welding edge width; α, opening angle of pinch-off pocket; *t*, width of pinch-off pocket. Courtesy of Hoechst AG.

parts of the container. The pinch-off must be designed to maximize the strength of the weld. Some different types of pinch-off designs are shown in Figure 23-5.

Due to the contact with both the bottle and the adjoining pinch-off section, which is typically thicker than the bottle walls, the heat load is higher in this region. For this reason, it is important to have a highly conductive pinch-off section material. However, due to the mechanical wear involved in the repeated cutting and compressing of the flash, high durability also is needed. For this combination, beryllium-copper is often the material of choice. For longer runs and higher durability, steel is used.

The width of the pinch-off edges is an important factor in the pinch-off design. The optimum width depends on material type, parison thickness, and part size. Too thin or too sharp an edge can result in cutting the parison rather than a solid welding. For most HDPE and PVC bottles of less than one gallon, the pinch-off land should be 0.010 in. PP requires a width of about 0.005 in. Another important parameter is the angle of the pinch edges. The results of a pinch-off optimization study for various sizes of polyethylene containers are listed in Table 23-3 and are shown in Figure 23-6.

The flash pockets are the areas in the parting surface of the mold bordering the pinch-offs that allow the mold to close without the parison holding it apart. The depth of these pockets should be large enough to slightly compress the parison so that it is cooled adequately.

Dams or beads are sometimes placed in the flash pockets parallel to the pinch-off to force some of the melted material into the mold cavity as the mold is being closed. This is done to reinforce the weld seam by making it thicker [6].

Flash Removal

Automatic trimming is essential in large-volume container production. This trimming can be performed in the mold or in stations separate from the blow molder.

One method for deflashing a part involves having the flash-pocket section of the mold hinged to or separated from the rest of the mold. Through the use of a hydraulic cylinder, or other such device, the section can be moved just before or after mold opening to separate the flash. The movement can be either rotational or axial.

By trimming in separate stations, the wear and tear associated with moving parts is isolated away from the mold, which eliminates the need to take the mold out of service simply to repair the cylinders or bearings of the flash pulling system.

The blow pin is easily designed to cut the neck and retain the shoulder flash until the bottle has been separated. For example, a small groove can

Table 23-3 Pinch-off design parameters (see Fig. 23-6)

Container Volume, V (liters)	Pocket Opening Angle, α	Welding Edge Width, b (mm)	Pocket Width, c	Compression Bar Width, L	Gap Between Compression Bars, x	Delayed Closing Travel, W_v (mm)	Delayed Closing Speed (mm/s)		
							HDPE		
							HMW	MMW	PP
≦1	30°	0.6–1	$2d$	$2b$–$3b$	$0.1c$	$2d$–$4d$	40–80	140–260	75–100
1–30	30°	1–3	$2d$	$2b$–$3b$	$0.1c$	$2d$–$4d$	10–40	75–140	20–75
30–250	30°	3–5	$1d$–$1.5d$	$2b$–$3b$	$0.1c$	$2d$–$4d$	3–10	10–75	10–20
250–2000	30°	5–7	$1d$–$1.5d$	$2b$–$3b$	$0.1c$	$2d$–$4d$	3–10	—	—

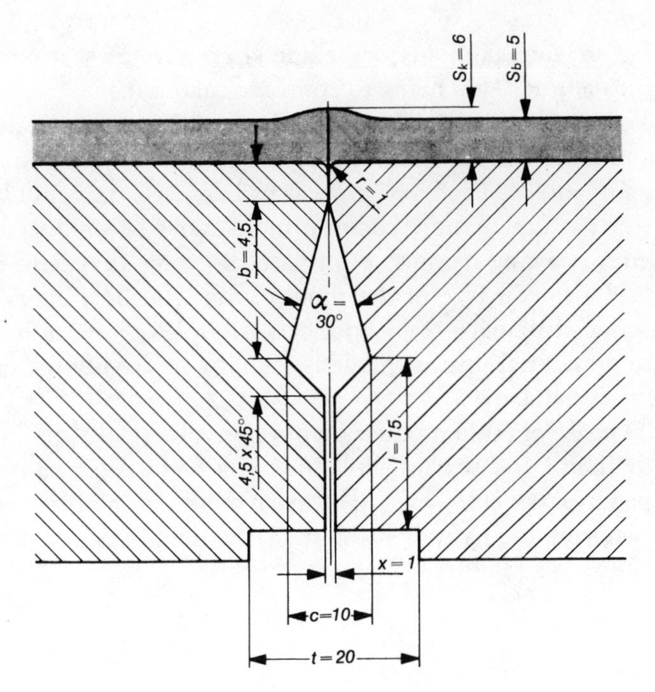

Figure 23-6 Optimized mold base for a 60-liter can made from HMW-HDPE. Dimensions are defined in Table 22-3. Courtesy of Hoechst AG.

be machined around its circumference to form a small undercut. The flash can then be stripped off the blow pin with an actuated sleeve.

Many large or unusually shaped parts are still manually trimmed. Torches can be used to diminish some of the edge roughness resulting from the flash removal.

MOLD CAVITY

The mold cavity is the exact representation of the blow molded object. It is produced in the mold making process by machining from a drawing or pattern, or by casting around a pattern. Pattern making is a very specialized craft. Key aspects of the cavity include the parting line orientation, the venting system, and the surface texture.

Production Techniques

Standard machining operations are used to fashion the cavity, including milling, turning, and grinding. Tracing machines are used to produce identical cavities, or to machine from a pattern. Most new machines are computer controlled. Machining is best for small numbers of simple configurations. Large machined molds are made from several sections that are assembled into a unit. This helps to conserve material.

Modern casting methods can accurately reproduce complicated configurations and unusual surface textures. Casting provides an economical means to produce many molds. The Shaw process utilizes a ceramic mold produced from a master pattern. A new ceramic mold must be produced for each casting. Vacuum casting helps to eliminate the voids that tend to be produced just under the surface of the mold cavity. Hot hobbing is a type of pressure casting utilizing a steel hob, or pattern. Pressure is held on the molten casting material until it hardens, helping to densify the crystalline structure of the metal.

The final phase of mold production often involves polishing, sandblasting, or photoetching the surface. Polishing is a manual operation and can be a very expensive part of the overall cost of a mold. The ability to reproduce a surface finish from the pattern without costly mold finishing is one of the advantages of casting [3].

Parting Line

Generally, the parting line of a mold is chosen to give two symmetric halves. The parting surface is usually a plane. The section of the parting plane should be chosen to avoid undercut features, and to yield the lowest blow ratio (ratio of part diameter to parison diameter) possible. Handles must be located along the parting plane; other locations require complex moving section molds, which are described later in this chapter [10].

For cylindrical containers, the parting line is not of critical importance due to radial symmetry. In oval containers (horizontal cross section) the parting line should run through the major axis. Rectangular containers can either have a parting plane orthogonally through the centers of the side faces, or diagonally through the corners. The corners at 90 degrees to the parting line for diagonally parted molds are usually very thin.

Some newer applications, typically for engineering resins, often utilize large flat surfaces, and double-wall construction. Some of these applications include business machine housing panels. These types of parts utilize prepinched and blown parisons that are slowly inflated as the mold closes. Sometimes, the parting line of the mold is irregular to allow for the unusual configuration of the part. Parting lines may then take an L, U, or irregular shape when viewed from the side.

The parting surface of the mold can be used to compression-mold flanges or extensions. This is done by cutting pockets into the plane that are the shape of the desired feature. Care must be taken to accurately match the thickness of the parison and the flange or extension depth so that the mold is not shimmed open during closing.

In the production of certain types of cases or contoured packages, hinges are produced between adjoining panel areas of the case. The hinges are pressed between the mold halves in recesses similar to the flash pockets. By controlling the amount of compression and the speed of the mold closing movement, the strength of the hinge can be increased through orientation of the polymer structure.

Venting

During the blow molding process, it is necessary to allow the air inside the cavity to escape as the parison inflates. This is accomplished by carefully providing passageways out of the cavity. These passages are called vents [4,11,12].

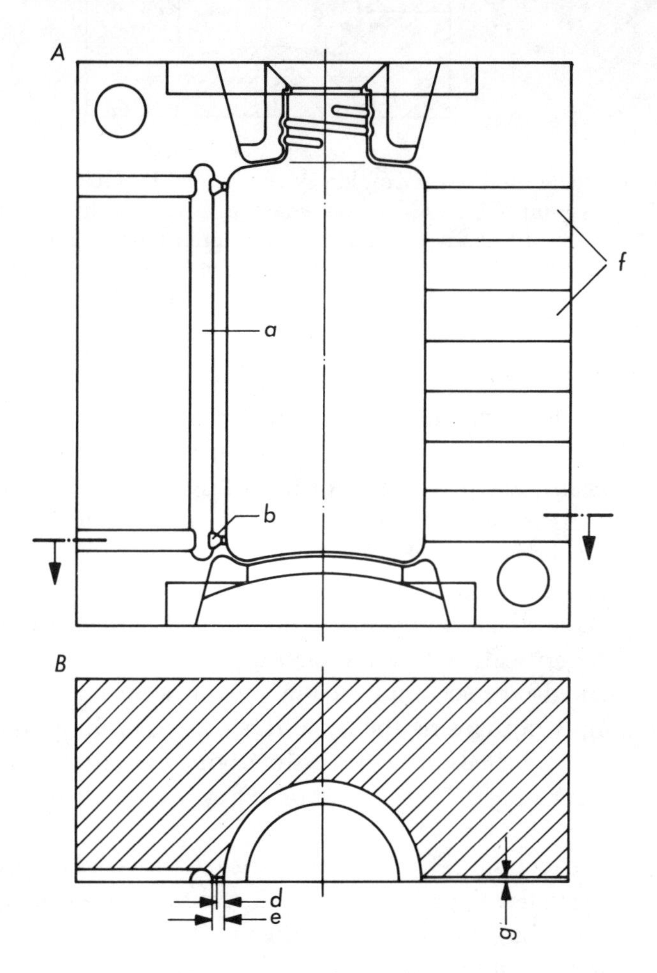

Figure 23-7 Venting of mold cavities in the mold parting line. A. Facilitating free escape by means of a groove *a* at a distance *e* from the cavity edge; venting in the base and shoulder edges by funnel-shaped venting channels *b*; $d = 0.5$ mm. B. Venting slits in the parting line *f* (0.1 mm deep, 25 mm wide); $g = 0.1$ mm. Courtesy of Hoechst AG.

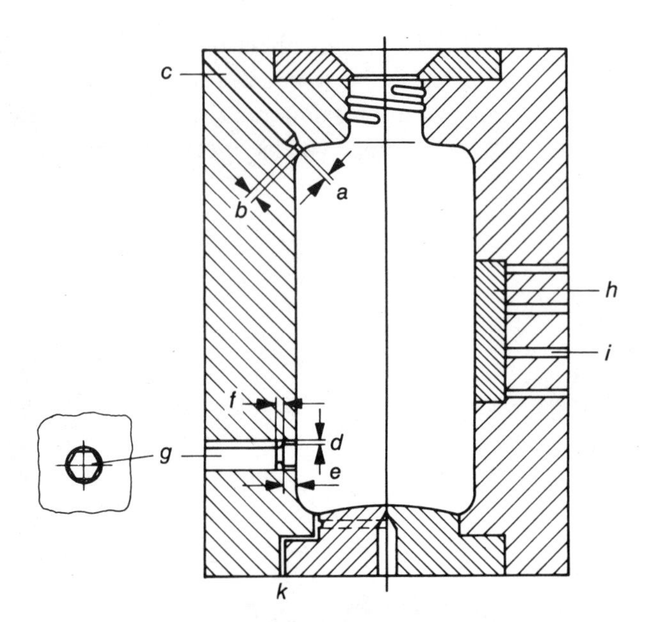

Figure 23-8 Suitable methods for mold cavity venting. Left half, vents: *a*, 0.1–0.3 mm dia.; *b*, 0.5–1.5 mm; *c*, blind hold; *d*, chamber height of the free circular section, 0.1–0.2 mm; *e*, slit depth, 0.5–1.5 mm; *f*, annular groove; *g*, bolt with venting slit; *k*, ventilation channel in the fitting point of an insert. Right half, mold cavity venting by means of plugs and plates made of sintered metal: *h*, plate made of sintered metal; *i*, vent. Courtesy of Hoechst AG.

There are several types of vents. Several common vent types are shown in Figures 23-7 and 23-8. One type utilizes flats approximately 0.002–0.004 in. deep that open into the cavity on the parting line. They are relieved to a depth of >.0.20 in. just outside the cavity. These vents can be placed around up to 50 percent of the parting line distance. On neck areas and around lips, concentric openings of from 0.002 up to 0.008 in. can be used. These vents are necessary to produce well-formed sharp edges free from air bubbles that are critical for leakfree sealing.

Venting can also be accomplished by putting shims between the mold halves and simply not closing the mold flush together. In deep corner sections, holes can be drilled through the the cavity and plugged with noncircular pins or vent plugs. These types of vents work well in large-volume parts.

Recently the technique for labeling bottles in the mold during the blowing phase has been developed. This procedure offers the advantages of synchronizing the labeling operation with the production cycle and improving bottle stiffness at a lighter weight.

Several vents in the label panel of the bottle are connected together and routed to a vacuum source. The label is placed in the mold by a mechanical arm during the portion of the cycle in which the mold is open. The vacuum holds the label in place until the expanding parison contacts it. Hot melt adhesives are used as fastening agents [13].

Cavity Surface

When processing polyethylene, the inner surface of the cavity must be slightly roughened. Without this surface texture, a poor surface finish can be encountered, similar to orange peel. This is caused by poor venting of the cavity. Tiny air bubbles are trapped and act as insulators that reduce the cooling of the plastic directly over them. The different cooling rates result in differential shrinkage, causing the poor surface. A rough surface texture provides a network of minute passageways to the vents, eliminating the air bubbles. Sandblasting with #80 or #120 grit is acceptable for most small polyethylene bottles. Larger parts may need grit sizes up to #30 for optimum efficiency [12].

When highly transparent or glossy bottles are to be produced, especially when using PET, PP, or PVC, it is necessary to polish the mold. Radial wiping using #360 grit can achieve an excellent surface and still allow air to escape to the parting line. Generous venting should be used on glossy containers. If desired, most materials can be plated.

Many objects have a textured surface that is intended to cover surface imperfections, or to provide a better grip. Different mold materials etch differently. Etching must be performed before hardening treatments. Adhesion of the part is higher in an etched mold, which can cause mold release difficulties or require the use of mold release agents. For texture depths of 0.10–0.012 in. in HDPE, it is necessary to have a mold texture of 0.025–0.030 in. in an aluminum alloy. When molding the stiffer engineering resins, it is desirable to add a draft angle of 1.5 degrees per 0.001 in. of texture depth. The textures are typically about 0.0025–0.003 in. in depth [14].

Inserts

It is often desirable to produce the pinch-off areas of the mold out of a tougher and more thermally conductive material than the rest of the mold. This is accomplished by the use of inserted sections, illustrated in Figure 23-9. Maintenance is also facilitated since these areas wear faster despite their higher toughness. It is easier to bolt on a new section or to repair a smaller piece than to refurbish the whole mold. Similarly, repairs can be approached in this manner [1].

Should a modification to the cavity be desired that would require adding metal (forming a depression or hollowed out area in the part), it is often easier to bolt in an insert rather than build up the metal by welding.

Figure 23-9 Interchangeable inserts are common features of blow molds. Courtesy of Wentworth Mould & Die Ltd.

Another use for an insert is to allow the rapid changing of brand name, contents designation, date of manufacture, or other information through the use of interchangeable engraved plates.

Small volume changes can be made with a series of different sized inserts. Undercut features can be achieved if loose inserts are placed in the mold between each cycle and removed with the part.

Handles

Molded handles must lie along the parting line of the mold to avoid complicated moving sections (see Figure 23-13). The central cutout, or eye, of the handle must be pinched out, and must therefore have a proper pinch-off/flash pocket design. Often this section is designed as an insert.

The cross section of the handle should be a rounded square or rectangle for optimal wall thickness [11]. The handle can be used to stiffen the container vertically, and to eliminate glugging while pouring, if it separately enters the neck.

Indented handholds are often located in the bases of large containers to facilitate pouring. They are best located across the parting line to avoid mold release difficulties, and should be well rounded and not too close to the rim to avoid thinning and breakage problems.

Handles can be injection molded separately and positioned as inserts in the mold so that they are welded onto the part during the blow cycle. Processes are currently being developed in which the handle is injected during the blow molding cycle itself. If this development proves successful, it will become possible to produce, in one step, PET containers with their handles in place.

ANCILLARY ELEMENTS

Base Plates

Blow molds are usually mounted onto base plates that are larger than the mold itself. A typical layout is shown in Figure 23-10. The molds are

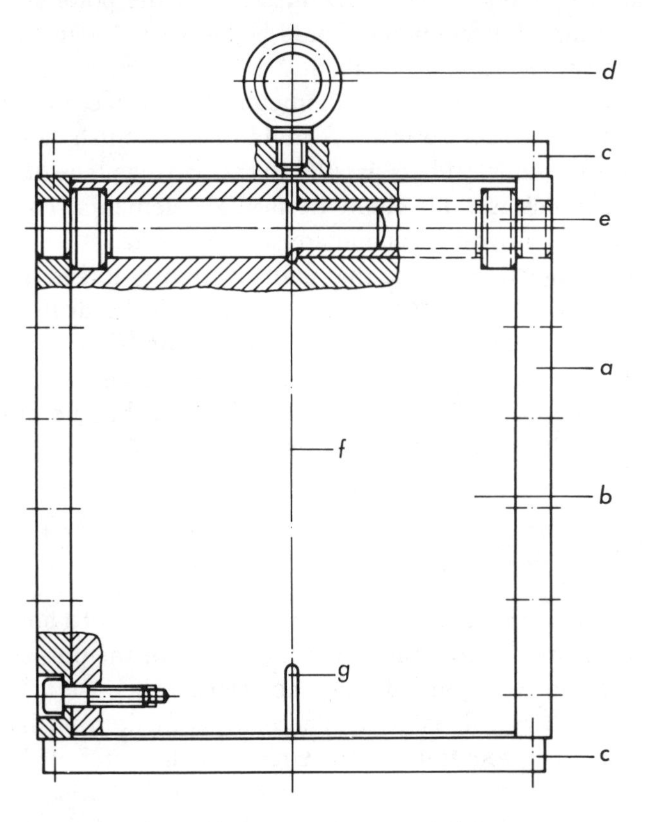

Figure 23-10 Mold block with transverse bar, *c*, and eyebolt, *d*, for mounting. *a*, mounting plate; *b*, blow mold; *e*, opening; *f*, mold parting line; *g*, slot for parting the mold halves on the bench. Courtesy of Hoechst AG.

attached to the backing plates by means of bolts or pins. The mold base can also consist of a beam or bar, especially with large molds. The unit is then mounted in the machine by attaching the base plate with clips or bolts to the platens.

The base plate can be used to mount multiple molds for easier installation. Spacers can also be mounted on the base plate when the molds are small in relation to platen size. This can help to spread the clamping force evenly on the platen, which insures square closing.

The base plate can also be used as a sort of manifold for the coolant lines. This eliminates the need to route individual hoses to each mold in a multiple mold arrangement. O-rings are used to seal the mating surfaces between the mold and base plate.

Alignment Pins

Alignment bolts and bushings are used to ensure accurate positioning of the mold halves as they are closed (see Figure 23-10). These steel parts are hardened to 55–60 Rockwell hardness, with the bolts slightly harder than the bushings. Bolt diameter should be between 3/8 and 1 1/4 in., and can penetrate the bushing up to about 2 1/2 times the diameter. For large molds bolts can be up to 3 in. in diameter. The surfaces of the bolt and bushing are usually recessed into the mold to allow flush closing.

Sometimes cylindrical bars parallel to the parting plane of the mold are used in larger molds as guide elements as guide elements. These bars have a larger tolerance for mismatch and are self-centering.

To allow removal of plastic that is inadvertently forced into the alignment bushing, the hole in which it is set should be drilled completely through the mold. The alignment bolts and bushings should be set as far as possible from the cavity. This is especially important when the part is completely pinched off. Too much pressure on too thin a section can lead to cracking or deformation in short time frames.

Striker Plates

Mold inserts are generally used to produce the neck of a bottle. A striker plate is usually fitted across the top of the insert for the purpose of cutting the flash from the neck and reducing the wear on the insert. Figures 23-11 and 23-12 show how this is done. The plate is made from hardened tool steel, and sharpened to a fine cutting edge. The diameter of the opening is matched to the diameter of the blow pin to shear the parison either as the blow pin is rammed down into the neck opening or as it is retracted.

The striker plate is typically produced with two or four semicircular shearing surfaces. When the shearing surface in use becomes too worn be-

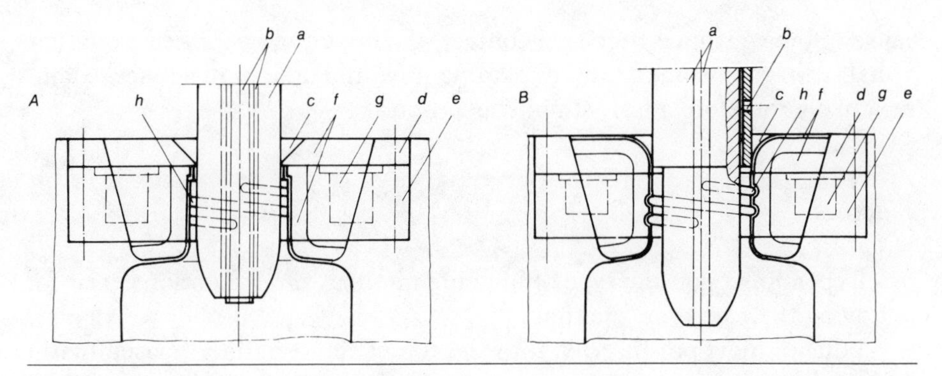

Figure 23-11 Mold and mandrel design for the calibration method. A. Internal calibration of a bottle neck while blowing through the mandrel. *a*, interchangeable calibration mandrel; *b*, blow nozzle; *c*, annual pinch-off groove; *d*, hardened surface plate of the neck insert with cutting edge; *e*, lower threaded part of the neck insert; *f*, pinch-off pocket in the neck/shoulder region; *g*, neck cooling; *h*, threaded bottle neck, which is reduced on OD at the neck top to ensure accurate calibration and prevent the escape of secondary air even with thin walls. B. Design for the calibration of a sealing surface that is free of flash. *a*, blow nozzle designed as calibration mandrel; *b*, cutting ring with narrow venting slit (0.5 mm) and vent holes; *c*, tulip-shaped pinch-off area; *d*, hardened surface plate of the neck insert; *e*, lower part of the neck insert with threads (*h*) that are interrupted in the parting line region; *f*, pinch-off pocket in the neck/shoulder region; *g*, neck cooling; *h*, threads which in A are interrupted along the parting line. Courtesy of Hoechst AG.

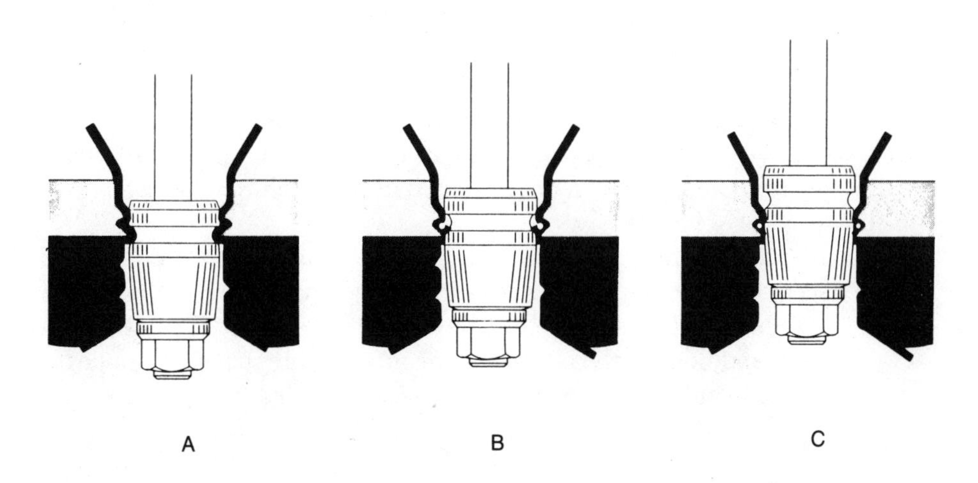

Figure 23-12 Sequence of operation for pull-up prefinish. A. Mold closed position. B. Shearing action of the blow pin through shear steel. C. Blow pin strokes up to provide prefinish for container neck. Courtesy of Johnson Controls, Plastics Machinery Division.

cause of the continual frictional contact, the plate can be rotated a quarter- or half-turn to put one of the remaining unworn surfaces in service. Maintenance is therefore a fast, simple operation.

Ejectors

Depending upon the type of blowing method, various options exist for removing the part from the mold.

When a blow pin inserted into the top of the container is used, withdrawal mechanisms can be arranged to grasp the tail flash through special slots in the base of the mold. The bottles can then be moved onto a cooling bed or conveyor belt for subsequent handling.

If a horizontally shuttling mold is used in conjunction with a vertical blow pin, the part is retained on the pin as the mold indexes over to grasp the next parison. The part is dropped onto the collection bed when the blow pin is vertically retracted.

After blowing, the article can be stripped off a bottom blow pin by means of a plate, sleeve, or gripper. This type of blowing technique is usually utilized for large articles such as 55-gallon drums.

In needle blow methods the needle is withdrawn slightly before full mold open. The part is retained in one mold half by a special undercut. Knockout rods or pins force the container out of the cavity.

Miscellaneous Features

A hole should be drilled and tapped into the top of heavy molds to allow insertion of an eyebolt. This allows the use of a crane or hoist to position the mold in the machine. Some arrangements should be provided to lock the halves of large molds together to prevent accidental opening during installations. A slot in the parting line can help in prying the halves apart for modification or repair when not mounted in the machine. These features are shown in Figure 23-10.

When molding large parts with extensive pinch-off lengths, landing pads can be used to reduce the loading on and increase the life of the pinch-off edges. These pads can be square and are preferably made from steel. When new, the pads should stop the closing of the mold before the pinch-off edges meet. The pads can later be worked down until the edges are the proper distance apart.

MOVING SECTION MOLDS

If it is desired to produce undercut features in a molded part that could not be released from a simple two-piece mold, moving sections must be used

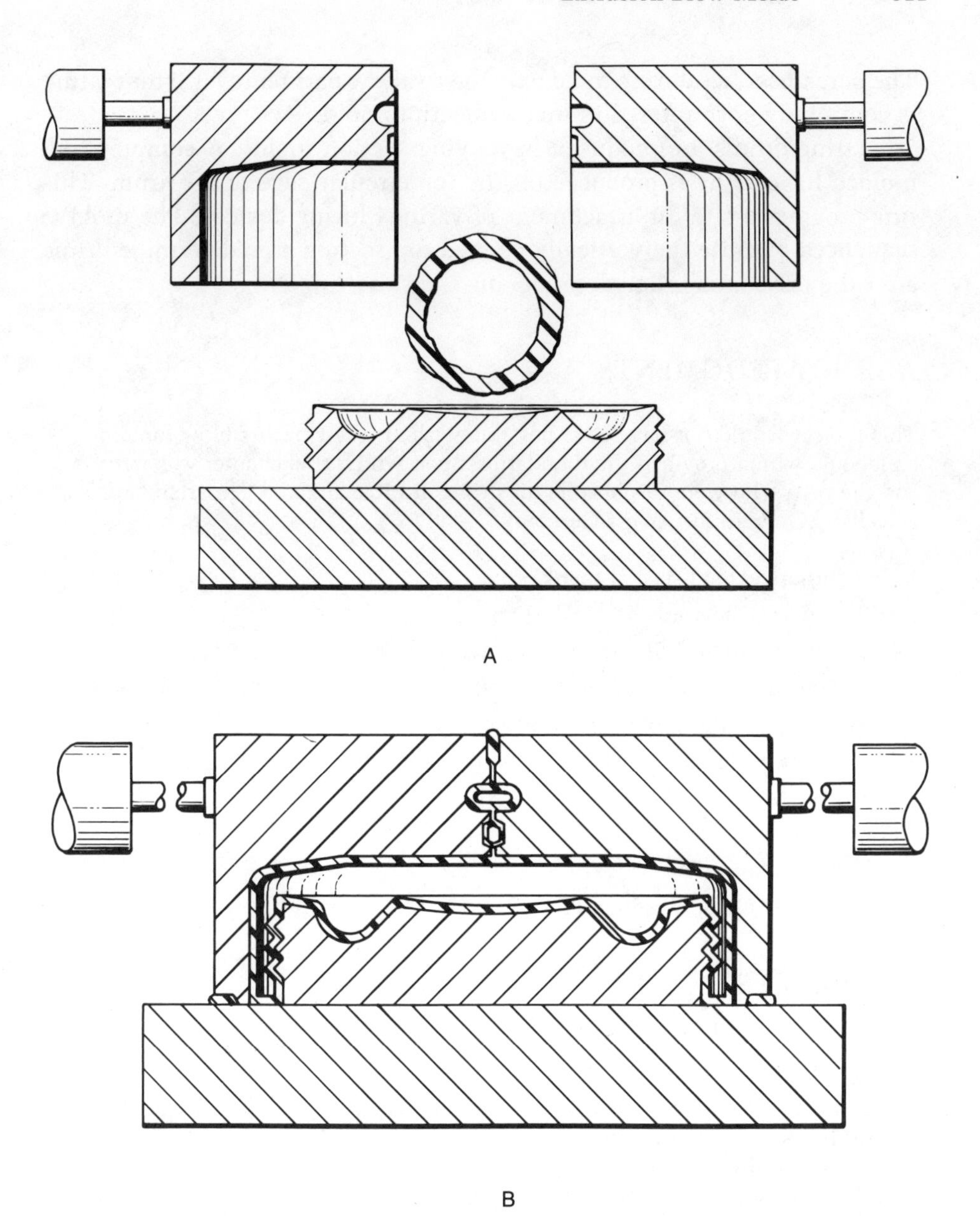

Figure 23-13 Moving section molds are used to produce parts that could not otherwise be molded. A. Open quarter-mold sections and thread forming core. B. Quarter-mold sections closed on thread forming core. Courtesy of Phillips 66 Company, Plastics Division.

[10,15]. This is how the handle is produced in some types of screw-on lids. The sequence begins with the moving sections each fully extended, as shown in Figure 23-13(a). The sections close together during the molding step to form the handle, as shown in Figure 23-13b. When molding, the slides move in sequence with mold opening, which allows part removal.

This same mold utilizes a retractable core to produce the internal threads.

The core unscrews during mold opening to allow part removal. This feature is commonly encountered in many injection molds.

L-ring drums are produced in moving section molds. A compression-molded lifting ring is produced on the top circular edge of the drum. This ring provides a site for attachment of various lifting devices. The mold is sequenced to close the vertically opened top section a predetermined time after the parison is inflated, producing the solid ring.

ACKNOWLEDGMENTS

This project would not have been possible without the Hoechst blow molding handbook, which provided the foundation upon which this chapter was written, and the men who were responsible for it: Eckard Raddatz and Christian Gondro, Hoechst AG, Frankfurt, West Germany.

The authors extend special thanks to:

H.R. Glover, Chevron Chemicals, Orange, TX

Dieter Wunderlich, FGH Systems, Denville, NJ

Scott Hartung, Seajay Inc., Neptune, NJ

Christopher Phillips, Johnson Controls, Manchester, MI

Don Peters, Phillips Chemical Co., Bartlesville, OK

Brush Wellman Inc., Warren, MI

Ampco Metals, New Berlin, WI

Uddeholm Corp., Totowa, NJ

NGK Metals Corp., Reading, PA.

REFERENCES

1. Suit, M., "Blow Molds." *Modern Plastics Encyclopedia.* New York: McGraw-Hill, 1986–1987. Pp. 331–333.
2. Cantillon, W. H. *Beryllium-Copper Molds for Blow Molding.* Brush Wellman Corp.
3. "Cast Beryllium Copper: Some Tips on Using It as a Mold Material." *Industrial Models and Patterns* (March–April 1962).
4. *A Short Course in Blow Molding High-Density Polyethylene.* Chevron Chemical Co., 1970.
5. Menges, G., and P. Mohren. *How to Make Injection Molds.* New York: Hanser, 1986.
6. Branscum, T. E., and R. Doyle. "Blow Mold Design for Large Irregular Shapes." *Modern Plastics* 40(7) pp. 144–148 (March 1963).
7. Gondro, C. *Modern Applications and Equipment for Blow Molding of Polyolefins.* Ausplas, 1982.
8. Segura, J. S. "Cryogenic Inner Cooling for Extrusion Blowmolding—The Extrublas Process." *Proc., 2n Ann. Blowmold Technical Conference.* Soc. Plastics Eng., November 1985.
9. Schubbach, R. "Optimierung des Quetschkantenbereichs und des Schliessvorgangs fuer Extrusionsblaswerkzeuge." *Plastverabeiter* 10 pp. 608–614 (1973).

10. Peters, D. L., and J. R. Rathman. "Blow Molding Highly Irregular Shaped Parts with Multiple Parting Lines." *Proceedings, ANTEC.* Soc. Plastics Eng., 1985.
11. Irwin, C. *Extrusion Blow Molding Tools.* Center for Professional Advancement (February 1985).
12. "Blow Molding Marlex." Phillips Chemical Co. V.2(2, 3, 5, 6); v.3(2, 3, 4); v.4(1, 4, 6). (March 1961–December 1963).
13. Hasl, H. V. M. "In-mold Labeling: The New Competitive Edge." *Proceedings, ANTEC.* Soc. Plastics Eng., 1986.
14. Rathgeber, J. "Blow Molding: The Evolution of an Imaginative Technology." *Proceedings, ANTEC.* Soc. Plastics Eng., 1987.
15. Peters, D. L. "Blow Molding Highly Irregular Shaped Parts with Moving Mold Sections." *Proceedings, ANTEC.* Soc. Plastics Eng., 1982.

CHAPTER

24

Injection Blow Molds

E. DIETER WUNDERLICH

Containers produced by injection blow molding have become the standard of comparison for quality packages. Intricate, precise neck finishes without flash, smooth bottoms, and skin-friendly surface finishes are the notable characteristics of these bottles. However, the tooling costs, which are generally two to three times higher than for extrusion blow molds, dictate economic limits to this process. Injection blow molds require:

- More complex mold engineering
- Preform molds
- Blow molds
- Support tooling
- Longer lead times

To overcome the economic drawbacks of injection blow mold tools, an average of eight to ten cavity molds must be run in one machine. Such molds consist of 200 to 250 components. To speed up lead time and reduce costs, an injection blow mold is built by several specialized toolmakers each performing a single critical task, such as turning, milling, grinding, polishing, or assembly. Most components not only have to be machined within 0.0005 in. tolerance, but must be concentric, parallel, and most important, square to each other. This is the only way to ascertain that all components will fit and line up during assembly and later will allow consistent production.

MOLD CONSTRUCTION

An injection blow mold is built in several steps, involving several different sections.

Engineering

The major task is the parison layout (Figure 24-1). It requires detailed knowledge of the injection blow process, material behavior, and swell and shrinkage factors (Table 24-1). The outside configuration of the parison is formed by the neck ring and the parison mold. The inside configuration is formed by the core rod.

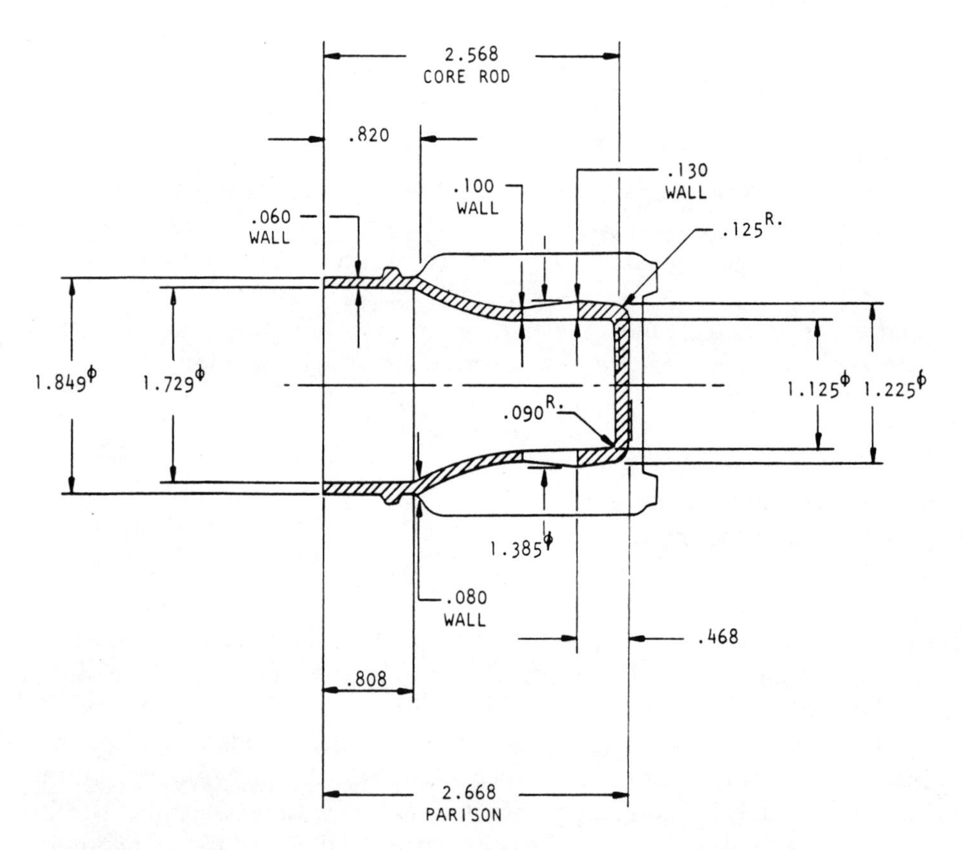

Figure 24-1 Typical parison core rod layout. Courtesy of Johnson Controls/Uniloy/Rainville, and Jomar Corporation.

Table 24-1 Tool shrinkage guide (as molded)

Material	*Recommended Factors and Sizes* E	T	*OH*	*Body*	*Hot Melt Factor*
LDPE	1.018	1.024	1.018	1.020	13.077
HDPE	1.020	1.026	1.018	1.018	12.5
PP	1.020	1.025	1.018	1.017	12.4
GPPS	1.003	1.005	1.005	1.005	16.1
PVC	1.001	1.005	1.005	1.005	20.1
Barex	1.003	1.005	1.005	1.005	18.35
Lexan	1.004	1.006	1.006	1.006	18.3
SAN	1.003	1.005	1.005	1.005	16.5

The above figures are only a guide. Actual shrinkage applied should be arrived by comparing containers of similar size and wall thickness.

Courtesy of Johnson Controls/Uniloy/Rainville and Jomar Corporation.

The parison volume is equal to the final container weight divided by its material density. The key to a good blowing parison is to calculate the volume at injection temperature and shape, within an optimal relationship between a minimum parison wall thickness of approximately .080 in. and a maximum blowup ratio of less than 3:1. This process requires extensive calculations.

Computer programs for calculating parison volume have been developed for regular shapes. However, for extreme oval bottles, and certain engineering materials, the individual calculations of an experienced designer will be required. A safer, but costlier and more time-consuming way is to build a unit cavity mold; purposely undersize the parison; run the mold; and make the necessary revisions by trial and error.

Once the parison shape has been established, the engineering of the mold components and machine tooling can begin.

Parison Mold

The parison mold (Figure 24-2) consists of two components, the body and the block neck ring. Some molders prefer to separate the nozzle area from the body to form a third component. Each component is kept in minimum contact with the other(s) by applying 0.020 in. deep air gaps. This helps to properly zone parison conditioning. A cooling line under the nozzle helps to set up this relative hot area. Often the same cooling line under the neck finish helps to set up the thread to ascertain *T, E,* and ID dimensions within specified tolerances. A third and/or fourth cooling line for circulation

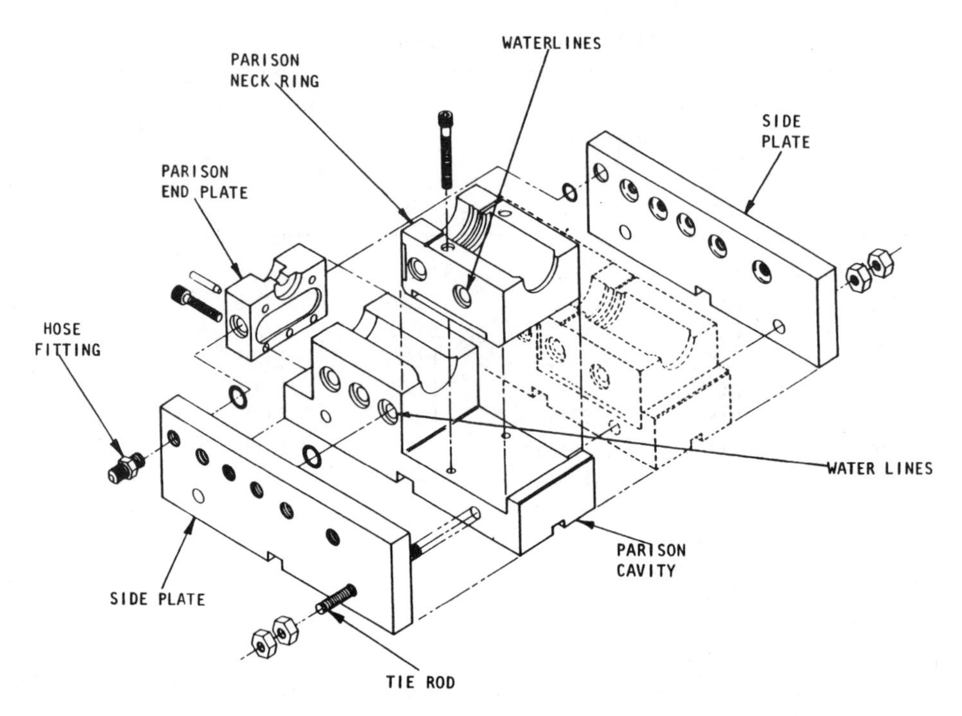

Figure 24-2 Parison mold assembly. Courtesy of Johnson Controls/Uniloy/Rainville, and Jomar Corporation.

of warmer fluid (water or oil) conditions the parison from its thermoplastic to its thermoelastic condition or blowing temperature.

Generally, molds are built with cooling lines in a V shape perpendicular to the parison. This allows proper zoning and parison conditioning all around as close as possible to the parting line. Cooling channels should be at least 0.4375 ($^7/_{16}$) in. diameter, located twice the diameter away from the parison surface to achieve uniform cooling, and spaced 0.875–1.25 in. (min/max) apart to leave room for mounting 0.375 ($^3/_8$) in. full flow fittings. The parison mold is 0.0005 in. narrower than the blow mold to compensate for heat expansion.

The block neck ring sits snug, but 0.0005 in. higher in the parison body. Often a key or dowel pin is used to maintain its center position. The hardened and ground neck ring forms the thread and locates the core rod in its pocket. Absolute concentricity and parallelism of all surfaces are critical to assure perfect centricity of the core rod. Because of wear and tear, parison neck rings are built from A2 air-hardened tool steel. In the case of molds for corrosive materials such as PVC, hardened stainless steel, at 55 to 56 RC, is used.

All surfaces that will be in contact with the plastic are highly polished and chrome plated for ease of fill and release.

Blow Molds

The blow mold (Figure 24-3) forms the final shape of the container. For polyolefin materials the blow mold body and bottom plug are made from 7075-T6 aircraft aluminum alloy. For hard resins such as PS or PC, A2 air-hardened tool steel is used, and for PVC, beryllium-copper alloy or 420 series stainless steel. The neck ring should be A2 air-hardened steel or stainless steel.

The cooling lines are drilled in a V shape to allow maximum circumferential cooling. For soft resins the bottom plug is fixed, but for hard resins it is retractable during mold opening to allow easy bottle release. Proper venting between the block neck ring and the blow mold body and the parting

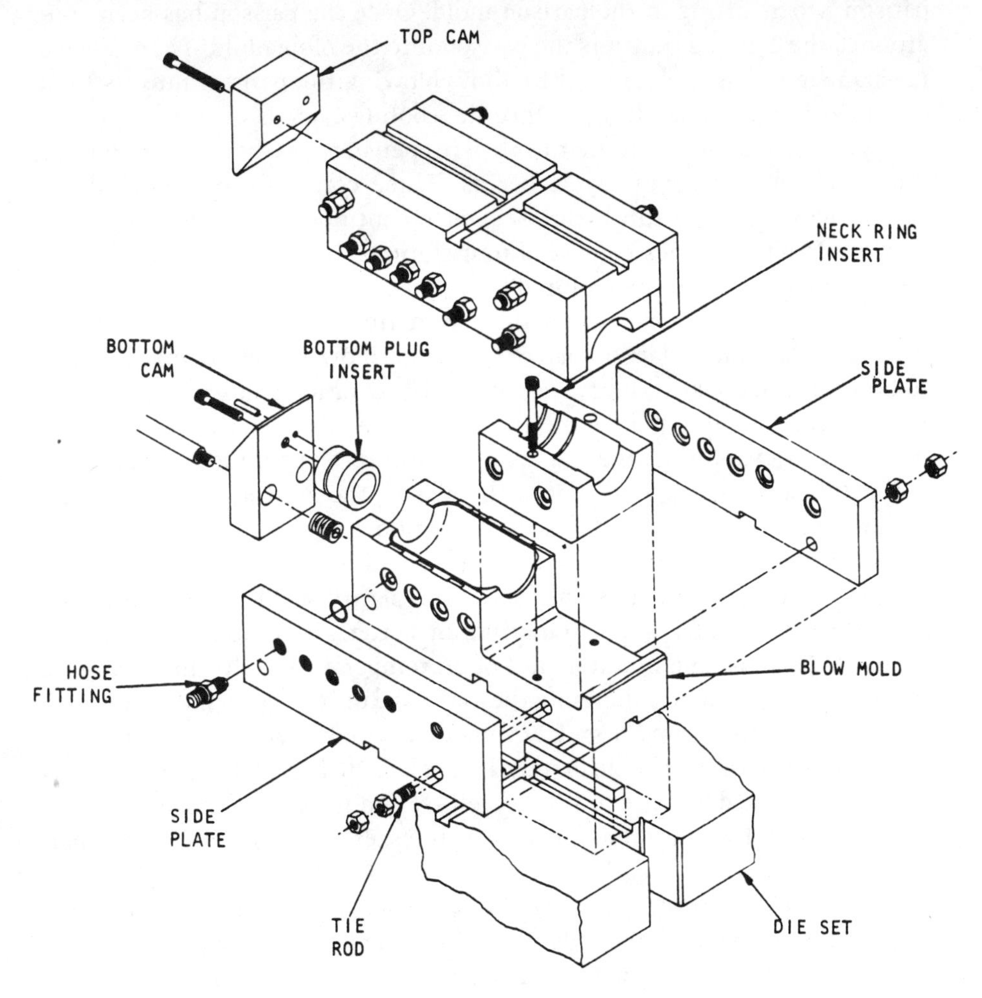

Figure 24-3 Blow mold assembly. Courtesy of Johnson Controls/Uniloy/Rainville, and Jomar Corporation.

line is important. Since the individual cavity blocks are tightly mounted next to each other, a 45-degree chamfer at each upper edge is necessary to let the air escape.

The parison and blow molds consist of individual cavity blocks. They are bolted tightly to each other with O-rings between. At each mold half end there is a steel side plate with $^1/_4$ in. NPT threads for $^3/_8$ in. full flow water fittings. The benefit of side plates is twofold: the outside cavities get full water flow and the threads can easily be repaired in case of damage.

Core Rods

The core rod (Figure 24-4) forms the internal diameter of the neck and parison, when sitting in the parison mold. Once the parison has been conditioned the core rod transfers the parison into the blow mold. There a valve mechanism opens the core rod to allow blowing the parison into its final container shape and cooling it. Once the cooling cycle is over, the core rod transfers the finished container to the stripper station where it is extracted from the rod, and a new cycle begins. Three core rods are required per parison/blow mold to fulfill this sequence continuously. (A few injection blow molding machines consist of four stations. In that case a fourth core rod/parison/blow cavity is required.)

Core rods blow either from the top or from the bottom. For narrow necks and *L/D* ratios larger than 8:1, bottom blow is the preferred choice to attain mechanical stability. For smaller *L/D* ratios and wide necks, top blow is preferred. In top blow the air does not pass through the entire length of the core rod and thus the rod is kept warmer. The core rod temperature is controlled by the applied heat of the plastic, the blow air, and internally or externally applied air in the stripper station.

If specific core rod temperatures are required to achieve certain processing conditions—such as enhancing the clarity in PP or ensuring non-crystallization of PET—air circulation, or better oil circulating, core rods are used. The cost and maintenance of oil temperature controlled core rods are substantially higher; therefore they are used only in special applications.

Core rods are ground from L2 oil-hardened tool steel. All areas that will be in contact with the plastic material must be polished in the flow direction and hard chromed for ease of fill and release. An important detail is perfect concentricity of the internal air passage to ascertain uniform core temperature.

Core rods are hardened to 52–54 RC. This is a bit less than the neck ring hardening, because the rods are easier to replace in the case of damage. The core rod holding diameter is approximately 0.004 in. smaller than the core rod holder diameter to allow proper alignment between the hot parison and the cold blow mold.

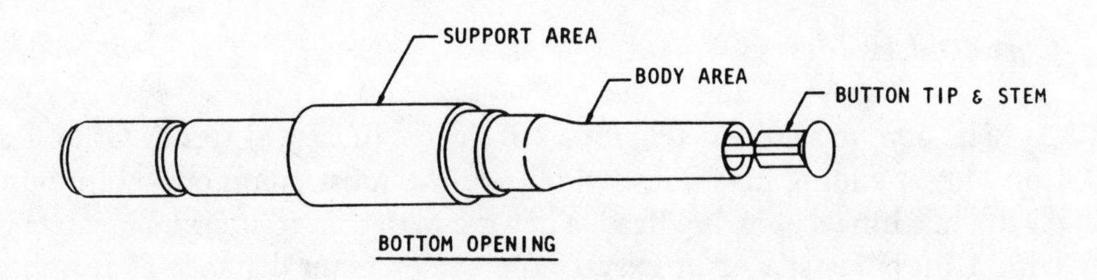

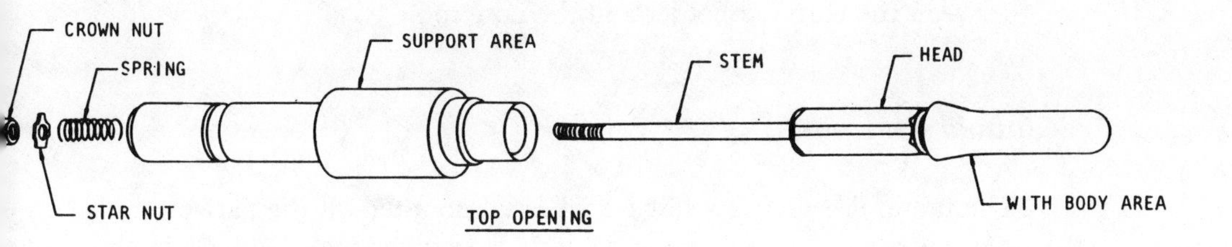

a

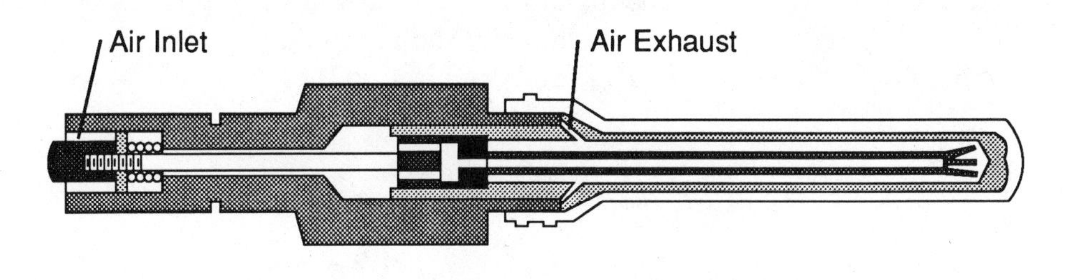

b

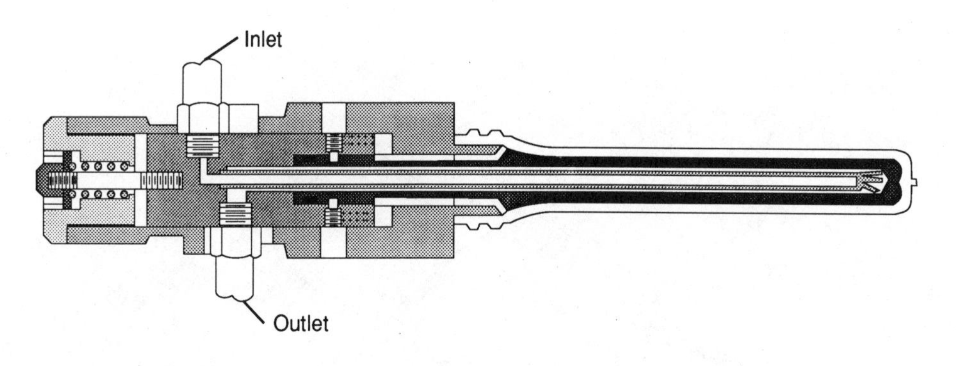

c

Figure 24-4 Core rods. (a) Typical assemblies. (b) Air-cooled rod. (c) Liquid-and-air-cooled core rod. Courtesy of Johnson Controls, Inc., Plastics Machinery Division and Jomar Corporation.

Core Rod Holder

The core rod holders (Figure 24-5)—three in case of three-station injection blow molding machines, which are the most common—are mounted on the machine's indexing head. They are perfectly ground square steel bars with 1 in. bores spaced at exactly the same center distance as in the corresponding mold. The core rods are located in these bores and held in place with core rod retainer rings. Care must be taken when storing these face bars: The back surface must be perfectly smooth to form a seal against blow air leakage between the index head and the core holder. An O-ring seals the gap between the core rod holder and the core rod.

Manifold Assembly

The manifold assembly (Figure 24-6) is mounted on the parison mold die set. The manifold itself must be designed low enough (3 in. high, generally) to clear the core rods with the parisons during indexing. The manifold clamps need to be strong enough to hold the manifold in a perpendicular position during injection. The manifold base is air-gapped to transfer minimal heat to the die set and subsequently the parison mold. Air-hardened nozzle retainers hold the nozzles firmly in place so they cannot leak.

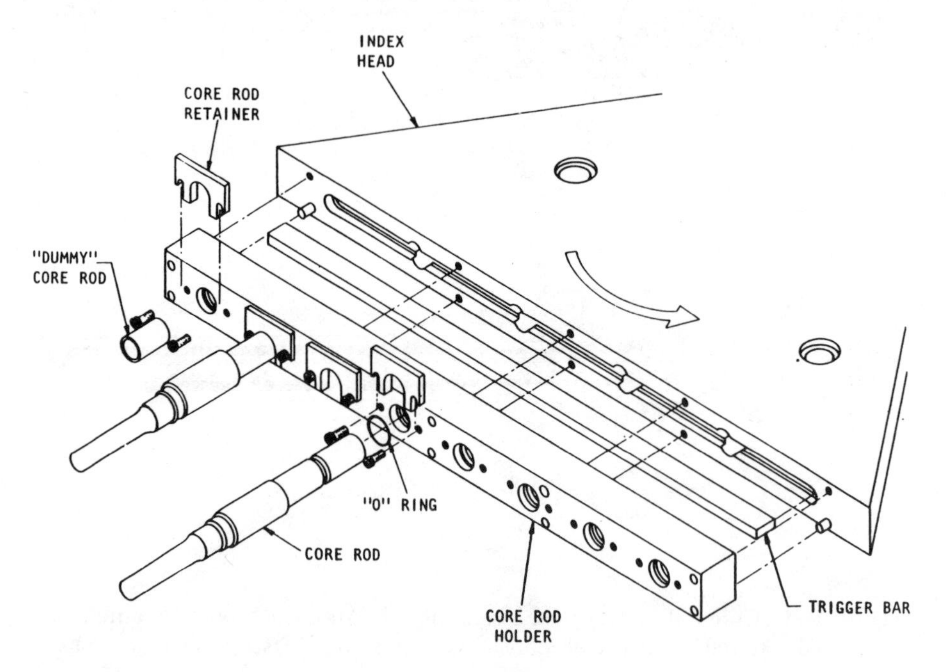

Figure 24-5 Core rod retainer assembly. Courtesy of Johnson Controls/Uniloy/Rainville, and Jomar Corporation.

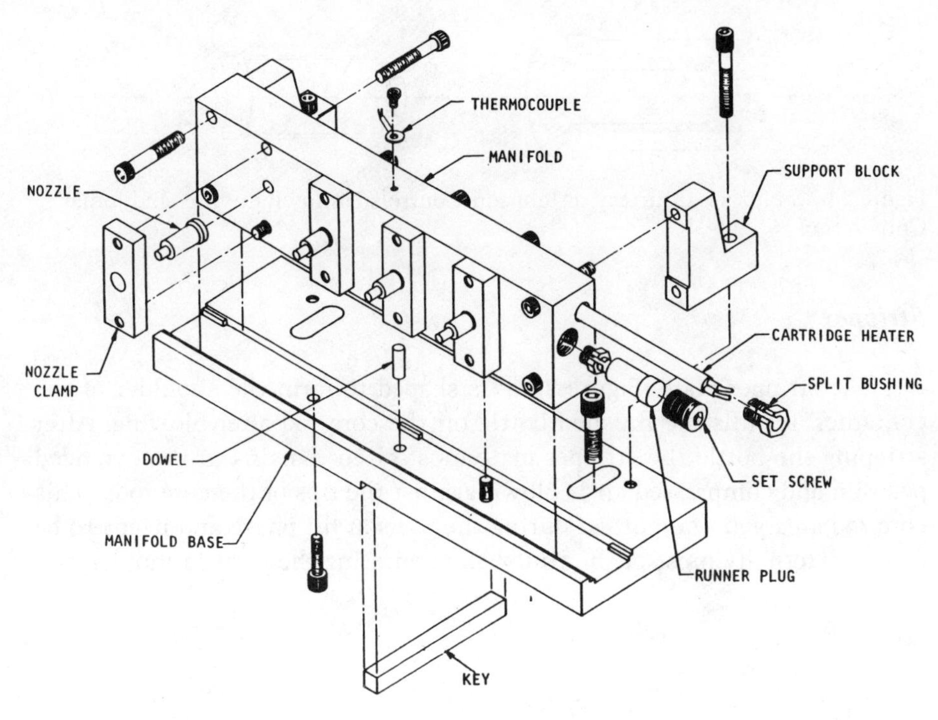

Figure 24-6 Manifold assembly. Courtesy of Johnson Controls/Uniloy/Rainville, and Jomar Corporation.

Electric cartridge heaters on each side should be a half inch longer than the manifold and of equal watt density to ensure uniform temperature throughout the width of the manifold. For conventional raw materials, a gun-drilled 0.625 ($^5/_8$) in. diameter runner is adequate. For heat-sensitive raw materials a two-piece, coat-hanger style manifold with highly polished runner and well-rounded corners is necessary to avoid degradation.

Nozzles

Nozzles (Figure 24-7) are a critical interface between the manifold and the parison mold. They are made from P21 pre-hardened tool steel for ease of machining and should be slightly softer than the parison mold, so that in case of overpacking the nozzle is damaged first. It is easier to replace.

Because the front surface of the nozzle comes in contact with the parison, it is important for it to be polished and chrome plated for easy release.

Nozzle gates for the center cavities are usually 0.040 to 0.060 in. diameter. Each successive nozzle is opened up by one drill size. The aim is to balance the parison filling within 0.125 ($^1/_8$) in. to avoid overheating of the parison that fills first. Nozzles should not be any longer than 1.5 in., to minimize residence time. Nozzles for heat-sensitive materials must be streamlined and polished internally to avoid hangups and thus degradation.

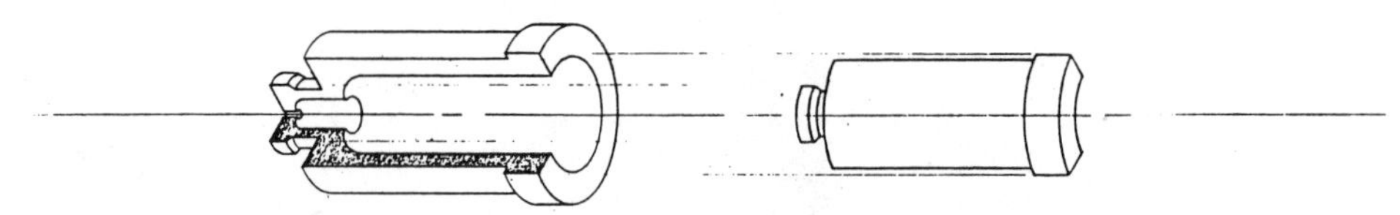

Figure 24-7 Nozzle. Courtesy of Johnson Controls/Uniloy/Rainville, and Jomar Corporation.

Stripper

The stripper plate (Figure 24-8) is shaped to form the shoulder of the container. It pulls the finished bottle off the core rod after blowing. After stripping the bottle the stripper plate most often remains in the extended position and compressed air is blown against the tips of the core rods. This core rod area gets the hottest during the parison fill phase and needs to be cooled externally most of the time before entering the parison mold again.

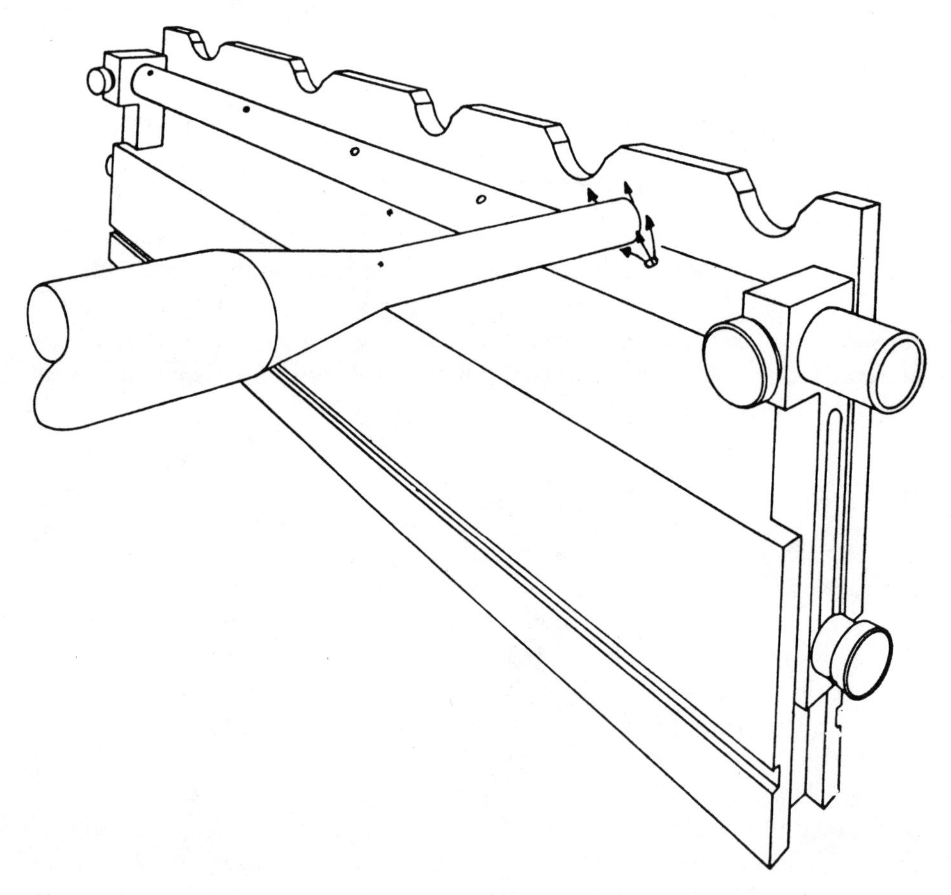

Figure 24-8 Stripper. Courtesy of Johnson Controls/Uniloy/Rainville, and Jomar Corporation.

A refinement of the stripper is called stripper-tipper. It grabs the finished containers by the neck finish, extracts them from the core rods, and releases them on to a conveyor in a stand-up position.

Die Set

The parison and blow mold cavities are assembled on individual die sets (Figure 24-9). Despite the key ways in two directions, the assembly requires a skilled craftsman to line up all cavities within 0.0005 in. For this reason the molds should remain on the die sets. During molding it is important to grease the die set leader pins and bushings daily. Die sets facilitate mold alignment in the machine and mold changeover, and protect the molds from damage during transport.

Before each mold changeover it is important to stone all external die set surfaces to eliminate any burrs. Even small burrs can cause mold misalignment.

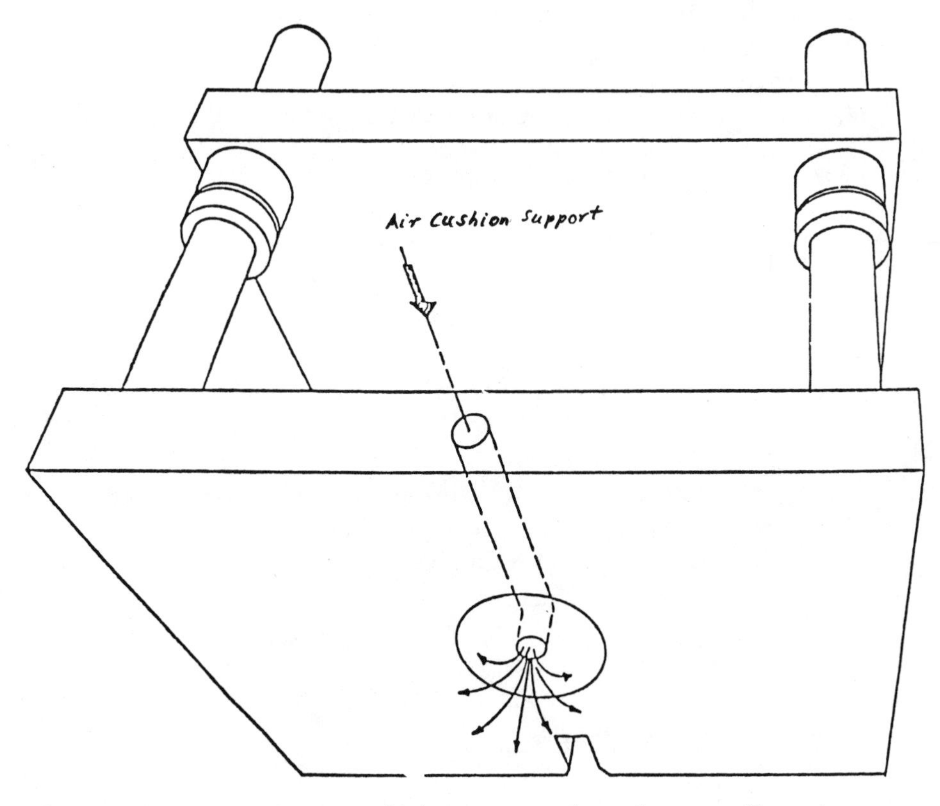

Figure 24-9 Die set. Courtesy of Johnson Controls/Uniloy/Rainville, and Jomar Corporation.

Checklist

The detailed checklist given as an appendix to this chapter covers all components of a blow mold. It can be used as a guide for designing or ordering molds, or as an inspection checklist before approving or accepting a new mold.

*CHECKLIST NEW MOLDS**

PART

1. Is part drawing approved?
2. Have you read all correspondence about job?
3. Is type of plastic, additives, and color specified?
4. Is function of part understood?
5. Can changes be made to make it simpler, better, easier to mold part?
6. Are number of cavities correct for annual volume?
7. Are tolerances indicated on critical dimensions?
8. Can these tolerances be maintained?
9. Is flame treating necessary?
10. Is cycle time realistic?
11. Are dimensions given including or excluding shrinkage?
12. What shrinkage factor was used?
13. Has adequate draft been specified?
14. Where does draft start and stop?
15. Have tapers been specified?
16. Has gate location been approved?
17. Has type of gate mark been specified? (telescoping-standard-ball-0.375, 0.250, or 0.187 in. dia.)
18. Is stripper mechanism sufficient?
19. Has mold finish been specified?
20. Is neck finish specified?
21. Has engraving master or art work been approved by customer?
22. Is engraving specified for size, type, and depth?
23. Are oil circulating core rods needed?
24. Is high-pressure blow needed?

MACHINE

1. Is clamp tonnage sufficient for parts?
2. Is swing radius sufficient?

*This list is courtesy of Johnson Controls/Uniloy/Rainville and of Jomar Corporation.

3. Will molds (blow and injection) physically fit in machine specified and clear safety gate?
4. Is trigger bar length sufficient?
5. Is opening stroke sufficient to transfer part?
6. Is injection capacity of plasticizer sufficient?
7. Are spacers needed for stack height?
8. Do water lines interfere with tie bars?
9. Is system for stripper-tipper and ejector pins needed?
10. Does machine need mixing nozzle?
11. Is 3-pressure inject needed?
12. Is 2-pressure blow needed?
13. Is continuous screw rotation (CRS) needed?
14. Is special screw needed?
15. Is high-pressure blow needed?
16. Is rotary union for circulating core rods needed?
17. Is process control recommended?
18. Do parts hit safety gate when stripped?

MOLD DESIGN

Injection

1. Is cavity steel specified right for part? (e.g., SS for PVC, P20, etc.)
2. Is type of plating specified?
3. Is nozzle size correct? 0.187, 0.250, or 0.375, or special?
4. Is nozzle seat inserted or air gapped?
5. Is heat isolation at neck ring or holding diameter required?
6. Is neck ring a special material (Be-Cu, SS, etc.)?
7. Are water ports located in right place?
8. Is there a sufficient number of water ports?
9. Are water ports drilled as we want?
10. Are molds vented proper amount for type of plastic to be molded?
11. Has parison/core rod layout been reviewed?
12. Does parison/core rod layout need any changes?
13. Is number of cavities correct?
14. Is cavity width correct?
15. Are cavities marked with numbers? Top or bottom? And job number?
16. Has heat treat been specified?
17. Is draft indicated sufficient?
18. Has heat expansion been calculated?
19. Is neck finish correct?

20. Are alignment pins needed?
21. Are alignment pins securely installed and spaced correctly?
22. Are alignment pins in correct half of mold?

Blow

1. Is cavity material specified suitable for part?
2. Is cavity finish specified? (e.g., polish, sandblast, etc.)
3. If cavity is sandblasted, is grit specified?
4. Is heat isolation at neck ring required?
5. Is neck ring a special material?
6. Are water ports located correctly?
7. Is there a sufficient number of water ports?
8. Are water ports drilled as we want?
9. Are molds *vented properly* and *sufficiently*?
10. Are alignment pins needed?
11. Are alignment pins securely installed and spaced correctly?
12. Are cavities marked with numbers? Top or bottom? And job number?
13. Is number of cavities correct?
14. Is width of cavities correct?
15. Is draft indicated and is it sufficient?
16. Is draft beginning and end specified?
17. Has customer approved art work or masters for engraving?
18. Is all engraving specified *exactly* for depth, size, type, and location?
19. Is bottom plug material specified?
20. Are bottom plugs correct? Push-up, shape, and/or special?
21. Is cooling of bottom plug necessary?
22. Are drive lugs needed in bottom plugs?

Core Rods

1. Has parison/core rod layout been reviewed?
2. Has core rod shape been reviewed?
3. Does core rod shape need any changes?
4. Is draft at neck finish adequate?
5. Are transition areas blended enough and/or smooth enough for material to be molded?
6. Is blow opening position best for part?
7. Is core rod holding diameter long and wide enough?
8. Is sealing diameter in correct spot for material to be molded?
9. Is dimple needed in bottom of tip?

10. Is steel specified?
11. Is plating specified for type and thickness?
12. Is heat treat/hardness specified?
13. Is grinding specified after heat treatment to straighten?
14. Are instructions for machining specified? (e.g., concentricity, straightness, and fit to neck ring)
15. Is air channel big enough? 0.281 standard.
16. Any deviation from 0.281 air channel *must be approved* before released.
17. Has air flow over tip been calculated?
18. Air flow over tip must match 0.281 air channel whenever possible.
19. Do any areas need vapor hone for release?
20. Are all radius areas sufficient?
21. Any radius changes suggested?
22. Is tip rod correct diameter?
23. If special material for core rod is specified, do we agree?
24. Are core rods temperature-controlled?
25. Are temperature-controlled core rods necessary?
26. Is circulating system adequate?
27. Are any changes suggested?
28. Are return springs right tension?
29. Are seals and O-rings right material for fluid?

Stripper

1. Is stripper sufficient for part?
2. Is contour right? And length and position to strip part correctly?
3. Is material specified?
4. Is center distance correct?
5. Is height correct?
6. Is number of cavities correct?
7. Is there enough adjustment for height?
8. Is tipper mechanism needed?
9. If tipper is used is there enough clearance not to crush neck finish?
10. Are ejector pins necessary?
11. Are bottle support rods needed?
12. Is tip cool fitting and tubing correct?
13. Is tip cool channel correct?
14. Is orifice size for tip cool holes specified?
15. Is holding slot correct size and location?
16. Is stripper marked for job?
17. Are all bolts specified?

MACHINE PARTS

Core Rod Holders

1. Number of holders specified: 3 or 4?
2. Is number of cavities correct?
3. Is center distance specified and correct?
4. Is hole size correct for core rods?
5. Are mounting holes correct and sufficient?
6. Are alignment pin holes correct (size and location)?
7. Are O-ring slots correct?
8. Is O-ring size specified and stamped on holder?
9. Are bolt sizes specified?
10. Are there any changes from normal?
11. Are the changes correct?
12. Is steel specified?
13. Is stress relieving, thickness, and height specified?
14. Stamped with center line distance, part number, and job number?

Core Rod Retainers

1. Is number of retainers specified?
2. Is thickness specified?
3. Are holes made to fit core rods (slotted-round)?
4. Are hole locations correct to core rod holder?
5. Are bolt holes correct size and/or slotted?
6. Is steel specified?
7. Are retainers individual units?
8. If retainers are not individual, why?

Manifold

1. Is manifold correct length?
2. Are center distances correct?
3. Is manifold height correct?
4. Will core rods transfer over manifold?
5. Is manifold to be air gapped from base?
6. Is RTD location hole in right place and tapped correctly?
7. Are double runners needed?
8. Are heater rod holes correct?
9. Are heater rods specified?
10. Are heater rods correct for length, diameter, voltage, and wattage?

11. Is manifold special for fluid circulation?
12. Is fluid porting correct?

Nozzle

1. Is steel specified?
2. Is steel specified right for application?
3. Is hardness specified?
4. Are all undercuts and tapers specified?
5. Is diameter correct for mold?
6. Is length correct for mold?
7. Is special finish (plating) required at orifice surface?
8. Are orifice sizes specified?
9. Are orifice sizes engraved on nozzles?
10. Does nozzle match retainer and manifold?

Nozzle Retainer

1. Are retainers right thickness?
2. Are thicker retainers necessary (excessive injection press)?
3. Is clearance correct for good seat against manifold?
4. Do holes match manifold drilling?
5. Is bolt size specified?
6. Will center hole match nozzle?
7. Is steel and hardness specified?

GENERAL

1. Have spare parts been specified?
2. Are quantities for all parts specified?
3. Chart for bolts, O-rings, nozzles, heaters, and special mechanisms?
4. Are all parts numbered and identified?
5. Scheduled completion dates for mold?
6. Is mold for export?
7. Are metric sizes specified for parts?

CHAPTER

25

General Mold Buying Practices

CIRO PETRUCELLI

Proper procedures for the procurement of molds to produce plastics products have evolved to become a standard practice known to all in the industry but followed by only a few. This is due mainly to the ever-increasing pace of a growing and changing industry trying to keep up to the demands of its customers to produce more sophisticated parts, with more sophisticated resins, at lower prices, and with as little a lead time as is possible.

As part of this industry, blow molding may well be the most rapidly growing and changing process today. Technological advancements in resins, molding equipment, and product applications have caused industrial designers to regard blow molding seriously as a mature process capable of producing products other than toys and trash cans.

The procurement of molds for the blow molding process is developing into a more technical responsibility than it was just a few short years ago. Although it may not take a master mold maker to buy molds, a good general knowledge of blow mold features and requirements is essential background for making the considerable investment that molds represent in overall equipment costs.

Blow molding encompasses three subindustries. The first is bottle molding—blowing bottles of all shapes and sizes for the packaging industry. This is the largest area of the blow molding industry in terms of the number of parts molded each year.

Second is industrial blow molding, which produces a vast range of products and is not limited to any particular form or size. Blow molded parts from small toys weighing just a few grams to large 90- and 330-gallon containers weighing 35 pounds or more are typical in this part of the industry. Automotive blow molded parts are also in this category. Plastic windshield washer jars, radiator overflow jars, and air conditioning and heating duct work make up a large volume of the industrial segment of the blow molding industry.

The third part of the industry is devoted to developments with blow moldable grades of engineering resins. This is in its infancy, but great strides have been taken and blow molded structural components are beginning to find their way into a variety of markets.

REQUEST FOR QUOTATION (RFQ)

Mold purchase begins with requests to one or more mold makers for price and delivery quotations. The initial request for a quotation may be the single most important phase of the mold procurement procedure in determining the magnitude of the investment to make.

Generally speaking, the accuracy of the mold quote in both price and delivery is directly related to the depth information, or lack of it, given to the mold maker. One of the most reliable methods of providing this information is to develop a standards document that not only spells out the requirements for the particular part, but also incorporates the purchasing company's standard mold requirements.

Because the dimensional envelope of a blow mold is based on the particular part to be molded, it is very important to address the molding press specifications of minimum and maximum shut heights and distances between tie bars. Unlike the standardization one finds on injection molding presses, blow molding presses of the same clamp or shot size capacities may vary greatly, depending on the press manufacturer.

The next most important piece of information the mold maker must know is the requirement for a machines mold or a cast mold. These are discussed in more detail later in this chapter to provide information that will help in making a choice.

In addition to the above information, the data in the following list of blow mold requirements should be included in an initial RFQ to help the mold maker provide an accurate price and to assure comparative quotations.

- Blowing Provisions: Top blow, bottom blow, or needle blow? (Location).
- Pinch-offs: Top and bottom only, or 100%? Pinch-off plates water cooled—Yes? No?. Pinch-off plate material, if inserts.
- Thread inserts: Material. Water cooled—Yes? No?

- Logo Areas: Artwork availability. Inserted—Yes? No? Interchangeable—Yes? No?
- Sectioned Mold: Indicate split(s).
- Prepinch Device: Approximate stroke.
- Slides: Method of movement.
- Knockout System: Method of actuation; location of pins.
- Texture: Defined area; specific pattern to be quoted. Sampling prior to texturing—Yes? No?

The above information is rather general. While it will serve as the basis for a preliminary quotation, the kind of detailed information a mold maker requires to make a final quotation and begin work is shown in Figure 25-1. The checklist at the end of chapter 23 is also a useful guide to the many details that must be considered.

Date ____

Company ____

Part Name ____

Part No. ____

Molding Press ____

Shut Height (min) ____________ Length × Width (max) ____________

Platen Layout ____

Plastic Shrink ______________"/" Model Furnished: Yes____ No____

No. of Cavities ________ Mold Sections ________Knife Insert ________

Type: Cast Alum. ________ Mach. Alum. ________ Other ________

Cooling: Drilled ________ Channel Back ________Cast Tubes ________

Type Blow: Top ______ Bot. ______ Blow Pin Size ______ Dia.______

Air Cyl. ________ Needle Size ________

Pinch-off: 100% _____ Ends Only _____ Inserted _____ H_2O _____

Inserts: Alum. ________ Steel ________ Bronze ________

Core Pulls: Mech. _______ Cyl. _______ Air _______ Hyd. _______

Prepinch: ______ Teflon Coated ______ Spring ______ Cyl. ______

Water Cooled ________

Thread Insert: Info ______ Steel ______ Alum. ______ H_2O ______

Engraving: Cast in ______ Inserted ______ Removable ______

Interchangeable ____________ Dial-A-Date ____________

Cavity Finish: Text Info ____ Polished ____ Blasted ____ Grit ____

Riser System: Rails _____ I-Beam _____ Alum. _____ Steel _____

Eject. System: Cyl. Act. ______ Spring Act. ______ Location ______

Mold Mounting: Flanged ______ Tapped Holes (size & layout) ______

Mold Identification: ______________ Trade Mark ______________

Cavity No. __________ Part No. __________ Location __________

Venting: Drilled Holes/Size ______ Core Vents ______ Other ______

Figure 25-1 Blow mold information sheet.

MOLD MATERIALS

The most commonly used material for producing blow molds is aluminum. The mold may either be machined from a solid block or cast. With the development of engineering resins, machined steel molds have also started to play an important part in the blow molding industry.

Machined Aluminum Molds

The two most commonly used aluminum alloys for machined cavities are 6061-T6 and 7075-T6. The 6061-T6 alloy, although not as hard as the 7075T6 Alloy—only 95 Brinell compared to 150 Brinell—has been used to a greater extent because of its better weldability. This can be an important factor for later changes and for mold reconditioning.

The method of construction using these alloys is to cut the cavity shapes directly into the blocks by conventional milling, one-to-one duplication, or by CNC machining centers.

Drilled water lines placed approximately 1 in. from the cavity and approximately 2½ in. on center are then drilled to surround the cavity with proper cooling.

Around the periphery of the cavities, the pinch-off is machined and left standing along with the pinch-off relief that allows the flashed areas to cool properly. Because of the mass of hot plastic in the flash area, attention to proper cooling is very important. Failure to cool this area properly could result in the hot flash curling up on the part proper and destroying the usefulness of the product being molded.

In addition to the lines for cooling, drilled lines must also be engineered for proper venting. The placement of vents is based on the shape of the product being molded. Vents are generally designed to provide an exhaust route for the air that is trapped on the back side of the parison as the parison is blown against the cavity.

Usually, vents are placed in the deepest parts of the cavity, in pockets that could trap air, and in areas such an engraved lettering to insure good definition. In many cases, pattern maker's core vents can be used. These are made of aluminum and have a number of 0.010-in.-wide slots milled in the face, which is only 0.060 in. thick. Depending on the area, these core vents can be ¼, ⅜, or ½ in. diameter. In textured areas, blank core vents without slots are installed in the cavity and after texturing a cluster of tiny holes is drilled through the face of these blanks to provide venting.

If pinch-off is on the ends of the part only, ½-in.-wide parting vents, 0.003–0.005 in. deep, along each side of the cavity, (one half only), should also be considered.

The overall cavity finish, if not textured, should have a sandblasted finish of approximately 60 grit. This aids in venting the trapped air and allows channeling to the main vents.

Other areas of venting to be concerned about are pin vents in a recessed boss, under inserts, in logos and lettering, and inserts in pinch-offs.

Cast Aluminum Molds

The most commonly used aluminum casting alloy for blow molds is 356A because of its good pouring characteristics in the foundry, and its good machining characteristics in the machine shop after casting.

Because blow molding is a low-pressure molding process, cast aluminum molds can stand up to even the highest production rate requirements for parts of all shapes and sizes.

The construction of cast aluminum molds starts with the development of a shrink model. The model incorporates two shrinkages, one for the casting, and one for the plastic. Usually these models will look exactly like a solid replica of the finished part with the exception of the expansion caused by the two shrinkages.

The cavity thickness of approximately $\frac{1}{2}$ in., the outer walls of approximately 1 in., and any support ribbing are all generated in the foundry during the course of producing the castings.

In cast aluminum molds cooling can be achieved by two methods: channel back (also referred to as flood cooling), and cast-in stainless steel tubing lines cast directly on the back side of the cavity metal thickness.

In channel back molds, back plates and any intrusion through the flooded back of the mold must be sealed in some fashion. Generally back plates are gasketed and screw holes are sealed with O-rings. Vent holes must be also designed around the water jacket.

In tubing molds, back plates may not be required. Screw holes and venting need not be sealed, but care must be taken not to drill or machine into the tubing.

Depending on the part and mold size, either $\frac{3}{8}$ or $\frac{1}{2}$ in. OD tubing can be used. The tubing is bent in a serpentine fashion on the back side of the cavity metal approximately 2–3 in. on center. The foundry procedure to accomplish this is done very carefully to make sure that the tubing stays in contact with the aluminum at all times and is surrounded sufficiently with aluminum to provide the contact required to dissipate heat quickly.

Correctly designed and constructed cast aluminum molds have proven to be good, reliable, and economically sound investments for producing industrial blow molded products.

OTHER AREAS OF CONCERN

One of the most important areas in any blow mold is the proper design of the pinch-off blade and its relation to the material being molded. Also, the proper depth of the pinch-off relief is critical to provide adequate cooling to the head or tail flash area.

If the pinch-off blade is too sharp, it will cut through the parison like scissors and the molded part will have a poor weld line. If the blade is too broad it may tend to hold the mold open, creating a thick web and hampering the trimming operation.

Proper blowing of the part is dependent on the geometry of the part and/or the molding press. Some molding presses only allow blowing from the head, commonly referred to as top blow. Others may only allow blowing through a blow pin positioned at the base of the machine; this is called bottom blowing. In either case, and depending on the part to be molded, both kinds of presses will allow blowing by use of a cylinder and needle through the side or back of the mold.

When top or bottom blow is used, steel inserts should be used to pinch off around the blow pin, rather than aluminum, due to the vulnerability to damage in that area of the mold.

CONCLUSION

The procurement of blow molds is an investment consideration that cannot be taken lightly. New developments are in process that must be evaluated carefully. These include computer-aided mold cooling and computer-aided preparison control of part thickness control. Laboratory testing by the major material suppliers is presently underway to develop new blow moldable grades of polymers, and equipment manufactures are preparing their machines to process these new polymers.

The blow molding industry is producing more sophisticated parts, in more sophisticated molds, in shorter lead times than ever before. And the industry has only begun to scratch the surface of its potential.

CHAPTER

26

CIM for Plastic Part Manufacturing and Mold Design

TIMOTHY WAGNER

Computer-integrated manufacturing (CIM) utilizes data from various computer systems throughout a company to perform the functions associated with all aspects of the design, manufacture, management, and support of plastic blow molding operations. A major feature of CIM is that manual reentry is not required in order for data from one system in the company to be used in another system; intercommunication, data transfer, and coordination are all computer controlled and facilitated. Thus CIM maximizes productivity through severe reductions in clerical tasks and paperwork, which allows management and staff to apply their efforts to improving product quality and customer service.

In creating a CIM environment in a blow mold factory, there are two characteristics of data that will often dictate CIM design considerations. First, purchase orders for blow molds can range from one cavity to several cavities per set, and each order may be for one or multiple sets. Second, voluminous amounts of data are used in individual files, and there may be a great number of individual files per order. CIM design considerations must take these characteristics into account.

In order to make the integrated system compatible with the people who will be using it, it is imperative to reduce all individual functions to simple menu selections or single tablet selection at the workstation. Long commands or detailed procedures will frustrate the users and reduce the utilization of company resources to meet production and sales goals. To maintain organization of the multitude of data files being generated and used, a standard file naming convention must be used throughout the company. The convention should be kept to an abbreviated form. Naming conventions utilizing more than ten characters for file names should be reexamined and perhaps simplified to minimize confusion.

Once the factory is integrated, data will proliferate at an exponential rate. Provision must be made to remove unneeded data while keeping current data and useful historical data active in the system. Ideally, there should be software on one of the computers in the factory that will allow the removal and reinstallation of all the data required for a job in an easy way. Removing sections of data for a job from different computers will make it confusing to reload a job when a repeat order is received. If a removal-reinstallation program is not used, automated cross references must be available to locate individual files for reloading into the individual computer systems.

This chapter examines techniques to share data between different computer systems by transporting and converting data to the correct formats. The key computer systems within a blow molding factory are:

- Computer-aided design (CAD)
- Computer-aided manufacturing (CAM)
- Process planning (PP)
- Direct numerical control (DNC)
- Coordinate measuring machines (CMM)
- Shop floor graphics (SFG)
- Data management system (DMS)
- Management information reporting (MIR)
  - Shop floor scheduling
  - Labor recording
  - Current job statusing
  - Analysis of company competitive performance

There is a wide variety of each of these systems on the market. As technology progresses, even more vendors will offer programs and systems covering additional applications, and for different industries. The process by which vendors are chosen is not covered here; however certain capabilities required for integration are addressed. This chapter deals specifically with how to share data in an integrated manufacturing operation.

DATA SHARING

To share data among the several technologies used in plastics blow molding production, applications software and command scripts must be developed to:

1. Extract the desired data from the application technology in ASCII format
2. Transmit the data to the target computer
3. Convert the data to the proper format for the target computer
4. Have the target computer read the data into the target software application

Because various combinations of computer hardware and software are possible, integration implementation will be explained in terms of the above steps for each possible interface.

Data Extraction

Each technology is executed with specific data relevant to the particular software used for that application. In order to share data, specific data must be segregated from the application and stored in ASCII format on the system. There are three basic ways to accomplish this: modification of software, redirection of standard output of reports to disk, or software development to perform the extraction.

If access is available to the source code for the software, it can be modified to output the data in an ASCII file, with a specific layout. This method is useful with an application that performs a specific function. By reading the source code, it is possible to output the data in a specific format, thus eliminating subsequent conversion for input to another technology.

The second method is to redirect a standard report to disk rather than the usual printout device. If the report contains all the data needed, a simple conversion program can be written to reformat the data for subsequent use.

The final method is to write software that will go directly to the data files and output the required data in the desired format. Since the data structure is usually proprietary information and difficult to decipher without vendor support, this method is the most difficult to use.

Data Transmission

Once the data are extracted, the ASCII data file may or may not have to be moved to another computer. With networking options growing, it may be possible to access the data file from another computer without physically moving it. If this option is not available, there are several key concepts to adhere to.

When sending data files between computers, it is highly probable that it will be desirable or necessary to transfer all the files ready for transmission to a specific computer. For example, there may be several different types of data files resident on the CAD/CAM system that must be moved to another computer. There are several ways to perform this task.

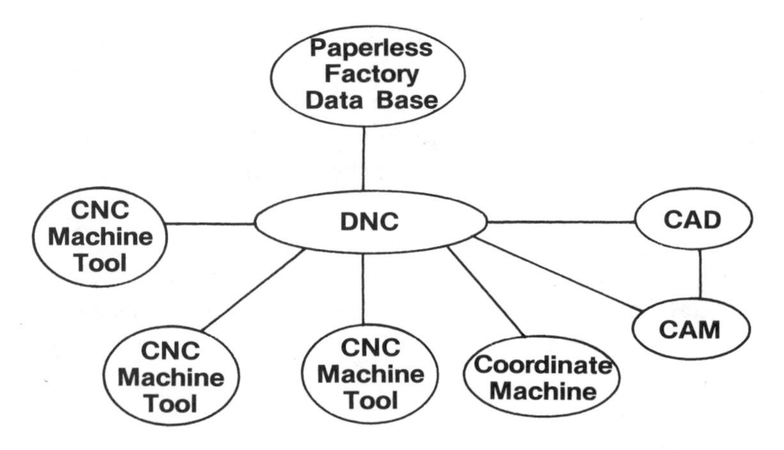

Figure 26-1 Factorywide computer-integrated communications network.

Assume that the DNC system can emulate a terminal on the CAD/CAM system (see Figure 26-1). Assume also that file transfer software is available on both systems; that at least one of the CAD/CAM terminals is multiuser and multitasking; and that all the data files of different types are stored in one accessible directory on the CAD/CAM system.

To make these files available to the system, the applications that generated them must have adhered to two specific procedures. First, the files of extracted data must have been named according to the established convention. That might be a six-character job or category identification, followed by a three-character extension that is a descriptive abbreviation, drawn from a standard list of abbreviations. For example, the file name of a drawing in plotter format for job number 123 would be JOB123.PLT. Second, all such files must have been stored in a particular CAD/CAM directory.

In order to make the transfer between the two computers, a command is issued from the DNC computer that instructs the CAD/CAM system to send a list of all the data files in the specified CAD/CAM directory to the DNC system. When this directory list is received on the DNC system, a program is executed to read it and output a macro that will process the data files when they are transmitted by CAD/CAM. The macro is executed on the DNC computer to receive and process data from all files called for on the basis of the naming convention, for example, all JOB123 files, or all *123.PLT files. Depending on the extension of the file name (i.e., PLT), each file is processed.

This technique may sound complicated, however, the basic concept is to tell the receiving computer what is to be sent, and then to send it. When the receiving computer knows what is coming, it can get ready to receive and process the data. If done correctly, intersystem file transfer can be done with little or no manual intervention.

Data Conversion

The receiving computer will generally have a standard input format to which data must conform. Data conversion is the process by which the extracted data are converted to the format acceptable to the receiving computer. This conversion can take place during data extraction, during data transmission, or as an independent step.

In order to minimize manual interaction, the data can be converted when they are received from the other computer. This conversion usually requires software to rearrange data, and because the vendor does not usually supply conversion information, frequently requires mathematical manipulations to mate with the target computer.

Data Acceptance

Data acceptance is the physical assimilation of data into the receiving computer so that is readily available to the user of the receiving computer. Again, this can be performed directly after the data are received by including the necessary loading procedures in the macro that receives the data.

INTERFACES

In a CIM plant, systems used in various operational areas must interface—be linked for intercommunication—with those in one or more other areas. The interfaces for efficient operation are discussed here.

Computer-aided Design (CAD) to Computer-aided Manufacturing (CAM)

At present there are three basic choices available in purchasing CAD/CAM computer hardware and software. First is the classic turnkey system provided by a single vendor. It has totally integrated hardware and software—nothing additional is required to pass geometry from CAD to CAM.

Second is the workstation solution. It is similar to the turnkey system except that CAD/CAM software is obtained from one vendor and the fully compatible computer hardware is obtained from a second vendor. Again, little or no additional integration between CAD and CAM is required.

The third choice is similar to the workstation solution except that there are two software vendors, one for CAD and one for CAM. Generally speaking, software is available to create an ASCII file of the CAD geometry and other software to read that ASCII file into the CAM software such as a IGES which is a standard in the CAD/CAM industry. Even though software exists to accomplish the task, it is necessary to set up a command or tablet selection to execute this software to create the ASCII file. The next tablet selection will be a means of logging off the CAD software and logging onto the CAM software to read in the ASCII CAD file for subsequent use in CAM.

If CAM is performed by different people in different areas, these tablet selections must file or recall the CAD ASCII file from a predetermined filing location. Here the naming convention used through the factory will facilitate recalling the file. With intervendor coordination, these utilities can be set up by the vendor's applications engineer during installation.

CAD to Shop Floor Display of Drawings (SFG)

The purpose of this technology is to provide people throughout the plant with paperless access to CAD drawings without spending a considerable amount on a workstation.

The system on which the drawings are to be seen must be able to create graphic displays from data stored in files, not just inputted locally. For the shop floor, this would be the direct numerical control or DNC system. The following steps delineate one possible implementation.

1. Designer completes drawing on CAD system.
2. Designer makes a tablet selection to create a data file called JOB123.PLT, which contains the drawing in ASCII plotter format. This plotter file is stored in a specific directory with all other data files being sent to the shop floor.
3. The plotter file, which is a record of a series of lines and arcs, travels to the DNC computer where the data are converted to the input format that the DNC system will accept. The conversion of the data from CAD/CAM system format to DNC system format can take place anywhere along the route the data must travel to get to the target computer. Conversion may not be necessary if the DNC system accepts the data format that the CAD system generates.
4. Once the data arrive at the target computer, provisions must be made to accept/assimilate/load the data into the target computer so that the user will have access to it with minimum delay.
5. When the user displays the drawing, the target computer must have display-only software to pan and zoom the drawing without modifying it.

When providing this capability, procedures must be instituted to maintain only current drawing revisions on the system.

CAM to Shop Floor Display of NC/CNC Cutting Instructions

The technology of providing a graphical display of numerical control/computer numerical control machine tool cutting instructions is a mature technology called backplotting. Backplotting provides the machine tool operator with a display on the shop floor terminal of the cutter motion to ensure that the quality of data will generate the correct part and that the integrity of the data on the shop floor is the same as when it was generated on the CAM system.

There are many kinds of backplotting displays. The conditions on the shop floor (coolant, dirt, grime, smoke) will dictate the type of terminal required. That in turn may limit what kind of backplotting display may be used. Assuming shop floor conditions are acceptable, there are technologies on the market that will show the cutter motion as a series of lines and arcs connected together. A more advanced display shows cutter motion with the tool diameter incorporated. There is also technology that shows the cutter moving across a block of material in solid form, to provide the machinist with a picture of the emerging, and finally finished part.

Depending on how the coordinate systems are established for the CAD drawing and the NC/CNC data, it is possible to overlay the two on the same display. This requires consistent matching of the drawing origin print (0,0) and the origin of the NC/CNC tool path in all files and documents.

Since the DNC terminal at the NC/CNC machine tool will also be required to send data to the machine tool during machining, the speed at which data are displayed is important. One possible scenario is for the operator to start the display while he is setting up the workpiece in the machine tool so that he can check the display before starting machining. Subsequently, the time required to establish the display must be less than the set-up time.

CAM to DNC System Interface

When NC/CNC cutting instructions for cavities are generated on the CAM system, it is possible to generate enough data to require the equivalent of five miles of paper tape. This quantity of data will occupy a considerable amount of space and require a significant amount of time to transmit from the CAM station to the DNC system.

If the CAM software calculates the length of the resulting paper tape, the equivalent disk space can be determined by calculating the number of characters on the tape and converting that to megabytes of data, as follows:

$$\text{No. char} = \text{Tape length, ft} \times 12\ \text{in./ft} \times 10\ \text{char/in.}$$

At one byte per character:

$$\text{Disk space, Mbyte} = \frac{\text{No. chars on tape}}{1{,}000{,}000}$$

Depending on how many cavity records are to be kept on disk at any particular time, a rough estimate can be made for data storage. In addition to cavity NC/CNC data, space must be allocated for software and other data files, which also may be large.

Once the cavity data have been generated, they can be sent directly to the machine tool from the workstation or can be sent to a DNC system. If the workstation is used to send the data to the machine tool, it may not be available for anything else. If multiple cavities are to be cut, the workstation could be useless for a considerable amount of time. If the data are transmitted to a DNC system, the workstation will be occupied for only one transmission. Furthermore, once the data are in the DNC system, a number of machine tools can access the data for multiple sets and multiple cavities per set.

If the workstation is multitasking, a background task to transmit the data to a machine tool or DNC system can be established. If the background task is used on a workstation that also feeds data to a machine tool, the station must have the power to accomplish both tasks without starving the machine tool for data. Under a multitasking environment, transmission to the machine tool should have the highest priority.

For a network of CAD/CAM workstations, the system should be set up so that the data can be sent out without interrupting anyone on the network. If designers are regularly interrupted to move data, their attitudes and performance may be severely impaired.

As the network between the CAD/CAM system and the DNC network is established, precautions must be taken not to send the data too fast. Even though speed is important, excessive speed will corrupt the integrity of the data by adding or changing the data. Although slow speeds require long transmissions, repeated transmissions take even longer. Speed is dependent on the communication technology used. Some allow very fast transmission, others require slower speeds. Data transmission speed is expressed in bits per second, or baud. (1 byte = 8 bits). When this speed is known, the time required to transmit a CAM record of a given tape length can be calculated as follows:

$$\text{No. bits} = \text{Tape length, ft} \times 12\ \text{in./ft} \times 10\ \text{char/in.} \times 8\ \text{bits/char}$$

$$\text{Transmission time, hours} = \frac{\text{No. bits} \div \text{baud [bits/sec]}}{3600\ \text{sec/hour}}$$

(Note that if the number of characters or the megabytes of data is known, No. bits = No. char × 8, or = Mbytes × 1,000,000 × 8.)

If errors do occur, reduce the speed of transmission (baud rate) on both the sending and receiving computer.

CMM to CAD Interface

Assuming that the communications network has been set up to allow passage of a ASCII data file from the coordinate measuring machines (CMM) to the CAD system, it is possible to use CMM data to augment the design process. As with CAM-generated cavity data, the amount of CMM data can be very large. Depending on the format of data generated by the CMM and the CAD macro capability, the CMM data may or not have to be converted as they are sent to the CAD system. The data can be displayed on CAD as lines and arcs in a two- or three-dimensional format depending on the application. If the data are to be used as the basis of a surface, the scanning process on the CMM may have to be unidirectional rather than bidirectional.

CMM to CAM Interface

CMM data can be passed to CAM through CAD or they can bypass CAD and go directly to CAM. Under CAM, the data can be used as boundary data for related tool paths. For example, when a cavity is cut using manual duplicating or CNC machining from CMM data, the pinch programs required around the neck and push-up can be generated by scanning the relevant area of the cavity. The data can be displayed on CAM and offset to accommodate probe diameter and pinch width to form the geometry to drive a cutter for pinch programs.

PROCESS PLANNING

Data sharing is just one aspect of computer-integrated manufacturing. Another key benefit of CIM is being able to determine at any time the status of a job or part of a job with regard to schedules and budgets.

To meet a promised delivery date, each job must be scheduled down to the day that each operation must be performed for each component of the blow mold. In scheduling, the mold body will probably be the controlling factor. Within the mold body, the cavity machining will be the key scheduling factor. Given an approximate machining time based on material type and cavity volume, the rest of the schedule can be determined in relation to that requirement.

Assuming that several jobs are active in the plant at the same time, there must be a method of organizing all the various steps, dates, and estimated times for each component for each job. One method of organizing the work load for the shop, assuming material availability, and keeping accurate labor records is to assign each individual task a job name, a component name, a step number, a completion date, and an estimated time for the task. With this structure, labor recording and project status can be tracked.

There are a number of process planning systems on the market. An important point in evaluating them is to determine the time required to input the data. Another point to consider is that order sizes of blow molds may require a considerable number of process plans to be written.

The key factors in meeting budget are material, labor, and equipment. With an integrated plant, Purchasing can be supplied with a bill of materials from the CAD system electronically, to speed processing. Furthermore, if software is used to perform the quoting function, the quoted price can be downloaded to the financial software for budget variation calculations and status reports. Finally, through process planning and labor recording, machine tool utilization can be tracked and analyzed.

JOB SUMMARY AND ANALYSIS

As a job progresses, data on adherence to budget and schedule will be collected from normal day to day operations. In order to complete the CIM picture, mathematical analysis of the collected data can identify production problem areas as well as areas of profitable performance. For this analysis, further streamlining of the factory procedures and processes can be performed to meet company goals.

APPENDIX

Mold Maintenance Program for PVC, PET, and PE Container Molds

MAX SUIT

The molds used to produce PVC and PET containers should always have highly polished cavities. It is therefore best to polish them once every two weeks, and certainly before they begin to oxidize. Soft paper tissue and a polishing compound should be used for polishing the cavities. A small amount of the mold polishing compound should be rubbed into each cavity, which should then be polished with a clean tissue until the cavity gives a mirror image. The tissue should be replaced constantly in order to prevent scratching while polishing.

The mold cavities used to produce PE containers should always be sandblasted, because sandblasting produces a smooth container finish, which helps proper venting. Sharp lines on the container finish indicate the need to have the cavities sandblasted. When sandblasting, extra care should be taken to protect the pinch-off edges on the parting line. After sandblasting, the mold should be completely disassembled and all cavity split vents and pin vents cleaned, and the parting line vents checked for proper depth and remachined if necessary. The cavity pin vents or vent plugs should be cleaned every four to five weeks, because of the wax buildup on narrow passages. Improper venting of the mold will result in a poor surface finish, drop test, and uneven wall thickness of the end product.

If PVC material is used, please note that PVC gas is corrosive and will attack the mold material more aggressively if proper venting is not maintained, thereby causing rapid mold deterioration.

Mold cooling lines should be checked for corrosion and any other obstructions that could prevent the flow of the coolant. If any corrosion is detected, the cooling lines should be cleaned immediately. The coolant liquid should also be checked to make sure that the glycol being used is compatible with aluminum, since many of the glycols will cause corrosion when used with aluminum. Reduced cooling would result in poor impact strength, deformation capacity change, and poor surface finish on the containers. In addition, it causes increased cycle time and reduced production.

Guide pins and bushings should be replaced at least once a year. However, if cycle times are increased, play is detected in mold halves, machine die locks are too loose, or pin and bushing wear is noticeable, the pins and bushings should be changed more frequently. New guide pins and bushings will improve mold life and prevent cavity and bottle mismatch.

Striker plates and blow pin plates are vital parts of the mold, and so should be kept in good condition or replaced when necessary. Any uneven wear in the striker-plate cutoff area would produce a poor surface and flash in a neck sealing area, resulting in leaky containers.

Pinch-off edges are designed to cut excessive plastic from the container every time the mold is closed if bottom detabbers are used. However, flash in neck, handle, and shoulder areas remains on the container until its removal as a secondary operation. Worn down pinch-off edges would produce a heavy wall in the pinch-off area, causing deformed containers and difficulty in trimming. Restoring mold pinch-off edges requires specially trained toolmakers, and should therefore be left to the qualified moldmaker.

Whenever the operation is shut down for any length of time or the molds put into storage, all waterlines should be blown out with compressed air and the cavities should be coated with a protective agent to prevent corrosion. We have used a product made by C.R.C. Chemicals known as 3-36. It is available through Carling Products in Mississauga, Ontario, Canada, or C.R.C. Chemicals in Westminster, Pennsylvania, U.S.A.

Index

Wentworth - Alumni Library